T0212507

Lecture Notes in Artificial Intelligence 9875

Subseries of Lecture Notes in Computer Science

Ngoc-Thanh Nguyen · Yannis Manolopoulos
Lazaros Iliadis · Bogdan Trawiński (Eds.)

Computational
Collective Intelligence

8th International Conference, ICCCI 2016
Halkidiki, Greece, September 28–30, 2016
Proceedings, Part I

 Springer

Editors
Ngoc-Thanh Nguyen
Wrocław University of Technology
Wrocław
Poland

Yannis Manolopoulos
Aristotle University of Thessaloniki
Thessaloniki
Greece

Lazaros Iliadis
Department of Forestry and Management
Democritus University of Thrace
Orestiada, Thrace
Greece

Bogdan Trawiński
Wrocław University of Technology
Wrocław
Poland

ISSN 0302-9743   ISSN 1611-3349 (electronic)
Lecture Notes in Artificial Intelligence
ISBN 978-3-319-45242-5   ISBN 978-3-319-45243-2 (eBook)
DOI 10.1007/978-3-319-45243-2

Library of Congress Control Number: 2016949588

LNCS Sublibrary: SL7 – Artificial Intelligence

Printed on acid-free paper

This Springer imprint is published by Springer Nature
The registered company is Springer International Publishing AG Switzerland

Preface

This volume contains the proceedings of the 9th International Conference on Computational Collective Intelligence (ICCCI 2016), held in Halkidiki, Greece, September 28–30, 2016. The conference was co-organized by the Aristotle University of Thessaloniki, Greece, the Democritus University of Thrace, Greece, and the Wrocław University of Science and Technology, Poland. The conference was run under the patronage of the IEEE SMC Technical Committee on Computational Collective Intelligence.

Following the successes of the First ICCCI (2009) held in Wrocław, Poland, the Second ICCCI (2010) in Kaohsiung, Taiwan, the Third ICCCI (2011) in Gdynia, Poland, the 4th ICCCI (2012) in Ho Chi Minh City, Vietnam, the 5th ICCCI (2013) in Craiova, Romania, the 6th ICCCI (2014) in Seoul, South Korea, and the 7th ICCCI (2015) in Madrid, Spain, this conference continues to provide an internationally respected forum for scientific research in the computer-based methods of collective intelligence and their applications.

Computational collective intelligence (CCI) is most often understood as a sub-field of artificial intelligence (AI) dealing with soft computing methods that enable making group decisions or processing knowledge among autonomous units acting in distributed environments. Methodological, theoretical, and practical aspects of CCI are considered as the form of intelligence that emerges from the collaboration and competition of many individuals (artificial and/or natural). The application of multiple computational intelligence technologies such as fuzzy systems, evolutionary computation, neural systems, consensus theory, etc., can support human and other collective intelligence, and create new forms of CCI in natural and/or artificial systems. Three subfields of the application of computational intelligence technologies to support various forms of collective intelligence are of special interest but are not exclusive: Semantic Web (as an advanced tool for increasing collective intelligence), social network analysis (as the field targeted to the emergence of new forms of CCI), and multi-agent systems (as a computational and modeling paradigm especially tailored to capture the nature of CCI emergence in populations of autonomous individuals).

The ICCCI 2016 conference featured a number of keynote talks and oral presentations, closely aligned to the theme of the conference. The conference attracted a substantial number of researchers and practitioners from all over the world, who submitted their papers for the main track and 12 special sessions.

The main track, covering the methodology and applications of CCI, included: multi-agent systems, knowledge engineering and Semantic Web, natural language and text processing, data-mining methods and applications, decision support and control systems, and innovations in intelligent systems. The special sessions, covering some specific topics of particular interest, included cooperative strategies for decision making and optimization, meta-heuristics techniques and applications, Web systems and human–computer interaction, applications of software agents, social media and the Web of linked data, computational swarm intelligence, ambient networks, information technology in biomedicine, impact of smart and intelligent technology on education, big data mining and searching, machine learning in medicine and biometrics, and low-resource language processing.

We received 277 submissions. Each paper was reviewed by two to four members of the international Program Committee of either the main track or one of the special sessions. We selected the 108 best papers for oral presentation and publication in two volumes of the *Lecture Notes in Artificial Intelligence* series.

We would like to express our thanks to the keynote speakers, Plamen Angelov, Heinz Koeppl, Manuel Núñez, and Leszek Rutkowski, for their world-class plenary speeches. Many people contributed toward the success of the conference. First, we would like to recognize the work of the PC co-chairs and special sessions organizers for taking good care of the organization of the reviewing process, an essential stage in ensuring the high quality of the accepted papers. The workshops and special sessions chairs deserve a special mention for the evaluation of the proposals and the organization and coordination of the work of the 12 special sessions. In addition, we would like to thank the PC members, of the main track and of the special sessions, for performing their reviewing work with diligence. We thank the Organizing Committee chairs, liaison chairs, publicity chair, special issues chair, financial chair, Web chair, and technical support chair for their fantastic work before and during the conference. Finally, we cordially thank all the authors, presenters, and delegates for their valuable contribution to this successful event. The conference would not have been possible without their support.

It is our pleasure to announce that the conferences of the ICCCI series continue their close cooperation with the Springer journal *Transactions on Computational Collective Intelligence*, and the IEEE SMC Technical Committee on Transactions on Computational Collective Intelligence.

Finally, we hope that ICCCI 2016 significantly contributes to the academic excellence of the field and leads to the even greater success of ICCCI events in the future.

September 2016

Ngoc-Thanh Nguyen
Yannis Manolopoulos
Lazaros Iliadis
Bogdan Trawiński

Conference Organization

Honorary Chairs

Pierre Levy — University of Ottawa, Canada
Tadeusz Więckowski — Wrocław University of Science and Technology, Poland

General Chairs

Yannis Manolopoulos — Aristotle University of Thessaloniki, Greece
Ngoc Thanh Nguyen — Wrocław University of Science and Technology, Poland

Program Chairs

Lazaros Iliadis — Democritus University of Thrace, Greece
Costin Badica — University of Craiova, Romania
Kazumi Nakamatsu — University of Hyogo, Japan
Piotr Jędrzejowicz — Gdynia Maritime University, Poland

Special Session Chairs

Bogdan Trawiński — Wrocław University of Science and Technology, Poland
Elias Pimenidis — University of the West of England, UK

Organizing Chairs

Apostolos Papadopoulos — Aristotle University of Thessaloniki, Greece

Keynote Speakers

Plamen Angelov — Lancaster University, UK
Heinz Koeppl — Technische Universität Darmstadt, Germany
Manuel Núñez — Universidad Complutense de Madrid, Spain
Leszek Rutkowski — Czestochowa University of Technology, Poland

Special Sessions Organizers

1. WASA 2016: 6th Workshop on Applications of Software Agents

Mirjana Ivanovic	University of Novi Sad, Serbia
Costin Badica	University of Craiova, Romania

2. CSDMO 2016: Special Session on Cooperative Strategies for Decision Making and Optimization

Piotr Jędrzejowicz	Gdynia Maritime University, Poland
Dariusz Barbucha	Gdynia Maritime University, Poland

3. RUMOUR 2016: Workshop on Social Media and the Web of Linked Data

Diana Trandabat	University "Al. I. Cuza" of Iasi, Romania
Daniela Gifu	University "Al. I. Cuza" of Iasi, Romania

4. WebSys 2016: Special Session on Web Systems and Human–Computer Interaction

Kazimierz Choroś	Wrocław University of Science and Technology, Poland
Maria Trocan	Institut Supérieur d'Électronique de Paris, France

5. MHTA 2016: Special Session on Meta-Heuristics Techniques and Applications

Pandian Vasant	Universiti Teknologi Petronas, Malaysia
Bharat Singh	Big Data Analyst, Hildesheim, Germany
Neel Mani	Dublin City University, Dublin, Ireland
Rajalingam Sokkalingam	Universiti Teknologi Petronas, Malaysia
Junzo Watada	Waseda University, Japan

6. CSI 2016: Special Session on Computational Swarm Intelligence

Urszula Boryczka	University of Silesia, Poland
Mariusz Boryczka	University of Silesia, Poland
Jan Kozak	University of Silesia, Poland

7. AMNET 2016: Special Session on Ambient Networks

Vladimir Sobeslav	University of Hradec Kralove, Czech Republik
Ondrej Krejcar	University of Hradec Kralove, Czech Republic
Peter Brida	University of Žilina, Czech Republic
Peter Mikulecky	University of Hradec Kralove, Czech Republic

8. ITiB 2016: Special Session on IT in Biomedicine

Ondrej Krejcar	University of Hradec Kralove, Czech Republic
Kamic Kuca	University of Hradec Kralove, Czech Republic
Dawit Assefa Halle	Addis Ababa University, Ethiopia
Tanos C.C. Franca	Military Institute of Engineering, Brazil

9. *ISITE 2016: Special Session on Impact of Smart and Intelligent Technology on Education*

Petra Poulova	University of Hradec Kralove, Czech Republic
Ivana Simonova	University of Hradec Kralove, Czech Republic
Katerina Kostolanyova	University of Ostrava, Czech Republic
Tiia Ruutmann	Tallinn University of Technology, Estonia

10. *BigDMS 2016: Special Session on Big Data Mining and Searching*

Rim Faiz	University of Carthage, Tunisia
Nadia Essoussi	University of Carthage, Tunisia

11. *MLMB 2016: Special Session on Machine Learning in Medicine and Biometrics*

Piotr Porwik	University of Silesia, Poland
Agnieszka Nowak-Brzezińska	University of Silesia, Poland
Robert Koprowski	University of Silesia, Poland
Janusz Jeżewski	Institute of Medical Technology and Equipment, Poland

12. *LRLP 2016: Special Session on Low-Resource Language Processing*

Ualsher Tukeyev	al-Farabi Kazakh National University, Kazakhstan
Zhandos Zhumanov	al-Farabi Kazakh National University, Kazakhstan

International Program Committee

Sharat Akhoury	University of Cape Town, South Africa
Ana Almeida	GECAD-ISEP-IPP, Portugal
Orcan Alpar	University of Hradec Kralove, Czech Republic
Bashar Al-Shboul	University of Jordan, Jordan
Thierry Badard	Laval University, Canada
Amelia Badica	University of Craiova, Romania
Costin Badica	University of Craiova, Romania
Hassan Badir	ENSAT, Morocco
Dariusz Barbucha	Gdynia Maritime University, Poland
Nick Bassiliades	Aristotle University of Thessaloniki, Greece
Artur Bąk	Polish-Japanese Academy of Information Technology, Poland
Narjes Bellamine	ISI & Laboratoire RIADI/ENSI, Tunisia
Maria Bielikova	Slovak University of Technology in Bratislava, Slovakia
Pavel Blazek	University of Defence, Czech Republic
Mariusz Boryczka	University of Silesia, Poland
Peter Brida	University of Žilina, Slovakia

Robert Burduk	Wrocław University of Science and Technology, Poland
Krisztian Buza	Budapest University of Technology and Economics, Hungary
Aleksander Byrski	AGH University Science and Technology, Poland
Jose Luis Calvo-Rolle	University of A Coruña, Spain
David Camacho	Universidad Autonoma de Madrid, Spain
Alberto Cano	Virginia Commonwealth University, USA
Frantisek Capkovic	Slovak Academy of Sciences, Slovakia
Dariusz Ceglarek	Poznan High School of Banking, Poland
Amine Chohra	Paris-East University (UPEC), France
Kazimierz Choroś	Wrocław University of Science and Technology, Poland
Mihaela Colhon	University of Craiova, Romania
Jose Alfredo Ferreira Costa	Universidade Federal do Rio Grande do Norte, Brazil
Ireneusz Czarnowski	Gdynia Maritime University, Poland
Paul Davidsson	Malmö University, Sweden
Gayo Diallo	University of Bordeaux, France
Tien V. Do	Budapest University of Technology and Economics, Hungary
Ivan Dolnak	University of Žilina, Slovakia
Olfa Belkahla Driss	Université de Tunis, Tunisia
Atilla Elci	Aksaray University, Turkey
Vadim Ermolayev	Zaporozhye National University, Ukraine
Nadia Essoussi	University of Carthage, Tunisia
Rim Faiz	University of Carthage, Tunisia
Faiez Gargouri	University of Sfax, Tunisia
Mauro Gaspari	University of Bologna, Italy
Antonio Gonzalez-Pardo	Universidad Autonoma de Madrid, Spain
Huu Hanh Hoang	Hue University, Vietnam
Tzung-Pei Hong	National University of Kaohsiung, Taiwan
Josef Horalek	University of Hradec Králové, Czech Republic
Frédéric Hubert	Laval University, Canada
Maciej Huk	Wrocław University of Science and Technology, Poland
Dosam Hwang	Yeungnam University, Korea
Lazaros Iliadis	Democritus University of Thrace, Greece
Agnieszka Indyka-Piasecka	Wrocław University of Science and Technology, Poland
Dan Istrate	Université de Technologie de Compiègne, France
Mirjana Ivanovic	University of Novi Sad, Serbia
Jaroslaw Jankowski	West Pomeranian University of Technology, Poland
Joanna Jędrzejowicz	University of Gdansk, Poland
Piotr Jędrzejowicz	Gdynia Maritime University, Poland
Jason Jung	Chung-Ang University, Korea
Przemysław Juszczuk	University of Silesia, Poland

Ioannis Karydis	Ionian University, Greece
Petros Kefalas	University of Sheffield International Faculty, CITY College, Greece
Rafał Kern	Wrocław University of Science and Technology, Poland
Marek Kisiel-Dorohinicki	AGH University Science and Technology, Poland
Attila Kiss	Eötvös Loránd University, Hungary
Jitka Komarkova	University of Pardubice, Czech Republic
Marek Kopel	Wrocław University of Science and Technology, Poland
Ivan Koychev	University of Sofia St. Kliment Ohridski, Bulgaria
Jan Kozak	University of Silesia, Poland
Adrianna Kozierkiewicz-Hetmańska	Wrocław University of Science and Technology, Poland
Ondrej Krejcar	University of Hradec Kralove, Czech Republic
Dariusz Król	Wrocław University of Science and Technology, Poland
Elżbieta Kukla	Wrocław University of Science and Technology, Poland
Julita Kulbacka	Wrocław Medical University, Poland
Marek Kulbacki	Polish-Japanese Academy of Information Technology, Poland
Piotr Kulczycki	Systems Research Institute of the Polish Academy of Science, Poland
Kazuhiro Kuwabara	Ritsumeikan University, Japan
Florin Leon	Technical University Gheorghe Asachi of Iasi, Romania
Edwin Lughofer	Johannes Kepler University Linz, Austria
José María Luna	University of Cordoba, Spain
Juraj Machaj	University of Žilina, Slovakia
Bernadetta Maleszka	Wrocław University of Science and Technology, Poland
Marcin Maleszka	Wrocław University of Science and Technology, Poland
Yannis Manolopoulos	Aristotle University of Thessaloniki, Greece
Antonio David Masegosa Arredondo	University of Granada, Spain
Adam Meissner	Poznań University of Technology, Poland
Ernestina Menasalvas	Universidad Politecnica de Madrid, Spain
Héctor Menéndez	Universidad Autonoma de Madrid, Spain
Jacek Mercik	Wrocław School of Banking, Poland
Peter Mikulecky	University of Hradec Kralove, Czech Republic
Alin Moldoveanu	University Politechnica of Bucharest, Romania
Javier Montero	Universidad Complutense de Madrid, Spain
Ahmed Moussa	Abdelmalek Essaadi University, Morocco
Grzegorz J. Nalepa	AGH University of Science and Technology, Poland

Filippo Neri	University of Naples Federico II, Italy
Linh Anh Nguyen	University of Warsaw, Poland
Ngoc-Thanh Nguyen	Wrocław University of Science and Technology, Poland
Adam Niewiadomski	Lodz University of Technology, Poland
Alberto Núñez	Universidad Complutense de Madrid, Spain
Manuel Núñez	Universidad Complutense de Madrid, Spain
Tarkko Oksala	Aalto University, Finland
Tomasz Orczyk	University of Silesia, Poland
Rafael Parpinelli	Santa Catarina State University, Brazil
Marek Penhaker	VSB – Technical University of Ostrava, Czech Republic
Dariusz Pierzchała	Military University of Technology, Poland
Marcin Pietranik	Wrocław University of Science and Technology, Poland
Elias Pimenidis	University of the West of England, UK
Bartłomiej Płaczek	Silesian University of Technology, Poland
Piotr Porwik	University of Silesia, Poland
Radu-Emil Precup	Politehnica University of Timisoara, Romania
Ales Prochazka	Institute of Chemical Technology, Czech Republic
Paulo Quaresma	Universidade de Evora, Portugal
Ewa Ratajczak-Ropel	Gdynia Maritime University, Poland
José Antonio Sáez	University of Granada, Spain
Virgilijus Sakalauskas	Vilnius University, Lithuania
Jose L. Salmeron	University Pablo de Olavide, Spain
Jakub Segen	Polish-Japanese Academy of Information Technology, Poland
Ali Selamat	Universiti Teknologi Malaysia, Malaysia
Natalya Shakhovska	Lviv Polytechnic National University, Ukraine
Andrzej Siemiński	Wrocław University of Science and Technology, Poland
Vladimir Sobeslav	University of Hradec Kralove, Czech Republic
Stanimir Stoyanov	University of Plovdiv Paisii Hilendarski, Bulgaria
Yasufumi Takama	Tokyo Metropolitan University, Japan
Bogdan Trawiński	Wrocław University of Science and Technology, Poland
Maria Trocan	Institut Superieur d'Electronique de Paris, France
Krzysztof Trojanowski	Polish Academy of Sciences, Poland
Ventzeslav Valev	Bulgarian Academy of Sciences, Bulgaria
Izabela Wierzbowska	Gdynia Maritime University, Poland
Michal Woźniak	Wrocław University of Science and Technology, Poland
Krzysztof Wróbel	Uniwersity of Silesia, Poland
Drago Žagar	University of Osijek, Croatia
Danuta Zakrzewska	Lodz University of Technology, Poland

| Constantin-Bala Zamfirescu | Lucian Blaga University of Sibiu, Romania |
| Aleksander Zgrzywa | Wrocław University of Science and Technology, Poland |

Program Committees of Special Sessions

WASA 2016: 6th Workshop on Applications of Software Agents

Amelia Badica	University of Craiova, Romania
Olivier Boissier	ENS Mines Saint-Etienne, France
Paolo Bresciani	FBK, Italy
Marius Brezovan	University of Craiova, Romania
Zoran Budimac	University of Novi Sad, Serbia
Mihaela Colhon	University of Craiova, Romania
Weihui Dai	Fudan University, China
Adina Magda Florea	University Politehnica of Bucharest, Romania
Giancarlo Fortino	University of Calabria, Italy
Daniela Gifu	Alexandru Ioan Cuza University of Iasi, Romania
Adrian Groza	Technical University of Cluj-Napoca, Romania
Sorin Ilie	University of Craiova, Romania
Galina Ilieva	University of Plovdiv Paisii Hilendarsky, Bulgaria
Nicolae Jascanu	Dunarea de Jos University of Galati, Romania
Gordan Jezic	University of Zagreb, Croatia
Systä Kari	Tampere University of Technology, Finland
Petros Kefalas	University of Sheffield International Faculty, Thessaloniki, Greece
Setsuya Kurahashi	University of Tsukuba, Japan
Mario Kusek	University of Zagreb, Croatia
Florin Leon	Technical University Gheorghe Asachi of Iasi, Romania
Marin Lujak	University Rey Juan Carlos, Spain
Viorel Negru	West University of Timisoara, Romania
Andrea Omicini	University of Bologna, Italy
Mihaela Oprea	University Petroleum-Gas of Ploiesti, Romania
Agostino Poggi	University of Parma, Italy
Ilias Sakellariou	University of Macedonia, Greece
Stanimir Stoyanov	University of Plovdiv Paisii Hilendarsky, Bulgaria
Denis Trcek	University of Ljubljana, Slovenia
George Vouros	University of Piraeus, Greece
Constantin-Bala Zamfirescu	University of Sibiu, Romania

CSDMO 2016: Special Session on Cooperative Strategies for Decision-Making and Optimization

Dariusz Barbucha	Gdynia Maritime University, Poland
Ireneusz Czarnowski	Gdynia Maritime University, Poland
Joanna Jędrzejowicz	University of Gdansk, Poland

Piotr Jędrzejowicz	Gdynia Maritime University, Poland
Edyta Kucharska	AGH University of Science and Technology, Poland
Antonio D. Masegosa	University of Deusto, Spain
Javier Montero	Universidad Complutense de Madrid, Spain
Ewa Ratajczak-Ropel	Gdynia Maritime University, Poland
Iza Wierzbowska	Gdynia Maritime University, Poland
Mahdi Zargayouna	IFSTTAR, France

RUMOUR 2016: Workshop on Social Media and the Web of Linked Data

Nuria Bel	Universitat Pompeu Fabra, Spain
Costin Badica	University of Craiova, Romania
Georgeta Bordea	National University of Ireland, Ireland
Steve Cassidy	Macquarie University, Australia
Dragoş Ciobanu	University of Leeds, UK
Mihaela Colhon	University of Craiova, Romania
Dan Cristea	Alexandru Ioan Cuza University of Iaşi, Romania
Thierry Declerck	Universitat des Saarlandes, Saarbrücken, Germany
Daniela Gîfu	Alexandru Ioan Cuza University of Iaşi, Romania
Jorge Gracia	Universidad Politecnica de Madrid, Spain
Radu Ion	Microsoft Ireland, Ireland
John McCray	National University of Ireland, Ireland
Rada Mihalcea	University of Michigan, USA
Andrei Olariu	University of Bucharest, Romania
Octavian Popescu	IBM Research, USA
Dan Ştefănescu	Vantage Labs, USA
Diana Trandabăţ	Alexandru Ioan Cuza University of Iaşi, Romania
Dan Tufiş	Romanian Academy Research Institute for Artificial Intelligence Mihai Drăgănescu, Romania
Piek Vossen	Vrije Universiteit, Amsterdam, The Netherlands
Gabriela Vulcu	National University of Ireland, Ireland
Michael Zock	Aix-Marseille University, France

WebSys 2016: Special Session on Web Systems and Human–Computer Interaction

František Čapkovič	Academy of Sciences, Slovakia
Kazimierz Choroś	Wrocław University of Science and Technology, Poland
Jarosław Jankowski	West Pomeranian University of Technology, Poland
Ondřej Krejcar	University of Hradec Kralove, Czech Republic
Matthieu Manceny	Institut Supérieur d'Électronique de Paris, France
Aleš Procházka	Institute of Chemical Technology, Czech Republic
Andrzej Siemiński	Wrocław University of Science and Technology, Poland
Maria Trocan	Institut Supérieur d'Électronique de Paris, France
Aleksander Zgrzywa	Wrocław University of Science and Technology, Poland

MHTA 2016: Special Session on Meta-Heuristics Techniques and Applications

Gerhard-Wilhelm Weber	Middle East Technical University, Turkey
Kwon-Hee Lee	Dong-A University, Korea
Igor Litvinchev	Nuevo Leon State University, Mexico
Mohammad Abdullah-Al-Wadud	King Saud University, Saudi Arabia
Vo Ngoc Dieu	HCMC University of Technology, Vietnam
Gerardo Maximiliano Mendez	Instituto Tecnologico de Nuevo Leon, Mexico
Leopoldo Eduardo Cárdenas Barrón	Tecnológico de Monterry, Mexico
Denis Sidorov	Irkutsk State University, Russia
Weerakorn Ongsakul	Asian Institute of Technology, Thailand
Goran Klepac	Raiffeisen Bank Austria, Croatia
Herman Mawengkang	The University of Sumatera Utara, Indonesia
Igor Tyukhov	Moscow State University of Mechanical Engineering, Russia
Hayato Ohwada	Tokyo University of Science, Japan
Ugo Fiore	Federico II University, Italy
Leo Mrsic	University College Effectus – College for Law and Finance, Croatia
Nguyen Trung Thang	Ton Duc Thang University, Vietnam
Nikolai Voropai	Energy Systems Institute, Russia
Shiferaw Jufar	Universiti Teknologi Petronas, Malaysia
Xueguan Song	Dalian University of Technology, China
Ruhul A. Sarker	UNSW, Australia
Vipul Sharma	Lovely Professional University, India

CSI 2016: Special Session on Computational Swarm Intelligence

Urszula Boryczka	University of Silesia, Poland
Mariusz Boryczka	University of Silesia, Poland
Miłosław Chodacki	University of Silesia, Poland
Diana Domańska	University of Oslo, Norway
Wojciech Froelich	University of Silesia, Poland
Przemysław Juszczuk	University of Silesia, Poland
Jan Kozak	University of Silesia, Poland
Dariusz Pierzchała	Military University of Technology, Poland
Rafał Skinderowicz	University of Silesia, Poland
Tomasz Staś	University of Economics in Katowice, Poland
Beata Zielosko	University of Silesia, Poland

AMNET 2016: Special Session on Ambient Networks

Ana Almeida	Porto Superior Institute of Engineering, Portugal
Peter Brida	University of Žilina, Slovakia
Ivan Dolnak	University of Žilina, Slovakia

Elsa Gomes	Porto Superior Institute of Engineering, Portugal
Josef Horalek	University of Hradec Kralove, Czech Republic
Josef Janitor	Technical University of Kosice, Slovakia
Ondrej Krejcar	University of Hradec Kralove, Czech Republic
Goreti Marreiros	Porto Superior Institute of Engineering, Portugal
Peter Mikulecký	University of Hradec Kralove, Czech Republic
Juraj Machaj	University of Žilina, Slovakia
Marek Penhaker	Technical University of Ostrava, Czech Republic
José Salmeron	Universidad Pablo de Olavide of Seville, Spain
Ali Selamat	Universiti Teknologi Malaysia, Malaysia
Vladimir Sobeslav	University of Hradec Kralove, Czech Republic
Stylianakis Vassilis	University of Patras, Greece

ITiB 2016: Special Session on IT in Biomedicine

Dawit Assafa Haile	Addis Ababa University, Ethiopia
Peter Brida	University of Žilina, Slovakia
Richard Cimler	University of Hradec Kralove, Czech Republic
Rafael Dolezal	University of Hradec Kralove, Czech Republic
Ricardo J. Ferrari	Federal University of Sao Carlos, Brazil
Tanos C.C. Franca	Military Institute of Engineering, Brazil
Ondrej Krejcar	University of Hradec Kralove, Czech Republic
Kamil Kuca	University of Hradec Kralove, Czech Republic
Juraj Machaj	University of Žilina, Slovakia
Petra Maresova	University of Hradec Kralove, Czech Republic
Marek Penhaker	Technical University of Ostrava, Czech Republic
Jan Plavka	Technical University of Kosice, Slovakia
Teodorico C. Ramalho	Federal University of Lavras, Brazil
Saber Salehi	Universiti Teknologi Malaysia, Malaysia
Ali Selamat	Universiti Teknologi Malaysia, Malaysia

ISITE 2016: Special Session on Impact of Smart and Intelligent Technology on Education

Pavel Doulik	Jan Evangelista Purkyne University, Czech Republic
Blanka Klimova	University of Hradec Kralove, Czech Republic
Katerina Kostolanyova	University of Ostrava, Czech Republic
Silvia Pokrivcakova	Constantine the Philosopher University, Slovakia
Tatiana Polyakova	Moscow State Technical University for Automobile and Road, Russia
Petra Poulova	University of Hradec Kralove, Czech Republic
Maria Teresa Restivo	Universidade do Porto, Portugal
Tiia Ruutmann	Tallinn University of Technology, Estonia
Jiri Skoda	Jan Evangelista Purkyne University, Czech Republic
Marcela Sokolova	University of Hradec Kralove, Czech Republic
Ivana Simonova	University of Hradec Kralove, Czech Republic

| Darina Tóthova | European University Information Systems Slovakia |
| Milan Turcani | Constantine the Philosopher University, Slovakia |

BigDMS 2016: Special Session on Big Data Mining and Searching

Ajith Abraham	Machine Intelligence Research Labs, USA
Thierry Badard	Laval University, Canada
Hassan Badir	ENSAT Tangier, Morocco
Chiheb Ben N'cir	Université de la Manouba, Tunisia
Ismaïl Biskri	Université du Québec à Trois-Rivières, Canada
Guillaume Cleuziou	Université d'Orléans, France
Ernesto Damiani	University of Milan, Italy
Gayo Diallo	University of Bordeaux, France
Aymen Elkhlifi	University of Paris Sorbonne, France
Nadia Essoussi	FSEG Nabeul, University of Carthage, Tunisia
Rim Faiz	IHEC, University of Carthage, Tunisia
Sami Faiz	ISAMM, University of Manouba, Tunisia
Riadh Farah	ISAMM, University of Manouba, Tunisia
Faiez Gargouri	ISIMS, University of Sfax, Tunisia
Lamia Hadrich Belguith	FSEGS, University of Sfax, Tunisia
Frédéric Hubert	Laval University, Canada
Ahmed Moussa	ENSA, Abdelmalek Essaadi University, Morocco
Maria Malek	EISTI, France
Gabriella Pasi	University of Milan Bicocca, Italy

MLMB 2016: Special Session on Machine Learning in Medicine and Biometrics

Nabendu Chaki	University of Calcutta, India
Robert Czabański	University of Silesia, Poland
Adam Gacek	Institute of Medical Technology and Equipment, Poland
Marina Gavrilova	University of Calgary, Canada
Manuel Graña	University of the Basque Country, Spain
Alicja Wakulicz-Deja	University of Silesia, Poland
Robert Koprowski	University of Silesia, Poland
Agnieszka Nowak-Brzezińska	University of Silesia, Poland
Nobuyuki Nishiuchi	Tokyo Metropolitan University, Japan
Małgorzata Przybyła-Kasperek	University of Silesia, Poland
Marek Kurzyński	Wrocław University of Science and Technology, Poland
Roman Simiński	University of Silesia, Poland
Janusz Jeżewski	Institute of Medical Technology and Equipment, Poland
Dragan Simic	University of Novi Sad, Serbia
Ewaryst Tkacz	Silesian University of Technology, Poland

Dariusz Mrozek	Silesian University of Technology, Poland
Bożena Małysiak-Mrozek	Silesian University of Technology, Poland
Michał Dramiński	Polish Academy of Sciences, Warsaw, Poland
Michał Kozielski	Silesian University of Technology, Poland
Rafał Deja	Academy of Business in Dabrowa Gornicza, Poland

LRLP 2016: Special Session on Low-Resource Language Processing

Ualsher Tukeyev	al-Farabi Kazakh National University, Kazakhstan
Zhandos Zhumanov	al-Farabi Kazakh National University, Kazakhstan
Francis Tyers	The Arctic University of Norway, Norway
Madina Mansurova	al-Farabi Kazakh National University, Kazakhstan
Altynbek Sharipbay	L.N. Gumilyov Eurasian National University, Kazakhstan
Orken Mamyrbayev	Institute of Information and Computational Technologies, Kazakhstan
Rustam Musabayev	Institute of Information and Computational Technologies, Kazakhstan
Zhenisbek Assylbekov	Nazarbayev University, Kazakhstan
Aibek Makazhanov	Nazarbayev University, Kazakhstan
Jonathan Washington	Indiana University, USA
Djavdet Suleimanov	Institute of Applied Semiotics, Tatarstan, Russia
Sergazy Narynov	Alem Research Company, Kazakhstan
Miquel Esplà-Gomis	University of Alicante, Spain
Altangerel Chagnaa	National University of Mongolia, Mongolia

Additional Reviewers

Barbieri, Francesco	Khusainov, Aydar	Mocanu, Andrei
Bosque-Gil, Julia	Kravari, Kalliopi	Montero, Javier
Cherichi, Soumaya	Kumar, Krishan	Ouamani, Fadoua
Eklund, Ulrik	Labba, Chahrazed	Rodríguez
Hooper, Paul Charles	Mani, Neel	Fernández, Víctor
Jain, Nikita	Martín, Alejandro	Singh, Meeta
Jemal, Dhouha	Missaoui, Sondess	Takáč, Peter

Contents – Part I

Natural Language and Text Processing

Data Mining Methods and Applications

Decision Support and Control Systems

Innovations in Intelligent Systems

Cooperative Strategies for Decision Making and Optimization

Meta-Heuristics Techniques and Applications

Web Systems and Human-Computer Interaction

Contents – Part II

Impact of Smart and Intelligent Technology on Education

Big Data Mining and Searching

Machine Learning in Medicine and Biometrics

Low Resource Language Processing

Multi-agent Systems

A Novel Interaction Protocol of a Multiagent System for the Study of Alternative Decisions

Florin Leon[✉]

Department of Computer Science and Engineering, "Gheorghe Asachi"
Technical University of Iaşi, Bd. Mangeron 27, 700050 Iaşi, Romania
fleon@cs.tuiasi.ro

Abstract. The process of decision making is one of the core components of a cognitive system. In this paper, a simple, deterministic protocol for agent interaction in a multiagent system is proposed, which is based on passive stigmergy and can exhibit complex interactions, although in the end it stabilizes. This system can be used to study the effect of alternative decisions. A statistical analysis of several scenarios is provided, in which a perturbation, i.e. a single or a small set of alternative decisions, can change the final utility of an agent from minimum to maximum.

Keywords: Multiagent systems · Decision making · Alternative decisions · Perturbations

1 Introduction

A very important issue in decision making is the "what if" question. Given a certain system, it is interesting to explore its entire range of possible expressions, which in decision theory would correspond to possible outcomes, given the individual decisions that the agents are able to make. Since time is linear, the exploration of all possibilities may involve the presence of an oracle that would reveal the best alternative required to reach a specific goal.

In order to perform such an analysis, our goal is to design a simple interaction protocol for a multiagent system whose overall behaviour should be complex enough so that the best way to reach a specific outcome should not be obvious, or should even be impossible to predict without running the system explicitly.

In previous works [5–8], such a multiagent system was described, whose behaviour was based on pairwise interactions between agents which exchanged resources and had internal models about their peers to try and estimate the benefit from the corresponding interactions. The analysis of the system revealed that it is capable of exhibiting a large range of behaviours, from asymptotically stable to periodic and chaotic (in the technical sense of chaos theory), where small perturbations in the initial state of the system can greatly impact its overall behaviour.

After studying it, several characteristics that seem important for the design of such a system have been identified:

© Springer International Publishing Switzerland 2016
N.T. Nguyen et al. (Eds.): ICCCI 2016, Part I, LNAI 9875, pp. 3–12, 2016.
DOI: 10.1007/978-3-319-45243-2_1

- a preference for greedy decisions that try to maximize the immediate reward without regard or knowledge about the long-term utility;
- incomplete information about the overall state of the system and therefore imperfect basis to make decisions;
- lack of reliable information about the actual consequences of a decision, because of the unpredictable long-term behaviour of the system;
- lack of control about the decisions of the other agents.

In this paper, a new, simpler, faster protocol will be described, where the idea of interaction between agents is still present but in an abstracted form of passive stigmergy. The focus is not on the types of behaviour, because in this case they are always asymptotically stable, but on the consequences of decisions and on the possibility of finding alternative decisions that can lead to different outcomes in the final utilities of the agents.

The article is organized as follows. Section 2 presents some related work on these issues. Section 3 describes the interaction protocol of the multiagent system and its mathematical formalization. Section 4 covers some experimental studies with a statistical analysis of the alternative decisions. The final section contains the conclusions.

2   Related Work

Decision making is a crucial human issue that has been extensively studied from different perspectives and has applications in many fields, e.g. computer science, psychology, economics, social science, politics, military etc.

In the agent field, the problem of deciding with limited information was addressed in [10], which provides a statistical model about which problem solving approach to take with limited a-priori information. Another important matter is how to combine knowledge from local distributed sources in order to facilitate global decision-making [9].

In psychology, [2] considers optimal decision making in two alternative forced choice tasks: six models of decision making are analysed and it is proven that there is always an optimal trade-off between speed and accuracy that maximizes various reward functions.

Further, it was found that the optimal observer model may provide a parsimonious account of both response time and error rate data following Hick's Law, i.e. that response time increases log-linearly with the number of alternatives, and that people approximate Bayesian inference in multi-alternative choice, except for some perceptual limitations [3].

In neuro-economics, [11] discusses whether the brain is unitary, i.e. a unified system as a single information processor, or dual, in which two or more competing "selves" can reach separate decisions and some sort of meta-control is needed to decide the final outcome, which can also involve the concept of Nash equilibrium.

In marketing, [12] presents an experiment to study the empirical applications of alternative choice rules with a model based on Simon's idea of a satisficing decision maker. The model strongly supports the satisficing stopping rule.

A review of several decision making models, with emphasis on time-critical decision making in a military context, is given in [1].

3 Multiagent System Design

In previous works [5–8], a multiagent system was proposed and thoroughly studied. It included internal models of agents that were used to choose other agents for interaction. The interactions were simple exchanges of resources, and the success or failure of these attempts gave the agents different rewards.

This paper aims to further abstract the agent choices by designing a new, simple system which nevertheless exhibits all the criteria of behaviour stated in the introduction. Instead of direct agent interactions, an approach based on *passive stigmergy* is used, which involves altering the environment so that the effects of another agent's actions change [4].

Formally, the system is composed of n agents and m states. Usually, m is greater or equal to n. Each state is defined by a combination of values. The number of values, v, is the same for all states. Each value is a random positive integer from a predefined domain: $v_i \sim U(0, v_{max})$, $v_i \in \mathbb{N}$, $\forall i = 1..v$. For example, if $v = 10$ and $v_{max} = 9$, a state can look like this: $\{7, 2, 0, 0, 9, 3, 7, 9, 6, 1\}$.

The agents have a partial view of the states: they can only perceive a random combination of v_p values, where $1 \leq v_p \leq v$. v_p is the same for all the agents, but the particular values they can perceive are different. For example, if $v_p = 3$ and $v = 10$, one agent can perceive the values with indices $\{1, 4, 2\}$, while another can perceive the values with indices $\{4, 5, 8\}$. The values and the order of the perceived values are also generated uniformly random from their allowed domain and initialized before the start of the simulation.

The multiagent system runs sequentially. In each step, the agents are activated in lexicographic order and perform the actions presented below. The reason for this ordering, as opposed to simulating parallelism by executing agents sequentially and in a random order, or just executing concurrently, is that we want the system to be deterministic. Randomness is only involved in the initialization of the system, i.e. the initial features of the states and agents. Afterwards, the execution of the system is completely deterministic. This is important when studying alternative decisions, because we can compare the behaviour of the system under different circumstances, and we can emphasise the situations when minor changes in the state of the system at one point can cause very different outcomes, eventually.

In each step of the simulation, each agent performs an elementary action: it chooses a state and a state value from the set of its perceived values, and tries to perform a unitary change on that value, i.e. adding or subtracting 1 unit.

The philosophy of the system is to allow indirect interactions between agents, through the states of the environment. Ideally, two agents can cooperate if they continuously choose the same state and state value, and one agent increments it, while the other one decrements it. The number of "adding" agents (type A) should be approximately equal to the number of "subtracting" agents (type S), in order to have a balanced system.

A key factor that introduces non-linearity into the system is the existence of two thresholds for the agent actions. A type A agent can only perform its action on a state value if the average of the state values is below a maximum threshold. This non-linearity also has a subtle side. The agent has only partial information about the states, therefore it cannot know the real average of the state values. It can only approximate it using the average of the values it can perceive. If the average of the perceived values is below the threshold, the agent attempts to act. If the average of the entire set of values is below the threshold, the agent receives a positive reward r_p, e.g. $r_p = 1$. Otherwise, it receives a negative reward r_n, e.g. $r_n = -1$.

Another condition is that the target state value is below the maximum allowed value v_{max}. In this case, it is ensured that the domain of the state values is never violated.

Similarly, a type S agent has to obey a corresponding minimum threshold and its target state values must be at least 1.

Overall, we can distinguish between fixed parameters, namely the number of agents, states, values, perceived values and thresholds, and the random initial configuration containing the actual state values, and perceived values for each agent.

In our case, with $v_{max} = 9$, the maximum threshold is $\theta_{max} = 5$ and the minimum threshold is $\theta_{min} = 4$. This is the most constrained setting that still allows the agents to cooperate, e.g. a type A agent can increment a state value and a type S agent can decrement it back. More distant thresholds would increase the probability of the agents being successful. However, the aim of the paper is to design a flexible system that allows both failure and success, in order to see under which circumstances failure can be changed into success.

Each agent has a utility computed out of successive rewards. Its initial value is 0 and then, after each action that receives reward r (either r_p or r_n), the utility u is updated as: $u_{t+1} = u_t \cdot \gamma + r \cdot (1 - \gamma)$, where γ is a discount factor, e.g. 0.9, which measures the importance of future rewards compared to the current utility. The idea of a discount factor is characteristic of most reinforcement learning algorithms.

There are situations when an agent has no visible options to act, i.e. there are no states which obey the conditions presented above. In this case, the agent does not take any action and its utility remains unchanged.

The agents are always executed in a lexicographic order, and the decisions to act are deterministic. They analyse the states also in a lexicographic order. For a certain state, the order of the state values they analyse is given by their sets of perceived values.

In Fig. 1, the evolution of two systems is presented, starting from different configurations. The graphs show the utility of the 6 agents comprising the system. One can see that after an initial unstable period, some agents stabilize to a utility equal to 1 while others stabilize to a negative utility of −1. The agents in the former category have *implicitly* found a way to cooperate with agents of the complementary type. The agents in the latter category can no longer find any suitable state values to act on, either because they have been changed by the other agents, or because the average of the perceived values gives them the incorrect information that the state is inaccessible, although it is not the case.

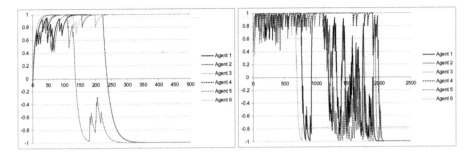

Fig. 1. The convergence of agent utilities: a) faster convergence; b) slower convergence

4 Experimental Study

This section presents a probabilistic analysis of the system running under different configurations. We study the kind of utilities the agents eventually obtain and some approaches that can be used to find most of the solutions to the alternative decision problem, i.e. whether there exists a single perturbation or small group of perturbations that can change the final utility of a specific agent from −1 to 1.

4.1 The Effect of Perceptual Capability

This study is concerned with the probability of an agent to correctly assess whether the average of the entire set of values of a state lies in the allowed domain (e.g. above the minimum threshold) when the average of the perceived values does. The problem of the average being less than a maximum threshold is analogous and if we assume that the thresholds are symmetrical, i.e. $\theta_{min} = v_{max} - \theta_{max}$, we can only consider a single problem with a minimum threshold θ.

Let V_A be a random variable that is true when $\sum_{v_i \in V} v_i/v \geq \theta$ and false otherwise, where V is the set of state values. Let V_P be a random variable that is true when $\sum_{v_i \in P} v_i/v_p \geq \theta$ and false otherwise, where P is the set of perceived state values. We can empirically compute the conditional probability $P(V_A \mid V_P)$ by generating a large number of states, i.e. 10 million, and using a frequentist approach to assess the probability. This can also be obtained analytically, but given the rather large number of fixed parameters, it was considered easier to use the Monte Carlo method.

Figure 2 displays this conditional probability when the number of perceived values varies, for different values of the threshold θ.

The statistical average of state values is $v_{max}/2$. When θ is high, above this value, it is more difficult to assess that the global average is feasible only if the local average happens to be feasible. Conversely, when θ is low, the chances of finding a feasible global average are higher. A good compromise between these extremes is given by a value for θ close to the $v_{max}/2$. Therefore, these were the values that were used in the following experiments: $\theta_{min} = 4$ and $\theta_{max} = 5$. Also, the number of perceived values

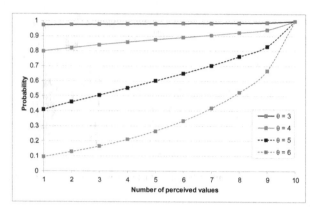

Fig. 2. The conditional probability $P(V_A \mid V_P)$ for different thresholds when the number of perceived values varies

should be small enough to have a clear contrast between the local and global perspectives, but high enough to allow sufficient overlapping in the perceived values of the agents, so that they have a chance to implicitly cooperate and have continuous interactions. That is why we chose $v_p = 3$ in the following statistical analysis.

4.2 The Probability of Finding Solutions

We define a "solution" to the alternative decision problem as a single change or a small set of changes to the initial decision making structure of an agent, such that it would obtain a positive utility, e.g. 1, instead of a negative utility, e.g. −1, if the system were executed from the initial conditions again and the agents chose their actions deterministically according to the presented protocol, except for the specific steps defined by the above-mentioned changes.

The changes can be seen as small perturbations that can "flip" the final state of an agent. In this paper, we only make a preliminary, statistical investigation of the existence of solutions.

Solutions are represented as a list of triples containing the index of the agent which makes the alternative decision, the simulation step in which it is performed and the actual decision index, from the set of possible decisions. It is reminded that the possible decisions are about the state and the perceived state value on which the agent acts.

Therefore, a stochastic approach was chosen to find the solutions (Pseudocode 1). There, *noTrials* represents the maximum number of attempts at finding a solution for a particular configuration, while *noMetaTrials* represents the number of repetitions of the same configuration, used for statistical analysis.

The GENERATEALTERNATIVEDECISIONS procedure depends on the type of perturbation. Three scenarios will be presented in the following sections, where the corresponding pseudocode will be given. Since the probability of finding a solution increases with the number of trials, we modelled this process as a probability distribution. The exponential negative distribution seems to capture the nature of the process.

Pseudocode 1. Stochastic search for solutions

```
procedure STOCHASTICSOLUTIONFINDER
begin
    for i = 1 to noMetaTrials do
        repeat
            env := GenerateEnvironment(RandomUniform(maxInt)) » random initial configuration
            RunToStabilization(env) » standard simulation
        until there exists targetAgent with u_final(targetAgent) = −1
        for j = 1 to noTrials do
            alternativeDecisionList := GENERATEALTERNATIVEDECISIONS(targetAgent)
            RunToStabilization(env, alternativeDecisionList) » simulation with alternative decisions
            if u_final(targetAgent) = 1 then
                break » solution found, record statistics
            end
        end
    end
end
```

4.3 Scenarios

a. Multiple Alternative Decisions of the Target Agent

In this scenario, we aim to find a variable number of alternative decisions, between 1 and a maximum number d_{max}, e.g. 10. The GENERATEALTERNATIVEDECISIONS procedure follows Pseudocode 2.

Pseudocode 2. Procedure for generating random alternative decisions

```
procedure GENERATEALTERNATIVEDECISIONS
begin
    for i = 1 to 1 + RandomUniform(d_max)  do » between 1 and d_max
        step := RandomUniform(maxSteps) » maxSteps is the number of steps in the standard simulation
        decision := RandomUniform(maxInt) modulo NoPossibleDecisions(step)
        alternativeDecisionList.Add((targetAgent, step, decision)) » the list is initially empty
    end
    return alternativeDecisionList
end
```

Table 1 presents the statistical results for the specified number of agents and states. The other fixed parameters, used for all these scenarios are: $v = 10$, $v_p = 3$, $\theta_{min} = 4$, $\theta_{max} = 5$. The recorded results are: the average number of trials after which a solution was found, the average number of steps of a simulation, the number of alternative decisions in *alternativeDecisionList*, and the percentage of the configurations for which a solution was found.

One can see that the number of problems with solutions increases with the number of agents and decreases with the number of states. With more agents present, a certain agent has more chances to cooperate, i.e. there are more chances to exit an "isolated" final state by choosing a different state value to act on, which in turn can benefit from the cooperation of other agents of the complementary type that also act on that state value.

Table 1. Statistical results for the first scenario

Agents	States	Trials	Steps	Alternative decisions	Configurations with solutions (%)
10	100	26.000	550.273	4.182	11
20		18.318	614.773	4.273	22
50		22.429	498.310	3.690	42
100		21.381	442.905	4.119	42
50	50	25.956	544.844	3.756	45
	100	22.429	498.310	3.690	42
	200	20.484	805.000	3.613	31
	500	16.276	1883.345	4.034	29

The latter effect is less intuitive. Although the problem space is larger, the decrease is not caused by the limited number of trials. Even with 1000 trials, the percentage of the solutions does not increase more than expected by following the negative exponential distribution.

This phenomenon remains to be further investigated, because one might think that a larger number of states increases the chance of an agent to find a feasible state value to act upon. However, this is not the case. Moreover, the larger number of simulation steps until stabilization reflects the fact that the interactions between the agents become more complex. The agents' ability to cooperate may be hampered in this way, but only a thorough investigation of the internal workings of the system may lead to a satisfactory explanation.

b. Single Alternative Decision of the Target Agent

This scenario is a simplification of the previous one, where we aim to find a single alternative decision that can reverse an unfavourable outcome for the target agent. The GENERATEALTERNATIVEDECISIONS follows Pseudocode 2, but without the *for* loop. The results are given in Table 2.

An interesting fact is that even the least invasive kind of perturbation, i.e. a single alternative decision, can change the final outcome in many cases. Compared with the first scenario, this case is more unfavourable to the heavily constrained configurations, such as those with a small number of agents, but it is more favourable to those with a higher number of agents.

Table 2. Statistical results for the second scenario

Agents	States	Trials	Steps	Configurations with solutions (%)
10	100	57.500	723.000	2
20		31.273	596.455	11
50		30.226	470.742	31
100		23.353	404.961	51
50	50	25.839	361.258	31
	100	30.226	470.742	31
	200	23.750	816.321	28
	500	26.259	2001.926	27

c. Multiple Alternative Decisions of All Agents

This scenario is the most general, where all the agents can make alternative deci-sions. The GENERATEALTERNATIVEDECISIONS follows Pseudocode 2, with the change of *alternativeDecisionList*.Add((RandomUniform(*noAgents*), *step*, *decision*)).

As expected, in Table 3 it can be seen that the number of solutions increases because of a greater diversity of influences.

Table 3. Statistical results for the third scenario

Agents	States	Trials	Steps	Alternative decisions	Configurations with solutions (%)
10	100	27.000	598.583	3.500	12
20		16.378	628.649	4.324	37
50		18.409	542.439	4.015	66
100		10.145	590.084	4.108	83
50	50	10.840	375.240	4.040	75
	100	18.409	542.439	4.015	66
	200	17.311	1078.295	4.311	61
	500	13.292	2148.292	4.231	65

5 Conclusions

The design of a multiagent system was presented, which despite simplicity can exhibit complex interactions that make it impossible to predict its final state, although its rules are completely deterministic. This system can be used to analyse the effect of alter-native decisions. In this paper, we performed a preliminary statistical investigation trying to assess the probability with which one or a small set of alternative decisions can change the final utility of an agent from minimum (-1) to maximum (1).

Many directions of future work can be envisioned. The internal configuration of the system should be analysed to see whether one can directly predict the existence of a solution or not. If an agent's set of perceived values is disjoint from those of the other agents, it cannot cooperate with them, i.e. it is isolated. In this case, the lack of a solution is easy to explain. But beside this case, the protocol can exhibit a very complex pattern of interactions that can lead to both positive and negative final utilities. This is worth investigating further, especially since the presented protocol is simpler and much faster than a previous one. When the solutions cluster together (as the steps in which they should be performed), it would be interesting to investigate whether some kind of compression is possible, i.e. whether there is a general model that defines the locations of the clusters.

References

1. Azuma, R., Daily, M., Furmanski, C.: A review of time critical decision making models and human cognitive processes. In: Proceedings of the 2006 IEEE Aerospace Conference, Big Sky, Montana, USA (2006). doi:10.1109/aero.2006.1656041
2. Bogacz, R., Brown, E., Moehlis, J., Holmes, P., Cohen, J.D.: The physics of optimal decision making: a formal analysis of models of performance in two-alternative forced choice tasks. Psychol. Rev. **113**(4), 700–765 (2006)
3. Hawkins, G., Brown, S.D., Steyvers, M., Wagenmakers, E.J.: Context effects in multi-alternative decision making: empirical data and a bayesian model. Cogn. Sci. **36**, 498–516 (2012). doi:10.1111/j.1551-6709.2011.01221.x
4. Holland, O.E.: Multiagent systems: lessons from social insects and collective robotics. Adaptation, Coevolution and Learning in Multiagent Systems: Papers from the 1996 AAAI Spring Symposium, pp. 57–62. AAAI Press, Menlo Park (1996)
5. Leon, F.: A multiagent system generating complex behaviours. In: Bădică, C., Nguyen, N. T., Brezovan, M. (eds.) ICCCI 2013. LNCS, vol. 8083, pp. 154–164. Springer, Heidelberg (2013). doi:10.1007/978-3-642-40495-5_16
6. Leon, F.: Analysis of behavior stability in a multiagent system. In: Frank, T.D. (ed.) New Research on Collective Behavior, Psychology Research Progress. Nova Publishers, New York (2016)
7. Leon, F.: Design and evaluation of a multiagent interaction protocol generating behaviours with different levels of complexity. Neurocomputing **146**(SI), 173–186 (2014). doi:10.1016/j.neucom.2014.04.058
8. Leon, F.: Stabilization methods for a multiagent system with complex behaviours. Comput. Intell. Neurosci., Article ID 236285, 20 pages (2015). doi:10.1155/2015/236285
9. Przybyła-Kasperek, M., Wakulicz-Deja, A.: Global decision-making in multi-agent decision-making system with dynamically generated disjoint clusters. Appl. Soft Comput. **40**, 603–615 (2016)
10. Reches, S., Talman, S., Kraus, S.: A statistical decision-making model for choosing among multiple alternatives. In: Proceedings of AAMAS 2007, Honolulu, Hawaii, USA (2007)
11. Rustichini, A.: Dual or unitary system? Two alternative models of decision making. Cogn. Affect. Behav. Neurosci. **8**(4), 355–362 (2008). doi:10.3758/cabn.8.4.355
12. Stüttgen, P., Boatwright, P., Monroe, R.T.: A satisficing choice model. Mark. Sci. **31**(6), 878–899 (2012)

Modeling Internalizing and Externalizing Behaviour in Autism Spectrum Disorders

Laura M. van der Lubbe, Jan Treur, and Willeke van Vught[✉]

Behavioural Informatics Group, Department of Computer Science, Vrije
Universiteit Amsterdam, Amsterdam, The Netherlands
lauravanderlubbe@hotmail.com, j.treur@vu.nl,
willekevanvught@hotmail.com

Abstract. This paper presents a neurologically inspired computational model
for Autism Spectrum Disorders addressing internalizing and externalizing
behaviour. The model has been verified by mathematical analysis and it is
shown how by parameter tuning the model can identify the characteristics of a
person based on empirical data.

1 Introduction

Over the years, much research has been performed in the area of Autism Spectrum
Disorders (ASD); e.g., [9, 19]. Most people think of people with autism as shy or
socially disabled, but ASD can occur in different forms. It is sometimes difficult to find
a suitable way of counseling somebody with ASD. Persons with ASD often need extra
counseling, for example during high school. For such counseling to be effective it is
important to have insight in the specific variant the person has, and which specific
mental processes and behaviours occur. A computational model can be a basis for
someone to get more understanding of such mental processes and behaviours.

One distinction that can be made within ASD is between persons who are internalizing versus those who are externalizing; e.g., [14]. The former type of persons may
show anxiety whereas the latter type may show aggression. The computational model
presented here addresses these two types of mental processes and behaviours, and
enables to model both internalizing persons and externalizing persons with ASD,
depending on the settings of certain parameters representing characteristics of the
person. Besides this, the model also covers other characteristics that can occur in
persons with ASD, such as enhanced sensory processing sensitivity (e.g., [1]), reduced
mirror neuron activation (e.g., [12]), imperfect self-other distinction (e.g., [5]), and
reduced emotion integration (e.g., [11], pp. 73−74). To cover the latter aspects as well,
and to obtain an integrative model, elements of an earlier model from [20, 21] were
incorporated.

This new model extends earlier models [20, 21] with different behavior types and
contributes to the understanding externalizing and internalizing behavior of persons

---

Authorships are based on comparable contribution.

© Springer International Publishing Switzerland 2016
N.T. Nguyen et al. (Eds.): ICCCI 2016, Part I, LNAI 9875, pp. 13–26, 2016.
DOI: 10.1007/978-3-319-45243-2_2

with ASD. This can be used as a basis for human-aware or socially aware computing applications within the field of ASD. In the Sect. 2 neurological background information is discussed. After that, in Sect. 3 the model is presented. Section 4 discusses some simulation experiments. Section 5 contributes verification of the model by mathematical analysis. In Sect. 6 it is shown how the model can be used to automatically identify the characteristics of a person, based on empirical behavioural data and a parameter tuning method. Finally, Sect. 7 is a discussion.

2   Neurological Background

The proposed computational model was designed on the basis of findings and theories from Cognitive and Social Neuroscience and Developmental Psychology. In this section these are briefly discussed. Persons with ASD often have an enhanced sensitivity of their sensory processing. Incoming stimuli easily result in a level of stress that has to be handled in some way. Being an internalizing or externalizing human being, is one of the differences between persons with ASD [14]. Internalizing feelings means that you do not show them, but you do feel them. In [14], internalizing persons are described as being withdrawn-depressed, anxious-depressed and having somatic complaints. One of the findings in [23] is that anxiety is often comorbid with ASD, as it is related to enhanced sensitivity and problems with emotion regulation.

Anxiety can be seen as an internalizing behaviour, as the behaviours are not clearly shown to the outside world. However, other persons with ASD are showing more externalizing behaviour; this means that those persons do express their feelings, mostly negative feelings like anger. [14] summarizes externalizing behaviour as aggressive and rule-breaking behaviour. Anxiety is not the only behaviour that can result from bad emotion regulation; [13] describes that aggressive behaviour of persons with ASD can be caused by poor emotion regulation as well. Since aggressive behaviour is an externalizing behaviour type this behaviour is linked to externalizing children. [18] showed that aggressive behaviour in children with ASD could be caused by a combination of poor emotion regulation and impaired understanding of emotions of others.

Anxiety can be seen as a defensive reaction to a potential threat [6], in this case the avoiding of the gaze of the other person is a defensive reaction on the threat of a stimulus for which the person is highly sensitive. She or he does not communicate with the other person, which can be interpreted as a flight response. The externalizing person shows more a fight response and expresses aggressiveness, looks the other person in the eye and communicates with the other person. Such a person does not express anxiety.

The model introduced in Sect. 3 takes into account the two opposite behaviour types, internalizing and externalizing behaviour, as discussed above. The model addresses how these behaviour types relate to the dynamics of processes involving a number of internal states and the expressions of the body, an avoiding gaze and communication.

3   The Computational Model

The computational model has been designed using the temporal-causal network modelling approach described in [22]. According to this general dynamic modeling approach a model is designed at a conceptual level, for example, in the form of a graphical conceptual representation or a conceptual matrix representation. A graphical conceptual representation displays nodes for *states* and arrows for *connections* indicating causal impacts from one state to another (e.g., as shown in Fig. 1 below), and includes some additional information in the form of a *connection weight* for each connection (for the strength of the impact), and for each state a *speed factor* (for the timing of the effect of the impact), and the type of *combination function* used (to aggregate multiple impacts on the state). In Table 1 the states used in the model are briefly explained. These states are depicted as nodes in Fig. 1. Sensory representation states srs_s, srs_{self}, srs_B and srs_b are used for stimulus s, the agent *self*, other agent B, and body states b that embody and label emotional states; for body state b two instances are considered: *anx* for an anxious and *agg* for aggressive. For some of these (s and B), which refer to the external world also sensor states ss_s and ss_B are used to incorporate sensing from the world states. Two types of preparation states are considered: ps_b for body states b and ps_B for communication to the other agent B. The preparation states

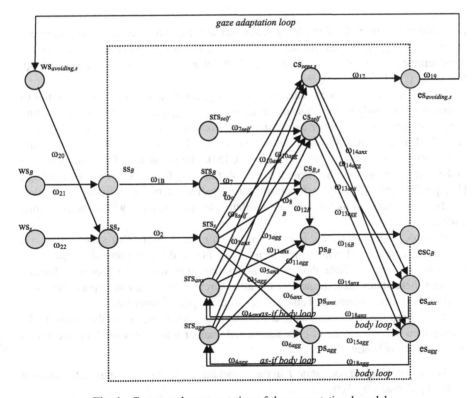

Fig. 1. Conceptual representation of the computational model

Table 1. States used in the model

State	Explanation
ss_s	sensor state for stimulus s
ss_B	sensor state for B
srs_s	sensory representation state of stimulus s
srs_{self}	sensory representation state of agent $self$
srs_B	sensory representation state of other agent B
srs_b	sensory representation state of body state b
ps_b	preparation state for body state b
ps_B	preparation state for communication to other agent B
$cs_{B,s}$	control state for self-other distinction concerning agent B
$cs_{sens,s}$	control state for enhanced sensory sensitivity for s
cs_{self}	control state for the agent itself
es_b	execution state for body state b
esc_B	execution state for communication to B
$es_{avoiding,s}$	execution state for avoidance of s
ws_s	world state for stimulus s
ws_B	world state for other agent B
$ws_{avoiding,s}$	world state for gaze avoiding stimulus s

ps_{anx} and ps_{agg} for each of the body states anx and agg are affected by the representation states srs_{anx} and srs_{agg} for these body states, and in turn affect in a cyclic manner these representation states srs_{anx} and srs_{agg}, both by an as-if body loop and a body loop, following [7, 8].

Execution (or expression) states es_{anx} and es_{agg} are included for these two types of preparations for body states, plus an execution state $es_{avoiding,s}$ for a stimulus s avoiding gaze. The actual execution or expression of preparations for body states and gaze is controlled by control states cs_{self} for $self$ and $cs_{sens,s}$ for enhanced sensory processing sensitivity for s (emotion regulation; e.g., [10, 15]). The preparation for communication to the other agent gets control from the control state $cs_{B,s}$ for self-other distinction (e.g., [12], pp. 201–202, [5]). By such control states specific internal monitoring and control functions are modeled that usually are attributed to specific areas within the prefrontal cortex.

In Table 2 for each state it is indicated which impacts from other states it gets, via which connections and with which weights. In Fig. 1 these weights are depicted as labels for the arrows. Note that as the nodes represent states, the processes happen between these states, as indicated by the arrows representing causal impact; the terms used in the fourth column in Table 2 refer to the types of processes.

The conceptual representation of the model as shown in Fig. 1 and the tables can be transformed in a systematic or even automated manner into a numerical representation of the model as follows [22]:

- At each time point t each state Y in the model has a real number value in the interval [0, 1], denoted by $Y(t)$

Table 2. Connections and their weights

From states	To state	Weight	Connection
ws_s	ss_s	ω_{22}	sensing stimulus s
$ws_{avoiding,s}$		ω_{20}	suppressing sensing of s
ws_B	ss_B	ω_{21}	sensing agent B
ss_s	srs_s	ω_2	representing s
ss_B	srs_B	ω_{1B}	representing B
ps_b	srs_b	ω_{4b}	predicting b
es_b		ω_{18b}	effectuating b
srs_s	ps_b	ω_{5b}	responding b
srs_b		ω_{6b}	amplifying b
srs_{anx}	ps_B	ω_{11anx}	responding communication
srs_{agg}		ω_{11agg}	to anxiety and aggression
$cs_{B,s}$		ω_{12B}	controlling communication
srs_B	$cs_{B,s}$	ω_{7B}	monitoring B for self-other
srs_s		ω_{8B}	monitoring s for self-other
srs_s	$cs_{sens,s}$	ω_9	monitoring s for sensitivity
srs_{anx}		ω_{10anx}	monitoring anxiety
srs_{agg}		ω_{10agg}	monitoring aggression
srs_{self}	cs_{self}	ω_{7self}	monitoring $self$
srs_s		ω_{8self}	monitoring s
srs_{anx}		ω_{3anx}	monitoring anx
srs_{agg}		ω_{3agg}	monitoring agg
cs_{self}	es_b	ω_{13b}	controlling response b
$cs_{sens,s}$		ω_{14b}	suppressing response b
ps_b		ω_{15b}	executing response b
ps_B	esc_B	ω_{16}	executing communication
$cs_{sens,s}$	$es_{avoiding,s}$	ω_{17}	executing avoidance of s
$es_{avoiding,s}$	$ws_{avoiding,s}$	ω_{19}	effectuating avoidance of s

- At each time point t each state X connected to state Y has an *impact* on Y defined as $\textbf{impact}_{X,Y}(t) = \omega_{X,Y} X(t)$ where $\omega_{X,Y}$ is the weight of the connection from X to Y
- The *aggregated impact* of multiple states X_i on Y at t is determined using a *combination function* $\mathbf{c}_Y(..)$:

$$\textbf{aggimpact}_Y(t) = \mathbf{c}_Y(\textbf{impact}_{X_1,Y}(t), \ldots, \textbf{impact}_{X_k,Y}(t)) = \mathbf{c}_Y(\omega_{X_1,Y}X_1(t), \ldots, \omega_{X_k,y}X_k(t))$$

where X_i are the states with connections to state Y
- The effect of $\textbf{aggimpact}_Y(t)$ on Y is exerted over time gradually, depending on *speed factor* η_Y:

$$Y(t + \Delta t) = Y(t) + \eta_Y[\mathbf{aggimpact}_Y(t) - Y(t)]\,\Delta t$$
$$\text{or} \quad \mathbf{d}Y(t)/\mathbf{d}t = \eta_Y[\mathbf{aggimpact}_Y(t) - Y(t)]$$

- Thus the following *difference* and *differential equation* for Y are obtained:

$$Y(t + \Delta t) = Y(t) + \eta_Y[\mathbf{c}_Y(\omega_{X_1,Y}X_1(t), \ldots, \omega_{X_k,Y}X_k(t)) - Y(t)]\,\Delta t$$
$$\mathbf{d}Y(t)/\mathbf{d}t = \eta_Y[\mathbf{c}_Y(\omega_{X_1,Y}X_1(t), \ldots, \omega_{X_k,Y}X_k(t)) - Y(t)]$$

As an example, according to the pattern described above the difference and differential equation for ps_{anx} are as follows:

$$ps_{anx}(t + \Delta t) = ps_{anx}(t) + \eta_{ps_{anx}}[\mathbf{c}_{ps_{anx}}(\omega_{5anx}\,srs_s(t), \omega_{6anx}srs_{anx}(t)) - ps_{anx}(t)]\,\Delta t$$
$$dps_{anx}/\mathbf{d}t = \eta_{ps_{anx}}[\mathbf{c}_{ps_{anx}}(\omega_{5anx}srs_s(t), \omega_{6anx}srs_{anx}(t)) - ps_{anx}(t)]$$

So, for any set of values for the connection weights, speed factors and any choice for combination functions, each state of the model gets a difference or differential equation assigned. For the model considered here this makes a set of 17 coupled difference or differential equations, that together, in mutual interaction describe the model's behaviour. Note that the speed factors enable to obtain a realistic timing of the different states in the model, for example, to tune the model to the timing of processes in the real world.

For all states except the sensor state ss_s, for the combination function either the *identity function* $\mathbf{id}(..)$ or the *advanced logistic sum combination function* $\mathbf{alogistic}_{\sigma,\tau}(\ldots)$ is used [22]:

$$\mathbf{c}_Y(V) = \mathbf{id}(V) = V$$
$$\mathbf{c}_Y(V_1, \ldots V_k) = \mathbf{alogistic}_{\sigma,\tau}(V_1, \ldots, V_k) = \left(\frac{1}{1+e^{-\sigma(V_1 + \ldots + V_k - \tau)}} - \frac{1}{1+e^{\sigma\tau}}\right)(1+e^{-\sigma\tau})$$

Here σ is a *steepness* parameter and τ a *threshold* parameter. The advanced logistic sum combination function has the property that activation levels 0 are mapped to 0 and keeps values below 1. The identity function $\mathbf{id}(..)$ is used for the 6 states with a single impact: ss_B, srs_B, srs_s, esc_B, $es_{avoiding,s}$, $ws_{avoiding,s}$. For example, the difference and differential equation for srs_s are as follows:

$$srs_s(t + \Delta t) = srs_s(t) + \eta_{srs_s}[\omega_2 ss_s(t) - srs_s(t)]\,\Delta t$$
$$dsrs_s(t)/\mathbf{d}t = \eta_{srs_s}[\omega_2 ss_s(t) - srs_s(t)]$$

The function $\mathbf{alogistic}_{\sigma,\tau}(\ldots)$ is used as combination function for the 10 states with multiple impacts, except the sensor state ss_s for stimulus s: srs_{anx}, srs_{agg}, $cs_{sens,s}$, cs_{self}, $cs_{B,s}$, ps_B, ps_{anx}, ps_{agg}, es_{anx}, es_{agg}. For example, the difference and differential equation for ps_{anx} are as follows:

$$ps_{anx}(t + \Delta t) = ps_{anx}(t) + \eta_{ps_{anx}}[\mathbf{alogistic}_{\sigma,\tau}(\omega_{5anx}srs_s(t), \omega_{6anx}srs_{anx}(t)) - ps_{anx}(t)]\,\Delta t$$
$$dps_{anx}/\mathbf{d}t = \eta_{ps_{anx}}[\mathbf{alogistic}_{\sigma,\tau}(\omega_{5anx}srs_s(t), \omega_{6anx}srs_{anx}(t)) - ps_{anx}(t)]$$

For the sensor state ss_s the effect of the avoiding gaze is modelled by the following combination function $c_{ss_s}(V_1, V_2)$, where V_1 refers to the impact $\omega_{22}\, ws_s(t)$ from ws_s on ss_s and V_2 to the impact $\omega_{20}\, ws_{avoiding,s}(t)$ from $ws_{avoiding,s}$ on ss_s:

$$c_{ss_s}(V_1, V_2) = V_1(1 - V_2)$$

This function makes the sensing of stimulus s inverse proportional to the extent of avoidance; e.g., sensing s becomes 0 when avoidance is 1, and V_1 when avoidance is 0. According to this combination function the difference and differential equation for ss_s are as follows:

$$ss_s(t + Dt) = ss_s(t) + \eta_{ss_s}[\omega_{22}ws_s(t)(1 - \omega_{20}ws_{avoiding,s}(t)) - ss_s(t)]\,\Delta t$$
$$dss_s/dt = \eta_{ss_s}[\omega_{22}ws_s(t)(1 - \omega_{20}ws_{avoiding,s}(t)) - ss_s(t)]$$

4 Simulation Experiments

The numerical representation of the model discussed above (in the form of the 17 difference equations for the 17 states) has been implemented in Python. In this section simulations are discussed that show the two different types of behaviours. In order to let the model show the behavior of an externalizing or internalizing person the parameters are constrained. Using the constraints shown in Table 3 the behavior of an externalizing or internalizing case will be shown in a realistic manner, according to the neurological background of Sect. 2. Internalizing persons suppress the aggressive response (high ω_{14agg}) more than the anxious response (low ω_{14anx}); for externalizing people this is opposite. Externalizing persons have a strong control over their communication (high ω_{12}), while internalizing persons don't (low ω_{12}). Externalizing persons have a low avoidance of stimuli (low ω_{17}), while internalizing people have a stronger tendency to avoid stimuli (high ω_{17}). The model also functions with parameters disregarding the constraints. However, in these cases the behavior cannot be classified as either internalizing or externalizing.

For the simulations discussed here the step size Δt was 0.5, all speed factors were 1, and all connection weights except the four in Table 3 were always 1. Moreover, for the states that use an advanced logistic sum combination function the threshold and steepness values were as shown in Table 4.

The initial values of all states were 0, except for the world states ws_s and ws_B for stimulus s and the other agent B, which as a form of input for the agent had constant

Table 3. Intervals for parameters for externalizing and internalizing

Weight	Externalizing	Internalizing
ω_{14agg}	[−0.3, 0]	[−1, −0.7]
ω_{14anx}	[−1, -0.7]	[−0.3, 0]
ω_{12}	[0.7, 1]	[0, 0.3]
ω_{17}	[0, 0.3]	[0.7, 1]

Table 4. Parameter values used for steepness σ and threshold τ

state	τ	σ	state	τ	σ	state	τ	σ
srs_{anx}	0.8	8	$cs_{B,s}$	1	40	es_{anx}	1.5	5
srs_{agg}	0.8	8	cs_{self}	1	40	es_{agg}	1.5	5
ps_{anx}	1	8	$cs_{sens,s}$	1.2	40	esc_B	0.5	2
ps_{agg}	1	8	ps_B	1.5	2	$es_{avoiding,s}$	0.5	40

value 1 for the whole time. The upper graph in Fig. 2 shows simulation results for an externalizing, aggressive person ($\omega_{12} = 1$, $\omega_{14agg} = 0$, $\omega_{14anx} = -1$ and $\omega_{17} = 0$), i.e., there is no suppression of the aggressive response, but there is suppression of the anxious response, there is no stimulus avoiding gaze, and there is communication. In the first few time steps all values go up: in the beginning, there is an input for the sensor states in the model and it takes some time to reach all other states, and in particular the control states.

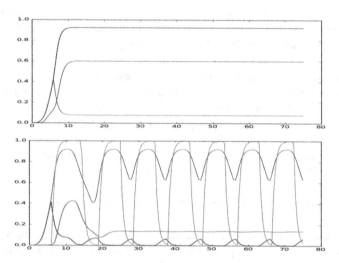

Fig. 2. Simulations of an externalizing person (upper graph) and an internalizing person (lower graph). Horizontal axis: time. Vertical axis: activation value. Light blue: $es_{avoiding,s}$ (gaze avoiding s). Green: es_{anx} (expressing anxiety). Red: esc_B (communication). Dark blue: es_{agg} (expressing aggressiveness). (Color figure online)

In time step 6, the anxious expression declines fast. This is due to the suppressive effect of ω_{14anx} (from $cs_{sens,s}$ to es_{anx}). The expressed aggression goes up and stays activated; the person is aggressive as long as the stimulus is present. Because in this simulation the stimulus never fades away, the aggressiveness stays too. When the person becomes aggressive, he/she faces the stimulus (no avoidance) and starts to communicate to the other agent (e.g., yelling at somebody). In the simulation the aggressiveness level never becomes 1; there is always a little bit of anxiousness present.

The lower graph in Fig. 4 shows a simulation of an internalizing, anxious person ($\omega_{12} = 0$, $\omega_{14agg} = -1$, $\omega_{14anx} = 0$, and $\omega_{17} = 1$); i.e., there is suppression of the aggressive response, but there is no suppression of the anxious response, there is an avoiding gaze, and there is no communication. In the beginning, the values of all states go up again, but in time step 6, it can be seen that the aggressiveness declines rapidly. As part of this drop, body representation state srs_{agg} drops, which has influence on the preparation state ps_B for communication. Also body representation state srs_{anx} (which is high) has an influence on ps_B. This causes that ps_B is not declining immediately. Because the control state $cs_{sens,s}$ is high, the execution of the avoiding gaze $es_{avoiding,s}$ becomes also high which causes the person to look away, this causes that the stimulus fades away for that person and so is the expression of anxiousness. Because the anxiousness drops, and body representation state srs_{agg} is still low, the communication preparation state ps_B becomes lower which causes the communication to stop. When the person is less anxious, the control state $cs_{sens,s}$ becomes lower which causes less suppression of aggressive expression es_{agg} and therefore the aggressiveness is shown a little. Because the stimulus is fading away for the person, there is no reason not to look at s anymore and the person looks at it again. The process goes on like this, ending up in a repeating (limit cycle) pattern. The communication is not coming back, this makes sense because if a person is internalizing, he or she withdraws him/herself socially and does not communicate anymore.

5   Verification by Mathematical Analysis

In this section, it is discussed how a mathematical analysis was performed of the equilibria of the model, in order to enable verification of (the implementation of) the model. A state Y has a *stationary point* at t if $dY(t)/dt = 0$. The model is in *equilibrium* a t if every state Y of the model has a stationary point at t. From the specific format of the differential or difference equations it follows that state Y has a stationary point at t if and only if

$$Y(t) = \mathbf{c}_Y(\omega_{X_1,Y} X_1(t), \ldots, \omega_{X_k,Y} X_k(t))$$

where X_i are the states with connections to state Y, and $\mathbf{c}_{Xi}(\ldots)$ is the combination function for Y. If the values of the states for an equilibrium are indicated by $\underline{X}_i$ then being in an equilibrium state is equivalent to a set of 17 equilibrium equations for the 17 states X_i of the model:

$$\underline{X}_i = \mathbf{c}_{X_i}(\omega_{X_1,X_i} \underline{X}_1, \ldots, \omega_{X_k,X_i} \underline{X}_k)$$

For example, for state srs_s the identity function is used as a combination function; then the above equilibrium equation is

$$\underline{\mathbf{srs}}_s = \omega_2 \, \underline{\mathbf{ss}}_s$$

The 5 equilibrium equations for ss_B, srs_B, esc_B, $es_{avoiding,s}$, $ws_{avoiding,s}$ are similar to this one. As another example, for state ps_{anx} the combination function is the advanced logistic function; then the equilibrium equation is

$$\underline{\mathbf{ps}}_{anx} = \mathbf{alogistic}_{\sigma,\tau}(\omega_{5anx}\,\underline{\mathbf{srs}}_s,\ \omega_{6anx}\,\underline{\mathbf{srs}}_{anx})$$

The 9 equilibrium equations for srs_{anx}, srs_{agg}, $cs_{sens,s}$, cs_{self}, $cs_{B,s}$, ps_B, ps_{agg}, es_{anx}, es_{agg} are similar to this more complicated one. Finally, the equilibrium equation for ss_s is:

$$\underline{\mathbf{ss}}_s = \omega_{22}\,\underline{\mathbf{ws}}_s\left(1 - \omega_{20}\,\underline{\mathbf{ws}}_{avoiding,s}\right)$$

These 17 equilibrium equations cannot be solved analytically in an explicit manner, due to the 10 equations among them involving a logistic function. However, they still can be used for verification of the model. This can be done by substituting the values found in a simulation at the end time in these equations, and then check whether the equations hold. This indeed has been done for a number of arbitrary cases (for different parameter values for ω_{12}, ω_{14agg}, ω_{14anx}, and ω_{17}) for the externalizing type of person, and the equations turned out to always hold (with an accuracy 10^{-15} or lower).

For the internalizing type the equilibrium equations never hold, as then the pattern becomes a limit cycle with state values changing all the time. However, in a limit cycle each state fluctuates between a minimum and a maximum value. At the time points for these minima and maxima the state has a stationary point, which means that the equation

$$Y(t) = \mathbf{c}_Y(\omega_{X_1,Y}\,X_1(t),\ldots,\omega_{X_k,Y}X_k(t))$$

should be fulfilled. This can be verified as well. This indeed has been done for the minima and maxima within the limit cycle of the internalizing type with $\omega_{12} = 0$, $\omega_{14agg} = -1$, $\omega_{14anx} = 0$, and $\omega_{17} = 1$ (using step size 0.05). It turned out that the stationary point equations were indeed fulfilled for all states of the model, with an average accuracy of 0.0041 over the minima and maxima of all states (the maximal deviation among the minima and maxima of the different states was 0.027, which is still reasonable, as in that case a very high steepness $\sigma = 40$ was applied, which can lead to sharp turning points).

The period of the limit cycles was also analyzed. The period was analyzed after the first drop, as only from there the graph becomes stable. Analyzing this, for the most clear internalizing type it was found that the period is constant for all states, with value 9.5. This is coherent with the theory behind the model that the same process is repeated all the time. The length of the period depends on the exact parameter values for ω_{12}, ω_{14agg}, ω_{14anx}, and ω_{17}. For other values within the constraints it can reach 17. Unfortunately, there are no methods known to analyse this period mathematically.

The verification outcomes provide evidence that the model (as implemented in Python) does what is expected.

6 Tuning Characteristics in the Model

The model can be used to model different types of persons, with different characteristics. These personal characteristics are represented by the values of parameters in the
model. For practical use of the model, for a given person these values have to be found.
These values could be based on questionnaires filled by the person, but it would be
more convenient if the characteristics can be determined automatically based on
observed behaviour of the person; this can be used as a form of automated diagnosis.
This section shows how this indeed can be done. In order to test this, some observable
empirical data are needed.

Because real empirical data were not available yet, the (pseudo-empirical) data used
to explain and test the approach were generated (by a third person) based on the model
with random parameter values for ω_{12}, ω_{14agg}, ω_{14anx}, and ω_{17} that satisfy the constraints
described in Table 3, after which noise was added to make the data more like realistic.
These parameters have been addressed as these parameters determine the type of
characteristics of a person as addressed in this paper. The observable states used are the
body and communication execution (or expression) states es_{agg}, es_{anx}, $es_{avoidance,s}$, and
esc_B. The dots in Fig. 3 show these expression states according to the pseudo-empirical
data. The dots show high values for es_{agg} and esc_B, and low values for es_{anx} and
$es_{avoidance,s}$. This already indicates that the person can be an externalizing person. To
find the parameter values characterizing the person represented by these data, two
parameter estimation methods were used: exhaustive search and simulated annealing.
For both cases an error function based on the sum of squares of the deviations was used.

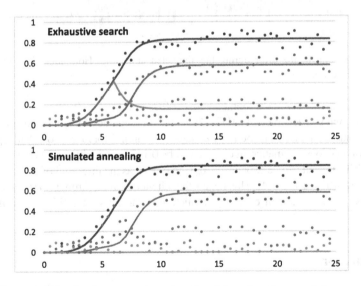

Fig. 3. The model (lines) using parameter values found by exhaustive search (upper graph) and
simulated annealing (lower graph) in comparison to the data (dots). Horizontal axis: time.
Vertical axis: activation value. Light blue: $es_{avoiding,s}$ (gaze avoiding s). Green: es_{anx} (expressing
anxiety). Red: esc_B (communication). Dark blue: es_{agg} (expressing aggressiveness). (Color figure
online)

The constraints for the intervals of the possible values of the parameters limit the set of possible parameter values. Therefore *exhaustive search* can be feasible. This method has the advantage that a set with all possibilities is created, which gives the certainty that the correct set (global optimum) is among them. Other methods may only come up with a local optimum. Using exhaustive search with grain size 0.01, the following weights were found: $\omega_{12} = 0.71$, $\omega_{14agg} = -0.18$, $\omega_{14anx} = -0.84$, and $\omega_{17} = 0.3$. The accuracy for these weights was found to be 0.0563. Such values indeed are expected to represent somebody who does externalizing. This shows that it is indeed possible to identify the characteristics of a person expressed in terms of the found parameter values, using exhaustive search applied to behavioural data. In Fig. 3 (upper graph) it is shown how for these parameter values the model fits to the data.

As an alternative parameter tuning method, also a *simulated annealing* approach was applied. The lower graph in Fig. 3 shows how for these parameter values the model fits to the data. It can be seen that over time (and decreasing temperature) the changes become smaller. The graph in Fig. 4 shows the plot of the error during this process.

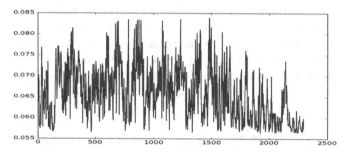

Fig. 4. Progression of the error during the simulated annealing. Horizontal axis: number of iterations. Vertical axis: error value

The best weights that were found are $\omega_{12} = 0.721$, $\omega_{14agg} = -0.195$, $\omega_{14anx} = -0.884$, $\omega_{17} = 0.117$, with an accuracy of 0.0573. Note that this accuracy for simulated annealing is just a bit worse than what was found by exhaustive search. Moreover, note that apparently the value of ω_{17} does not matter much, as it can be 0.117 or 0.3 without much difference in accuracy. This indeed can be explained from the model, as the threshold of $es_{avoiding,s}$ is 0.5 and steepness 40 (see Table 4). This means that all values of ω_{17} from 0 to 0.3 will lead to (practically) no activation of $es_{avoiding,s}$ (which also can be seen in Fig. 3: the flat light blue line).

7 Discussion

This paper presented a computational model for persons with ASD enabling to distinguish two different behaviour types that are prevalent in ASD: internalizing and externalizing behaviour [14]. The model was inspired by findings and theories from Cognitive and Social Neuroscience and designed as a network of mental states according to the temporal-causal network modeling approach presented in [22]. By simulation experiments and mathematical analysis the model was verified.

The presented model specifically addresses findings and theories concerning internalizing and externalizing behaviour types within ASD. However, as elements from the model described in [20, 21] were adopted as well, the model also integrates some other aspects of ASD, addressed by theories: reduced mirror function or poor self-other distinction; e.g., [5, 12].

The model represents specific personal characteristics by specific values of parameters included in the model, such as connection weights. By using proper choices for these connection weights, the model can either simulate an internalizing person or an externalizing person. Moreover, it was shown how based on a given data set concerning a person's behaviour, by parameter estimation methods the behaviour type of this person can be identified automatically.

The computational model proposed here can be used as an ingredient to develop human-aware or socially aware computing applications (e.g. [16, 17, 19]) that can provide help in getting more understanding of the different behaviour types and their influence on the behaviour of a person with ASD. More specifically, in [2, 19] it is shown how such applications can be designed with knowledge of human and/or social processes as a main ingredient represented by a computational model of these processes which is embedded within the application. Such computational models can have the form, for example, of qualitative causal models, or of dynamical numerical models. As an example, in [3, 4] this design approach is illustrated to obtain a human-aware software agent supporting professionals in attention-demanding tasks, based on an embedded dynamical numerical model for attention. The computational model for ASD proposed here can be used in a similar manner to design a human-aware or socially aware software agent to support persons with ASD. This might be helpful in particular for those who are counseling or supervising persons with ASD. The method shown to identify the behaviour type of a person based on empirical behavioural data can be useful, for example, in choosing a counseling approach.

References

1. Baker, A.E.Z., Lane, A.E., Angley, M.T., Young, R.L.: The relationship between sensory processing patterns and behavioural responsiveness in autistic disorder: a pilot study. J. Autism Dev. Disord. **38**, 867–875 (2008)
2. Bosse, T., Hoogendoorn, M., Klein, M., Treur, J.: A generic agent architecture for human-aware ambient computing. In: Mangina, E., Carbo, J., Molina, J.M. (eds.) Agent-Based Ubiquitous Computing, pp. 35–62. World Scientific Publishers: Atlantis Press, Amsterdam (2009)
3. Bosse, T., Memon, Z.A., Oorburg, R., Treur, J., Umair, M., de Vos, M.: A software environment for an adaptive human-aware software agent supporting attention-demanding tasks. J. AI Tools **20**, 819–846 (2011)
4. Bosse, T., Memon, Z.A., Treur, J., Umair, M.: An adaptive human-aware software agent supporting attention-demanding tasks. In: Yang, J.-J., Yokoo, M., Ito, T., Jin, Z., Scerri, P. (eds.) PRIMA 2009. LNCS, vol. 5925, pp. 292–307. Springer, Heidelberg (2009)

5. Brass, M., Spengler, S.: The inhibition of imitative behaviour and attribution of mental states. In: Striano, T., Reid, V. (eds.) Social Cognition: Development, Neuroscience, and Autism, pp. 52–66. Wiley-Blackwell, Oxford (2009)

6. Cisler, J.M., Olantunji, B.O., Feldner, M.T., Forsyth, J.P.: Emotion regulation and the anxiety disorders: an integrative review. Psychopathol. Behav. Assess **32**, 68–82 (2010)

7. Damasio, A.R.: The Feeling of What Happens: Body and Emotion in the Making of Consciousness. Harcourt Brace, New York (1999)

8. Damasio, A.R.: Looking for Spinoza. Harcourt, Orlando (2003)

9. Frith, U.: Autism: Explaining the Enigma. Blackwell, Oxford (2003)

10. Goldin, P.R., McRae, K., Ramel, W., Gross, J.J.: The neural bases of emotion regulation: reappraisal and supression of negative emotion. Biol. Psychiatry **63**, 577–586 (2008)

11. Grèzes, J., de Gelder, B.: Social perception: understanding other people's intentions and emotions through their actions. In: Striano, T., Reid, V. (eds.) Social Cognition: Development, Neuroscience, and Autism, pp. 67–78. Wiley-Blackwell, Oxford (2009)

12. Iacoboni, M.: Mirroring People: The New Science of How We Connect with Others. Farrar, Straus & Giroux, New York (2008)

13. Marsee, M.A., Frick, P.J.: Exploring the cognitive and emotional correlates to proactive and reactive aggression in a sample of detained girls. Abnorm. Child Psychol. **35**, 969–981 (2007)

14. Noordhof, A., Krueger, R.F., Ormel, J., Oldehinkel, A.J., Hartman, C.A.: Integrating autism-related symptoms into the dimensional internalizing and externalizing model of psychopathology. TRAILS Study Abnorm. Child Psychol. **43**, 577–587 (2015)

15. Ochsner, K.N., Gross, J.J.: The neural bases of emotion and emotion regulation: a valuation perspective. In: Handbook of Emotional Regulation, 2nd edn., pp. 23–41. Guilford, New York (2014)

16. Pantic, M., Pentland, A., Nijholt, A., Huang, T.S.: Human computing and machine understanding of human behavior: a survey. In: Proceedings of the International Conference on Multimodal Interfaces, pp. 239–248 (2006)

17. Pentland, A.: Socially aware computation and communication. IEEE Comput. **38**, 33–40 (2005)

18. Pouw, L.B.C., Rieffe, C., Oosterveld, P., Huskens, B., Stockmann, L.: Reactive/proactive aggression and affective/cognitive empathy in children with ASD. Res. Dev. Disabil. **34**(4), 1256–1266 (2013)

19. Treur, J.: On human aspects in ambient intelligence. In: Mühlhäuser, M., Ferscha, A., Aitenbichler, E. (eds.) AmI 2007 Workshops. CCIS, vol. 11, pp. 262–267. Springer, Heidelberg (2008)

20. Treur, J.: A cognitive agent model displaying and regulating different social response patterns. In: Walsh, T. (ed.) Proceedings of the Twenty-Second International Joint Conference on Artificial Intelligence, IJCAI 2011, pp. 1735–1742 (2011)

21. Treur, J.: Displaying and regulating different social response patterns: a computational agent model. Cogn. Comput. **6**, 182–199 (2014)

22. Treur, J.: Dynamic modeling based on a temporal-causal network modeling approach. Biol. Inspired Cogn. Architectures **16**, 131–168 (2016)

23. White, S.W., Mazefsky, C.A., Dichter, G.S., Chiu, P.H., Richey, J.A., Ollendick, T.H.: Social-cognitive, physiological, and neural mechanism underlying emotion regulation impairments: understanding anxiety in autism spectrum disorder. Int. J. Dev. Neurosci. **39**, 22–36 (2014)

Analysis and Refinement of a Temporal-Causal Network Model for Absorption of Emotions

Eric Fernandes de Mello Araújo[(⊠)] and Jan Treur

Behavioural Informatics Group, VU Amsterdam, Amsterdam, The Netherlands
{e.araujo, j.treur}@vu.nl

Abstract. For an equilibrium for any member t An earlier proposed temporal-causal network model for mutual absorption of emotions aims to model emotion contagion in networks using characteristics such as traits of openness and expressiveness of the members of the network, and the strengths of the connections between them. The speed factors describing how fast emotional states change, were modeled based on these characteristics according to a fixed dependence relation. In this paper, particular implications of this choice are analyzed. Based on this analysis, a refinement of the model is proposed, offering alternative ways of modeling speed factors. This refinement is also analyzed and evaluated.

Keywords: Temporal-causal network model · Absorption model · Emotion contagion · Computational modeling

1 Introduction

The social phenomenon called emotion contagion indicates the process by which emotions of a person are affected by emotions of other persons when they are interacting in a social network. This concept has a foundation in neurological findings on mirror neurons [1], and can be used to understand emotions, for example in situations where decisions can be affected by the emotional state of a person. This can occur in urgent situations, when events with a short duration can create disturbances in decisions, but also in processes with longer durations, like mood and depression, commitment with work, et cetera.

Different computational models have been proposed to model emotion contagion. Among them are temporal-causal network models [2] such as the absorption model, introduced in [3] and the amplification model introduced in [4]. The current paper focuses on the absorption model. In this model emotion contagion is modeled using personal characteristics (or traits) such as openness (how a person is open to be influenced by others) and expressiveness (how a person expresses him or herself in the social network), and the strength of the connection between persons. This paper presents an analysis and refinement of the absorption model, in particular by considering multiple options for the way in which the speed factor is modeled. In the original absorption model, a fixed dependence relation is used for the speed factor, describing how the speed factor relates to the traits and connections in the network. In the proposed refined absorption model, in addition two alternative ways are offered that

© Springer International Publishing Switzerland 2016
N.T. Nguyen et al. (Eds.): ICCCI 2016, Part I, LNAI 9875, pp. 27–39, 2016.
DOI: 10.1007/978-3-319-45243-2_3

relate the speed factor in different ways to these network characteristics. The effects and improvements that are obtained from these alternative options are analyzed and evaluated as well. This work also shows a more in depth mathematical analysis to better understand convergence and stability in the model. These analyses show that the presented temporal-causal network model is trustworthy and can be very useful to understand different contexts of emotions in social networks.

The paper has the following structure: Sect. 2 will explain the original absorption model in detail. In Sect. 3 an analysis is made in particular concerning the speed factor in the model. Section 4 presents two possible alternative ways to model the speed factor. A scaled approach and an advanced logistic function approach are the options explored in this section. Section 5 presents mathematical analysis of the model regarding monotonicity and equilibria. Section 6 presents results using the new approach for the model, and Sect. 7 presents the conclusions and future works.

2　Emotion Absorption: The Temporal-Causal Network Model

In this section, the computational model for mutual absorption of emotions is presented [3, 5]. This model has been developed as a temporal-causal network model; see [2]. First, the most important concepts used in the model are explained, both in terms of a conceptual representation and a numerical representation. The section concludes with examples of applications of this computational model of emotion contagion.

The model distinguishes some characteristics of persons and the connections between them, represented by parameters. These characteristics affect emotion contagion in the network. The model describes how internal emotion states q_A of persons A affect each other. However, internal states do not affect each other in a direct manner. First, they have to be expressed, after which they can be observed by another person, and in turn such an observation can affect the internal state of this other person. So, internal emotion states q_A affect each other by *contagion as a three-step process*, for which each step has its own characteristics (indicated by ε_B, α_{BA}, δ_A, respectively):

- from internal emotion state q_B of B to expressed emotion by B ε_B
- from expressed emotion by B to observed emotion by A α_{BA}
- from observed emotion by A to internal emotion state q_A of A δ_A

The characteristic for the extent to which a person B expresses him or herself within the network is captured by the concept of *expressiveness*, modeled by parameter ε_B. Similarly, the characteristic for the extent to which a person A is open to be influenced is represented by the *openness*, modeled by parameter δ_A. The strength of the relation between two people in the network is described by the *channel strength*, modeled by parameter α_{BA}. They are formalized by the numerical representations ε_B, α_{BA}, and δ_A as real numbers between 0 and 1.

Based on the above steps, the overall contagion process is modeled in terms of the *connection weight* ω_{BA} from sender B to receiver A. This represents the resulting influence of the internal emotion state of sender B on the internal emotion state of receiver A and depends on the above three parameters as shown in (1).

$$\omega_{BA} = \varepsilon_B \alpha_{BA} \delta_A \tag{1}$$

In the model this ω_{BA} is used to determine the strength of the impact from the emotion state of B to the emotion state of A at some time point t:

$$\textbf{impact}_{BA}(t) = \omega_{BA} q_B(t)$$

where $q_B(t)$ is the emotion level of B at time t. The overall contagion strength ω_A to q_A represents the total effect from all nodes that are connected to emotion state q_A of person A; it is modeled as in (2).

$$\omega_A = \sum_{B \neq A} \omega_{BA} \tag{2}$$

The aggregated impact $\textbf{aggimpact}_A(t)$ at time t of all connected emotion states q_{Bi} on emotion state q_A is modeled by a *scaled sum function* (see [2]) $\textbf{ssum}_{\omega A}(...)$ with the overall connection strength ω_A as scaling factor, as shown in (3).

$$
\begin{aligned}
\textbf{aggimpact}_A(t) &= \textbf{ssum}_{\omega_A}\left(\textbf{impact}_{B_1A}(t), \ldots, \textbf{impact}_{B_kA}(t)\right) \\
&= \left(\textbf{impact}_{B_1A}(t) + \ldots + \textbf{impact}_{B_kA}(t)\right)/\omega_A \\
&= \left(\omega_{B_1A}q_{B_1}(t) + \ldots + \omega_{B_kA}q_{B_k}(t)\right)/\omega_A \\
&= \left(\omega_{B_1A}/\omega_A\right)q_{B_1}(t) + \ldots + \left(\omega_{B_kA}/\omega_A\right)q_{B_k}(t)
\end{aligned}
\tag{3}
$$

From this it follows that $\textbf{aggimpact}_A(t)$ is calculated as a weighted average of the emotion levels of the connected states q_B as in (4).

$$\textbf{aggimpact}_A(t) = \sum_{B \neq A} w_{BA} q_B(t) \tag{4}$$

with weights

$$w_{BA} = \omega_{BA}/\omega_A = \varepsilon_B \alpha_{BA} \delta_A \bigg/ \sum_{C \neq A} \varepsilon_C \alpha_{CA} \delta_A = \varepsilon_B \alpha_{BA} \bigg/ \sum_{C \neq A} \varepsilon_C \alpha_{CA}.$$

The sum of these weights is 1. The dynamics for the contagion for this temporal-causal network model (see also [2]) is described in (5).

$$\Delta q_A(t + \Delta t) = q_A(t) + \eta_A[\textbf{aggimpact}_A(t) - q_A(t)]\Delta t \tag{5}$$

We denote with η_A the speed factor of A, which is chosen $\eta_A = \omega_A$ here. Sometimes the aggregated impact $\textbf{aggimpact}_A(t)$ on A is denoted by the shorter notation $q_A*(t)$.

This temporal-causal network model of emotion contagion has been investigated further and applied in a number of studies. For example, it was applied to predict the emotion levels of team members, in order to maintain emotional balance within a team [6]. When the teams' emotion level was found to become deficient, the model, which was embedded in an ambient agent, provided support to the team by proposing the team leader to give his employees a pep talk [6]. The pep talk is an example of an intervention strategy. Another study experimented with simulations of changes in the social network structure in order to guide the contagion process in a certain direction [7]. Yet another study used the model to predict changes on Physical Activity levels of a group of friends/acquaintances from the same course, applying the model to behavior contagion [8].

3 Analysis of the Absorption Model

The absorption model is based on two main assumptions, one of which addresses the *level* of the emotions and the other one the *speed* of change of the emotion levels:

(1) The emotion level $q_A(t)$ of a person A is affected linearly by the weighted average $\sum_{B \neq A} w_{BA} q_B(t)$ of the emotion levels of the connected persons B.
(2) For each person A the speed of change η_A of his or her emotion level linearly depends on the overall connection strength ω_A within the network: $\eta_A = \omega_A$

Roughly spoken, assumption (1) entails that the members of the network adapt to an average emotion level in the network. As a consequence, the emotion levels will converge to a common emotion level, which is between the minimal and maximal initial emotion levels of the connected members (see Sect. 3.1 for example simulations showing this, and Sect. 5 for a mathematical proof). This is in contrast to, for example, the amplification model introduced in [4] where assumption (1) is not taken as a point as departure, and as a result emotion contagion spirals can be modeled that reach levels higher (or lower) than any of the initial levels.

The second assumption (2) makes that the more connected members in the network, the higher the speed of change will be, in a proportional manner. In the current paper, assumption (1) is kept, but assumption (2) is critically analyzed in more depth and loosened in order to create room for alternatives. This second assumption (2) is an answer on the more open question:

> How does the speed of change of the emotion level depend
> on the network structure and size?

Specific variants of this question are the following. If a person has more connections to members with a given average emotion level, will he or she adapt faster to this average emotion level? Has the number of relations in real life effect on your speed of change for adapting to them? If a person has more friends, will his or her emotions be affected faster than the emotions of another person with fewer friends? If so, to which extent? Is this relation linear or proportional, or is it inherently nonlinear? Is this increase of the speed going on indefinitely, or is there some bound for it?

In [4] these questions were answered in a most simple, linear, proportional manner, as expressed by assumption (2) above. However, it is doubtful whether this most simple linear option is the most plausible option for realistic networks. The initial studies of the absorption model itself in [4] already highlighted two constraints: (a) In dynamic property P3 in Sect. 5 (referring to Theorem 5 in Sect. 4) in [4] it is stated that for some initial emotion values the emotion values eventually can run out of their boundaries 0 or 1. Also, (b) ω_A is a cumulative value based on the number of connections and their weights; when this number increases, because of the assumption (2) $\eta_A = \omega_A$, also the speed factor η_A increases in a proportional manner, without any limitation. For (b) Bosse et al. [4] used adaptations to the choice of the step size Δt to control that the model stays within the boundaries. That can work well for a few nodes, but this entails that all the time a smaller value for Δt has to be chosen, when the number of nodes becomes larger. This is possible, but not very practical. The hypothesis is that problem (a) relates to the strongly increasing value for the speed factor η_A for larger networks entailed by the choice of taking it equal to ω_A. Some experiments were run for analysis keeping the same characteristics of the experiment done by Bosse et al. [4] but with more nodes. The idea is to analyze if the choice for ω_A as a speed factor η_A indeed is the bottleneck for issues (a) and (b) of the absorption model.

3.1   Analysis of the Original Model

The same simulations in [4] were run again, with more nodes added to the scenario in order to better understand how the model works, and what alternatives are possible. 7 scenarios were created according to the Appendix A of [4]. The scenarios are:

1. All members have $\omega = 1$ – fully open channels (1a)
2. All members have $\omega > 0$ – big openness for all (1b)
3. All members have $\omega > 0$ – small openness for all (1c)
4. All members have $\omega = 0$ – no changes on emotional levels (2)
5. One member has $\omega = 0$ ($\delta = 0$) (3)
6. Only one member has $\omega \neq 0$ (all other members have $\delta = 0$) (4)
7. One member has $\omega = 0$ ($\delta = 0$ and $\varepsilon = 0$) (5)

Below a brief analysis of the effects on scenario 1a is made, showing what happens when the number of nodes is increased. The results of the other scenarios can be found at Appendix A (http://www.few.vu.nl/~efo600/iccci16/ICCCI16_A.pdf). The first scenario (1a) considers the maximal contagion that can happen. For that, all the parameters (expressiveness, openness and channel strengths) are set to 1. Moreover, Δt is set to 0.1. Figure 1 shows the differences between graphs for different numbers of members, 3, 9, and 18 nodes. As the initial values for the emotion levels for the tests have been generated at random, different convergence points, according to the average of the initial values are shown in the graphs. The convergence value for all the nodes that emerges is an average of the initial emotion levels, as shown in [4].

As all parameters are equal to 1, the speed factor η_A for each member A is the in-degree of the nodes minus 1: $\eta_A = n - 1$, with $n = $ number of nodes. As the speed factor η_A determines the next emotion level for all the nodes (Eq. (5)), the emotion levels converge faster, and at some network size (after 12 members) oscillation in the emotional levels occurs due to the sudden changes caused by the high speed factor. Note that to see this effect Δt was not decreased, what normally would be a measure taken; it was kept at 0.1. Such a decrease would be possible; however, decreasing Δt with the size of the network indefinitely is neither practical nor desirable.

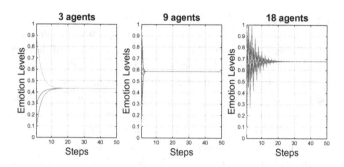

Fig. 1. Full channel connections for 3 to 18 members

3.2 Mathematical Analysis of the Problem

This section addresses mathematical analysis of the problem concerning the oscillation of the emotion levels. The sudden changes will be explained showing the effect of the increase in the value of the choice of the speed factor $\eta_A = \omega_A$ for larger networks. The equation for ω_A is the sum of all the connection strengths generated by the nodes in-connected to q_A. The strength by which the emotion from each B is received by A is calculated by $\omega_{BA} = \varepsilon_B \alpha_{BA} \delta_A$, as seen before. If all nodes have ε and δ higher than zero, and if the network is fully connected (in other words, there is no $\delta_{BA} = 0$ to any pair B, A), and the number of neighbors in the network increases, the value of ω_A also increases proportionally and by the assumption (2) $\eta_A = \omega_A$ the same holds for η_A. While the speed factor increases, the changes from $q_A(t)$ to at the next emotion level $q_A(t + \Delta t)$ become more sudden, and less realistic. As a matter of illustration, imagine a fully connected network (all the channel strengths α_{BA} equal to 1), where every member has openness and expressiveness equals 0.5. So, in that case

$$\omega_{BA} = \varepsilon_B \alpha_{BA} \delta_A = 0.5 \times 1 \times 0.5 = 0.25$$

Therefore, $\omega_{BA} = 0.25$ is the connection strength between the emotion states q_A for each of the members that are connected. So, if n is the number of nodes, ω_A for any of the fully connected network will be $(n - 1) \times 0.25$. If the number of nodes is increased 10 times, the speed factor $\eta_A = \omega_A$ will be increased around 10 times as well. For an increase to 1000 nodes, the speed factor will be 100 times bigger. Figure 2 (left graph) shows that this increase follows a linear tendency. As it may be doubted that this indefinite increase of η_A is realistic, in Sect. 4 alternative options for the speed factor η_A will be considered, with patterns corresponding to the other graphs in Fig. 2, where there is some bound in the increase of the speed factor.

So assuming the speed factor $\eta_A = \omega_A$ (assumption (2) of the original absorption model) causes unbalanced behaviour of the model. In order to avoid abrupt changes, usually the value of Δt is made smaller and smaller for larger networks. This approach is conceptually and practically inadequate as the speed factor refers to the velocity of the changes in the emotions, whereas Δt refers to the time steps of the model, and has nothing to do with the speed factor η_A. That incompatibility is a reason to consider alternative answers on the main question concerning speed factors, as discussed next.

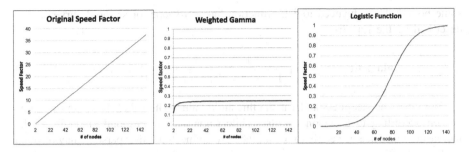

Fig. 2. Three types of relations for the speed factor depending on number of nodes left: η_A increasing linearly as the number of members increases middle: η_A increases up to a limit, due to a weighted speed factor right: η_A increases according to an advanced logistic function

4   Alternative Ways to Model the Speed Factor η_A

In this section two alternative ways of modeling the speed factor η_A are explored. In both cases the speed factor increases with the size of the network, but stays under a certain bound, according to patterns as shown in Fig. 2 middle and right graph. The first option is by modeling the speed factor as a *scaled* ω_A, with scaling factor the number n of nodes in the network:

$$\eta_A = \omega_A/n$$

This option avoids the effect caused by the increasing in the number of members. In this case, and using the same network as in Fig. 2, it can be seen that the value for η_A converges to the ω_{BA} which is the same for all nodes in this scenario. Figure 2 middle graph shows the new situation when for η_A the above scaled model is used.

Using this option will assure that the speed factor η_A has boundaries defined according to the following mathematical analysis. For the new calculation for η_A, it holds for all A, $0 \leq \eta_A < 1$.

This can be verified as follows. It holds

$$\sum_{B \neq A} \varepsilon_B \times \alpha_{BA} \times \delta_A \leq n - 1$$

as each of the terms of this sum is ≤ 1. Therefore

$$\eta_A = \omega_A/n = \sum_{B \neq A} \varepsilon_B \times \alpha_{BA} \times \delta_A/n \leq (n-1)/n < 1$$

In other words the speed factor η_A is now bounded by 1. Note that by a slight modification this bound can be set to any number η by multiplying this by an extra parameter η (the same holds for the second alternative discussed below): $\eta_A = \eta\,\omega_A/n$.

A second alternative is to use an advanced logistic function in order to gradually increase the speed with network size but keep the values for the speed factors within some bound. The advanced logistic function has a S shape, or sigmoid curve, and is described by the Eq. (6).

$$\textbf{alogistic}_{\sigma,\tau}(\eta_A) = \left[\frac{1}{(1+e^{-\sigma(\eta_A-\tau)})} - \frac{1}{(1+e^{\sigma,\tau})}\right](1+e^{-\sigma,\tau}) \qquad (6)$$

Here σ is the steepness and τ is the threshold value. The values for σ and τ can be chosen according to a person's traits. While some persons respond gradually to the increasing influence of people to whom they are connected, other persons may respond by flare-ups. For the situation of a person that responds linearly to the increasing on their cumulative ω_A, a low steepness value such as $\sigma = 0.3$ can be chosen, and, for example, $\tau = 20$. The results for this situation can be seen at Fig. 2, right graph.

As can be seen, the logistic function also keeps the speed factor values between 0 and 1, and if the parameters of the function are well adjusted, the equation can give more realistic outcomes.

5 Mathematical Analysis

This section presents some of the results of a mathematical analysis for the model after the changes at the speed factor calculation.

Definition 1. A network is called strongly connected if for every two nodes A and B there is a directed path from A to B and vice versa.

Lemma 1. Let a temporal-causal network model be given based on scaled sum functions for states q_A:

$$\textbf{d}_{q_A}/\textbf{d}t = \eta_A\left[\sum_{B\neq A} \omega_{BA}q_A/\omega_A - q_A\right]$$

Then the following holds.

(a) If for some state q_A at time t for all states q_B connected toward q_A it holds $q_B(t) \geq q_A(t)$, then $q_A(t)$ is increasing at t: $\textbf{d}q_A(t)/\textbf{d}t \geq 0$; if for all states B connected toward A it holds $q_B(t) \leq q_A(t)$, then $q_A(t)$ is decreasing at t: $\textbf{d}q_A(t)/\textbf{d}t \leq 0$.

(b) If for all states q_B connected toward q_A it holds $q_B(t) \geq q_A(t)$, and at least one state q_B connected toward q_A exists with $q_C(t) > q_A(t)$ then $q_A(t)$ is strictly increasing at t: $\textbf{d}q_A(t)/\textbf{d}t > 0$. If for all states q_B connected toward q_A it holds $q_B(t) \leq q_A(t)$, and at least one state q_B connected toward q_A exists with $q_C(t) < q_A(t)$ then $q_A(t)$ is strictly decreasing at t: $\textbf{d}q_A(t)/\textbf{d}t < 0$.

Proof of Lemma 1. (a) From the differential equation for $q_A(t)$

$$\mathbf{d}_{q_A}/\mathbf{dt} = \eta_A \left[\sum_{B \neq A} \omega_{BA} q_B / \omega_A \, q_A \right]$$

$$= \eta_A \left[\sum_{B \neq A} \omega_{BA} q_B - \omega_A - q_A \right] / \omega_A$$

$$= \eta_A \left[\sum_{B \neq A} \omega_{BA} q_B - \sum_{B \neq A} \omega_{BA} \, q_B \right] / \omega_A$$

$$= \eta_A \sum_{B \neq A} \omega_{BA} [q_B - q_A] / \omega_A$$

it follows that $\mathbf{d}q_A(t)/\mathbf{dt} \geq 0$, so $q_A(t)$ is increasing at t. Similar for decreasing.
(b) In this case it follows that $\mathbf{d}q_A(t)/\mathbf{dt} > 0$, so $q_A(t)$ is strictly increasing. Similar for decreasing. ■

Theorem 1 (convergence to one value). Let a strongly connected temporal-causal network model be given based on scaled sum functions for the states q_A

$$\mathbf{d}_{q_A}/\mathbf{dt} = \eta_A \left[\sum_{B \neq A} \omega_{BA} q_B / \omega_A - q_A \right]$$

and with equilibrium values q_A. Then for all A and B the equilibrium values q_A and q_B are equal: $q_A = q_B$. Moreover, this equilibrium state is attracting.

Proof of Theorem 1. Take a state q_A with highest value q_A. Then for all states q_C it holds $q_C \leq q_A$. Suppose for some state q_B connected toward q_A it holds $q_B < q_A$. Take a time point t and assume $q_C(t) = q_C$ for all states q_C. Now apply Lemma 1b) to state q_A. It follows that $\mathbf{d}q_A(t)/\mathbf{dt} < 0$, so $q_A(t)$ is not in equilibrium for this value q_A. This contradicts that this q_A is an equilibrium value for q_A. Therefore, the assumption that for some state q_B connected toward q_A it holds $q_B < q_A$ cannot be true. This shows that $q_B = q_A$ for all states connected towards q_A. Now this argument can be repeated for all states connected toward q_A. By iteration every other state in the network is reached, due to the strong connectivity assumption; it follows that all other states in the temporal causal network model have the same equilibrium value as q_A. From Lemma 1b) it follows that such an equilibrium state is attracting: if for any state the value is deviating it will move to the equilibrium value. ■

Proposition 1 (Monotonicity Conditions). (a) If $q_A^*(t) \leq q_A(t)$ then $q_A(t)$ is monotonically decreasing; it is strictly decreasing when $q_A^*(t) > q_A(t)$.
(b) If $q_R^*(t) \leq q_R(t)$ then $q_R(t)$ is monotonically increasing; it is strictly increasing when $q_R^*(t) > q_R(t)$.

Proof. This follows from the differential equation

$$\mathbf{d}_{q_C}(t)/\mathbf{dt} = \eta_C \big(q_C^*(t) - q_C(t) \big)$$

and the fact that $0 \leq q_C(t) \leq 1$ and $0 \leq q_C^*(t) \leq 1$. ■

Lemma 2. Suppose all ω_{CD} are nonzero. Then for an equilibrium the following holds:
(a) $q_A^* = 0$ if and only if $q_C = 0$ for all $C \neq A$
(b) $q_B^* = 1$ if and only if $q_C = 1$ for all $C \neq B$

Proof. (a) From

$$q_A^* = \sum_{C \in G \setminus \{A\}} w_{CA} q_C = 0$$

and the fact that all terms are nonnegative it follows that $w_{CA} q_C = 0$ for all $C \neq A$ and conversely.

(b) From

$$q_B^* = \sum_{C \in G \setminus \{B\}} w_{CB} q_c = 1$$

and the fact that

$$\sum_{C \in G \setminus \{B\}} w_{CB} = 1$$

it follows that $q_C = 1$ for all $C \neq B$ and conversely.   ∎

Lemma 3. For an equilibrium for any member the following holds:
  (a) If $q_A = 0$ then $q_A^* = 0$
  (b) If $q_B = 1$ then $q_B^* = 1$

Proof. (a) From

$$q_A^* - q_A = 0$$

with $q_A = 0$ it follows

$$q_A^* = 0$$

(b) From

$$q_B^* - q_B = 0$$

with $q_B = 1$ it follows

$$q_B^* = 1$$ ∎

Proposition 2. Suppose some A is given and all w_{BA} are nonzero. Then for an equilibrium the following holds:
  (a) If $q_A = 0$ then $q_C = 0$ for all C
  (b) If $q_B = 1$ then $q_C = 1$ for all C

Proof. This immediately follows from Lemmas 2 and 3.   ∎

Proposition 3. Suppose all w_{DC} are nonzero. Then for an equilibrium it holds
  (i) If $q_A = 0$ for some A then $q_C = 0$ for all $C \in G$.
  (ii) If $q_B = 1$ for some then $q_C = 0$ for all $C \in G$.

Proof. This immediately follows from Proposition 2.   ∎

6 Results

Using the alternative models for the speed factors η_A and comparing them with the outcomes for the original model, it turns out that the oscillation is not present anymore in any of the new approaches. For scenario 1(a), Fig. 3, it is possible to observe that for 3 members the logistic function delays the convergence point, especially because the logistic function will give a lower value when ω_A is lower.

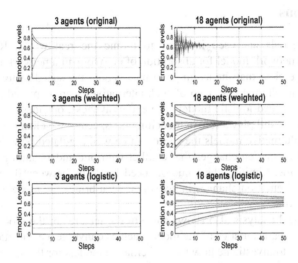

Fig. 3. Comparison for scenario 1(a) between the 3 speed factors

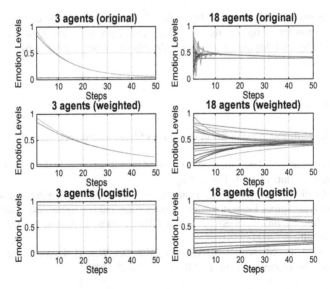

Fig. 4. Scenario 3 and the different speed factors used to calculate the emotion levels

For scenario 3, it is clear how the use of both the scaled and or logistic model for the speed factor corrects the awkward slopes from the original model without needing any change on the time step Δt used (Fig. 4). As noticed at scenario 1(a), for fewer nodes, the logistic function still keeps the convergence point later. This can be handled at the logistic function itself through steepness and threshold adjustments.

More results and analysis for the other scenarios and graphs can be found at Appendix B (http://www.few.vu.nl/~efo600/iccci16/ICCCI16_B.pdf).

7 Conclusions

Mathematical models are used in order to mimic the real world. Regarding the temporal-causal network model for absorption of emotions in a network introduced by [6] it has become clear that the assumption made about the speed factor isn't perfect, and gives room to alternatives. Two of such alternatives were explored here: a scaled model and an advanced logistic model. The expressiveness, openness to changes, and the strength of links still play a role in modelling the speed of the change of the emotion level. By these alternative models the speed can be well regulated between boundaries and do not lead to sudden changes that conflict with our understanding of emotional evolution over time. Limiting the value of speed factor η_A between 0 and 1 creates a stable slope in the emotion changes in networks, what brings the model closer to what it is expected to do.

A mathematical analysis also shows some of the features of the model. Part of the analysis explains characteristics of the model such as convergence and stability. Future work can be done to investigate how these alternative models for the speed factor affect the results of previous research, and how they can be combined with the model for emotion contagion spirals from [3].

Acknowledgements. E.F.M. Araujo's stay at the VU Amsterdam was funded by the Brazilian Science without Borders Program, through a fellowship given by the Coordination for the Improvement of Higher Education Personnel – CAPES (reference 13538-13-6).

References

1. Iacoboni, M.: Mirroring people: the new science of how we connect with others. Macmillan, New York (2009)
2. Treur, J.: Dynamic modeling based on a temporal-causal network modeling approach. Biol. Inspired Cogn. Archit. **16**, 131–168 (2016)
3. Bosse, T., Duell, R., Memon, Z.A., Treur, J., van der Wal, C.: A multi-agent model for emotion contagion spirals integrated within a supporting ambient agent model. In: Yang, J.-J., Yokoo, M., Ito, T., Jin, Z., Scerri, P. (eds.) PRIMA 2009. LNCS, vol. 5925, pp. 48–67. Springer, Heidelberg (2009)
4. Bosse, T., Duell, R., Memon, Z.A., Treur, J., Van Der Wal, C.N.: Multi-agent model for mutual absorption of emotions. In: ECMS, pp. 212–218 (2009)

5. Bosse, T., Duell, R., Memon, Z.A., Treur, J., van der Wal, C.N.: Agent-based modeling of emotion contagion in groups. Cogn. Comput. **7**(1), 111–136 (2015)
6. Duell, R., Memon, Z.A., Treur, J., Van Der Wal, C.N.: An ambient agent model for group emotion support. In: 2009 3rd International Conference on Affective Computing and Intelligent Interaction and Workshops, ACII 2009, pp. 1–8. IEEE (2009)
7. Klein, M., Manzoor, A., Mollee, J., Treur, J.: Effect of changes in the structure of a social network on emotion contagion. In: 2014 IEEE/WIC/ACM International Joint Conferences on Web Intelligence (WI) and Intelligent Agent Technologies (IAT), vol. 3, pp. 270–277. IEEE (2014)
8. Araújo, E.F.M., Tran, A.V., Mollee, J.S., Klein, M.C.: Analysis and evaluation of social contagion of physical activity in a group of young adults. In: Proceedings of the ASE BigData & SocialInformatics 2015. ACM (2015)

Cognitive Modelling of Emotion Contagion in a Crowd of Soccer Supporter Agents

Berend Jutte and C. Natalie van der Wal[✉]

Department of Computer Science, Vrije Universiteit,
Amsterdam, Netherlands
c.n.vander.wal@vu.nl

Abstract. This paper introduces a cognitive computational model of emotion contagion in a crowd of soccer supporters. It is useful for: (1) better understanding of the emotion contagion processes and (2) further development into a predictive and advising application for soccer stadium managers to enhance and improve the ambiance during the soccer game for safety or economic reasons. The model is neurologically grounded and focuses on the emotions "pleasure" and "sadness". Structured simulations showed the following four emergent patterns of emotion contagion: (1) hooligans are very impulsive and are not fully open for other emotions, (2) fanatic supporters are very impulsive and open for other emotions, (3) family members are very easily influenced and are not very extravert, (4) the media is less sensible to the ambiance in the stadium. For validation of the model, the model outcomes were compared to the heart rate of 100 supporters and reported emotions. The model produced similar heart rate and emotional patterns. Further implications of the model are discussed.

Keywords: Cognitive modelling · Crowd behaviour · Emotion contagion

1 Introduction

Do you remember the soccer game of the Champions League Final in 2005? This match between the soccer teams Liverpool F.C. from England and AC Milan from Italy, showed why soccer is played and enjoyed by people all over the world. In the first half of the match the score was 0–3 in favour of AC Milan. Within fifteen minutes after the break the score was 3–3. After 90 min the score was still 3–3. During the extra time, none of the teams scored a goal. After the penalties, Liverpool F.C. won the Champions League. During such a match all kinds of emotions arise. Especially the switch between pleasure at first and sadness in a later stage for the supporters from AC Milan, and the other way around for the supporters from Liverpool F.C. is very interesting. How these emotions arise and influence the ambiance in the stadium is not only very fascinating to understand in general, but can also be important for economic and safety reasons. For example, during almost every soccer game some supporters sing racist passages or show racist activities. The mood of the crowd has a big influence on this behaviour [1]. Another interesting effect of the crowd's mood is on the sales of beverages and food. It is known that when the supporters enjoy their game more, sales in and around the stadium rise [2]. With these reasons in mind, the focus of the current

© Springer International Publishing Switzerland 2016
N.T. Nguyen et al. (Eds.): ICCCI 2016, Part I, LNAI 9875, pp. 40–52, 2016.
DOI: 10.1007/978-3-319-45243-2_4

work is on understanding how sadness and pleasure arise and distribute through a crowd of soccer supporters. The aim is to build a cognitive computational model of this emotion contagion process to better understand the dynamics and with a future application in mind where an ambient intelligent system can monitor the crowd's behavior and emotions and to provide advice and support to soccer stadium managers on possible interventions.

During a soccer game, supporters are mainly feeling sadness or pleasure [3]. The feeling of "pleasure" and "sadness" are processed and controlled in the brain. Human pleasure reactions occur across a distributed system of brain regions. Furthermore, emotions are spread amongst people through emotion contagion. Emotion contagion is part of the cognitive system and categorized as: (1) automatic subconscious contagion through mimicry and feedback or (2) conscious transfer through social comparison of moods and appropriate response in groups, mediated by attention [4]. Automatic emotion contagion is represented in this work, based on the principle of mirror neurons. [5, 6]. The conscious emotion contagion is represented in this work as a social phenomenon, where emotions of group members can be absorbed by the other group members or can be amplified, in the way that the real emotion of other group members will be reinforced. [7]. This social phenomenon is based on the social connections the supporters have with each other. Our work is based upon [7–9]. We have chosen for modeling a social process with a neurologically grounded model for two reasons: (1) to model human mental processes in relation to reality, namely representing the continuous dynamic processes of firing neurons and (2) because this modeling approach with cognitive and affective mental states can be more effective in predicting human behaviour than a model without them. [8].

This paper examines how the dynamical pattern of pleasure and sadness can be represented and modelled based on neuro-scientific concepts and theories, such as reward mechanisms in the brain, mirror neurons and somatic marking. The hypothesis of this research is that the emotions sadness and pleasure show specific dynamical patterns during a soccer game and these patterns are dependent on anticipation and people influencing other people with their emotions and cognitions. More specifically, we postulate that: (1) on an individual level, a supporter will experience a higher intensity of pleasure if its preferred team scores a goal. In the same fashion, a supporter will experience a higher intensity of sadness if the non-preferred team scores a goal; (2) on a group level, supporters are sensitive for the emotions of other supporters (For instance, when the preferred team has scored a goal for the home supporters, the away supporters will experience a higher intensity of pleasure over time, because the whole stadium more or less feels the emotion 'pleasure'). As a first step in validating this model, the simulation results will be compared to the heartbeats of 100 persons, measured during an entire soccer game. Also, the model outcomes are compared with real-time reported emotions and in hindsight reported emotions of soccer supporters. The rest of this paper is organised as follows. In Sect. 2 the underlying neurological underlying theories and the proposed cognitive computational model are described. Next, in Sect. 3 the results from the structured simulations are shown. Section 4 presents the validation of the model. The paper concludes with a discussion in Sect. 5.

2   Cognitive Model for Emotion Contagion in Agents

Neurological background. Human emotions occur across a distributed system of brain regions. The Nucleus Accumbens (NAcc) is related to the reward system. The reward system is a group of brain structures that mediates the reinforcement. The most important pathway of the reward system is the dopamergic pathway (Ventral Tegmental Area (VTA) – NAcc). The feeling of pleasure or desire to pleasure is regulated among this pathway. Furthermore, the Orbitofrontal Cortex (OC) is responsible for the controlling of the emotion state and the Anterior Cingulate Cortex (ACC) is responsible for the progressing and expression of the emotion state [10]. This ACC has a very strong and important connection with the Amygdala. The Amygdala is responsible for processing both positive and negative emotions. Until recently it was thought that it was only associated with negative processing. [11]. The Amygdala links different information, that comes from different senses, to different emotions. In every different situation, the Amygdala determines which emotional reaction is the most useful in that particular situation. In most of the situations the emotional response is fast and automatic (e.g. fear; fight or flight reaction). The connections between these brain regions that are involved in the processing of pleasure and sadness can be mapped in an abstract relation mapping.

There is little scientific knowledge about the connections between the related brain regions at the lower level of the brain (on the level of neurons). There is, however, neurological research available about the connections at a higher level (the abstract idea). For example, as described earlier, a dopaminergic pathway connects the VTA and the NAcc with each other. This connection is related to the Amygdala. The Amygdala (emotion processing) has an important connection with the ACC (emotion expression) [12, 13]. Besides different brain regions that are related to the emotions "pleasure" and "sadness", hormones and somatic markers are important as well. The ACC has an important function in the regulation of blood pressure and heart frequency. These two body processes are modelled in this work as well and used for validation of the model.

Agent-based Computational Model. The previously described theories are modelled in concepts and their relations, expressed in numerical values. Figure 1 gives an overview of the structure and organization of all states and relations, for each agent. Tables 1 and 2 show all the states and connections that are included in the model. Every state will be described below.

Input States. The Context state (Eq. 1) represents the situational context, which can be the current score, the importance or the current stage of the match. Its value is either 0 or 1; 0 means the state is not active (no importance or urgency) and 1 means the state is active (high importance or urgency). The Goal state (Eq. 2) is active when a goal is made by one of the two teams. It remains active for 100 time steps. Its value is either 0 or 1: 0 means not active and 1 means active. The Emotion Sensor state (Eq. 3) is the state where the emotions of other supporters are aggregated, as a weighted sum, to influence the agent. Its range is [0,1], whereby 0 means there is no incoming emotion

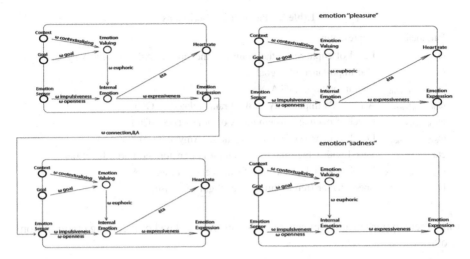

Fig. 1. Meta-model: cognitive model emotion contagion soccer supporters agents (left); concepts and relations specified per emotion pleasure and sadness (right)

Table 1. Description of all states

State	Description	Value
Context	Representation of context during the game. The current situation of the match at that moment (for example; qualification, current score, period of the match)	0 or 1 (0 representing a context of low importance and low urgence and 1 representing a highly important or urgent context)
Goal	A goal is made by one of the two teams	0 or 1 (0 representing no goal is scored and 1 representing a goal is scored)
Emotion Sensor	Level of all incoming emotions from other (connected) supporters.	[0,1] (0 representing no incoming emotion and 1 representing a maximal incoming emotion)
Emotion Valuing	This state represents the incoming internal valuing of the game related to the emotions	0 or 1 (0 representing no incoming values from context state and goal state and 1 representing incoming values from both context state and goal state or only from value state)
Internal Emotion	This state represents the processing part of both positive and negative emotions	[0,1] (0 representing no incoming emotion and 1 representing a maximal incoming emotion)
Emotion Expression	The expression of the emotion from the processing state	[0,1] (0 representing no incoming emotion and 1 representing a maximal incoming emotion)
Heart rate	This state mimics the heart rate of the agent	[70,180] (70 representing the minimal heart rate and 180 representing the maximal heart rate)

Table 2. Parameter descriptions

Parameter	Representing:
$\omega_{euphoric}$	Level of euphoria in the brain (Connection VTA-Nacc with Amygdala)
ω_{goal}	The importance of a goal
$\omega_{contextualizing}$	Internal valuing (VTA-Nacc)
$\omega_{impulsiveness}$	A person's impulsivity in the brain (related to OC)
$\omega_{openness}$	Level of openness to emotions of others (DS-ACC)
$\omega_{expressiveness}$	Level of expressiveness of agent A (Amygdala-ACC)
eta	Speed with which internal emotion is transformed into a heart rate
$\omega_{connection,B,A}$	Level of social connection from agent B to agent A
beta	Speed factor (representing fight-flight response)

from other supporters and 1 means there is a maximum level of incoming emotion from all others.

$$\text{Context } (t) = 0 \text{ OR } 1 \tag{1}$$

$$\text{Goal } (t) = 0 \text{ OR } 1 \tag{2}$$

$$\text{Emotion Sensor}(t) = \text{Emotion Sensor}(t) + \left(\sum\nolimits_{B \in G \backslash \{A\}} (\omega_{connection,B,A} * \text{Emotion Expression}_B(t)) \right) / (\text{total} \tag{3}$$
$number of\ \omega_{connection,B,A})$

Internal States. The $\omega_{contextualizing}$ parameter mimics the connection between the VTA and the NAcc in the brain, better known as the dopaminergic pathway. The range of Emotional Valuing state is [0 or 1]; 0 means there is no incoming value from Context and Goal, 1 means either there is an incoming value from Context and an incoming value from Goal (4a) or there is an incoming value from Context and no incoming value from Goal (4b). Formula 4a is used for input Goal(t) = 1, and 4b for Goal(t) = 0. This choice was made, because in Eq. 4a, w hen a goal is scored, both Context and Goal and therefore Emotion Valuing will be 1. Thereafter, Context will stay 1, Goal will become 0, and Emotion Valuing will be 0.5, which means a decrease of Emotion Valuing, and eventually a decrease of the Emotion Expression. However, this is not supposed to happen after a goal is scored, since the emotion has to be steady. Therefore, when Goal(t) = 0, Emotion Valuing is only dependent on Context. In this way, Emotion Valuing will stay 1. The $\omega_{euphoric}$ portrays the connection between the VTA - NAcc and the Amygdala in the brain [14] (Limbic system; NAcc). The $\omega_{impulsiveness}$ mimics the impulsivity of a person in the brain, which is related to the OC [15] (OC). This is the connection between the Emotion Sensor state and Internal Emotion state. The Internal Emotion state portrays the Amygdala [11] (Amygdala). The $\omega_{openness}$ portrays the openness from a person to emotions from other persons. This connection is related to the Dorsal Striatum and ACC [16] (Dorsal and ventral striatum). The Dorsal Striatum is involved in anticipating emotion expression and regulation of the heart rate. The Ventral Striatum is more related to anticipating the reward mechanism [17]. The range Internal Emotion for "pleasure" (5a) or "sadness" (5b) is [0,1]; 0 means there is no incoming emotion ("sadness" or "pleasure") from Emotion Valuing and Emotion Sensor and 1 means there is a maximum level of incoming emotion ("sadness" or "pleasure") from Emotion Valuing and Emotion Sensor. The parameter beta in formula 5b is based on the evolutional neurological theory about fear and the 'fight or flight' theory. Stress and negative feelings are processed faster compared to positive feelings, based on the flight and fight reactions [18]. Furthermore, people with a negative feeling act more narrow-minded and will most likely not open up completely to other people. In this way, their own emotion disrupts the incoming emotion from other people [19].

$$\text{Emotion Valuing}(t + \Delta t) = \text{Emotion Valuing}(t) + \eta \left(\left(\left(\text{Goal}(t) * \omega_{goal} \right) + \left(\text{Context}(t) * \omega_{contextualizing} \right) \right) / 2 \right) - \text{Emotion Valuing }(t) \right) \Delta t \tag{4a}$$

$$\text{Emotion Valuing}(t + \Delta t) = \text{Emotion Valuing}(t) + \eta \left(\left(\text{Context}(t) * \omega_{contextualizing} \right) - \text{Emotion Valuing}(t) \right) \Delta t \tag{4b}$$

$$\text{Internal Emotion}(t + \Delta t) = \text{Internal Emotion}(t) + \eta \left(\text{EmotionSensor}(t) * \omega_{impulsive} * \omega_{openess} + \left(\text{InternalValuing}(t) * \omega_{euphoric} \right) / 2 \right) - \text{Internal Emotion}(t) \right) \Delta t \tag{5a}$$

$$\text{Internal Emotion}(t + \Delta t) = \text{Internal Emotion}(t) + \eta \left(\left(\text{Emotion Sensor}(t) * \omega_{impulsive} * \omega_{openess} + (1 - beta) \right) + \left(\left(\text{Internal Valuing}(t) * \omega_{euphoric} \right) * beta \right) \right) - \text{Internal emotion}(t) \right) \Delta t \tag{5b}$$

Output States. The $\omega_{expressiveness}$ is the connection between Internal Emotion state and Emotion Expression. This connection portrays the relation between the Amygdala and the ACC with the related functions. [Knutson et al., 2001a,b] (ACC; Amygdala). The Emotion Expression state mimics the function of the ACC [20, 21] (ACC). The range of Emotion Expression is [0,1]; 0 means there is no incoming emotion ("sadness" or "pleasure") from Internal Emotion and 0 means there is a maximum level of incoming emotion ("sadness" or "pleasure") from Internal Emotion. The heart rate is connected and correlated with the ACC. The ACC plays a role in the regulation of heart rate frequency [21]. For Heart Rate a range of [70,180] was chosen, representing a common minimum heart rate of 70 and a maximum of 180, calculated as follows.

$$\text{Emotion Expression }(t + \Delta t) = \text{Emotion Expression }(t) + \eta \left(\text{Internal Emotion }(t) * \omega expressiveness - \text{Emotion Expression }(t) \right) \Delta t \tag{6}$$

$$\text{Heart rate}(t + \Delta t) = \text{Heart rate}(t) + eta * \left(\left(\text{Internal Emotion }(t + \Delta t) - \text{Internal Emotion }(t) / \max(\text{Internal Emotion}(t)) * \left(\left(\max(\text{Heartrate}(t) - \text{Heart rate}(t) / \left(\left(\max(\text{Heart rate}(t) - \text{Heart rate}(t) / \text{Heart rate}(t)) \right) \right) * \Delta t \tag{7}$$

Agents and their Environment. The agents representing supporters, are placed in a virtual stadium, see Fig. 2. In every stadium section, five supporter agents are placed: one agent from each of the five supporter subgroups explained below. Only when mentioned, the amount of supporters in a section is increased to 20. Agents in every section are connected with agents from three other sections. In this way all the supporters are directly or indirectly connected with each other, which is based on reality

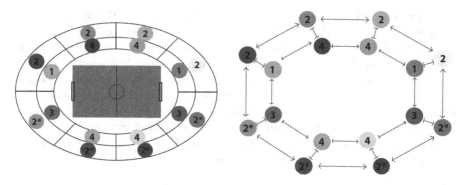

Fig. 2. Schematic overview of the agent environment, with the related connections between them

and meant for emotion contagion to arise. The five different supporter subgroups in the model are: (1) hooligans; (2) Fanatic supporters; (3) family members; (4) neutral supporters; (5) away supporters. Hooligans always have a believe to win the game and the thought that they can influence the game by their enthusiasm. They think that supporting their own players and provoking the opponents will help their team win the game (songs, choruses and choreographies). Hooligans are a group of hostile inter-acting people [21]. Often, hooligans have cognitive and mental problems. This is related with changes in function of the reward system and the limbic system. It is suggested that they have a very active reward system in combination with other cog-nitive problems. The desire to feel pleasure by visiting a soccer game is high (relatively high $\omega_{euphoric}$). Besides this, they have a reduced control system for emotions (rela-tively high $\omega_{impulsiveness}$). In addition, they are not fully open for emotions of other supporters (relatively low $\omega_{openness}$). However, they are very sensitive to emotions evoked by the game, like a goal. Fanatic supporters do not show the behavioural characteristics of a hooligan (e.g. violence), but have clearly visible emotions. They try to maintain a good ambiance in the stadium by singing and screaming during the game. They are also very sensitive to emotions evoked by the game and other supporters. Most of the time there is excessive use of alcohol (though less excessive than the hooligans' use of alcohol). Family members are stereotypically represented by a father who is visiting the game with his son. This is the largest group of supporters. They are quite silent during the game (compared with the fanatic supporters and hooligans). They are very happy when their favourite team scores a goal, and are sad when the opposite team scores a goal. Furthermore, this group does not sing during the game and does not drink any alcohol. Neutral supporters represent either the press, media, sponsors or business partners. They do not have any relatable emotion during the game and do not have any preference for one of the two teams. This is the smallest group of supporters. Away supporters have the same characteristics as the fanatic supporters as described above. However, their team of preference is the opposite team.

3 Results

To examine if the proposed model exhibits the patterns that can be expected from literature (described in Sect. 1), four different scenarios were simulated: matches (1) that are different in contexts (importance of the match), (2) with difference in distribution of the supporters within the stadium, (3) that are different in the final score and by (4) with and without emotion contagion. Figures 3 and 4 show the results of the simulations of a match with a 'non-important' context (representing a standard random friendly match) and a match with a 'highly important' context (representing the very important Champions League final from 2005). Parameter settings for the simulations are shown in Table 3. During the rest period, there is no emotion contagion between the supporters. Figure 3 shows that supporters feel some positive stress (emotion 'pleasure') at different periods during the match: at the beginning, before the rest, after the rest, shortly before the end of the match and after a goal has been scored. In the 'high' importance scenario, positive stress is felt during almost the entire match. See Fig. 4. Here, supporters are more euphoric. Also, higher levels of both emotions 'pleasure' and 'sadness' are visible, representing more intense emotions and 'more being at stake'. In both Figs. 3 and 4, there is clear emotion contagion between the supporters at the start: all the emotions converge/are averaging out. In this way, general emotional patterns arise for every different subgroup within different simulations. After this first period of the game, the process of emotion contagion is clearly present as well in different values of emotions between the supporters groups. In general, the sub-groups 'hooligans' and 'fanatic supporters' (both home and away supporters) have the highest values of the emotions 'pleasure' and 'sadness'. These two subgroups are generally the most loudly represented supporters during a soccer game. When the preferred team has scored a goal, the fanatic supporters and hooligans show the highest increase of the emotion 'pleasure'. The subgroups media and family members all show only a small increase in the emotion 'pleasure'. Regarding the emotion 'sadness', this emotion shows the same yet opposite pattern compared with the emotion 'pleasure'. Overall, the subgroups 'family members' and 'media' have a lower value of the emotions 'pleasure' and 'sadness' than the other subgroups. Family members mostly behave less extravert than hooligans and fanatic supporters. The subgroup media is the neutral group, which has no preference for neither one of the two playing teams. They are fully influenced by the ambiance in the soccer stadium. However, taken into account the difference of the values of the emotions 'pleasure' and 'sadness' between the different subgroups, all the emotional patterns are more or less the same. This finding reflects the general and collective ambiance during a soccer game. Figure 5 shows the simulation results of the Champions League final in 2005 without emotion contagion. It can be seen that the supporters do not influence each other's emotions. Most prominently the emotions do not converge from the start of the simulations and are generally a bit more intense or less intense on average. From the simulations with different distributions of supporters similar patterns were found. Especially in the hooligans and fanatic supporters subgroups the influence of emotion contagion was shown in the results by stronger convergence of emotions. In the different scoring

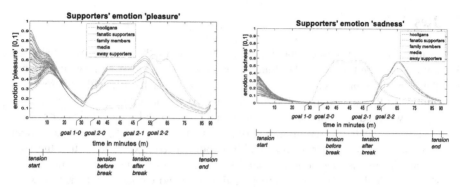

Fig. 3. Pleasure (left) and sadness (right) levels during a non-important context match

Table 3. Simulation settings

Parameter	Value
Delta time; eta; length	0,01; 1; 6300 time steps (representing 6300 s)
$\omega_{euphoric}$	Hooligans, fanatic and away supporters:1; family members: 0,8; media: 0,6
$\omega_{impulsiveness}$	Hooligans, fanatic and away supporters: 0,9; family members and media: 0,6
$\omega_{openness}$	Hooligans: 0,7; Fanatic, away supporers: 0,9; family members and media: 1
$\omega_{expressiveness}$, $\omega_{contextualizing}$, ω_{goal}	All supporters: 1

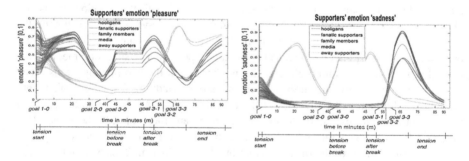

Fig. 4. 'Pleasure' (left) and 'sadness' (right) levels during a highly important context match

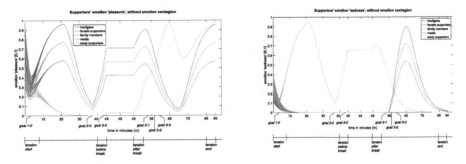

Fig. 5. 'Pleasure' (left) and 'sadness'(right) during a highly important context match without emotion contagion

matches, again the most prominent changes in emotion contagion were found in the hooligans and fanatic supporters and also pleasure and sadness reach more intense levels based on the context of the match.

4 Validation

As first steps in validating the model, the following data are compared with the outcomes of the model: (1) the heart rates from 100 supporters during the match between Feyenoord and AFC Ajax on November 8th, 2015; (2) and (3) 41 subjective reportings, in hindsight, of experienced emotions during a match gathered through questionnaires subjective reportings of three soccer supporters gathered in real time during a match. The result of step one are shown in Fig. 6. The model outcome (left) shows the same pattern as the average measurement of the heart rate of 100 supporters (right). The heart rate peak after a goal has been scored in the model outcome is not as high as in the real measurements, but the overall pattern is similar. The results of the pilot study, are shown in Fig. 7. The bold line in the figures, for both "pleasure" and "sadness", show the average reported emotion. It shows, more or less, the same pattern as the model outcomes. The results of the questionnaire study, are shown in Fig. 8. The bold line in the figures, shows the average reported emotions. Again, it shows the same pattern as the model outcomes. These validation tests confirm that the model can show real patterns of heart rate and emotions.

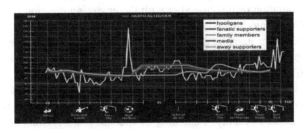

Fig. 6. Heart rates of soccer supporters; model outcomes (colour) and real data (yellow) (Color figure online)

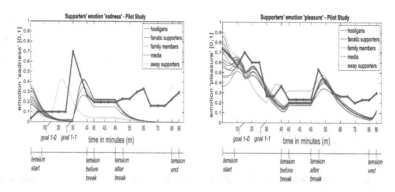

Fig. 7. Model Outcomes and reported emotions (thick lines) of pilot study

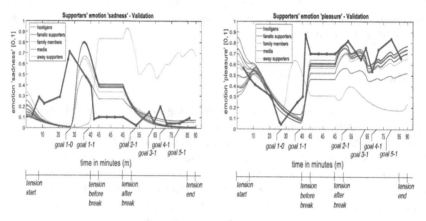

Fig. 8. Model outcomes and reported emotions (thick lines) of questionnaire study

5   Conclusions and Discussions

In this work the emotion contagion and patterns of the emotions "pleasure" and "sadness" were simulated during a soccer game in a virtual stadium, by a dynamic agent-based computational model. The main research question was "how does emotion contagion work in different situations during the game?" All stated phenomenon in Sect. 1, were verified in the simulation results, together with the emerging emotion contagion mechanisms. The main findings are that the emotions "pleasure" and "sadness" have different patterns during a soccer game and amongst different types of supporter. Furthermore, emotion contagion starts very clearly at the beginning of the match, and continues, yet to a lesser extent, during different activities during the soccer game.

One of the strengths of the current work is that the model is based on real life activities. The cognitive model is based on neurological theories and thus mimics a mechanism that exists in reality. Functions of specific brain regions and connections between those regions are related to specific functions and parts of the model. Furthermore, the different supporter groups are based on real supporters with their related characteristics that are visiting actual soccer games. The model heart rates outcomes show a similar pattern as the real life measurements, implying that it is reasonable to believe the model is based on relevant concepts. Another strength is that during the simulations different contexts were taken into account. Furthermore, to the author's knowledge, this paper presents the first model that simulates the ambiance during an activity that is related to crowd behaviour in sport stadiums. There is a lot of literature on different brain regions and their related functions and a large amount of models that model and predict human behaviour. Yet these models are mainly based on calculations derived from social interactions theories. The current work is based on existing neurological theories and social connections, unique in research on emotion contagion.

Aspects that need further development are the following. Firstly, there are more contextual situations during a soccer game, besides goals, like yellow cards, red cards and tackles that can also be included in the model. Secondly, parameter tuning can be improved; such as parameter 'eta' could have a value between 0 and 1, taken the new contextual situations into account. Also, the characteristics of the different subgroups, the omega parameters, can for now only be based on educative guesses, based on the neurological theories. Thirdly, interactions between the emotions 'pleasure' and 'sadness' could be modelled. Apart from that, a person's expectation of the game might influence the parameters and the emotion of that particular person. For example, when a supporter expects an easy win against the opponent, the value of the emotion could be lower when winning the game compared to the value of the emotion at low expectation of winning the game. Lastly, when looking at the validation, it becomes clear that the heart rate patterns are more or less the same, except for the peak after the moment when a goal is scored. Improvements are planned for future work.

In conclusion, the proposed model effectively simulates the emotions 'pleasure' and 'sadness' during a soccer game. It helps us understand the emotion contagion process of soccer supporters better. An envisioned future application of this model uses the current and predicted positive and negative emotions of the supporters, to advise on enhancing positive emotions and reducing or preventing negative emotions in order to improve the ambiance in a soccer stadium. This can be beneficial for both economic and safety reasons.

References

1. Back, L., Crabbe, T., Solomos, J.: Beyond the racist/hooligan couplet race, social theory and football culture. Br. J. Soc. **50**, 419–442 (1999)
2. Aknin, L.B., Barrington-Leigh, C.P., Dunn, E.W., Helliwell, J.F., Burns, J., Biswar-Diener, R., Kemeza, I., Nyende, P., Ashton-James, C.E., Norton, M.I.: Prosocial spending and well-being: cross-cultural evidence for a psychological universal. J. Pers. Soc. Psychol. **104**, 635 (2013)
3. Football Passions: report of research conducted by the social issues of research centre commissioned (2008). http://www.sirc.org/football/football_passions.pdf
4. Okaya, M., Yotsukura, S., Sato, K., Takahashi, T.: Agent evacuation simulation using a hybrid network and free space models. In: Yang, J.-J., Yokoo, M., Ito, T., Jin, Z., Scerri, P. (eds.) PRIMA 2009. LNCS, vol. 5925, pp. 563–570. Springer, Heidelberg (2009)

5. Lacoboni, M., Molnar-Szakacs, I., Gallese, V., Buccino, G., Mazziotta, J.C., Rizzolatti, G.: Grasping the Intentions of Others with One's Own Mirror Neuron System (2005)

6. Rizzolatti, G.: The mirror-neuron system and imitation. In: Hurley, S., Chater, N. (eds.) Perspectives on Imitation: From Cognitive Neuroscience to Social Science, vol. 1, pp. 55–76. MIT Press (2005)

7. Bosse, T., Duell, R., Memon, Z.A., Treur, J., van der Wal, C.N.: A multi- agent model for emotion contagion spirals integrated within a supporting ambient agent model. Cogn. Comput. **7**(1), 111–136 (2015)

8. Bosse, T., Hoogendoorn, M., Klein, M.C., Treur, J., Van Der Wal, C.N., Van Wissen, A.: Modelling collective decision making in groups and crowds: Integrating social contagion and interacting emotions, beliefs and intentions. Aut. Agents Multi-Agent Syst. **27**(1), 52–84 (2013)

9. Sharpanskykh, A., Treur, J.: Modelling and analysis of social contagion in dynamic networks. Neurocomputing **146**, 140–150 (2014)

10. Adang, O.M.J.: Systematic observations of violent interactions between football hooligans. In: Thienpont, K., Cliquet, R. (eds.) In-Group/Out- Group Behaviour In Modern Societies, pp. 1–172. Vlaamse Gemeenschap, Brussel (1999)

11. McLean, J., Brennan, D., Wyper, D., Condon, B., Hadley, D., Cavanagh, J.: Localisation of regions of intense pleasure response evoked by soccer goals. Psychiatry Res. **17**(1), 33–43 (2009)

12. Knutson, B., Adams, C.M., Fong, G.W., Hommer, D.: Anticipation of Increasing monetary reward selectively recruits nucleus accumbens. J. Neurosci. **21**, 1–5 (2001)

13. Knutson, B., Fong, G.W., Adams, C.M., Varner, J.L., Hommer, D.: Dissociation of reward anticipation and outcome with event- related fMRI. NeuroReport **12**, 3683–3687 (2001)

14. Schacter, D.L.: Psychology (2012)

15. Kringelbach, M.L.: The Pleasure Center: Trust Your Animal Instincts. Oxford University Press, New York (2009)

16. Seymour, B., Daw, N., Dayan, P., Singer, T., Dolan, R.: Differential encoding of losses and gains in the human striatum. J. Neurosci. **27**, 4826–4831 (2007)

17. O'Doherty, J.P., Buchanan, T.W., Seymour, B., Dolan, R.J.: Predictive neural coding of reward preference involves dissociable responses in human ventral midbrain and ventral striatum. Neuron **49**, 157–166 (2006)

18. Goligorsky, M.S.: The concept of cellular "flight-or-fight" reaction to stress. Am. J. Physiol. **280**(4), F551–F561 (2001)

19. Frederickson, B.L., Branigan, C.: Positive emotions broaden the scope of attention and thought–action repertoires. Cogn. Emot. **19**(3), 313–332 (2005)

20. Shima, K., Tanji, J.: Science **282**, 1335–1338 (1998)

21. Bush, G., Luu, P., Posner, M.I.: Cognitive and emotional influences in anterior cingulate cortex. Trends Cogn. Sci. **4**(6), 215–222 (2000)

The Role of Mood on Emotional Agents Behaviour

Petros Kefalas[1]([✉]), Ilias Sakellariou[2], Suzie Savvidou[1],
Ioanna Stamatopoulou[1], and Marina Ntika[3]

[1] The University of Sheffield International Faculty, CITY College,
Thessaloniki, Greece
{kefalas,ssavidou,istamatopoulou}@city.academic.gr
[2] University of Macedonia, Thessaloniki, Greece
iliass@uom.edu.gr
[3] South-East European Research Centre, Thessaloniki, Greece
mantika@seerc.org

Abstract. Formal modelling of emotions in agents is a challenging task. This is mainly due to the absence of a widely accepted theory of emotions as well as the two-way interaction of emotions with mood, perception, personality, communication, etc. In this work, we use the widely accepted dimensional theory of emotions according to which mood is a significant integrated factor for emotional state change and therefore behaviour. The theory is formally modelled as part of a state-based specification which naturally leads towards simulation of multi-agent systems. We demonstrate how moods of individual agents affect the overall behaviour of the crowd in a well-known example, that of the El Farol problem.

Keywords: Emotional agents · Formal modelling · Crowd behaviour · El Farol · Multi-agent systems simulation

1 Introduction

It is widely known that emotions can change crowd behaviour from rational to emotional. Modelling of artificial emotions in multi-agent systems (MAS) is challenging by itself, let alone the fact that emotions are affected by a number of other features, such as mood, appraisal, personality, emotion contagion, etc. The definition of emotion has always been accompanied by disagreement, with dominating "movements" in Psychology having different definitions [2]. For the purposes of this work, the most suitable definition of *emotions* is *"passions–as defined as event-instigated or object-instigated states of action readiness with control precedence"* [7]. The dimensional approach of emotion [15] attracts considerable interest and is widely used for measuring emotion.

Mood is a long-lasting emotion, which affects and is affected by personality traits. As an emotion, it can also be represented and measured in two dimensions. In its extremes *High Positive Affect* is characterised by positivity, full energy,

© Springer International Publishing Switzerland 2016
N.T. Nguyen et al. (Eds.): ICCCI 2016, Part I, LNAI 9875, pp. 53–63, 2016.
DOI: 10.1007/978-3-319-45243-2_5

concentration, and engagement, while *Low Negative Affect* can be described as a negative and passive state, potentially involving anger, fear, guilt and nervousness [19].

Personality traits determine the way individuals emotionally react to particular circumstances. Ideas about the connection between emotion, mood and particular personality traits have reached a point of convergence, placing all three components in direct relation to the dimensional approach.

There are several attempts to model emotion in MAS. A number of computational models, logically formalised around BDI have been proposed [8,14], and a number of applications reported on emergency evacuation [13,18,21], context-aware decision support systems [11], negotiation [17], interconnection with socials norms [5], teamwork and cooperation between agents [12], etc. Unlike any of the related approaches, we attempt to formally model the dimensional theory of emotions including mood and other aspects that affect emotions in MAS.

The aim of this paper is to present a formal model of artificial agents infused with artificial emotions based on the dimensional theory as well as to demonstrate in practice that the model can result into believable simulations that take into account different mood distributions over a crowd. The main contribution is a new formal model for emotions that includes the interaction of emotions, perception, mood, personality and emotion contagion. This is an enhancement of an already reported approach [10] that used basic emotion theory as a first step towards the modelling and simulation of emotional agents.

The paper is organised as follows: Sect. 2 presents the dimensional approach theory to emotions as described in Psychology. In Sect. 3, we discuss how artificial emotions and mood can be formally represented in an agent. Section 4 briefly discusses the Netlogo simulation of a MAS derived from a formal state-based model, and results that demonstrate how mood affects the behaviour of the crowd in a well-known example, namely the El Farol problem. Finally, Sect. 5 concludes the paper and suggests directions for future work.

2   Emotions and Moods: The Dimensional Approach

In Psychology there are many theories of emotion. The main issue is that emotions are extremely hard to measure. The dimensional approach is based on the idea that emotion can be represented and measured in two dimensions: (a) *valence*, representing how pleasurable it is for the individual to experience this state (feeling "good" or "bad"); and *arousal*, representing how likely the person is to take some action due to their particular state (feeling "energized" or "down"). These two dimensions represent what has also been called as "core affect", and can be either free floating (mood) or it can be caused by some stimulus initiating an emotional episode. Core affect is always consciously present. It is closely linked to emotion: feeling happy leads to perceiving objects in a congruent way and overestimating their pleasantness as more pleasant than real. The more positive the core affect is, the most pleasant the stimuli are going to be perceived and vice versa.

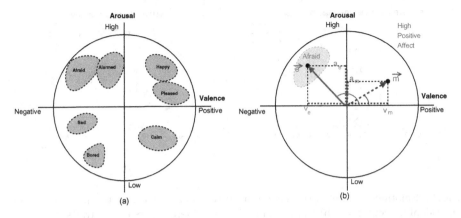

Fig. 1. (a) The circumplex of the dimensional theory of emotions with example emotions located in the two dimensions: arousal and valence. (b) Emotions and moods as vectors.

The dimensional approach was mainly used as an alternative to the discrete emotions theory, according to which there are 5–7 basic emotions [4]. There were several problems associated with the basic emotion theory, among which were difficulties in measuring changes in the emotion in a continuous way. Also, there are several emotional episodes involving more than the basic emotions, which therefore cannot be measured with the basic terms. In contrast, the dimensional theory uses a circumplex to depict emotions (Fig. 1a). The circular space of the circumplex both adequately and reliably represents the emotional space on two dimensions: pleasure and arousal [15]. The circumplex is intended for measuring the dominant emotions and not all emotions experienced concurrently. It is agreed that this model is practical and easy to use for capturing every day natural emotion, and also suitable for measuring both object-less (i.e. mood or core affect) emotions as well as object-directed (basic emotions) states. Despite the controversies and theoretical arguments doubting the suitability of the dimensional approach, it is nowadays widely used, mainly because it allows collection of unambiguous, culturally- and linguistically-free data, and most importantly because it can capture the development of an emotional episode over time. Researchers have occasionally used more than two dimensions by adding a new one, for instance control [6], but such approaches are not widely accepted in contemporary psychological research.

3   A Formal Model for Emotions and Emotional Agents with Mood

Emotions can be represented as a vector $\overrightarrow{e} = (\hat{\omega}, |\overrightarrow{e}|)$ in the two-dimensional space, i.e. with the angle and the magnitude of the emotion vector within the circumplex, with $\hat{\omega} \in [0, 360)$ and $|\overrightarrow{e}| \in [0, 1]$. Alternatively, emotions are

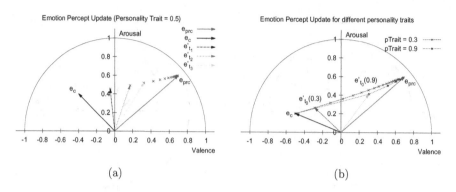

Fig. 2. Emotion updates. (a) Change in current emotion e_c, in the presence of an emotional percept e_{prc}. (b) Change in e_c for two different personality traits.

coordinates in the circumplex determined by the tuple (v_e, a_e), where v_e is the valence measure and a_e the arousal measure, respectively, with $v_e, a_e \in [-1, 1]$ and $\sqrt{v_e^2 + a_e^2} \leq 1$ (Fig. 1b). *Moods* can be considered as medium term emotions and, for the purpose of this study, will not change over time. Moods can also be seen as vectors $\overrightarrow{m}$ within the circumplex, or equivalently as tuples (v_m, a_m), where v_m is the valence measure and a_m the arousal measure of the mood, respectively, with $v_m, a_m \in [-1, 1]$ and $\sqrt{v_m^2 + a_m^2} \leq 1$ (Fig. 1b).

Agents' emotions change over time through stimuli perceived in the environment or other agents. A perceived stimulus may be positive or negative in the sense that it shifts the emotion vector towards positive or negative valence, high or low arousal. Each input is associated with an emotional percept e_{prc} affecting the agent's emotional state. The personality trait f_p also plays an important role in the emotion update process. Thus, the higher the value of the latter is, the more "perceptive" the agent is to emotional percepts.

However, according to the theory, the new emotion vector has to remain within the circumplex. Thus, given a stimulus and an associated emotional percept e_{prc} in terms of valence and arousal (v_{prc}, a_{prc}), the personality trait f_p with $f_p \in (0, 1]$, and the current emotion $\overrightarrow{e_c} = (v_e, a_e)$ of the agent, the updated emotion vector $\overrightarrow{e_c}' = (v_e', a_e')$ is given by the following two equations:

$$v_e' = v_e + \frac{f_p \cdot \Delta v}{1 + e^{-f_p \cdot (\Delta v)}} \quad \text{where } \Delta v = v_{prc} - v_e \tag{1}$$

$$a_e' = a_e + \frac{f_p \cdot \Delta a}{1 + e^{-f_p \cdot (\Delta a)}} \quad \text{where } \Delta a = a_{prc} - a_e \tag{2}$$

The use of a logistic function allows for a shift of the current emotion vector closer to the emotional percept vector, while preserving the properties imposed by the circumplex. If the specific input persists in the subsequent time points, then the emotion vector will eventually align with the corresponding emotional percept vector. In Fig. 2a, $\overrightarrow{e_{t_1}}' \dots \overrightarrow{e_{t_3}}'$ are the updated emotion vectors if the stimulus persists in the time points that follow.

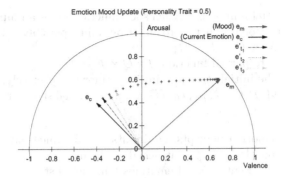

Fig. 3. The role of mood and personality in the change of emotional state.

The personality trait value determines how quickly the emotion vector converges to the emotional percept. In Fig. 2b, $\overrightarrow{e_{t_0}}'(0.3)$ is the updated emotion under $f_p = 0.3$, whereas $\overrightarrow{e_{t_0}}'(0.9)$ is the updated emotion under $f_p = 0.9$.

Mood, $\overrightarrow{e_m} = (v_m, a_m)$, also affects the current agent emotion $\overrightarrow{e_c}$, though to a lesser extent, in each iteration. Following the same approach, the updated $\overrightarrow{e_c}' = (v_e', a_e')$ is given by the following equations:

$$v_e' = v_e + \frac{f_p \cdot \Delta v_{me}}{1 + e^{2 \cdot (\Delta v_{me})}} \quad \text{where } \Delta v_{me} = v_m - v_e \tag{3}$$

$$a_e' = a_e + \frac{f_p \cdot \Delta a_{me}}{1 + e^{2 \cdot (\Delta a_{me})}} \quad \text{where } \Delta a_{me} = a_m - a_e \tag{4}$$

The two pairs of logistic functions defined differ in the steepness of the curve. This reflects the different impact of the two updates on the current emotion vector. In the absence of stimuli, the agent emotion vector will align with its mood vector after a number of iterations. This number is again determined by the personality trait of the agent. Figure 3 depicts the change of the emotion vector $\overrightarrow{e_c}$ under the influence of the mood $\overrightarrow{e_m}$.

Our main research theme is on formal modelling and simulation of crowd behaviour. We have been successfully using a formal method, namely X-machines, which are state-based machines extended with a memory structure, an n-tuple of values. This makes the formal model more compact as compared to memory-less state machines. Transitions between states are triggered by functions, which accept inputs and memory values and produce an output and new memory values. In a previous work, we have made an initial attempt to integrate artificial emotions as plug-ins within X-machines [9]. For the purposes of this work, an *emotions X-machine* is defined as: $^e\mathcal{X} = (\Sigma, \Gamma, Q, M, \Phi, F, q_0, m_0, E)$ where:

- Σ and Γ are the input and output alphabets, respectively;
- Q is the finite set of states;
- M is the (possibly) infinite set called memory;

- Φ is a set of partial functions φ; each such function maps an input, a memory value and an emotional structure to an output and a possibly different memory value, $\varphi : \Sigma \times M \times E \rightarrow \Gamma \times M$;
- F is the next state partial function, $F : Q \times \Phi \rightarrow 2^Q$;
- q_0 and m_0 are the initial state and initial memory respectively;
- $E = ({}^e\Sigma, {}^eQ, \mathcal{M}, P, C, {}^e\Phi, \rho_\sigma, \overrightarrow{e_0}, \overrightarrow{m_0})$ is an emotional structure formalisation with:

  - ${}^e\Sigma$ is a set of emotion percepts, i.e. inputs $\sigma \in \Sigma$ which have an emotional valence and arousal (v_{prc}, a_{prc}) attached to them;
  - ${}^eQ$ is a set of representations of emotions (emotional state);
  - $\mathcal{M}$ is a set of moods;
  - P is a personality trait type;
  - C is a contagion model type;
  - ${}^e\Phi : {}^eQ \times \mathcal{M} \times P \times C \times M \times \Sigma \rightarrow {}^eQ$ is the set of emotions revision functions ${}^e\varphi$, that given an emotional state $\overrightarrow{e} \in {}^eQ$, a mood $\overrightarrow{m} \in \mathcal{M}$, a contagion model $c \in C$, a personality trait $p \in P$ and a memory tuple $m \in M$ returns a new emotional state $\overrightarrow{e}' \in {}^eQ$;
  - ρ_σ is an input revision function, which given an input σ transforms it into an emotional percept taking into account the current emotional state, the mood and the personality, that is, $\rho_\sigma : \Sigma \times {}^eQ \times P \times \mathcal{M} \rightarrow {}^e\Sigma$;
  - $\overrightarrow{e_0}$ is the initial emotional state and $\overrightarrow{m_0}$ is the initial mood.

With regards to previous definitions of ${}^e\mathcal{X}$ [9], in this work we attempt to use the dimensional model of emotions instead of a set of basic emotions. In addition, we integrate mood $\mathcal{M}$ into the emotional structure formalisation E in order to investigate the role that it plays in the rate of change of the emotion state ${}^eQ$. *Emotional Contagion*, C, is a result of interaction between agents, which can also affect their emotions. The *personality trait* of an agent P is defined by the Big Five [3]: (O, C, E, A, N), values for the different personality factors of Openness, Consciousness, Extraversion, Agreeableness, Neuroticism. Both contagion and personality trait types were studied extensively in [13].

4   Feel Like Going to the Bar?

The El Farol problem was originally introduced as a problem in game theory [1]. The El Farol bar in Santa Fe hosts Irish music sessions on Thursday nights. However, since the bar is rather small, attracting a large crowd results in visitors not having a good time. The population consists of N individuals, and if more than 60 % attend, then they would all be better off if they had stayed at home. In the case that less attend, then their decision for going to the bar pays off. Unfortunately, individuals have to decide without knowing the intentions of others, knowing, however, the attendance of previous weeks. The problem is introduced as an example of reasoning based on patterns in "ill-defined situations" [1].

The approaches so far attempt to find strategies that predict the current week's attendance based on historical data. We adopt a different approach, where

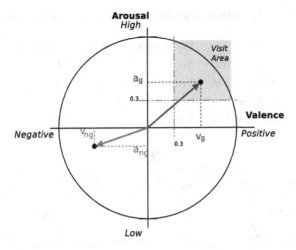

Fig. 4. Threshold for deciding to visit El Farol. Vector (v_g, a_g) is well in the visiting area, whereas an agent with an emotional state (v_{ng}, a_{ng}) will decide to stay at home.

each individual agent bases its decision on its current emotional state. Since the emotional state of the agent is expressed in terms of valence (pleasure) and arousal (how likely the agent is to take action), an agent decides to visit the bar if both components of the emotion vector exceed a threshold, i.e. $v_e > 0.3 \wedge a_e > 0.3$. The intuition behind the approach is that, in order to decide to attend, they have to be in a "pleasurable state" towards the visit and willing to take action (visit). On the contrary, if they are in a negative valence towards visiting and unwilling to take action, it is clear that they will not decide to go (Fig. 4).

As such, decision is based only on the agents's current emotional state. What motivates people to visit El Farol is mood, which reflects the long term tendency to visit the bar. The population is initially given a random mood that lies in the visit area depicted in Fig. 4, invariant during the course of the experiment. Since the emotional state converges to the mood in the absence of emotional percepts, all agents would decide to visit El Farol, if they did not perceive their environment. However, during a crowded evening, visiting agents receive negative emotional percept, whereas in a non crowded night a positive one. If they decide to stay at home and the bar is crowded, then, although their emotional state does not change with respect to the valence component, their attitude towards action is affected. For instance, if the agent stayed at home and the bar was crowded, then the valence component of the stimulus remains unaffected (value 0), but it is less willing to take the action of visiting the bar, and thus the arousal component of the stimulus is -0.2. This reduction in arousal is less than the one received if the agent had actually visited the bar on a crowded night (-0.4), i.e. the impact of experiencing a crowded night is more significant than that of hearing about it. Thus, we formulated emotional percept as depicted in Table 1.

In this simple scenario, the $^e\mathcal{X}$ model that describes the agent behaviour is rather simple with the functions and the state diagram as depicted in Fig. 5.

Table 1. Agent input and the associated emotional percepts.

Input	Bar crowded	Valence	Arousal
Night at bar	No	0.4	0.4
Night at bar	Yes	−0.4	−0.4
Night at home	No	0	0.2
Night at home	Yes	0	−0.2

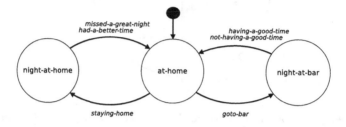

Fig. 5. The State Diagram of the agent in the El Farol prolem.

The $^e\mathcal{X}$ model was implemented in NetLogo [20], a well-established platform for MAS simulation, extensively used by the research community. The implementation was based on the TXStates DSL [16] that allows easily encoding $^e\mathcal{X}$ agents in the platform. Although the experimentation is preliminary, the results obtained seem to support the use of emotions as an alternative to other solutions provided to the problem. In all experiments reported in this work, agents start with a random emotional state and have a random personality trait. As stated, to represent the tendency agents have to visit the bar, agents have a random mood that lies within the "Visit Area" (Fig. 4).

The evolution of bar attendance with respect to time in a random single run of the experiment is shown in Fig. 6. Initially, since agents have a random

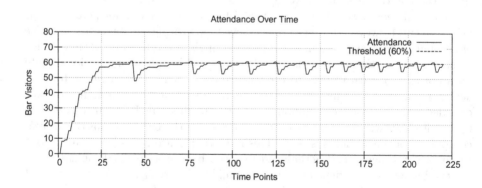

Fig. 6. Attendance in the El Farol prolem.

emotional state, but a positive mood toward attending the bar, the number of bar visitors increases steadily, until it reaches the 60 % threshold. When the attendance reaches the latter, a negative emotional percept is generated, that "lowers" the emotional state of the agents: this negative stimulus has an effect on the agent's decision regarding visiting the bar in the next round. Since agents have different personality traits and different moods, the negative percept pushes only a percentage of them not to visit the bar.

Agents do not "recall" historical data on attendance, this is encoded in their emotional state. The behaviour described is common in all experiments. The average attendance of several runs of the simulation is 56.9 with a standard deviation of 4.4. What affects attendance is the distributions of mood and personality traits. The relation between these and the evolution of attendance over time needs to be more carefully investigated, however such a task is outside the scope of this paper. These preliminary results are an indication of how emotions can play a significant role in regulating MAS societies.

5 Conclusions and Further Work

Without doubt emotions affect agent action in real societies. Consequently, modelling emotional agents has attracted significant attention. We introduced a formal model of emotional agents based on the well-established dimensional theory. The model does not only deal with short term agent emotion change due to agent input (percepts) but hosts medium term agent emotional states (moods), and their effect in the agent decision process.

Such a formal model leads to the development of interesting MAS. The El-Farol problem served as an initial example application to show how emotions can affect agent actions and even regulate societies. Preliminary results show that directly modelling the problem description as to what the agents "feel", and not introducing a utility that describes that, can lead to an agent action pattern that seems to be self-regulated. These results have to be investigated in depth and associated to the initial emotion, personality trait and mood distributions.

The El Farol model can act as a testbed for testing a number of issues that have not been discussed, such as emotion contagion, i.e. the change in the emotion caused by the emotional state of other agents. This aspect has not been investigated in this work and could be interesting; for instance, since agents "live" in specific neighbourhoods, the emotional states of agents living close by could play a role in their decision process, i.e. modelling a common social reputation mechanism.

References

1. Arthur, W.B.: Inductive reasoning and bounded rationality. Am. Econ. Rev. **84**(2), 406–411 (1994)
2. Cornelius, R.: The Science of Emotion: Research and Tradition in the Psychology of Emotion. Prentice Hall, USA (1996)

3. Costa Jr., P.T., McCrae, R.: Revised NEO Personality Inventory (NEO-PI-R) and NEO Five-Factor Inventory (NEO-FFI) manual (1992)
4. Ekman, P.: An argument for basic emotions. Cogn. Emot. **6**(3/4), 169–200 (1992)
5. Fix, J., von Scheve, C., Moldt, D.: Emotion-based norm enforcement and maintenance in multi-agent systems: foundations and petri net modeling. In: Proceedings of the 5th International Joint Conference on Autonomous Agents and Multiagent Systems, pp. 105–107. ACM (2006)
6. Fontaine, J.R., Scherer, K.R., Roesch, E.B., Ellsworth, P.C.: The world of emotions is not two-dimensional. Psychol. Sci. **18**(12), 1050–1057 (2007)
7. Fridja, N.: The psychologists point of view. In: Lewis, M., Haviland-Jones, J., Feldman-Barrett, L. (eds.) Handbook of Emotions, 3rd edn. The Guildford Press, New York (2008)
8. Jiang, H., Vidal, J.M., Huhns, M.N.: EBDI: an architecture for emotional agents. In: Proceedings of the 6th International Joint Conference on Autonomous Agents and Multiagent Systems, AAMAS 2007, pp. 1–3. ACM, New York (2007)
9. Kefalas, P., Sakellariou, I., Basakos, D., Stamatopoulou, I.: A formal approach to model emotional agents behaviour in disaster management situations. In: Likas, A., Blekas, K., Kalles, D. (eds.) SETN 2014. LNCS, vol. 8445, pp. 237–250. Springer, Heidelberg (2014)
10. Kefalas, P., Stamatopoulou, I., Basakos, D.: Formal modelling of agents acting under artificial emotions. In: Balkan Conference in Informatics, BCI 2012, Novi Sad, Serbia, 16–20 September 2012, pp. 40–45 (2012)
11. Marreiros, G., Santos, R., Ramos, C., Neves, J.: Context-aware emotion-based model for group decision making. IEEE Intell. Syst. **25**(2), 31–39 (2010)
12. Nair, R., Tambe, M., Marsella, S.: The role of emotions in multiagent teamwork. Who Needs Emotions (2005)
13. Ntika, M., Sakellariou, I., Kefalas, P., Stamatopoulou, I.: Experiments with emotion contagion in emergency evacuation simulation. In: Proceedings of the 4th International Conference on Web Intelligence, Mining and Semantics (WIMS 2014), pp. 49:1–49:11 (2014)
14. Pereira, D., Oliveira, E., Moreira, N., Sarmento, L.: Towards an architecture for emotional BDI agents. In: Proceedings of the Portuguese Conference on Artificial intelligence (EPIA 2005), pp. 40–46, December 2005
15. Russell, J.: A circumplex model of affect. J. Pers. Soc. Psychol. **39**(6), 1161–1178 (1980)
16. Sakellariou, I., Dranidis, D., Ntika, M., Kefalas, P.: From formal modelling to agent simulation execution and testing. In: Proceedings of the 7th International Conference on Agents and Artificial Intelligence (ICAART 2015), pp. 87–98 (2015)
17. Santos, R., Marreiros, G., Ramos, C., Neves, J., Bulas-Cruz, J.: Personality, emotion, and mood in agent-based group decision making. IEEE Intell. Syst. **26**(6), 58–66 (2011)
18. Tsai, J., Fridman, N., Bowring, E., Brown, M., Epstein, S., Kaminka, G.A., Marsella, S., Ogden, A., Rika, I., Sheel, A., Taylor, M.E., Wang, X., Zilka, A., Tambe, M.: ESCAPES: evacuation simulation with children, authorities, parents, emotions, and social comparison. In: Sonenberg, L., Stone, P., Tumer, K., Yolum, P. (eds.) AAMAS, pp. 457–464. IFAAMAS (2011)
19. Watson, P., Clark, L., Tellegen, A.: Development and validation of brief measures of positive and negative affect: the PANAS scales. J. Pers. Soc. Psychol. **54**(6), 1063–1070 (1988)

20. Wilensky, U.: NetLogo (1999). http://ccl.northwestern.edu/netlogo/. Center for Connected Learning and Computer-Based Modeling, Northwestern University, Evanston, IL
21. Zoumpoulaki, A., Avradinis, N., Vosinakis, S.: A multi-agent simulation framework for emergency evacuations incorporating personality and emotions. In: Konstantopoulos, S., Perantonis, S., Karkaletsis, V., Spyropoulos, C.D., Vouros, G. (eds.) SETN 2010. LNCS, vol. 6040, pp. 423–428. Springer, Heidelberg (2010)

Modelling a Mutual Support Network for Coping with Stress

Lenin Medeiros[✉], Ruben Sikkes, and Jan Treur

Behavioural Informatics Group, Vrije Universiteit Amsterdam,
De Boelelaan 1081, 1081 HV Amsterdam, Netherlands
{l.medeiros,j.treur}@vu.nl, rubensikkes@gmail.com

Abstract. The emotional state of an individual is continuously affected by daily events. Stressful periods can be coped with by support from a person's social environment. Support can for example reduce stress and social disengagement. Before improvements on the process of support are however made, it is essential to understand the actual real world process. In this paper a computational model of a network for mutual support is presented. The dynamic model quantifies the change in the network over time of stressors and support. The model predicts that more support is provided when more stress is experienced and when more people are capable of support. Moreover, the model is able to distinguish personal characteristics. The model behaves according to predictions and is evaluated by simulation experiments and mathematical analysis. The proposed model can be important in development of a software agent which aims to improve coping with stress through social connections.

Keywords: Stress · Coping · Mutual support · Network · Computational model

1 Introduction

It is a fact that the emotional state of an individual is continuously influenced by events from the environment. A loved one passing away might have a negative influence while a wedding might have a positive influence on the emotional state. All these events combined can be considered the basis for an individual's overall stress level. Too many stressful life events could lead to depression [13]. Therefore, the reduction of the overall stress level can lead to many health benefits [8].

Mutual support networks are social networks in which its members provide psychological and material support to each other in order to help such individuals to cope with stress [2]. In this scenario, stress responses would be perceived via, for example, things said in a conversation between friends in such a network. Regarding to the relation between stress and social networks, there is the concept of stress buffering. This concept is originated from the hypothesis formulated by John Cassel and Sidney Cobb [1] by which they argued that people with strong social ties, i.e. close friends, parents and sons, brothers and sisters, etc. could be protected from the negative effects related to negative life stressful events. As what is stated in [12], better social integration (doing social activities, etc.) leads to better resilience to post traumatic stress.

© Springer International Publishing Switzerland 2016
N.T. Nguyen et al. (Eds.): ICCCI 2016, Part I, LNAI 9875, pp. 64–77, 2016.
DOI: 10.1007/978-3-319-45243-2_6

Therefore, in this work, we consider stress buffering as the capacity of facing stressful events without getting into a stressed state of mind. Besides that, here in this work we are typically dealing with conversation, via a social network, between people with strong social ties as a type of social integration in order to help people to cope with stress. The advantages of social support for stress buffering is already known in the literature; e.g., [8, 11, 12, 14]. A dynamic model could be used to develop a tool (a software agent) that would help people in coping with stress by their social connections with people with strong social ties, since the goals of such a tool could be, for example: to select automatically a set of a given member's friends or relatives who can provide better support, to monitor who are seeking support in a given period due to, for example, a case of death in his or her family, etc.

In this paper, a dynamic model is presented for a mutual support network addressing the process of providing and receiving support. Such a model deals with the timing of positive and negative life events, and induced stress levels. To this model we pass, as input, values representing the intensity of positive and negative life events faced by the users of the network as well as the period of time they last. The model then returns, as output, values over time representing the amount of: the stress faced by the users, the support they receive and provide from and to each other, the perception that each of them have about the stress levels of their friends and the capacity they have to provide support. Note that the term dynamic came from the fact that through such a model we are able to represent the change of its states values over time. The computational model proposed here can be used as an ingredient to develop human-aware or socially aware computing applications; e.g. [9, 10, 15]. More specifically, in [5, 15] it is shown how such applications can be designed with knowledge of human and/or social processes as a main ingredient represented by a computational model of these processes which is embedded within the application. As an example, in [6, 7] this design approach is illustrated to obtain a human-aware software agent supporting professionals in attention-demanding tasks, based on an embedded dynamical numerical model for attention. The computational model for mutual support proposed here can be used in a similar way to design a human-aware or socially aware software agent to support persons suffering from stress effects.

This document is organized as follows; Sect. 2 formally describes our proposed model and shows the mathematical details of it. Section 3 shows the model working in order to describe some behavioral characteristics of it. In Sect. 4 the tuning process of our model is shown in order to make clear how we adjusted our model to keep it as close as possible to reality. Finally, Sect. 4 concludes our paper.

2   The Computational Model

There are a few requirements for the model that have been identified previously during the review of the literature about people coping with stress and social network models:

- A person should be supported more when there are more resources available by the others;
- The users of such a social network typically have strong social ties with each other;

- More help should be given when more help is needed;
- More support from people inside this network should lead to lower stress levels rather than receiving no support at all;
- Individual personal characteristics can be represented by different strengths of connections within persons;
- Social network characteristics can be represented by different strengths of connections between persons.

Simulations to test these requirements will be shown in the next section. The mutual support social network that will be analysed will consist of a number of persons.

The computational model has been designed using the temporal-causal network modelling approach described in [16]. According to this general dynamic modeling approach, a model is designed at a conceptual level, for example, in the form of a graphical conceptual representation or a conceptual matrix representation. A graphical conceptual representation displays nodes for states S and arrows for connections indicating causal impacts (shown by the nodes and arrows in Fig. 1) from one state to another like, for example, when the occurrence of a negative life event leads to the growth of an individual's stress level, and includes some additional information in the form of:

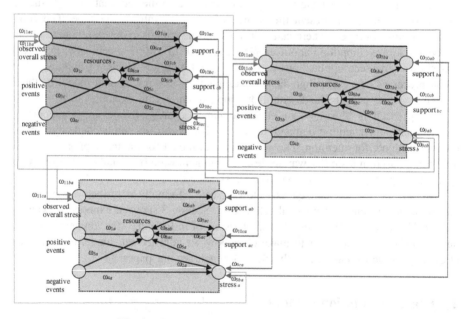

Fig. 1. Conceptual representation of the model

- For each connection from a state X to a state Y, a connection weight $\omega_{X,Y}$ (for the strength of the impact, i.e. the impact of a state X in a state Y is directly proportional to the value $\omega_{X,Y}$ when it is positive or inversely proportional when such a value is negative);

- For each state Y a speed factor η_Y (which represents how fast the effect will take place somehow that it controls if the impact of the effect will be slow or quick);
- For each state Y the type of combination function $c_Y(..)$ used (to aggregate multiple impacts on a state).

Regarding the states of the model, for a given person X, ne_X and pe_X are the negative and positive life events for this individual, s_X represents his or her stress level, r_X is the amount of resources available for this person, $sup_{X,Y}$ is the support provided by X to Y (another person), and oos_X is the overall stress level within the part of the network observed by X. These states are depicted as nodes in Fig. 1 for three particular persons a, b and c. In Table 1 the different connections are summarized with their weights, which are also depicted as labels of the arrows in Fig. 1. For the sake of simplicity in Fig. 1 such a network of only three persons is shown. The three coloured boxes represent the different persons that are involved in this model. All of them use the same states and architecture. At the core of each social network member stands the individual experienced stress level of such a person, represented by state s_X. The perceived stress of the other persons in the social network is the observed overall stress state oos_X. This state is a perception of how much stress the other persons in the network are experiencing and is therefore directly affected by the stress levels of the other persons. In our model, users could perceive stressed friends by the things they are telling in the social network via posts or direct messages. For sure a given user could hide your stressed state of mind, but we are not dealing with this situation; this could be, perhaps, studied in more details during a future work.

Table 1. Connections and respective weights

from	weight	+/-	to	from	weight	+/-	to
pe_X	ω_{1X}	+	r_X	pe_X	ω_{2X}	−	s_X
ne_X	ω_{3X}	−		ne_X	ω_{4X}	+	
s_X	ω_{5X}	−		$sup_{Y,X}$	$\omega_{9Y,X}$	−	
$sup_{X,Y}$	$\omega_{8X,Y}$	−		r_X	$\omega_{6X,Y}$	+	$sup_{X,Y}$
s_Y	$\omega_{11Y,X}$	+	oos_X	oos_X	$\omega_{7X,Y}$	+	
				s_Y	$\omega_{10Y,X}$	+	

This observed stress of the others is used by each person to form a perception of the overall amount of support that would be needed. Besides the perception of experienced stress of the others, state r_x indicates the resource (in terms of emotion, energy, time) a person has. It is important to mention that emotion is important for the resources because we expect that a person feeling sad, stressed, etc. is not indicated to provide support. If a person has no resources, no support can be given. Giving support drains these resources since it costs energy, time and it can change the emotional state of the support provider due to empathy: an individual can be sad because one of his best friends is facing a stressful situation. The relationship between giving support and available resource is therefore reciprocal.

Resources are also affected by the experienced stress level of the person. When a lot of stress is experienced this will negatively influence the amount of resources a person will have. The amounts of negative and positive events (indicated by pe_X and ne_X) have an influence on both the experienced stress and the available resources. The positive and negative events do not have the same values for the different persons because a real world process is modelled where persons are living their own lives. An event taking place in one person's life does not mean that a different person experiences this event as well. Finally a reciprocal connection exists between person X giving support to another person Y and this effect of this support by decreasing stress levels in person Y due to their strong social tie. A graphical conceptual representation of the model is shown in more detail in Fig. 1. Note that some of the arrows are bidirectional, where each direction has its own connection weight. The states pe_X and ne_X indicate the positive and negative events encountered by person X in his or her environment over time. Note that these events could be detected via emotions shared by things the users can tell in posts or direct messages, i.e. the same case of the state oos_X. For specific scenarios these are given as input for the persons.

In Table 1 for each state (in the 'to'column) it is indicated which impacts from other states (in the 'from'column) it gets, with which weights, and with which sign +/− indicating whether the weight value is positive (strengthening) or negative (suppressing). In Fig. 1 these weights are depicted as labels for the arrows. Note that as the nodes represent states, the processes happen between these states, as indicated by the arrows representing causal impact.

The conceptual representation of the model as shown in Fig. 1 and the tables can be transformed in a systematic or even automated manner into a numerical representation of the model as follows [16]:

- at each time point t each state Y in the model has a real number value in the interval $[0, 1]$, denoted by $Y(t)$
- at each time point t each state X connected to state Y has an impact on Y defined as **impact**$_{X,Y}(t) = \omega_{X,Y}X(t)$ where $\omega_{X,Y}$ is the weight of the connection from X to Y
- The *aggregated impact* of multiple states X_i on Y at t is determined using a *combination function* $\mathbf{c}_Y(..)$:

$$\mathbf{aggimpact}_Y(t) = \mathbf{c}_Y(\mathbf{impact}_{X_1,Y}(t), \ldots, \mathbf{impact}_{X_k,Y}(t))$$
$$= \mathbf{c}_Y(\omega_{X_1,Y}X_1(t), \ldots, \omega_{X_k,Y}X_k(t))$$

where X_i are the states with connections to state Y
- The effect of **aggimpact**$_Y(t)$ on Y is exerted over time gradually, depending on speed factor η_Y:

$$Y(t + \Delta t) = Y(t) + \eta_Y[\mathbf{aggimpact}_Y(t) - Y(t)]\,\Delta t$$
$$\text{or} \quad \mathbf{d}Y(t)/\mathbf{d}t = \eta_Y[\mathbf{aggimpact}_Y(t) - Y(t)]$$

- Thus, the following *difference* and *differential equation* for Y are obtained:

$$Y(t+\Delta t) = Y(t) + \eta_Y[\mathbf{c}_Y(\omega_{X_1,Y}X_1(t), \ldots, \omega_{X_k,Y}X_k(t)) - Y(t)]\,\Delta t$$
$$dY(t)/dt = \eta_Y[\mathbf{c}_Y(\omega_{X_1,Y}X_1(t), \ldots, \omega_{X_k,Y}X_k(t)) - Y(t)]$$

As an example, according to the pattern described above the difference and differential equation for the resource state r_Y for person Y providing support to persons X_1, ..., X_k are as follows:

$$r_Y(t+\Delta t) = r_Y(t) +$$
$$\eta_{r_Y}[\mathbf{c}_{r_X}(\omega_{1Y}\mathrm{pe}_Y(t), \omega_{3Y}\mathrm{ne}_Y(t), \omega_{5Y}s_Y(t), \omega_{8Y,X_1}\mathrm{sup}_{Y,X_1}(t), \ldots, \omega_{8Y,X_k}\mathrm{sup}_{Y,X_k}(t)) - r_Y(t)]\,\Delta t$$
$$dr_Y(t)/dt =$$
$$\eta_{r_Y}[\mathbf{c}_{r_Y}(\omega_{1Y}\mathrm{pe}_Y(t), \omega_{3Y}\mathrm{ne}_Y(t), \omega_{5Y}s_Y(t), \omega_{8Y,X_1}\mathrm{sup}_{Y,X_1}(t), \ldots, \omega_{8Y,X_k}\mathrm{sup}_{Y,X_k}(t)) - r_Y(t)]$$

So, for any set of values for the connection weights, speed factors and any choice for combination functions, each state of the model (as shown in Sect. 2) gets a difference and differential equation assigned, except the positive and negative events, which are used as input. For the model considered here this makes a set of 4 coupled difference or differential equations per person, that together, are 12 equations that in mutual interaction describe the model's behaviour. Note that the speed factors enable to obtain a realistic timing of the different states in the model, for example, to tune the model to the timing of processes in the real world. An often used combination function [16] is the *advanced logistic sum function* **alogistic**$_{\sigma,\tau}(\ldots)$:

$$\mathbf{c}_Y(V_1, \ldots V_k) = \mathbf{alogistic}_{\sigma,\tau}(V_1, \ldots, V_k) = \left(\frac{1}{1+e^{-\sigma(V_1+\ldots+V_k-\tau)}} - \frac{1}{1+e^{\sigma\tau}}\right)(1+e^{-\sigma\tau})$$

Here σ and τ are *steepness* and *threshold* parameters. This function is symmetric in its arguments: its result is independent of the order of the arguments. Moreover, the function is monotonic for its arguments, maps activation levels 0 to 0 and keeps values below1. For all states except the support states $\mathrm{sup}_{X,Y}$ the advanced logistic sum function was used. For example for the resource state r_Y:

$$\mathbf{c}_{r_Y}(V_1, V_2, V_3, V_4, V_5) = \mathbf{alogistic}_{\sigma,\tau}(V_1, V_2, V_3, V_4, V_5)$$

Then:

$$dr_Y(t)/dt = \eta_{r_Y}[\mathbf{alogistic}_{\sigma,\tau}(\omega_{1Y}\mathrm{pe}_Y(t), \omega_{3Y}\mathrm{ne}_Y(t), \omega_{5Y}s_Y(t), \omega_{8Y,X_1}\mathrm{sup}_{Y,X_1}(t), \ldots, \omega_{8Y,X_k}$$
$$\mathrm{sup}_{Y,X_k}(t)) - r_Y(t)]$$

Similarly for the states oos_X and s_X. For the support state $\mathrm{sup}_{Y,X}$ from Y to X the following combination function is used:

$$\mathbf{c}_{\mathrm{sup}_{Y,X}}(V_1, V_2, V_3) = \min(V_1, V_2)V_3/V_2 \text{ if } V_2 > 0, \text{ else } 0$$

where V_1 refers to impact **impact**$_{r_Y,\mathrm{sup}_{Y,X}}(t)$ from the resource state r_Y, V_2 to impact **impact**$_{\mathrm{oos}_Y,\mathrm{sup}_{Y,X}}(t)$ from the observed overall stress state oos_Y, and V_3 to impact **impact**$_{s_X,\mathrm{sup}_{Y,X}}(t)$ from the other person's own stress state s_X. Note that this combination function makes the provided support (when needed) proportional to the own resource level and also proportional to the fraction of the other person's stress level from the overall stress level. Moreover, note that in contrast to the other combination function, this function is not symmetric in its arguments. For example, based on the above, assuming $oos_Y(t) > 0$, the difference and differential equation for $sup_{Y,X}$ are as follows:

$$sup_{Y,X}(t + \Delta t) = sup_{Y,X}(t) +$$
$$\eta_{sup_{Y,X}}[\min(\omega_{6Y,X} r_Y(t), \omega_{7Y,X} oos_Y(t)) \omega_{10Y,X} s_X(t)/(\omega_{7Y,X} oos_Y(t)) - sup_{Y,X}(t)] \Delta t$$
$$\mathbf{d}sup_{Y,X}(t)/\mathbf{dt} =$$
$$\eta_{sup_{Y,X}}[\min(\omega_{6Y,X} r_Y(t), \omega_{7Y,X} oos_Y(t)) \omega_{10Y,X} s_X(t)/(\omega_{7Y,X} oos_Y(t)) - sup_{Y,X}(t)]$$

3 Verification of the Model

In this section it is described whether the proposed model acts according to what was expected. To find this out, both a mathematical and experimental analysis have been performed.

3.1 Mathematical Analysis

To verify (the implementation of) the model, a mathematical analysis was performed of the equilibria of the model. A state Y has a *stationary point* at t if $\mathbf{d}Y(t)/\mathbf{dt} = 0$. The model is in *equilibrium* at t if every state Y of the model has a stationary point at t. See Fig. 2 for an example of an equilibrium state that is reached. From the specific format of the differential or difference equations it follows that state Y has a stationary point at t if and only if

$$Y(t) = \mathbf{c}_Y(\omega_{X_1,Y}X_1(t), \ldots, \omega_{X_k,Y}X_k(t))$$

where X_i are the states with connections to state Y, and $\mathbf{c}_{X_i}(\ldots)$ is the combination function for Y. If the values of the states for an equilibrium are indicated by $\underline{\mathbf{X}}_i$ then for

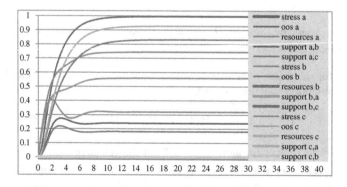

Fig. 2. The final equilibrium state values for the first scenario

a community of three persons being in an equilibrium state is equivalent to a set of 12 equilibrium equations for the 12 states X_i of the model:

$$\underline{\mathbf{X}}_i = \mathbf{c}_{X_i}(\omega_{X_1,X_i}\underline{\mathbf{X}}_1, \ldots, \omega_{X_k,X_i}\underline{\mathbf{X}}_k)$$

Most states have the advanced logistic sum function as combination function. For example, for state r_Y the equilibrium equation is:

$$\underline{\mathbf{r}}_Y(t) = \mathbf{alogistic}_{\sigma,\tau}(\omega_{1Y}\,\underline{\mathbf{pe}}_Y, \omega_{3Y}\,\underline{\mathbf{ne}}_Y, \omega_{5Y}\,\underline{\mathbf{s}}_Y, \omega_{8Y,X_1}\,\underline{\mathbf{sup}}_{Y,X_1}, , \omega_{8Y,X_2}\,\underline{\mathbf{sup}}_{Y,X_2})$$

An exception is state $\sup_{Y,X}$; in that case the equilibrium equation is

$$\underline{\mathbf{sup}}_{Y,X} = \min(\omega_{6Y,X}\,\underline{\mathbf{r}}_Y, \omega_{7Y,X}\,\underline{\mathbf{oos}}_Y)\,\omega_{6Y,X}\,\underline{\mathbf{r}}_Y\,\omega_{10Y,X}\,\underline{\mathbf{s}}_X/\omega_{7Y,X}\,\underline{\mathbf{oos}}_Y$$

or

$$\underline{\mathbf{sup}}_{Y,X}\,\omega_{7Y,X}\,\underline{\mathbf{oos}}_Y = \min(\omega_{6Y,X}\,\underline{\mathbf{r}}_Y, \omega_{7Y,X}\,\underline{\mathbf{oos}}_Y)\,\omega_{6Y,X}\,\underline{\mathbf{r}}_Y\,\omega_{10Y,X}\,\underline{\mathbf{s}}_X$$

The equilibrium equations cannot be solved in an explicit analytical manner, due to the logistic functions. Therefore the verification approach as sometimes used, by first solving the equations and then comparing the values to values found in simulations does not work here. However, for the purpose of verification of the model, solving the equations is actually not needed. The equations can also be used themselves for verification by just substituting the equilibrium values found in a simulation in them and then checking whether they are fulfilled (and with which accuracy). This indeed has been done and the equations turned out to always hold (with an accuracy $< 10^{-6}$).

A second manner in which mathematical analysis was performed was by verifying the requirements listed at the start of Sect. 2; in a more exact formulation they are:

(1) a person X should be supported by $\sup_{Y,X}$ more when there are higher levels of resources r_Y available by the others Y
(2) the connections in such a social network typically have high weights ω
(3) more help $\sup_{Y,X}$ should be given to X when more help is needed due to a higher stress level s_X
(4) more support $\sup_{Y,X}$ to a person X should result in a lower stress level s_X since we are assuming that Y and X have a strong social tie;
(5) individual personal characteristics can be represented by different strengths of connections within persons
(6) social network characteristics can be represented by different strengths of connections between persons

Requirement (2) was fulfilled in the considered scenarios. The last two requirements (5) and (6) indeed are fulfilled by the temporal-causal network modelling approach followed. By choosing values for the weights different characteristics of persons and network are obtained. Now consider requirement (1). Given the definition of the combination function $\mathbf{c}_{sup_{Y,X}}(\ldots)$ it holds:

$$\mathbf{aggimpact}_{\sup_{Y,X}}(\mathbf{t}) = \mathbf{c}_{\sup_{Y,X}}(\mathbf{impact}_{r_Y,\sup_{Y,X}}(t), \mathbf{impact}_{\mathrm{oos}_Y,\sup_{Y,X}}(t), \mathbf{impact}_{s_X,\sup_{Y,X}}(t))$$

$$= \min\left(\mathbf{impact}_{r_Y,\sup_{Y,X}}(t), \mathbf{impact}_{\mathrm{oos}_Y,\sup_{Y,X}}(t)\right)\mathbf{impact}_{s_X,\sup_{Y,X}}(t) \,/$$

$$\mathbf{impact}_{\mathrm{oos}_Y,\sup_{Y,X}}(t) \qquad\qquad \text{if } \mathbf{impact}_{\mathrm{oos}_Y,\sup_{Y,X}}(t) > 0, \text{ else } 0$$

From this formula it follows that when r_Y is higher, then $\mathbf{impact}_{r_Y,\sup_{Y,X}}(t)$ is higher and therefore $\mathbf{aggimpact}_{\sup_{Y,X}}(t)$ will be higher, which results in a higher level for support $\sup_{Y,X}$. Therefore requirement (1) is fulfilled. In a similar way it has been verified that (3) is fulfilled: from the above formula it follows that when s_X is higher, then $\mathbf{impact}_{s_X,\sup_{Y,X}}(t)$ is higher and therefore $\mathbf{aggimpact}_{\sup_{Y,X}}(t)$ will be higher, which results in a higher level for support $\sup_{Y,X}$. Finally, for requirement (4) the combination function for s_X has to be considered. The weight of the connection from $\sup_{Y,X}$ to s_X is negative, so for higher $\sup_{Y,X}(t)$ the impact

$$\mathbf{impact}_{\sup_{Y,X},s_X}(t) = \omega_{9Y,X}\sup_{Y,X}(t)$$

is lower (more negative). Now the combination function $\mathbf{alogistic}_{\sigma,\tau}(..)$ is monotonic for its arguments, while $\mathbf{impact}_{s_{Y,X},s_X}$ is one of its arguments in the formation of $\mathbf{aggimpact}_{s_X}(t)$. Therefore $\mathbf{aggimpact}_{s_X}(t)$ is lower when $\sup_{Y,X}$ is higher, so also requirement (4) is fulfilled. It turns out that the verification outcomes provide evidence that the model does what is expected.

3.2   Results of Simulation Experiments

For the simulated scenarios discussed in this section, we assumed that the 3 users involved are very close friends. We did not use real people facing real situations, but we imagine that the events that occurred were announced by them via Facebook posts and all of them could read it through the Facebook timeline. They provided support to each other with conversations by direct messages via the Facebook messenger and, finally, all of them did tell how they were feeling after each conversation session. The proposed model can show a wide variety of possible behavioural outcomes according

Table 2. Inputs and equilibrium outcomes of the first simulation

Events (static)		
$pe_A = 0.8$	$pe_B = 0.2$	$pe_C = 0.2$
$ne_A = 0.2$	$ne_B = 0.9$	$ne_C = 0.7$
Final equilibrium state values		
$s_A = 0$	$s_B = 0.743060$	$s_C = 0.559141$
$oos_A = 0.995380$	$oos_B = 0.8314$	$oos_C = 0.9279$
$r_A = 0.320417$	$r_B = 0$	$r_C = 0$
$sup_{A,B} = 0.239196$	$sup_{B,A} = 0$	$sup_{C,A} = 0$
$sup_{A,C} = 0.17999$	$sup_{B,C} = 0$	$sup_{C,B} = 0$

to different inputs over time. Only two of them are discussed here. In both cases we performed our model from time $t = 0$ until $t = 40$ with $\Delta t = 0.02$, and for all states speed factor $\eta = 0.5$, steepness $\sigma = 5$ and threshold $\tau = 0.15$. The first scenario concerns static input from the world and was done for the analysis of equilibria. Table 2 describes the stimuli and the final values for all states for this simulation. Note that, in

Table 3. Inputs and outcomes per phase for the second scenario

Events from $t = 0$ to $t = 10$		
$pe_A = 0.8$	$pe_B = 0.8$	$pe_C = 0.8$
$ne_A = 0.5$	$ne_B = 0.5$	$ne_C = 0.5$
State values at time $t = 10$		
$r_A = 0.52$	$r_B = 0.52$	$r_C = 0.52$
Events from $t = 10$ to $t = 20$		
$pe_A = 0.8$	$pe_B = 0.5$	$pe_C = 0.5$
$ne_A = 0.5$	$ne_B = 0.7$	$ne_C = 0.8$
State values at time $t = 20$		
$s_A = 0$	$s_B = 0.189204$	$s_C = 0.292407$
$oos_A = 0.675918$	$oos_B = 0.445901$	$oos_C = 0.281336$
$r_A = 0.23196$	$r_B = 0.005085$	$r_C = 0.003409$
$sup_{A,B} = 0.064991$	$sup_{B,A} = 0$	$sup_{C,A} = 0$
$sup_{A,C} = 0.102943$	$sup_{B,C} = 0.011371$	$sup_{C,B} = 0.010635$
Events from $t = 20$ to $t = 30$		
$pe_A = 0.6$	$pe_A = 0.6$	$pe_A = 0.6$
$ne_A = 0.3$	$ne_A = 0.3$	$ne_A = 0.3$
State values at time $t = 30$		
$s_A = 0$	$s_B = 0.001243$	$s_C = 0.001921$
$oos_A = 0.031166$	$oos_B = 0.019070$	$oos_C = 0.0122$
$r_A = 0.470451$	$r_B = 0.485435$	$r_C = 0.489418$
$sup_{A,B} = 0.00306$	$sup_{B,A} = 0$	$sup_{C,A} = 0$
$sup_{A,C} = 0.008533$	$sup_{B,C} = 0.008062$	$sup_{C,B} = 0.005258$
Events from $t = 30$ to $t = 40$		
$pe_A = 0.3$	$pe_A = 0.6$	$pe_A = 0.6$
$ne_A = 0.6$	$ne_A = 0.3$	$ne_A = 0.3$
State values at time $t = 40$		
$s_A = 0.116481$	$s_B = 8.168*10^{-6}$	$s_C = 1.262 * 10^{-5}$
$oos_A = 0.000373$	$oos_B = 0.202912$	$oos_C = 0.202826$
$r_A = 0.0030602$	$r_B = 0.32747792$	$r_C = 0.327735$
$sup_{A,B} = 7.746*10^{-5}$	$sup_{B,A} = 0.116647$	$sup_{C,A} = 0.116647$
$sup_{A,C} = 0.000120$	$sup_{B,C} = 0.000117$	$sup_{C,B} = 7.58 * 10^{-5}$

this case, person A only faces minor stress and substantially more positive events. Because of that A has resources to provide support to the other persons.

Besides that, persons B and C both face relatively high amounts of stressful events, B a bit more than C. That is the reason why B receives more support from A rather than C. Figure 2 shows the evolution of the state values of all persons (A, B and C, respectively) over time. Here the horizontal axis represents time and the vertical axis the state values. By looking at such a chart it is clear that all state values reach an equilibrium: after a sufficiently large amount of time point *t* the states do not change over time anymore.

The second scenario chosen is much more challenging. For the same time interval and step of the first scenario simulated, the positive and negative events were changed 3 times during the execution (each 500 time steps). These changing environmental conditions starting at time points 0, 10, 20, and 30 are shown in Table 3, together with the resulting state values after each period (see also Fig. 3). In the first period all three can handle the stressful events themselves, so no mutual help is needed. All states get level 0 except the resources that reach level 0.52 at time point 10. In the second period

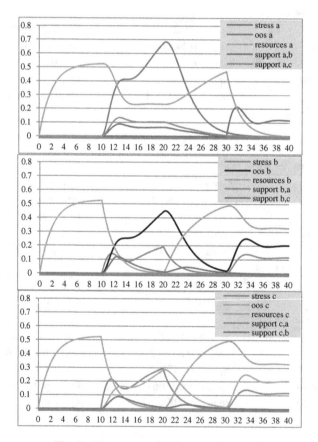

Fig. 3. The patterns for the second scenario

starting at time $t = 10$ person A starts to provide support to both persons B and C respecting the following condition: $sup_{A,B} < sup_{A,C}$. That occurs because C has a higher stress level than B: $s_B < s_C$. In the third period, starting at $t = 20$ the same amounts of positive and negative events occur for all persons, respecting $pe_X > ne_X$. From this moment on, the stress levels of persons B and C start to decrease considerably until their values get close to 0. The same happens with the states $sup_{A,X}$.

The resource levels of persons B and C start to increase as well. Such a behaviour was again expected since, with higher levels of positive events rather than the negative ones, lower stress levels are expected (both s_X and oos_X) which leads to less needs for support, so low values for the states $sup_{X,Y}$ can be expected. Finally, at $t = 30$ the amounts of positive and negative events were changed again. Persons B and C have the same configuration which respects the condition $pe_X > ne_X$. On the other hand, person A encounters a more stressful period: $pe_A < ne_A$. From that moment on, it can be seen that the state s_A starts to increase as expected, leading to the growth of the values of the states $sup_{X,A}$. Note that in all simulations it can be observed that the states $sup_{X,Y}$ always reduce the growth of s_Y.

Moreover, note that, due to abrupt changes on the environmental conditions, a given person may give support to another one even though he or she is facing stressful events. This situation occurs because it takes a while until such a negative event consumes all the resources of this person: the person has a buffer for the encountered stress. This indeed can happen in the real world and therefore the computational model should also show such a behaviour.

4   Discussion

A dynamic computational model was presented to simulate a mutual support network with persons that may face stress. In such a context, persons who are connected with each other in a social network can provide support to their friends who are facing stress and receive support as well when it is necessary. It is important to make clear that this process might be seen as different from an emotion contagion process since, in our case, the emotional states values of every individual tends to be different along time if they are receiving different stimuli. According to what was found in the literature, it was assumed that positive and negative events can affect a given person's stress level, leading to a stressful state of mind. However, receiving support from others, such as friends, is a good help in coping with stress.

The model was designed based on the temporal-causal network modelling approach described in [16]. The mathematical analysis as well as the results from the simulations that were performed showed that the model acts as we expect: the social network members feel the positive and negative effects of encountered events on their stress levels and, additionally, they also are able to provide support to others when it is necessary, leading to a reduction in other's stress levels.

In the literature not much work can be found on computational models for mutual support networks to handle stress. Some exceptions are [2–4]. Here [3, 4] have a different scope in that they focus on more complex internal cognitive structures and explicit interactions initiated by requesting actions for support.

However, [3] is more related to the perspective followed here. In this work a global configuration approach for support provision is followed, based on information for the whole network on need for support and possibilities to give support for all members of the network. Based on such an overall picture of all relevant information for the network a configuration of support provision is determined in a sequential manner by a kind of sequential generate and test (and backtrack) method. The approach proposed in the current paper is different in that it works at a more local level, and not in a global manner for the network as a whole. At each point in time each member of the network determines in a dynamic and autonomous manner the support to be provided only for his or her contacts, independent of other nonconnected members. Because this happens in a dynamic manner, the process is emergent and highly adaptive; for example if, for a given member support is provided by more members, this will contribute to effectiveness, and therefore soon the level of support can be adapted.

The presented model can be a basis to develop a human-aware or socially aware software agent application (e.g., [5, 9, 10, 15]) that can provide support in the social interaction, for example, by including a smart social media application helping in monitoring the states of the members of the network, and giving signals when somebody's stress levels are becoming too high. In other future work the aim is to study more factors that could play some role on this process despite the ones described here. For instance, can different types of connections between persons be incorporated? Additionally, it could be interesting to expand the model to cope with a network containing a significant number of members. Another future challenge is to find out how to collect real data in order to see how the model would act in a real world scenario. It is also important to state that two different individuals could need different amount of support due to the same stressful situation, but we don't take it into account in our model as a matter of simplification. It could be implemented in a future work as well. Finally, we believe that the simulated scenarios described could be executed using human users facing real stressful events in an extension of this work.

Acknowledgements. The authors would like to state that Lenin Medeiros' stay at Vrije Universtiteit Amsterdam was funded by the Brazilian Science without Borders program. This work was performed with the support from CNPq, National Council for Scientific and Technological Development - Brazil, through a scholarship which reference number is 235134/2014-7.

References

1. Anderson, N.B.: Encyclopedia of Health and Behavior, vol. 1. Sage (2004)
2. Aziz, A.A., Ahmad, F.: A Multi-agent model for supporting exchange dynamics in social support networks during stress. In: Imamura, K., Usui, S., Shirao, T., Kasamatsu, T., Schwabe, L., Zhong, N. (eds.) BHI 2013. LNCS, vol. 8211, pp. 103–114. Springer, Heidelberg (2013)
3. Aziz, A., Klein, M.C.A., Treur, J.: Intelligent configuration of social support networks around depressed persons. In: Peleg, M., Lavrač, N., Combi, C. (eds.) AIME 2011. LNCS, vol. 6747, pp. 24–34. Springer, Heidelberg (2011)

4. Aziz, A.A., Treur, J.: Modelling dynamics of social support networks for mutual support in coping with stress. In: Nguyen, N.T., Katarzyniak, R.P., Janiak, A. (eds.) New Challenges in Computational Collective Intelligence. SCI, vol. 244, pp. 167–179. Springer, Heidelberg (2009)
5. Bosse, T., Hoogendoorn, M., Klein, M.C.A., Treur, J.: A generic agent architecture for human-aware ambient computing. In: Mangina, E., Carbo, J., Molina, J.M. (eds.) Agent-Based Ubiquitous Computing, pp. 35–62. Atlantis Press, World Scientific Publishers (2009)
6. Bosse, T., Memom, Z.A., Treur, J., Umair, M.: An adaptive human-aware software agent supporting attention-demanding tasks. In: Yang, J.-J., Yokoo, M., Ito, T., Jin, Z., Scerri, P. (eds.) PRIMA 2009. LNCS, vol. 5925, pp. 292–307. Springer, Heidelberg (2009)
7. Bosse, T., Memon, Z.A., Oorburg, R., Treur, J., Umair, M., de Vos, M.: A software environment for an adaptive human-aware software agent supporting attention-demanding tasks. J. AI Tools 20, 819–846 (2011)
8. Ditzen, B., Heinrichs, M.: Psychobiology of social support: the social dimension of stress buffering. Restor. Neurol. Neurosci. 32, 149–162 (2014)
9. Pantic, M., Pentland, A., Nijholt, A., Huang, T.S.: Human computing and machine understanding of human behavior: a survey. In: Huang, T.S., Nijholt, A., Pantic, M., Pentland, A. (eds.) ICMI/IJCAI Workshops 2007. LNCS (LNAI), vol. 4451, pp. 47–71. Springer, Heidelberg (2007)
10. Pentland, A.: Socially aware computation and communication. IEEE Comput. 38, 33–40 (2005)
11. Schalkwijk, F.J., Blessinga, A., Willemen, A., Van Der Werf, Y.D., Schuengel, C.: Social support moderates the effects of stress on sleep in adolescents. J. Sleep Res. 24, 407–413 (2015)
12. Schwarzer, R., Bowler, R., Cone, J.: Social integration buffers stress in New York police after the 9/11 terrorist attack. Anxiety, Stress & Coping 27(1), 18–26 (2014)
13. Smith, A., Wang, Z.: Hypothalamic oxytocin mediates social buffering of the stress response. Biol. Psychiatry 76(4), 281–288 (2014)
14. Stein, E., Smith, B.: Social support attenuates the harmful effects of stress in healthy adult women. Soc. Sci. Med. 146, 129–136 (2015)
15. Treur, J.: On human aspects in ambient intelligence. In: Muehlhauser, M., Ferscha, A., Aitenbichler, E. (eds.) AmI 2007 Workshops. CCIS, vol. 11, pp. 262–267. Springer, Heidelberg (2008)
16. Treur, J.: Dynamic modeling based on a temporal–causal network modeling approach. Biologically Inspired Cogn. Architect. 16, 131–168 (2016)

Knowledge Engineering
and Semantic Web

Knowledge Integration Method for Supply Chain Management Module in a Cognitive Integrated Management Information System

Marcin Hernes[(✉)]

Wrocław University of Economics,
ul. Komandorska 118/120, 53-345 Wrocław, Poland
marcin.hernes@ue.wroc.pl

Abstract. The diversity of the criteria and methods of analysis used in Supply Chain Management systems leads to a situation in which the system generates a lot of variants of solutions. However, the user expects the system one final variant. Therefore, the integration of knowledge is necessary. To resolve this problem, the consensus method is proposed in this paper. The aim of this paper is to develop and verify a method for knowledge integration in the SCM module in Cognitive Integrated Management Information System (CIMIS).

The first part characterizes a SCM module in CIMIS. Next, a method for knowledge integration has been described. The last part of paper presents results of verification of developed method.

Keywords: Supply Chain Management · Integrated management information systems · Knowledge integration · Consensus method

1 Introduction

Nowadays, one can observe an increased interest of organizational units in systems focusing on integrating supply chain management (SCM). The systems enable strict coordination of activities of business partners, which is usually achieved by electronic data interchange (EDI). Systems of the class may function independently, however more and more often such systems function as a module of an integrated management information system. Reference works [8, 12, 13] point out that the SCM module shall include two solutions enabling dynamic reaction to emerging needs, which in turn may result in an increased value of all companies participating in a given supply chain. The SCM module then shall, on the basis of up-to-date and reliable information gathered from the surrounding environment, present users, in real time, with a variant of a solution related to the flow of products. It is however a difficult task due to the turbulent nature of the environment and due to a high computational complexity of algorithms used in the SCM module. Consequently, it leads to the module generating various variants of products flow between particular cooperating partners. Each of such variants may have different values of attributes (features) describing the flow of products (one of such attributes can be for example a delivery time). A user, however, in order to make a decision, needs to get from the system just one, final variant which

© Springer International Publishing Switzerland 2016
N.T. Nguyen et al. (Eds.): ICCCI 2016, Part I, LNAI 9875, pp. 81–89, 2016.
DOI: 10.1007/978-3-319-45243-2_7

will bring satisfactory benefits. It should be a variant which ensures a delivery of products in proper quantity and in an adequate time, while keeping costs and risk at the lowest possible level. Thus, knowledge in the SCM module shall be integrated.

In reference papers, as well as in practice, various methods of knowledge integration, have been presented and employed, for example negotiations [2], or deduction and computational methods [1]. However, it needs to be stressed that negotiations enable good integration of knowledge by reaching a compromise, but they require exchanging a large number of communications between elements of a system (e.g. which makes it difficult, or sometimes even impossible for the SCM module to function in real time). The deduction and computational methods, however, (for example ones based on the theory of games, classical mechanics, or the method of choice), enable obtaining a high computational efficiency of a system, but they do not guarantee proper integration of knowledge as it often happens that selecting one variant involves a high level of risk which has been previously discussed [7].

So far, in the practical realization of SCM modules (systems) no attention has been paid to the method which enables integration of knowledge in near to real time [7], and also guarantee reaching a proper compromise [10]. This is the consensus method [5, 10].

The aim of this paper is to develop and verify a method for knowledge integration in the SCM module in Cognitive Integrated Management Information System (CIMIS).

The paper has been divided as follows: the first part characterizes a SCM module in CIMIS. Next, a method for knowledge integration has been described. The last part of paper presents results of verification of developed method.

2 Supply Chain Management Module in CIMIS

The CIMIS has been detailed described in [6]. This is a multi-agent (based on LIDA cognitive agent architecture [4]) system consists of following sub-systems: fixed assets, logistics, manufacturing management, human resources management, financial and accounting, controlling, CRM, business intelligence. The SCM module is placed in logistic sub-system. Agents are grouped depending on an enterprise's position in a supply chain, for example supplier-serving agents, producer-serving agents, or retailer-serving agents.

However, solutions which have been offered so far [3, 9, 11], despite the fact that agents use various methods of managing the supply chain, it is the system user who has to decide independently which of the decisions generated by agents shall be executed. In the SCM module presented in the article however, the process of selecting final decisions is performed by an integration component.

It is assumed that SCM module components related to suppliers, producers, wholesalers, retailers, and individual customers, on the basis of information from transaction systems (for example Enterprise Resource Planning - ERP), analytical systems (for example Manufacturing Execution Systems - MES, Customer Relationship Management - CRM) and from internet sources, due to different criteria or methods of analyzing the information, generate various variants of solutions to individual elements of a supply chain. These variants are represented by means of information structures.

For the need of the paper, taking into account first of all its size, the producers-related SCM module has been used. The module consists of following elements (Fig. 1):

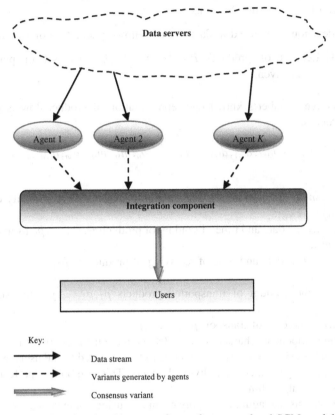

Key:

→ Data stream

--→ Variants generated by agents

⇒ Consensus variant

Fig. 1. A functional architecture of manufacturers-related SCM module.

- data servers,
- agents,
- integration component,
- users.

Data servers consist of data from Internet sources and transactional and analytical databases systems.

The agents are intelligent programs that are based on the data read from the servers carry out the process of calculation and reasoning. The result of these processes is decision related to supply management. Each agent uses a different method for supply chain management. In the present prototype used only a few commonly used methods in practice (described in details in [14]), which include:

- Fixed Order Quantity,
- Fixed Order Period,

- Optional Replenishment,
- Combined Replenishment
- Two Bin,
- Material Requirements Planning,
- Distribution Requirements Planning.

Agents' decisions are stored in database as following structure of variant:

Definition 1. Let set of products $P = \{p_1, p_2, \ldots, p_n\}$ and set of places $M = \{m_1, m_2, \ldots, m_g\}$ are given.

The collective member's knowledge representation of product flow is called the following structure:

$$x = \{\langle p_1, m_{s1}, m_{r1}, dt_{s1}, dt_{r1}, i_1, c_1 \rangle \ldots, \langle p_n, m_{sn}, m_{rn}, dt_{sn}, dt_{rn}, i_n, c_n \rangle\} \tag{1}$$

where:

$(s1, s2, \ldots, sn) \in [1 .. g]$, where g denotes number of places (cardinality of set M), $(r1, r2, \ldots, rn) \in [1 .. g]$,

$dt_{s1}, dt_{s2}, \ldots, dt_{sn}$- date and time of sending of product $p_1, p_2, \ldots, p_n$ from the places $m_{s1}, m_{s2}, \ldots, m_{sn}$,

$dt_{r1}, dt_{r2}, \ldots, dt_{rn}$- date and time of receiving of product $p_1, p_2, \ldots, p_n$ at the places $m_{r1}, m_{r2}, \ldots, m_{rn}$,

$i_1, i_2, \ldots, i_n$- the amount of transported products $p_1, p_2, \ldots, p_n$ (the size of the batch),

$c_1, c_2, \ldots, c_n$- the cost of transport $p_1, p_2, \ldots, p_n$.

Integration component (characterized in details in the next section of this paper), in turn, allows for agreeing to a variant (which is to be presented to the user) on the basis of different variants generated by individual agents. This component performs on the basis of consensus algorithm.

Users are persons managing the supply chain that using computers connected to the Internet read variants determined by the module consensus.

The next part of paper presents the method for agents' knowledge integration performing by integration component.

3   Method for Knowledge Integration

Knowledge integration is performed by consensus algorithm. The structures of variants constitute a profile on the basis of which a consensus is calculated, and within a given module they should have the same attributes.

Definition 2. The profile w = { $w^{(1)}$, $w^{(2)}$, ..., $w^{(K)}$} is called set of K structures of variants, such that:

$$W^{(1)} = \left\{ \left\langle p_1^{(1)}, m_{s1}^{(1)}, m_{r1}^{(1)}, dt_{s1}^{(1)}, dt_{r1}^{(1)}, i_1^{(1)}, c_1^{(1)} \right\rangle, \ldots, \left\langle p_n^{(1)}, m_{sn}^{(1)}, m_{rn}^{(1)}, dt_{sn}^{(1)}, dt_{rn}^{(1)}, i_n^{(1)}, c_n^{(1)} \right\rangle \right\}$$

$$W^{(2)} = \left\{ \left\langle p_1^{(2)}, m_{s1}^{(2)}, m_{r1}^{(2)}, dt_{s1}^{(2)}, dt_{r1}^{(2)}, i_1^{(2)}, c_1^{(2)} \right\rangle, \ldots, \left\langle p_n^{(2)}, m_{sn}^{(2)}, m_{rn}^{(2)}, dt_{sn}^{(2)}, dt_{rn}^{(2)}, i_n^{(2)}, c_n^{(2)} \right\rangle \right\}$$

(2)

$$W^{(K)} = \left\{ \left\langle p_1^{(K)}, m_{s1}^{(K)}, m_{r1}^{(K)}, dt_{s1}^{(K)}, dt_{r1}^{(K)}, i_1^{(K)}, c_1^{(K)} \right\rangle, \ldots, \left\langle p_n^{(K)}, m_{sn}^{(K)}, m_{rn}^{(K)}, dt_{sn}^{(K)}, dt_{rn}^{(K)}, i_n^{(K)}, c_n^{(K)} \right\rangle \right\}$$

The algorithm of consensus determining running in such a way that an ascending order of dt_{xy} value is set from all variants, and the same is done with values i_y and k_y. Then, calculations are done to determine between which values in the orders a value which is the consensus has been placed. The next step is to determine values of consensus of products, sending and receiving place by selecting from a profile values of attributes from a variant in which the distance between the cost of such a variant and the cost of a variant selected in the consensus is minimal. The algorithm finishes once all elements of a variant have been verified and a consensus has been found. Formal definition of the algorithm is as follows:

Algorithm 1

```
Data: Profile W= {W⁽¹⁾, W⁽²⁾, .... W⁽ᴷ⁾ }consists of K
structures of collective member`s knowledge related to
product flow.
Result: Consensus
```

$$CON = \left\{ \left\langle CON(p_1), CON(m_{s1}), CON(m_{r1}), CON(dt_{s1}), CON(dt_{r1}), CON(i_1), CON(c_1) \right\rangle, \ldots, \right.$$
$$\left. \left\langle CON(p_n), CON(m_{sn}), CON(m_{rn}), CON(dt_{sn}), CON(dt_{rn}), CON(i_n), CON(c_n) \right\rangle \right\}$$

```
for profile W.

START
1: Let   ∀CON(x) ∈ CON.CON(x) = ∅.
2: j:=1.
```
3: $g := dt_{rj}$ and $G := \left\langle g^{(1)}, g^{(2)}, \ldots, g^{(K)} \right\rangle$.
```
4: Determine pr(G).//ascending order of G.
```
5: $l_i^1 = (K+1)/2$, $l_i^2 = (K+2)/2$.

6: Determine $g^{(y)}$ where $l_j^1 \le g^{(y)} \le l_j^2$.

   $CON(g) = g^{(y)}$;

7: If $g := dt_{rj}$ then $g := dt_{sj}$, go to: 4.

   If $g := dt_{sj}$ then $g := i_j$, go to: 4.

   If $g := i_j$ then $g := c_j$, go to: 4.

   If $g := c_j$ then go to: 8.
```
8: g:=pj.
```
9: $g \in CON(g) \Leftrightarrow \min\left(\chi(CON(c_j), c_j) \right)$.
```
10: If g=pj then g:=msj, go to: 9.
    If g=msj then g:=mrj, go to: 9.
    If g=mrj then j:=j+1.
11: If j ≤ n then go to: 3.
    If j > n then END.
END.
Computational complexity is O(n²K).
```

Implementation of this algorithm in the SCM module allows for omitting mentioned earlier, the analysis by human of the various variants of product flow.

4   Research Experiment

In order to verify of the developed method for knowledge integration in SCM module,
the research experiment has been carried out. The aim of the experiment was to
compare the variants generated by the integration component with variants generated
by agents performed on the basis of the various another methods for supply chain
management supporting (mentioned in Sect. 2). The following conditions have been
assumed:

1. The initial stock value is 100, the demand for the next day is determined at random.
2. As chronon one day has been assumed. The test was performed over a period of 100
   days (each agent generated 100 knowledge structures on the basis of the structures
   of all the agents on any given day a consensus is determined).
3. In order to knowledge evaluation the following measures have been assumed:
   storage cost, delivery cost, delivery time and average coefficient of variation (risk
   level measure).
4. The storage cost was calculated based on the number of products stored in the
   storehouse (proportional relationship), and the holding time (proportional
   relationship).
5. The delivery cost is calculated based on the quantity of transported products
   (proportional relationship) and the delivery time (inverse proportion). Delivery time
   is determined by the deadlines specified in the planning of the demand by individual
   agents.
6. As a measure of risk the average coefficient of variation is used, because it is a
   relative measure, calculated as follows:

$$V = \frac{s}{|E(r)|} * 100\% \tag{3}$$

where:
   V – average coefficient of variation,
   s – average deviation of measured value,
   $E(r)$ – arithmetic average of measured value.
   The knowledge structures generated by individual agents on each day have been
saved in a database. Next, the consensus has been determined on the basis of these
structures. Table 1 presents achieved results.
   Analyzing the results of verification one can notice that the lowest average storage
costs and the shortest delivery times in the studied period have been obtained by
variants generated by agents a6 and a7 (229 and 227 respectively). The result obtained
by integration module (I), i.e. 239, ranks as the third in terms of the amount of average
costs of storing and delivery times in the analyzed period. So two agents have

Table 1. Results of the research experiment

Specification	Storage costs							
Number of agent	a1	a2	a3	a4	a5	a6	a7	I
Average	283	269	265	266	279	229	227	239
aver. coef. of variation [%]	5,5	9,1	6,1	7,5	5,9	6,1	4,9	1,9
Average of all the agents	260							
Specification	Delivery costs							
Number of agent	a1	a2	a3	a4	a5	a6	a7	I
Average	199	201	199	197	101	234	232	190
aver. coef. of variation [%]	3,0	16,2	8,3	7,2	97	3,1	4,6	2,2
Average of all the agents	194							
Specification	Delivery time							
Number of agent	a1	a2	a3	a4	a5	a6	a7	I
Average	2,0	1,5	2,1	2,1	2,71	0,4	0,2	1,34
aver. coef. of variation [%]	34,3	33,3	31,0	31,0	30,1	120	150	27,9
Average of all the agents	1,57							

generated better results than results generated using the consensus algorithm. However five agents have generated even worse results. Thanks to the use of the consensus algorithm it was possible to obtain lower average storage cost and shorter delivery times in comparison to average storage costs and delivery times of all agents which in the studied period amounted to 260 and 1,56 respectively.

It needs to be noticed, however, that even though decisions of agents a6 and a7 enabled obtaining low average storage cost and short average delivery times, they also generated high average delivery costs. The lowest average delivery costs could be obtained as a result of decisions generated by agent a5 (101), however they also generated high average storage costs (279) and long delivery times (2,71). The result obtained using the consensus algorithm, i.e. 190, also ranks as the third, in terms of average delivery costs, in the analyzed period. Application of the consensus algorithm has also enabled obtaining lower average delivery costs compared to average deliveries costs calculated on the basis of variants of all agents, which in the analyzed period amounted to 194.

While analyzing the risk connected with managing a supply chain, it has been observed that the use of the consensus algorithm enables executing the process with the lowest level of risk (average ratio of change was, in case of storage costs −1,9 %, 2,2 % in case of delivery costs, and 27,9 % in case of delivery times) among analyzed methods of supply chain management (for the remaining methods of supply chain management in the studied period the value of an average ratio of change in case of costs of storing ranged between 5,1 % and 9,1 %, between 3,0 % and 97 % in case of delivery costs, and between 31 % and 150 % in case of delivery times).

It can be said then that variants generated by the integration component are characterized by a low level of fluctuation of storage costs, delivery costs, and delivery times. The phenomenon may positively affect the stability of a company's financial liquidity (if costs do not fluctuate so much it is easier to plan them), and it help maintain continuity of production (low fluctuation of delivery times lowers the risk of downtime).

To sum up, it needs to be stressed that variants generated by the integration component enable, to obtain lower costs of storing and deliveries, and shorter delivery times in a given period of supply chain management while keeping the risk lower compared to the situation when we each time use a single method of supply chain management. The possibility of generating target variants in real time as opposed to a situation where a decision maker has to independently select from variants generated by individual methods is of great importance too. Consequently, the level of usefulness of a selected variant increases, which results in satisfactory benefits such as: timeliness, adequate volumes of a given batch, or decreased costs of deliveries, which may in turn lead to a company obtaining good financial results.

5   Conclusions

The SCM module enables integration and coordination of the flow of products, information and money between individual organizations within a supply chain, which of course affects their capacity to properly adapt to market demands. The use of the consensus algorithm in order to integrate knowledge, and to select one variant presented then to a user, based on variants suggested by a system, may lead to shortening time required to select such a variant, and to lowering the risk of choosing the worst variant. Consensus method cannot guarantees that a given decision will be optimal, however it does ensure an adequate level of satisfaction. Results of verification of the method of knowledge integration presented in the paper help to draw a conclusion that application of the consensus method enable to generate, in near to real time, variants which bring satisfactory benefits.

Further research may focus, among other things, on developing the function of assessment of knowledge of agents functioning within the SCM module, and on developing consensus algorithms enabling improvement of the agents' knowledge.

References

1. Barthlemy, J.P.: Dictatorial consensus function on n-trees. Math. Soc. Sci. **25**, 59–64 (1992)
2. Dyk, P., Lenar, M.: Applying negotiation methods to resolve conflicts in multi-agent environments. In: Zgrzywa A. (red.) Multimedia and Network Information systems, MISSI 2006. Oficyna Wydawnicza PWr, Wrocław (2006)
3. Farrell, B., Loffredo, D.: A Simple Agent for Supply Chain Management. Department of Computer Science, The University of Texas at Austin (2006)
4. Franklin, S., Patterson, F.G.: The LIDA architecture: adding new modes of learning to an intelligent, autonomous, software agent. In: Proceedings of the International Conference on Integrated Design and Process Technology. Society for Design and Process Science, San Diego (2006)
5. Hernes, M., Nguyen, N.T.: Deriving consensus for hierarchical incomplete ordered partitions and coverings. J. Universal Computer Science **13**(2), 317–328 (2007)

6. Hernes, M.: A cognitive integrated management support system for enterprises. In: Hwang, D., Jung, J.J., Nguyen, N.-T. (eds.) ICCCI 2014. LNCS, vol. 8733, pp. 252–261. Springer, Heidelberg (2014)
7. Hernes, M., Sobieska-Karpińska, J.: Application of the consensus method in a multi-agent financial decision support system. IseB **14**(1), 167−185 (2016). Springer, Heidelberg
8. Lu, D.: Fundamentals of Supply Chain Management. Dr. Dawei Lu & Ventus Publishing ApS, bookboon.com (2011)
9. Moyaux, T., Chaib-draa, B., D'Amours, S.: Supply chain management and multiagent systems: an overview. In: B. Chaib-draa, J.P. Müller (eds.), Multiagent-Based Supply Chain Management. SCI, vol. 28, pp. 1−27. Springer, Heidelberg (2006)
10. Maleszka, M., Mianowska, B., Nguyen, N.T.: A method for collaborative recommendation using knowledge integration tools and hierarchical structure of user profiles. Knowl. Based Syst. **47**, 1–13 (2013)
11. Podobnik, V., Petric, A., Jezic, G.: An agent-based solution for dynamic supply chain management. J. Univers. Comput. Sci **14**(7), 1080–1104 (2008)
12. Rutkowski, K.: Best practices in logistics and supply chain management. the case of central and eastern europe. In: Waters, D. (ed.) Global Logistics and Distribution Planning. Kogan Page, London (2010)
13. Sitek, P., Wikarek, J.: Cost optimization of supply chain with multimodal transport. In: Proceedings of the Federated Conference on Computer Science and Information Systems (2012). http://fedcsis.org/proceedings/fedcsis2012/pliks/182.pdf
14. Siurdyban, A., Møller, C.: Towards intelligent supply chains: a unified framework for business process design. Int. J. Inform. Syst. Supply Chain Manag. 5(1), 1−19 (2012). IGI Global, New York

An Asymmetric Approach to Discover the Complex Matching Between Ontologies

Fatma Kaabi[1(✉)] and Faiez Gargouri[2]

[1] Laboratory MIRACL, Faculty of Economic Sciences and Management,
Sfax, Tunisia
kaabifatma@yahoo.fr

[2] Laboratory MIRACL, The Higher Institute of Computer Science
and Multimedia of Sfax, BP 242, Sakiet Ezzit, 3021 Sfax, Tunisia

Abstract. This paper introduces an extensional and asymmetric alignment approach capable of identifying complex mappings between OWL ontologies. This approach employ the association rule to detect implicative and conjunctive mapping containing complex correspondences. Method for extracting the complex mappings is presented and results of experiments carried out on the large biomedical ontologies and the anatomy track available to Test library of Ontology Alignment Evaluation Initiative show the efficiency of the approach proposed.

1 Introduction

Ontology mapping is a well studied problem, several matching approaches have been proposed [10]. These methods aim at finding correspondences between the semantically related entities of those ontologies. From this approaches we can identify: the extensional approaches, and the intentional approaches. The majority of these methods finds only equivalence relations (CIDER-CL [6], YAM++ [7], LogMap [8]) and do not consider also the asymmetric relations like the subsumption. Most of the proposed approach are symmetrical and intentional. The only extensional and asymmetric method is the AROMA method [11]. Therefore, this method discovers only simple relationships.

Most existing matching approach concentrates on finding 1-1 mappings between two given ontologies. However, complex mappings are very useful in practice. Simple correspondences are not sufficient to express relationships that represent correspondences between entities since it (1) may be difficult to discover simple correspondences (or they do not exist) in certain cases, or (2) simple correspondences do not allow for expressing accurately relationships between entities.

As a motivating example, consider two ontologies $\mathcal{O}_1$ and $\mathcal{O}_2$ (Fig. 1) describing cell types. $\mathcal{O}_1$ is a part of the ontology CL^1 and $\mathcal{O}_2$ is an extract of the ontology $BCGO^2$. The proposed methods [7,8] can't find the most similar entity

---

[1] Cell Ontology (CL), http://www.cellontology.org/.
[2] Beta Cell Genomics Ontology, https://github.com/obi-bcgo/bcgo.

© Springer International Publishing Switzerland 2016
N.T. Nguyen et al. (Eds.): ICCCI 2016, Part I, LNAI 9875, pp. 90–97, 2016.
DOI: 10.1007/978-3-319-45243-2_8

node in $\mathcal{O}_2$ that maps to the entity node *phagocyte* in $\mathcal{O}_1$. But, the entity *phagocyte* can match to the intersection of the three entities *motile cell, native cell* et *stuff accumalating cell* dans $\mathcal{O}_2$. The terms describing the *phagocyte* concept are belonged in the dataset of the three concepts *motile cell, native cell* and *stuff accumalating cell.*

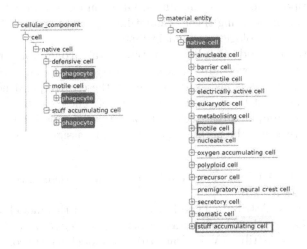

Fig. 1. Ontology O1 and ontology O2

The rest of this paper is organized as follows: First we review related work and we illustrate the limitations of existing complexes mapping approaches. Next we introduce the proposed method. Finally we present experimental results and concluding remarks.

2   Related Work

In order to find complex correspondences some approaches have been considered. We present in the following the most interesting ones. Doan and colleagues [12] developed a system **CGLUE** that uses machine learning techniques to semi-automatically generate semantic matching. This system finds disjunctions and equivalence relations between concepts. It finds complex matching between taxonomies. CGLUE is based on the notions of semantic similarity, expressed in terms of the joint probability distribution of the concepts involved. This system calculates the joint distribution of the concepts and use the joint distribution to compute any appropriate similarity measure.

CGLUE is based on the assumption that the children of any ontology entity are mutually exclusive and exhaustive. We note that the assumption maintains for many real ontology, in which the further specialization of an entity usually gives a partition of the instances of that entity. However, in many real ontologies, very sibling entities share instances. Hence, for these domains this approximating assumption is not hold.

The two approaches [13,14] are based on the inductive logic programming, ILP, and attempts at creating alignments by using the learning theory. These approaches take complex correspondences into account and not only equivalence correspondences. But here it is not possible to create complex mappings without learning correspondences out of instances. Often ontologies do not contain any instances. Hence the learning theory cannot be applied in order to find complex correspondences in ontologies without instances.

The pattern-based ontology matching approach presented in [15] define patterns to discover automatically complex correspondences. A master alignment of ontologies is necessary. To detect these correspondences, a set of simple conditions must be satisfied for each model. These conditions are a combination of structural, linguistic techniques and types compatibility. The defined models are (notice that the notation $i\#C$ is used to assign to an entity C from ontology O_i):

1. CAT (Class by Attribute Type Pattern): this model detects correspondences as $1\#A \equiv \exists 2\#R.2\#B$;
2. Class by Inverse Attribute Type Pattern (CAT^{-1}): this model allows correspondences as which are written as $1\#A \equiv 2\#B \cap 2\#R_1.T$, to be detected;
3. CAV (Class by Attribute Value Pattern): this model detects correspondences as $1\#A \equiv \exists 2\#R.\{...\}$, $(where \{...\}$ is a set of concrete data values)
4. PC (Property Chain Pattern): this model allows correspondences as $1\#R \equiv 2\#P \circ 2\#Q$.

This method can find a lot of complex correspondences. However, the used patterns cover only peculiar domains of ontologies.

After analysing these approaches, we note that the above mentioned methods only consider the equivalence relations between concepts and do not take into account the asymmetric relations such as the subsumption. To overcome these significant limitation, we have developed a new complex mapping methodology named ARCMA [4,5] (Association Rules Complex Matching Approach) which permit to map subsumption relations between entities.

3   A New Method for Complex Matching

The alignment method ARCMA [4], aims at finding complex correspondences between two OWL ontologies.

ARCMA follows three consecutive steps: (1) the term or data sets extraction (The pre-processing step), (2) the detection of association rules between entities of two ontologies and (3) the post-processing of results.

In the pre-processing step, a set of relevant terms embedded in the descriptions and entities instances is generated by using a natural language processing tools. We represent the entities (concepts and proprieties) by set of terms and data generated from their description and instances. We extract the name and the terms contained in the annotations (labels, comments, etc.). We also add the local name, the annotations and the values of its instances [5].

In the second step, ARCMA detects the complex matching between two OWL ontologies using the association rule model and a statical measure, the implication intensity [2]. A valid association rule $x \rightarrow y_1 \wedge \ldots \wedge y_i .. \wedge y_n$ means that the vocabulary associated to a source entity x aims to be included in the intersection between the relevant terms of set of entities y_i. For example, the valid rule *phagocyte → motile cell ∧ native cell ∧ stuff accumalating cell* could be interpret: The entity *phagocyte* corresponds to intersection of the three entities *motile cell, native cell* and *stuff accumalating cell*. The post-processing eliminates the redundancies in matcher found.

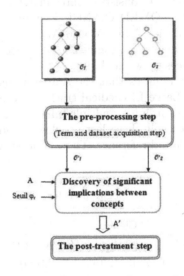

Fig. 2. The ARCMA process

Figure 2 illustrate the process of our method to discover the complex mappings between OWL ontologies. First, we use two OWL multiple inheritance ontologies. Then we apply a pretreatment process to define their relationship on a common extension. We also consider a reference alignment between these two ontologies. Next, we utilize the association rules to find complex correspondences type $x \Rightarrow y_1 \wedge \ldots \wedge y_i \ldots \wedge y_n$. Finally, we reduce the redundancy in the extracted rule set. A rule will be selected if none of its generative rules have a value of the implication intensity (φ) greater than or equals to its φ value.

4 Evaluation

To estimate the performance of our approach, a prototype is realized in Java. Our system supports input two OWL ontologies and a reference alignment, then comparing the correspondence obtained by our tool and those by a manual mapping.

This evaluation is carried out by exploiting the two metrics alignment quality: precision and recall [16]. Precision measures the ratio of correctly found correspondences over the total number of returned correspondences. Recall compute the ratio of correctly found correspondences over the total number of expected correspondences.

The experiment is performed on the large biomedical ontologies and the anatomy track available to Test library of Ontology Alignment Evaluation Initiative OAEI[3]. The Large Biomedical track contains the mapping of FMA (78,989 classes), NCI Thesaurus (66,724 classes) and SNOMED CT (306,591 classes) and uses the UMLS Metathesaurus as the basis for the track's reference mappings. The reference mappings only include subsumption and equivalence relations between classes. The track consists of three matching problems: FMA-NCI, FMA-SNOMED CT and SNOMED CT-NCI. The anatomy track includes the mapping of the two ontologies Adult Mouse Anatomy (AM) and part of the NCI thesaurus describing human anatomy. The reference mapping includes only equivalence correspondences between classes.

Our method ARCMA requires that the source ontology supports multiple inheritances. Among the Large Biomedical track and the anatomy track, there are only three ontologies containing multiple inheritances (SNOMED, AM and human). Hence, we will exploit these last ontologies and two references alignments: SNOMED CT-NCI and reference. The characteristics of these ontologies are shown in Table 1.

Table 1. Description of the ontologies used for the evaluation of ARCMA

Ontologies	Classes	Properties
Large SNOMED	122464 (40 % SNOMED)	55
Small SNOMED_fma	13412 (5 % SNOMED)	18
Small SNOMED_nci	51128 (17 % SNOMED)	63
Whole NCI	66724	190
Small NCI_fma	6488 (10 % NCI)	63
Small NCI_snomed	23958 (36 % NCI)	83
mouse	2744	3
human	3304	2

The Table 2 illustrates the results obtained by the alignment method ARCMA, with the rule selection threshold $\varphi_r = 0,9$.

In this table we note that in some tests such as $SmallSNOMED\_nci - SmallNCI\_fma$, the value of precision is 1, that means that the results of our method are the same given by an expert, and for the many other tests, the precision value is higher than 0.75, therefore, our system gives good results which are

---

[3] Ontology Alignment Evaluation Initiative Test library (2015), http://oaei.ontology matching.org/2015/seals-eval.html.

Table 2. Performance measures of ARCMA

Tests	Precision	Recall
mouse-human	0,8	0,571
Large SNOMED-Small NCI_fma	0,844	0,776
Large SNOMED-Small NCI_snomed	0,813	0,765
Large SNOMED-Whole NCI	0,808	0,778
Small SNOMED_fma-Small NCI_fma	0,927	0,731
Small SNOMED_fma-Small NCI_snomed	0,729	0,714
Small SNOMED_fma-Whole NCI_whole	0,811	0,860
Small SNOMED_nci-Small NCI_fma	1	0,667
Small SNOMED_nci-Small NCI_snomed	0,6	0,429
Small SNOMED_nci-Whole NCI	0,667	0,286

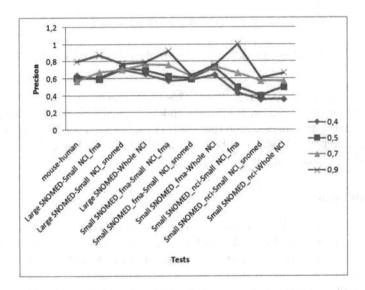

Fig. 3. Values of precision as a function of the threshold value φ_r

encouraging. For example, ARCMA discovered the following meaningful impli-
cations from SNOMED_small_overlapping_nci to NCI_small_overlapping_fma:
R1.CENTRAL_NERVOUS_SYSTEM_TRACT_STRUCTURE→CENTRAL_NERVOUS_SYSTEM_
PART AND NERVE
R2.DUODENAL_PAPILLA_STRUCTURE→BILIARY_TRACT AND DUODENUM AND
PANCREATIC_DUCT
R3.COLONIC_MUSCULARIS_PROPRIA_STRUCTURE→COLON AND
MUSCULARIS_PROPRIA
    Figures 3 and 4 show the influence of rule selection threshold φ_r on the pre-
cision and recall of ARCMA. We note that the value of the precision increases

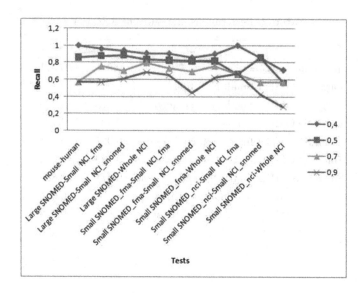

Fig. 4. Values of recall as a function of the threshold value φ_r

with the higher level of the threshold. This phenomenon clearly shows a correlation between the deviation from independence situation and the relevance of rules. In general, we can conclude that ARCMA achieved a good precision/recall values. The high recall value can be explained by the fact that UMLS thesaurus contains definitions of highly technical medical terms.

5 Conclusion

In this paper, we proposed a new approach for discovering complex mappings between two OWL ontologies. We utilized the association rule and the statical measure, the implication intensity, to detect implicative and conjunctive mapping containing complex correspondences. We implemented the approach and experimentally evaluated it on the large biomedical ontologies and the anatomy track, which demonstrated the high precision of the discovered correspondences. The principal advantage of this approach is that it is simple. Besides, the use of the implication intensity measure permit to approve the validity of the complex correspondences and justifies the good precision values obtained by ARCMA.

References

1. Agrawal, R., Imielinski, T., Swami, A.: Mining association rules between sets of items in large databases. In: The Proceedings of the 1993 ACM SIGMOD International Conference on Management of Data, pp. 207–216 (1993)
2. Blanchard, J., Kuntz, P., Guillet, F., Gras, R.: Implication intensity: from the basic statistical definition to the entropic version, chap. 28, pp. 473–485. CRC Press (2003)

3. Do, H., Rahm, E.: A system for flexible combination of schema matching approaches. In: The International Conference on Very Large Data Bases (VLDB 2002), pp. 610–621 (2002)
4. Kaâbi, F., Gargouri, F.: An approach to find complex matching between conceptual hierarchies. In: Proceedings of the 2012 IEEE 21st International Workshop on Enabling Technologies: Infrastructure for Collaborative Enterprises (WETICE 2012), pp. 205–210. IEEE Computer Society, Washington, DC (2012)
5. Kaabi, F., Gargouri, F.: A new approach to discover the complex mappings between ontologies. Int. J. Web Sci. **1**(3) (2012)
6. Gracia, J., Asooja, K.: Monolingual, cross-lingual ontology matching with CIDER-CL: evaluation report for OAEI 2013. In: Proceedings of the 8th International Conference on Ontology Matching, vol. 1111, pp. 109–116. CEUR-WS.org (2013)
7. Nago, D.D., Bellehsene, Z.: YAM++: a multi-strategy based approach for ontology matching task. In: Proceedings of the 8th International Conference on Ontology Matching, OM 2013, vol. 1111, pp. 211–218. CEUR-WS.org (2013)
8. Jiménez-Ruiz, E., Grau, B.C., Solimando, A., Cross, V.V.: LogMap family results for OAEI 2015. In: Proceedings of the 10th International Workshop on Ontology Matching collocated with the 14th International Semantic Web Conference (ISWC 2015), Bethlehem, PA, USA, 12 October 2015, pp. 171–175 (2015)
9. Kalfoglou, Y., Schorlemmer, M.: Ontology mapping: the state of the art. Knowl. Eng. Rev. **18**(1), 1–31 (2003)
10. Euzenat, J., Shvaiko, P.: Ontology Matching. Springer, Heidelberg (2007)
11. David, J., Guillet, F., Briand, H.: Association rule ontology matching approach. Int. J. Semant. Web Inf. Syst. **3**(2), 27–49 (2007)
12. Doan, A., Madhavan, J., Dhamankar, R., Domingos, P., Halevy, A.: Learning to match ontologies on the semantic web. VLDB J. **12**(4), 303–319 (2003)
13. Stuckenschmidt, H., Preu, L., Meilicke, C.: Learning complex ontology alignments a challenge for ILP research. In: Proceedings of the 18th International Conference on Inductive Logic Programming (2008)
14. Qin, H., Dou, D., LePendu, P.: Discovering executable semantic mappings between ontologies. In: Meersman, R., Tari, Z. (eds.) OTM 2007, Part I. LNCS, vol. 4803, pp. 832–849. Springer, Heidelberg (2007)
15. Ritze, D., Meilicke, C., Zamazal, O.S., Enschmidt, H.S.: A pattern-based ontology matching approach for detecting complex correspondences. In: Proceedings of the ISWC 2009 Workshop on Ontology Matching (2009)
16. Euzenat, J.: Semantic precision and recall for ontology alignment evaluation. In: Proceedings of the 20th International Joint Conference on Artificial Intelligence (IJCAI), Hyderabad (IN), pp. 348–353. AAAI Press, Menlo Park (CAUS) (2007)

An Evidential Approach for Managing Temporal Relations Uncertainty

Nessrine El Hadj Salem[1], Allel Hadjali[2], Aymen Gammoudi[1,2(✉)], and Boutheina Ben Yaghlane[3]

[1] ISGT, University of Tunis, 92 Boulevard 9 Avril 1938, 1007 Tunis, Tunisia
nessrine.hajsalem@gmail.com, aymen.gammoudi@ensma.fr
[2] LIAS/ENSMA, 1 Avenue Clement Ader, 86960 Futuroscope Cedex, France
allel.hadjali@ensma.fr
[3] IHEC, University of Carthage, IHEC-Carthage Presidence, 2016 Tunis, Tunisia
boutheina.yaghlane@ihec.rnu.tn

Abstract. Temporal information can be often perceived in a vague way as infected with imprecision and uncertainty. Therefore, we need to find some way of handling the invaluable temporal information. This paper presents a belief functions-based approach to represent temporal relations uncertainty in the point algebra context. We would like to show the concept of mass function is suitable for modeling the uncertain knowledge about possible relations between dates. The temporal uncertainty can also be expressed thanks to a vector of belief measures. A set of rules that allows some reasoning about evidential temporal relations, is then established.

Keywords: Temporal relations · Uncertainty · Evidence theory

1 Introduction

Temporal representation and reasoning is a central problem in Artificial Intelligence like planning, scheduling, causal reasoning and diagnosis. Temporal information can be often perceived in a vague way as infected with imprecision and uncertainty. There is not much work based on the treatment of information infected with imprecision, uncertainty and fuzziness. In the temporal context, imprecision or fuzziness means that temporal specifications are not rigid but flexible. This is due to the fact that knowledge of the time is expressed in a natural language. For example, one can define the primitive notion dates in a vague manner or use gradual descriptions to express linguistics relations between time intervals. As for the second imperfection, uncertainty, it is the result of a lack of information on the state of the world. It is impossible to determine if certain definite knowledge about the world is right or wrong. For example, we are not completely sure of the exact relative position of two temporal entities (expressed as intervals of time or moments).

Most work in temporal reasoning adopts a weakly expressive representation of uncertainty [3,15]. Indeed, temporal uncertainty is modeled using a disjunction

© Springer International Publishing Switzerland 2016
N.T. Nguyen et al. (Eds.): ICCCI 2016, Part I, LNAI 9875, pp. 98–107, 2016.
DOI: 10.1007/978-3-319-45243-2_9

of possible primitive relations between two temporal entities. For example, if an agent is not sure of the relation among relations $>$ and $=$, which exists between two dates t_1 and t_2, he expresses that $t_1 \geq t_2$ (where $\geq = \{>, =\}$). More the disjunction is wide, more the temporal relation is uncertain. The major disadvantage of this approach is that it does not quantify the uncertainty inherent in the temporal available knowledge. A second considerable limitation of this approach occurs when an order is required between the relations of each disjunction. This approach does not allow so to express that a relation is more plausible than another. That's why many theories have been used to represent temporal information. The most used is the probability theory [9] where the uncertainty is modeled by a vector of probabilities values associated to the three primitive relations. Other theories have been introduced to offer other models like possibility theory [7] when the information available is inconsistent, is completely taken into account by this approach, fuzzy sets theory [6]. The theory of evidence specialised to model the uncertainty more than the possibilistic and the probabilistic. This theory treat jointly temporal information through an explicit representation of uncertainty, imprecision, inconsistency. This theory allows probability and possibility measures based on the mass functions. In order to managing imperfect information, we proposed in our work a formalism of modelling and reasoning about temporal points. According to our knowledge, there is no much work that managing temporal relations used the evidence theory. That's why we aim proposing through this paper, new approaches to model and to infer new temporal information.

In this paper, we would discuss an approach based on the belief functions for the representation and management of uncertainty temporal relations in the case of algebra points (i.e. temporal entities are modeled in terms of dates/moments). The belief functions is a model of the uncertain more general than possibilistic and probabilistic models and allows to treat jointly the uncertainty and imprecision (due to lack of information). The paper is structured as follow. Section 2 recalls some basics notions needed to relay to the following sections. Section 3 presents a review of related work. In Sect. 4, we discuss more explicitly, first, the representation of the uncertainty between temporal relations and, secondly, a set of operations allowing to reason about the temporal relations and propagate uncertainty. Finally, Sect. 5 concludes the paper and sketches some perspectives that seem very interesting.

2   Background

We would like to give a brief recall; firstly, on temporal relations between dates, secondly, on the theory of belief functions.

2.1   Temporal Relations

Several qualitative formalisms have been proposed in the literature to represent temporal relations like [14]. One of the most known and used is the algebra points is in [15]. In this context, the temporal entities considered are the points of the real line (we also speak of moments or dates). The qualitative temporal relations that allow to compare two instants are in number of three and correspond to the following situations: a moment before ($<$) another, a moment is equal ($=$) to another, a moment after ($>$) another.

If a moment a before or equal to a moment b, This relation is denoted $a \leq b$ where $\leq$ represents the disjunction of two atomic relations $\{<, =\}$. The composition of two atomic relations is not necessarily an atomic relation. For example, if $t_1 < t_2$ and $t_2 > t_3$ then t_1 r t_3 where r $= \{<, =, >\}$. Algebra points includes a total of 8 temporal relations ($\emptyset$, $\{<\}$, $\{>\}$, $\{=\}$, $\{<, =\}$, $\{>, =\}$, $\{<, >\}$, $\{<, >, =\}$).

2.2 Evidence Theory

The theory of evidence (also known as the theory of belief functions) was developed by Shafer in 1976 [11] following the work of Dempster [4] on the upper and lower probabilities. This theory allow to model the belief in an event with mass functions. So, it allows to represent the uncertainty, but also the imprecise.

This theory supposes, first, the definition of a set of alternatives Θ, called frame of discernment, defined as follows:

$$\Theta = \{\theta_1, \theta_2, ..., \theta_n\} \tag{1}$$

The key element of this theory is belief assignment concept m defined as the following:

$$m : 2^\Theta \longrightarrow [0, 1] \tag{2}$$

$$m(\emptyset) = 0 \text{ and } \sum_{X \subseteq \Theta} m(X) = 1 \tag{3}$$

The elements X of Θ such as $m(X) > 0$ are called focal elements of m. The number $m(A)$ can be interpreted as the fraction of the mass unit allocated to X based on our state of knowledge. At the difference of probabilities, we see that it is possible to allocate mass to subsets of Θ and not only for singletons. This possibility provides a great flexibility representation to the model. The complete ignorance corresponds to $m(\Theta) = 1$, then a precise and safe knowledge corresponds to the allocation of the entire mass to a singleton Θ (m is called a certain mass). An imprecise and reliable knowledge will result in the allocation of the mass unit to a not singleton focal element. Uncertain knowledge is the allowance of the mass fractions in several focal elements.

A mass m can be equivalently represented by two no-additive measures: a belief function (or credibility) $Bel : 2^\Theta \longrightarrow [0, 1]$, defined by

$$Bel(X) = \sum_{\emptyset \neq Y \subseteq X} m(Y), \tag{4}$$

and a function of plausibility $Pl : 2^{\Theta} \longrightarrow [0,1]$, defined by

$$Pl(X) = \sum_{Y \cap X \neq \emptyset} m(Y). \tag{5}$$

The relation between Bel and Pl is given by the relation: $Pl(X) = 1 - Bel(\overline{X})$ where $\overline{X}$ refers to the complementary alternative of X. The couple $(Bel(A), Pl(A))$ characterizes the knowledge that we have on A.

One of the cornerstones of the theory of evidence make the ability to combine two sources of belief m_1 and m_2. A standard way to combine these two masses is the conjunctive sum operation defined by [12] (with m_1 and m_2 reliable, defined in the same frame of discernment Θ and provided by two independent experts):

$$m_1 \textcircled{\cap} m_2(Z) = \sum_{X,Y \subseteq \Theta : X \cap Y = Z} m_1(X) \times m_2(Y) \tag{6}$$

A particular property, such as the vacuous mass function. Permit us an only one focal set which is Ω, i.e. $m^{\Omega}(\Omega) = 1$. Such a basic belief assignment used in the case of total ignorance, i.e. we have no information about the actual state of our system. It's a particular case of multi-valued mapping [5]. To transfer the basic belief assignment of a frame of discernment Ω towards the cartesian product [13] of frames of discernment $\Omega \times \Theta$. This operation of vacuous extension, noted $\uparrow$, is defined by:

$$m^{\Omega \uparrow \Omega \times \Theta}(Y) = \begin{cases} m^{\Omega}(X) & \text{if } Y = X \times \Theta \\ 0 & \text{otherwise} \end{cases} \tag{7}$$

3   Related Works

What is noticeable is that, there are not many works on the modeling and treatment of the uncertainty of temporal relations. Most similar (and consider the algebra of points) to our approach are those proposed in [7–10]. In [9], the authors proposed a probabilistic model for processing with uncertain relations between dates. They define an uncertain relation between two dates a and b as any possible disjunction between the three primitive temporal relations (i.e., $<, =, >$). This disjunction is a composite relation of type $\leq = \{<, =\}$, $\geq = \{>, =\}$, $\neq = \{<, >\}$, or total ignorance $? = \{<, =, >\}$. The uncertainty is modeled by a vector $P_{ab} = (e^{<}, e^{=}, e^{>})_{a,b}$ where $e^{<}_{a,b}$ (resp. $e^{=}_{a,b}$, $e^{>}_{a,b}$) is the probability that $a < b$, i.e., a before b, (resp. $a = b$, $a > b$), and the sum $e^{<} + e^{=} + e^{>} = 1$. Operations, which involve preserving the semantics of probability, are proposed [9]. They allow to manage the propagation of uncertainty when two relations are composed, or when two temporal information on the same dates are merged. The major problem of this approach concerns the way in which the state of total ignorance is processed. When no information is known about the relation between any two dates, Ryabov and Puuronen suggest the use of probability values of the area related to the application. These values, denoted by $e^{<}_{D}$, $e^{=}_{D}$ and $e^{>}_{D}$, represent the probabilities of three primitive relations between two dates in this situation. This proposal makes sense only if a priori probability distribution is available. However, in practice it is difficult to produce new relation in such a priori probability. In [7], an approach based on the theory of possibilities for the representation and management of uncertainty in the temporal relations between dates, is proposed. An uncertain relation $r_{a,b}$ between a and b is represented by a normalized vector $\Pi_{ab} = (\Pi_{ab}^<, \Pi_{ab}^=, \Pi_{ab}^>)$ such as $\max(\Pi_{ab}^<, \Pi_{ab}^=, \Pi_{ab}^>) = 1$ where $\Pi_{ab}^<$ (resp. $\Pi_{ab}^=$, $\Pi_{ab}^>$) is the possibility that $a < b$ (resp. $a = b$, $a > b$). Standardization means that at least one relation of the three primitive relations is satisfied with a possibility equal to 1. The established possibilistic inference rules are simple and complete, even if they are qualitative. The case where the information available is inconsistent, is fully taken into account by this approach. In this work, we adopt the model based on the belief functions to represent and treat the uncertainty of the temporal relations between moments. This model is rich and can represent a subjective and imperfect (uncertain and imprecise) knowledge of a rational agent in an adequate and accurate way. As far as we know, this is the first time that the belief functions are used in the representation of temporal uncertainty.

4   Evidential Approach

Models of evidential reasoning temporal knowledge base as proposed are based on the interogation of the indefinite temporal information. Those models studied the inference rules like inverse, composition and combinaison. To show if these models are strong representation systems, the defined reasonment should give new result of temporal information. We defined the notion of point relations PR for an evidential approach, we presented the steps to generate point relations from imprecise temporal knowledge.

Such a new representation opens the way to the definition of a strong relational system for temporal knowledge base. A temporal knowledge base model is a strong evidential reasoning system if the future work we adopted an algorithm to processed a performing result over the set of its point relations. The point relations representation is the only way to define the temporal informations using the evidence theory.

In this section, models inherent uncertainty between dates through the theory of evidence. Then, a set of rules proposed to reason about evidential temporal relations.

4.1   Representation of Uncertain Temporal Relations

To model the indefinite time, we represented by a mass function on the frame of discernment Ω. An point relation represented by such a mass function is called an evidential points. The evidential representations of temporal relations are mass functions on $R = \{<, >, =\}$ between points. The temporal relation is written as $R_{A,B}$ that can be hold between two temporal points A and B. We refer to an element of this set as $r \in R$. An uncertain temporal relations represented by such a mass function is called an evidential temporal relations. In case of totally ignorance has the evidential form $\{\{R\}/1\}$ $m(\{R\}) = 1$, when it's no information. For example, $R_{A,B} = (m^{<} = 0.3, m^{=} = 0.3, m^{>} = 0.4)_{A,B}$. The state knowledge of an agent on the relations $R_{A,B}$ described in the referential and defined as follow;

$$2^{\Omega} = \{\emptyset, \{<\}, \{=\}, \{>\}, \{<,=\}, \{=,>\}, \{<,>\}, \{<,=,>\}\}.$$

The function of masses $m_{a,b}$ is defined: $2^{\Omega} \longrightarrow [0,1]$ and satisfied:

$$m_{a,b}(\emptyset) = 0 \text{ and } \Sigma_{r \subseteq \Omega} m_{a,b}(r) = 1.$$

For example: an agent can express the mass allocated to the possible relation between two dates a and b is $m(\{>\}) = 0,2$ and $m(\{=,<\}) = 0,8$.

Note that the state of knowledge of an agent on the uncertain relation $r_{a,b}$ can also be represented by a credibility vector of the form:

$$Bel_{a,b} = (Bel_{a,b}^{<}, Bel_{a,b}^{=}, Bel_{a,b}^{>}),$$

with $Bel_{a,b}^{<}$ (resp. $Bel_{a,b}^{=}, Bel_{a,b}^{>}$) measure the belief in the relation $a < b$ (resp. $a = b$, $a > b$).

If we take the previous example, it is easy to see that $Bel_{a,b}^{<} = 0$, $Bel_{a,b}^{=} = 0$ and $Bel_{a,b}^{>} = 0.2$. See Table 1.

Table 1. Table of belief measures.

Relation r	Bel(r)
$\emptyset$	0
$\{<\}$	0
$\{>\}$	0,2
$\{=\}$	0
$\{<,=\}$	0,8
$\{>,=\}$	0,2
$\{<,>\}$	0,2
Ω	1

It should also be noted that from the uncertainty vector $(Bel^{<}, Bel^{=}, Bel^{>})_{a,b}$, we can calculate the plausibility measure of the uncertain relation $\geq_{a,b}$ (resp. $\leq_{a,b}$, $\neq_{a,b}$) using the dual relation between both functions of belief and plausibility. Thus, we obtain:

$$Pl_{a,b}^{\geq} = 1 - Bel_{a,b}^{<}$$
$$Pl_{a,b}^{\leq} = 1 - Bel_{a,b}^{>}$$
$$Pl_{a,b}^{\neq} = 1 - Bel_{a,b}^{=}$$

4.2 Evidential Temporal Networks

To managing indefinite time in TED system, we propose to develop an evidential model of extension, called a network Evidential Temporal Network (ETN) which is a directed graph where nodes represent a time point (E = $\{E_1, E_2,..., E_n\}$) and the edges are the uncertain temporal relations (R_{E_i}, $E_j =\{[Bel^x,Pl^x] - \text{x} \in \text{X}\}$ with Bel^x and Pl^x designed the confidence level between two time points. An Temporal Evidential Database TED, is a set of N objects (tuples) and D domaine of attributes A_i, denoted by $D(A_i)$ (Table 2).

Table 2. A temporal evidential Table.

PatientNum	Disease	Temporal relation
1	Older 0.8	<
	{Less active, Overweight} 0.3	
2	Overweight	= 0.5 {=,>} 0.5

The evidential points of Temporal Evidential Database TED is a finite set of uncertain relations such that TED = $\{UR_1, UR_2,...,UR_k\}$. Each uncertain relations includes N objects, where each object contains one focal element for the constant time. Our contribution is summarized into 3 steps: First, their evidential points values are extended to the joint frame of discernment thanks to vacuous extension operation as mensionned in Table 3. Then, the extended evidential points values are inferred using the three inference rules as shown above. Finally, we can deduct the equation as follows $r_{a,c} =\sim (r1_{a,b} \oplus r2_{a,c}) = ((r_{a,b} \otimes r2_{b,c}))$.

Table 3. Uncertain relations of TDB.

UR_1	UR_2
(1, Older, <)	(1, Older, <)
(2, Overwight, =)	(2, Overwight,{=, >})
(1, {Less active, Overwight}, <)	(1, {Less active, Overwight}, <)
(2, {Overwight}, =)	(2, {Overwight}, =)

5 Reasoning Possible Relations

In this section, we present a set of rules composed of three basic operations (inversion, composition and combination). These rules allow to manage the propagation of uncertainty and reasoning about temporal evidential relations.

Operation of inverse:
The inversion operation ($\sim$) is an operation allows us to deduce the two vector $[Bel_{b,a}\!-\!Pl_{b,a}]$, of the relation $R_{b,a}$ from the two vector $[Bel_{a,b}\!-\!Pl_{a,b}]$ of the known relations $R_{a,b}$. This operation exchange <with>, = in the transitivity table of Allen [1].

Table 4 is the Temporal Evidential Database having three attribute.

Table 4. Extension of bbas from the evidential Table.

PatientNum	Symption	Temporal relation
1	{Diabetes, Stroke} 0.2	{<}
2	{Diabetes} 0.3	{=}
3	{Anemia} 0.5	{>}

- $Bel_{b,a} = (m^<, m^=, m^>)_{b,a}$
  $= (m^>, m^=, m^<)_{a,b}.$
- $Pl_{b,a} = (m^<, m^=, m^>)_{b,a}$
  $= (m^>, m^=, m^<)_{a,b}.$

Operation of composition:
The composition operation ($\otimes$) can derive a new temporal relation $R_{A,C}$ of the known relations $R_{A,B}$ and $R_{B,C}$, $R_{A,C} = R_{A,B} \otimes R_{B,C}$. We assume that the evidential interval values, $m(r)_{a,b}$, $m(r)_{b,c}$, where r $\in$ R, are known and defined on $\Omega_{TR} \uparrow \Omega_{TR} \times \Omega_D$. The evidential points values are calculated using the conjunctive rule of combination.

Table 5 is the conjunctive combination of all evidential objects detailed in Table 3. The combined bba is defined on a frame of discernment UR = {UR_1, UR_2, UR_3, UR_4} such that each UR_i is an uncertain relations.

Table 5. Example of composition operation of the extended bbas using the conjunctive rule of combination.

$m_1 \textcircled{\cap} m_2(C)$
(1, Older, <) 0.8 = 1 × 0.8 × 1
(1, {Less active, Overwight}, <) 0.2 = 1 × 0.2 × 1
(2, Less active, >) 0.6 = 1 × 0.6 × 1
(2, Less active, {<,=}) 0.4 = 1 × 0.4 × 1

Table 6. Example of combinaison operation of the extended bbas using the conjunctive rule of combination.

$m_1 ⓞ m_2(Z)$	
UR1	UR2
(1, Older, <) 0.8	(1, Older, <) 0.8
(2, Less active, >) 0.48	(2, Less active, $\{<,=\}$) 0.4
0.38	0.32
UR3	UR4
(1, {Less active, Overwight}, <) 0.2	(1, {Less active, Overwight}, <) 0.2
(2, Less active, >) 0.6	(2, Less active, $\{<,=\}$) 0.4
0.12	0.08

5.1 Operation of Combination

The combination operation ($\oplus$) can derive a new temporal relation $R_{A,C}$ given by two independent uncertain relations between two intervals a and b, noted $UR1_{a,b}$ and $UR2_{a,b}$. We assume that the evidential interval values, $m(ur1)_{a,b}$ and $m(ur2)_{a,b}^{\Omega}$, where r1, r2 $\in$ R, are known and defined on $\Omega_{TR} \uparrow \Omega_{TR} \times \Omega_D$. The evidential points values are calculated using the disjunctive rule of combination of all evidential objects. Table 6 combined the bba is demontred above.

Each uncertain relation UR_i can be expanded into different certain states called certain temporal relations R_j, as shown in Table 5. A same relation R_j can be a certain degree level of several uncertain temporal relations. For example, the certain relation R_1 is derived from uncertain temporal relations UR_1 and UR_2. Each uncertain relation is a candidate to represent the evidential relation. The degree of belief on a temporal relation to be candidate is given from the belief of its uncertain temporal relations.

6 Conclusion

In this article, we discussed the modeling and treatment of uncertainty of temporal relations in the framework of points algebra. The theoretic model used is the belief functions whose advantage is to allow to express uncertain and subjective knowledges as well imprecise ones. The key concept of our proposal is the mass function used to represent the belief of an agent on the possible relations between two dates. The uncertainty vector expressed in terms of belief measures, can also be calculated from mass functions. A set of rules controlling the propagation of uncertainty and governing evidential temporal reasoning relations, were also established. As future work, firstly, we plan to complete the reasoning part by the proposal of other rules, as the negation rule which allows to take into account the negative information of an agent on a possible temporal relation. Secondly, extend this study to temporal relations in the framework of intervals algebra [2].

Acknowledgments. This paper is supported by the project CNRS Defi Mastodons 2016: QDoSSi under the number 159982.

References

1. Al-Khatib, W., Day, Y.F., Ghafoor, A., Berra, P.B.: Semantic modeling and knowledge representation in multimedia databases. IEEE Trans. Knowl. Data Eng. **11**(1), 64–80 (1999)
2. Allen, J.F.: Maintaining knowledge about temporal intervals. Commun. ACM **26**(11), 832–843 (1983)
3. Chittaro, L., Montanari, A.: Temporal representation and reasoning in artificial intelligence: issues and approaches. Ann. Math. Artif. Intell. **28**(1–4), 47–106 (2000)
4. Dempster, A.P.: Upper and lower probabilities induced by a multiple valued mapping. Ann. Math. Stat. **38**(2), 325–339 (1967)
5. Denœux, T., Ben Yaghlane, A.: Approximating the combination of belief functions using the fast moebius transform in a coarsened frame. Int. J. Approximate Reasoning **31**(1), 77–101 (2002)
6. Guil, F., Marin, R.: Extracting uncertain temporal relations from mined frequent sequences. In: Thirteenth International Symposium on Temporal Representation and Reasoning, TIME 2006, pp. 152–159. IEEE (2006)
7. Hadjali, A., Dubois, D., Prade, H.: A possibility theory-based approach to the handling of uncertain relations between temporal points. In: Proceedings of the 11th International Symposium on Temporal Representation and Reasoning, TIME 2004, pp. 36–43. IEEE (2004)
8. Petridis, S., Paliouras, G., Perantonis, S.J.: Allen's hourglass: probabilistic treatment of interval relations. In: 17th International Symposium on Temporal Representation and Reasoning (TIME 2010), pp. 87–94. IEEE (2010)
9. Ryabov, V., Puuronen, S.: Probabilistic reasoning about uncertain relations between temporal points. In: Proceedings of the Eighth International Symposium on Temporal Representation and Reasoning, TIME 2001, pp. 35–40. IEEE (2001)
10. Ryabov, V., Puuronen, S., Terziyan, V.Y., et al.: Representation and reasoning with uncertain temporal relations. In: FLAIRS Conference, pp. 449–453 (1999)
11. Shafer, G.: A Mathematical Theory of Evidence. Princeton University Press, Princeton (1976)
12. Shafer, G.: The combination of evidence. Int. J. Intell. Syst. **1**(3), 155–179 (1986)
13. Smets, P.: Belief functions: the disjunctive rule of combination and the generalized bayesian theorem. Int. J. Approximate Reasoning **9**(1), 1–35 (1993)
14. Vila, L.: A survey on temporal reasoning in artificial intelligence. AI Commun. **7**(1), 4–28 (1994)
15. Vilain, M.B., Kautz, H.A.: Constraint propagation algorithms for temporal reasoning. In: AAAI, vol. 86, pp. 377–382 (1986)

An Improvement of the Two-Stage Consensus-Based Approach for Determining the Knowledge of a Collective

Van Du Nguyen[1,2(✉)], Ngoc Thanh Nguyen[1], and Dosam Hwang[2]

[1] Department of Information Systems, Faculty of Computer Science
and Management, Wrocław University of Technology, Wrocław, Poland
{van.du.nguyen,ngoc-thanh.nguyen}@pwr.edu.pl
[2] Department of Computer Engineering,
Yeungnam University, Gyeongsan, Korea
dshwang@yu.ac.kr

Abstract. Generally the knowledge of a collective, which is considered as a representative of the knowledge states in a collective, is often determined based on a single-stage approach. For big data, however, a collective is often very large, a multi-stage approach can be used. In this paper we present an improvement of the two-stage consensus-based approach for determining the knowledge of a large collective. For this aim, clustering methods are used to classify a large collective into smaller ones. The first stage of consensus choice aims at determining the representatives of these smaller collectives. Then these representatives will be treated as the knowledge states of a new collective which will be the subject for the second stage of consensus choice. In addition, all the collectives will be checked for susceptibility to consensus in both stages of consensus choice process. Through experiments analysis, the improvement method is useful in minimizing the difference between single-stage and two-stage consensus choice approaches in determining the knowledge of a large collective.

Keywords: Collective knowledge · Inconsistency knowledge · Two-stage consensus choice

1 Introduction

In recent years the problem of referring knowledge from autonomous and distributed sources for solving some common subjects in the real world is being more and more popular. For many subjects, people often ask opinions not one but many sources such as Internet search engines, forums, friends, etc. This phenomenon seems to be useful in giving a better solution because a large number of sources may have additional knowledge that a collective with smaller ones do not possess on their own and this knowledge may be relevant to subjects solved [1]. However, it also causes conflict because the knowledge given by these sources can conflict with each other. With a conflict we have in mind some inconsistency between knowledge states referred from autonomous sources (collective members) for the same subject in the real world.

© Springer International Publishing Switzerland 2016
N.T. Nguyen et al. (Eds.): ICCCI 2016, Part I, LNAI 9875, pp. 108–118, 2016.
DOI: 10.1007/978-3-319-45243-2_10

Thus solving such conflict is one of the most important tasks in the process of collective knowledge determination.

In collective knowledge determination objective case is the case in which the real knowledge state of the subject exists independently of the knowledge given by collective members. The real knowledge state is not known by the collective members when they are asking for given their knowledge states about the subject such as the problem of weather forecasts for a future day, the prediction of the currency rate for next month, etc. Thus the knowledge states in a collective reflect the real knowledge state to some degree because of incompleteness and uncertainty. The knowledge of a collective as a whole (collective knowledge) is determined on the basis of the knowledge states in a collective [2]. In [3] the author has proposed many consensus-based algorithms for determining collective knowledge for different knowledge representations such as logical, relational structures, ontology, etc. Consensus choice has usually been understood as a general agreement in situations where members have not agreed on some matters [4]. In practice a unanimous agreement is difficult to achieve [5]. Thus it is often not required in solving some common problems by using consensus choice. In this paper, however, consensus model is different from the others [6–10] whose aim at proposing a model to achieve a high level agreement between members. Instead, it is used for solving conflicts which arisen in the process of collective knowledge determination. In this case consensus model has been proved to be useful [4, 11–15].

Generally the knowledge of a collective, which is considered as a representative of the knowledge states in a collective, is often determined based on a single-stage approach. So far, almost all of consensus-based algorithms serving for collective knowledge determination are based on single-stage approach as in [3]. In case of big data, however, in which a collective is often very large and its knowledge is often determined based on multi-stage approach [16]. In this case one of the most important issues in the process of collective knowledge determination is how to minimize the difference between the knowledge determined based on multi-stage and single-stage approaches. Since the results of clustering methods can cause the number of members of each smaller collective to be different from each other. Then each representative can represent for a different number of members in each smaller collective. All of these representatives are treated equivalently in the next stage of consensus choice. This can cause the knowledge of the collective determined based on multi-stage consensus choice to be different from that based on single-stage consensus choice. In the previous work we have proposed a method to determine the knowledge of a large collective based on two-stage consensus choice [17]. In that work, through experiments simulation the proposed method is useful in minimizing the difference between these approaches. However, the problem of susceptibility to consensus has not been investigated. A collective is called susceptibility to consensus if the consensus of its knowledge states is sensible and acceptable as a representative of the whole collective [3]. In other words, whether the knowledge of a collective is good enough and can be acceptable as a compromise? Thus, in this work, we present an improvement method of the two-stage consensus choice in determining the knowledge of a large collective by taking into consideration the problem of susceptibility to consensus. To the best of our knowledge, this approach is novel and missing in the literature.

The remaining part of the paper is organized as follows. Section 2 presents some basic concepts about collective of knowledge states, knowledge of a collective, criterion of susceptibility to consensus, the problem of two-stage consensus choice. The improvement method, some experimental results are presented in Sect. 3. Finally, conclusion and future work are pointed out in Sect. 4.

2 Basic Notions

2.1 Collective of Knowledge States

In this study a collective is considered as a set of knowledge states given by collective members on the same subject. Firstly, let U be a set of objects representing the potential elements of knowledge referring to a concrete real-world subject. The elements of set U can represent logic expressions, tuples, etc. Set U can contain elements which are inconsistent with each other. Let 2^U be the powerset of set U that is the set of all subsets of U. Then the set of all k-element subsets (with repetitions) of set U is $\Pi_k(U)$ (for $k \in N$, N is the set of natural numbers), and let

$$\prod(U) = \bigcup_{k=1}^{\infty} \prod{}_k(U)$$

In this case $\Pi(U)$ is the set of all non-empty finite subsets with repetitions of set U. A set $X \in \Pi(U)$ is called a collective representing the knowledge states in a collective. Thus each element $x \in X$ represents the knowledge state of a collective member. In this work a collective is described as follows:

$$X = \{x_i = (x_{i1}, x_{i2}, \ldots, x_{im}) : i = 1, 2, \ldots, n\}$$

where $x_{ik} \in R$, $k = 1, 2, \ldots, m$; n is the number of collective members, m is the number of dimensions of each element in collective X. Thus each x_i is a multi-dimensional vector representing the collective members' opinion about a real-world subject.

2.2 Knowledge of a Collective

The knowledge of a collective (or collective knowledge) is understood as the consensus of the knowledge states in a collective. It is considered as the representative of a collective. So far, criteria 1-Optimality (O_1) and 2-Optimality (O_2) are the most popular criteria used for collective knowledge determination. This is due to the fact that satisfying these criteria it implies satisfying the majority of other criteria [3, 13]. In collective knowledge determination the knowledge of a collective is defined as follows:

Definition 1. *For a given collective X, the knowledge of X is determined by:*

- *criterion O_1 if:* $d(x^*, X) = min_{y \in U} d(y, X)$
- *criterion O_2 if:* $d^2(x^*, X) = min_{y \in U} d^2(y, X)$

where x^* is the collective knowledge X, $d(x^*, X)/d^2(x^*, X)$ is the sum of distances/squared distances from x^* to elements in collective X.

In case of satisfying criterion O_2, x^* has the following form:

$$x^* = \frac{1}{n}\left(\sum_{i=1}^{n} x_{i1}, \sum_{i=1}^{n} x_{i2}, \ldots, \sum_{i=1}^{n} x_{im}\right)$$

2.3   Susceptibility to Consensus

As aforementioned a collective is called susceptible to consensus if the consensus of its knowledge states is good enough and can be accepted as a representative of the knowledge states in a collective. In [3] the author has investigated the problem of consensus susceptibility for collectives involving real numbers. Also in that work some methods for improving the susceptibility to consensus have been proposed. In [18] the author has devoted the problem of consensus susceptibility for some representatives of collective knowledge such as: real number, binary vector, ordered partitions, or ordered coverings. The criterion for susceptibility to consensus is defined as follows [3]:

Definition 2. *A collective X is susceptible to consensus if and only if it satisfies*

$$d_{t\_mean}(X) \geq d_{min}(X)$$

where $d_{t\_mean}(X) = \frac{1}{n(n+1)}\sum_{i=1}^{n} d(x_i, X)$, $d_{min}(X) = \frac{1}{n}minD(X)$, $D(X) = \{d(x^*, X) :$ $x^* \in U\}$, *is the sum of distances from an element* x^* *to the elements in collective* X. *That is:*

$$d(x^*, X) = \sum_{i=1}^{n} d(x^*, x_i)$$

A collective satisfying the criterion in Definition 2 means that its knowledge states are dense enough for determining a good consensus. In other words, they are consistent enough for determining a good compromise which can be accepted as a representative for the whole collective. The consistency presents the density and coherence degree of the knowledge states in a collective [3].

2.4   Two-Stage Consensus Choice

Let $X = \{x_1, x_2, \ldots, x_n\}$ be a collective, $K(X)$ be the knowledge function which aims at determining the knowledge of collective X. Generally, the problem of collective knowledge determination based on two-stage consensus choice is described as follows:

- **Step 1.** Cluster collective X into smaller collectives $X_1, X_2, \ldots, X_k$
- **Step 2.** Determine the knowledge of each smaller collective

$$K(X_1), K(X_2), \ldots, K(X_k), \ Y = \{K(X_1), K(X_2), \ldots, K(X_k)\}$$

- **Step 3.** Determine the knowledge of collective $Y : K(Y)$

In this case clustering methods are used to classify a large collective into smaller ones. The first stage of consensus choice serves for determining the knowledge of these smaller collectives which will be treated as the knowledge states of a new collective. This new collective will be the subject for the second stage of consensus choice in the process of collective knowledge determination. In the next section we will present an improvement of the two-stage consensus choice approach in determining the knowledge of a large collective by taking into account the problem of susceptibility to consensus.

3 The Influence of Two-Stage Consensus-Based Approach on the Knowledge of a Collective

3.1 The Proposed Method

In this section we present a proposed method to determine the influence of two-stage consensus choice approach on determining the knowledge of a large collective. The detail procedure of the proposed method is described in the following figure.

According to Fig. 1, firstly, large collective X will be classified into smaller collectives $(X_1, X_2, \ldots, X_k)$ by clustering methods. In this study, *k-means clustering* and *random clustering* are applied. K-means clustering, is one of the most popular clustering methods in data mining, which aims at partitioning n elements into k clusters which may have different number of members [19]. The elements within a cluster are similar to one another and different from the elements in the other clusters. However, the second clustering method, we call it *random clustering* because of randomly dividing a large collective into smaller collectives (which nearly have the same number of members). After the clustering process finished, the susceptibility to consensus control will be invoked to check whether these smaller collectives are susceptible to consensus or not? In this process insusceptible collectives will be considered as outlier and will be eliminated. This causes the number of members in collective Y to be less than or equal to the number of smaller collectives $(j \le k)$. The knowledge of susceptible collectives will be considered as the knowledge states of a new collective. Then this new collective (Y) will be the subject of the second stage of consensus choice. If this new collective is susceptible to consensus, then the consensus choice is applied to determine a representative of the collective. Otherwise, some methods for improving the susceptibility to consensus are applied such as adding or removing a knowledge state or modifying weights of collective members [3, 20]. The numbers of members in smaller collectives are described in Table 1.

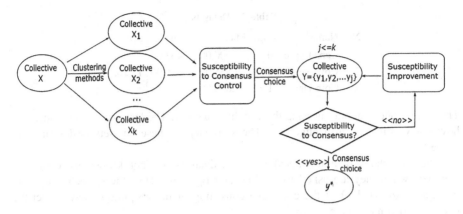

Fig. 1. The proposed method

Table 1. Number of members of smaller collectives

Collectives	Members
X_1	n_1
X_2	n_2
...	...
X_k	n_k

The new collective (Y) has the following form:

$$Y = \{y_1, y_2, \ldots, y_j\} : j \leq k$$

where $y_j = \min_{z \in U} d(z, X_j)$, or $y_j = \min_{z \in U} d^2(z, X_j)$.

According to [17], in case of weighted approach the weight values of the knowledge states in collective Y are as follows:

$$w(y_1) = \frac{n_1}{n}, w(y_2) = \frac{n_2}{n}, \ldots, w(y_j) = \frac{n_j}{n}$$

where n is the number of members in collective X.

3.2 Experimental Results

Datasets

In this study we use three separate datasets from the field of wisdom of crowds to determine the influence of two-stage consensus choice on determining the knowledge of a large collective knowledge. The first dataset is about guessing the total pieces of cereal in a weird-looking glass vase, which has been done by Gideon Rosenblatt[1].

---

[1] http://www.the-vital-edge.com/wisdom-of-crowds/.

Table 2. Datasets

No	Name	Min	Max	Real	Size
1	Cereal vase	0	7,000.00	467.00	436
2	Cow weight	1	14,555.00	1,355.00	17,185
3	Coin jar	1	4,000.00	739.54	602

The second one is about guessing the weight of a cow[2]. The last one is about guessing how many coins are in a huge jar[3]. The summary of these datasets is described in Table 2.

The real states (actual values) of these datasets are not known by collective members when they are asked for giving their opinions. Thus the collective opinion (collective knowledge) of the collective consisting of these opinions may reflect the real state to some degree.

Experimental Results

In the following section we present the experiments of the improvement method with the datasets mentioned in the previous section. First, experiments of applying Ckmeans.1d.dp method (which is based on dynamic programming for optimal one-dimensional clustering) to cluster a large collective into smaller ones [21]. The result of this algorithm also reveals the optimal number of clusters should be used for clustering a dataset. The results of using k-means clustering in the first stage are described in Tables 3, 4 and 5.

Table 3. The result in the first stage of cereal vase dataset

Cereal vase	Size	First stage	
		O_1	O_2
Collective 1	288	345.0	341.32
Collective 2	115	800.0	845.63
Collective 3	26	1,686.31	1,746.19
Collective 4	5	3,528.0	3,487.2
Collective 5	2	7,000.0	6,500.5

In the next step which collectives are insusceptible to consensus will be considered as outliers and will be removed. Therefore, in case of dataset cereal vase, the knowledge of collectives 4 and 5 will not be taken into account in the next stage of collective knowledge determination because of insusceptibility to consensus.

In case of using the random clustering method the results of collective knowledge determination in the first stage are described in Tables 6, 7 and 8.

In this case the number of smaller collectives is the same as that in case of using Ckmeans.1d.dp method. In other words, the number of smaller collectives (k) for each dataset is the same for both k-means and random clustering methods. According to [17],

[2] http://www.npr.org/sections/money/2015/08/07/429720443/17-205-people-guessed-the-weight-of-a-cow-heres-how-they-did.

[3] http://www.wired.com/2015/01/coin-jar-crowd-wisdom-experiment-results/.

Table 4. The result in the first stage of cow weight dataset

Cow weight	Size	First stage	
		O_1	O_2
Collective 1	2,079	600.00	557.85
Collective 2	3,941	895.00	897.19
Collective 3	5,506	1,250.00	1,256.63
Collective 4	4,215	1,623.00	1,625.79
Collective 5	1,250	2,200.00	2,233.66
Collective 6	159	3,456.00	3,501.42
Collective 7	19	6,500.00	6,326.42
Collective 8	15	12,121.00	12,058.87

Table 5. The result in the first stage of coin jar dataset

Coin jar	Size	First stage	
		O_1	O_2
Collective 1	266	300.00	557.92
Collective 2	158	356.56	747.91
Collective 3	94	261.65	599.92
Collective 4	42	349.98	669.35
Collective 5	25	346.86	604.07
Collective 6	17	313.01	711.95

Table 6. The result in the stage of cereal vase dataset

Cereal vase	Size	First stage	
		O_1	O_2
Collective 1	88	554.94	669.27
Collective 2	87	450.00	613.71
Collective 3	87	465.00	563.38
Collective 4	87	480.00	706.11
Collective 5	87	377.00	549.33

Table 7. The result in the first stage of cow weight dataset

Cow weight	Size	First stage	
		O_1	O_2
Collective 1	2148	1260.00	1310.68
Collective 2	2148	1250.00	1285.96
Collective 3	2148	1244.95	1274.99
Collective 4	2148	1253.84	1298.52
Collective 5	2148	1250.00	1291.52
Collective 6	2148	1234.00	1270.48
Collective 7	2148	1257.93	1297.89
Collective 8	2148	1238.00	1284.53

Table 8. The result in the first stage of coin jar dataset.

Coin jar	Size	First stage	
		O1	O2
Collective 1	101	331.75	595.83
Collective 2	101	347.35	616.56
Collective 3	100	313.97	526.49
Collective 4	100	328.52	578.95
Collective 5	100	313.03	598.08
Collective 6	100	343.73	691.45

however, non-weighted approach can cause the difference between the collective knowledge determined based on two-stage and single-stage approaches to be too large. Thus, in this study, in case of using k-means clustering method we only present weighted approach which is proved helpful in minimizing this difference [17]. The results of collective knowledge determination for both single-stage and two-stage approaches are presented in Table 9. In the second stage, again susceptibility to consensus will be checked for collectives involving the knowledge of the corresponding smaller collectives. However, the process of susceptibility improvement is not invoked because these collectives are susceptible to consensus. The susceptibility improvement in the second stage of consensus choice will be the subject of the future work. For this aim, we will investigate with other datasets and different numbers of smaller collectives using in both k-means and random clustering methods. For the problem of susceptibility improvement, in case of using k-means clustering method we intend to use the method of adding a knowledge state to the insusceptible collective. The added knowledge state is chosen from the collective with the higher number of members than the others. However in case of using random clustering method a formal model will be proposed to determine optimal weight values such that the collective is satisfied the criterion of susceptibility to consensus.

Table 9. The results in the second stage

		Cereal vase	Cow weight	Coin jar
Single-stage	S_O_1	450.00	1245.00	345.06
	S_O_2	622.00	1287.00	596.12
Two-stage (Ckmeans.1d.dp)	W_O_1	345.00	1,250.00	420.69
	W_O_2	548.00	1,286.25	596.12
Two-stage (random clustering)	Ran_O_1	450.00	1,244.95	343.59
	Ran_O_2	622.08	1,286.98	576.47

We notice that in this study the quality of collective knowledge is not taken into consideration because the most important issue in the problem of collective knowledge determination based on multi-stage consensus choice is how to minimizing the difference between collective knowledge determined based on single-stage and multi-stage approaches. Figure 2 presents a comparison of using k-means and random clustering methods in the process of collective knowledge determination based on two-stage consensus choice. According to this figure, the proposed method is helpful in minimizing the difference between the knowledge determined based on single-stage and two-stage approaches.

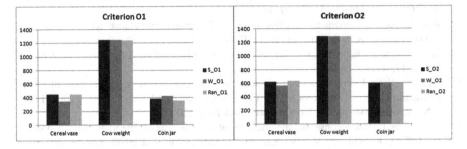

Fig. 2. The results in the second stage

4   Conclusion and Future Work

In this paper we have presented an improvement method of the two-stage consensus choice for determining the knowledge of a large collective by taking into account the problem of susceptibility to consensus. Through experiments, the improvement method is useful in minimizing the difference between collective knowledge determined based on two-stage and single-stage consensus-based approaches.

The future works should investigate this research problem with other structures such as binary vector, relational structure, etc. In addition, the problem of susceptibility improvement in the second stage of consensus choice also should be taken into consideration. To the best of our knowledge, paraconsistent logics could be useful [22].

Acknowledgement. We would like to thank Gideon Rosenblatt, Quoc Trung Bui, Erik Steiner for sharing datasets.

References

1. Vroom, V.H.: Leadership and the decision-making process. Organ. Dyn. **28**(4), 82–94 (2000)
2. Nguyen, N.T.: Processing inconsistency of knowledge in determining knowledge of collective. Cybern. Syst. **40**(8), 670–688 (2009)
3. Nguyen, N.T.: Advanced Methods for Inconsistent Knowledge Management. Springer, London (2008)
4. Day, W.H.E. The consensus methods as tools for data analysis. In: IFC 1987: Classifications and Related Methods of Data Analysis. North-Holland, Springer, Heidelberg (1988)
5. Kline, J.A.: Orientation and group consensus. Cent. States Speech J. **23**(1), 44–47 (1972)

6. Saint, S., Lawson, J.R.: Rules for Reaching Consensus: A Modern Approach to Decision Making. Jossey-Bass, San Francisco (1994)
7. Herrera-Viedma, E., Herrera, F., Chiclana, F.: A consensus model for multiperson decision making with different preference structures. IEEE Trans. Syst. Man Cybern. Part A Syst. Hum. **32**(3), 394–402 (2002)
8. Eklund, P., Rusinowska, A., de Swart, H.: A consensus model of political decision-making. Ann. Oper. Res. **158**(1), 5–20 (2008)
9. Wu, Z., Xu, J.: Consensus reaching models of linguistic preference relations based on distance functions. Soft. Comput. **16**(4), 577–589 (2012)
10. Herrera-Viedma, E., et al.: A review of soft consensus models in a fuzzy environment. Inf. Fusion **17**, 4–13 (2014)
11. Nguyen, N.T.: Using consensus methods for determining the representation of expert information in distributed systems. In: Cerri, S.A., Dochev, D. (eds.) AIMSA 2000. LNCS (LNAI), vol. 1904, pp. 11–20. Springer, Heidelberg (2000)
12. Nguyen, N.T.: Using consensus methods for solving conflicts of data in distributed systems. In: Jeffery, K., Hlaváč, V., Wiedermann, J. (eds.) SOFSEM 2000. LNCS, vol. 1963, pp. 411–419. Springer, Heidelberg (2000)
13. Nguyen, N.T.: Using consensus for solving conflict situations in fault-tolerant distributed systems. In: Proceedings of the First IEEE/ACM International Symposium on Cluster Computing and the Grid (2001)
14. Barthelemy, J.P., Guenoche, A., Hudry, O.: Median linear orders: heuristics and a branch and bound algorithm. Eur. J. Oper. Res. **42**(3), 313–325 (1989)
15. Nguyen, T.N., Sobecki, J.: Using consensus methods to construct adaptive interfaces in multimodal web-based systems. Univ. Access Inf. Soc. **2**(4), 342–358 (2003)
16. Maleszka, M., Nguyen, N.T.: Integration computing and collective intelligence. Expert Syst. Appl. **42**(1), 332–340 (2015)
17. Du Nguyen, V., Nguyen, N.T.: A two-stage consensus-based approach for determining collective knowledge. In: Le Thi, H.A., Nguyen, N.T., Do, T.V. (eds.) Advanced Computational Methods for Knowledge Engineering. AISC, vol. 358, pp. 301–310. Springer, Heidelberg (2015)
18. Kozierkiewicz-Hetmańska, A.: Analysis of susceptibility to the consensus for a few representations of collective knowledge. Int. J. Softw. Eng. Knowl. Eng. **24**(05), 759–775 (2014)
19. Kanungo, T., et al.: An efficient k-means clustering algorithm: analysis and implementation. IEEE Trans. Pattern Anal. Mach. Intell. **24**(7), 881–892 (2002)
20. Nguyen, N.T.: Methods for achieving susceptibility to consensus for conflict profiles. J. Intell. Fuzzy Syst. **17**(3), 219–229 (2006)
21. Wang, H., Song, M.: Ckmeans.1d.dp: optimal k-means clustering in one dimension by dynamic programming. R J. **3**(2), 29–33 (2011)
22. Nakamatsu, K., Abe, J.: The paraconsistent process order control method. Vietnam J. Comput. Sci. **1**(1), 29–37 (2014)

Natural Language and Text Processing

An Approach to Subjectivity Detection on Twitter Using the Structured Information

Juan Sixto$^{(\boxtimes)}$, Aitor Almeida, and Diego López-de-Ipiña

DeustoTech-Deusto Institute of Technology, Universidad de Deusto,
Avenida de las Universidades 24, 48007 Bilbao, Spain
{jsixto,aitor.almeida,dipina}@deusto.es

Abstract. In this paper, we propose an approach to the subjectivity detection on Twitter micro texts that explores the uses of the structured information of the social network framework. The sentiment analysis on Twitter has been usually performed through the automatic processing of the texts. However, the established limit of 140 characters and the particular characteristics of the texts reduce drastically the accuracy of Natural Language Processing (NLP) techniques. Under these circumstances, it becomes necessary to study new data sources that allow us to extract new useful knowledge to represent and classify the texts. The structured information, also called meta-information or meta-data, provide us with alternative features of the texts that can improve the classification tasks. In this study we have analysed the use of features extracted from the structured information in the subjectivity detection task, as a first step of the polarity detection task, and their integration with classical features.

Keywords: Twitter · Text categorization · Data mining for social networks · Subjectivity detection · Social networks

1 Introduction

Since the Twitter social network was created in 2006, it has experienced a substantial growth, having more than 100 million of daily active users and 500 million tweets every day [24] nowadays. Currently Twitter is one of the largest textual data sources used in the data mining and knowledge extraction fields of research. As a part of these fields, sentiment analysis is the computational study of people's opinions, appraisals, attitudes, and emotions toward entities, individuals, issues, events, topics and their attributes [15]. Several research groups have used sentiment analysis techniques [17] over the Twitter micro-texts with an acceptable grade of success. However, the particular characteristics of Twitter (Hashtags, user references, inclusion of URLs, maximum of 140 characters) generate loosely formated texts that are difficult to analyse. Addressing this challenge requires an adaptation of the classical techniques and tools to Twitter's unique requirements, that often results in a relevant decrease of their performance.

© Springer International Publishing Switzerland 2016
N.T. Nguyen et al. (Eds.): ICCCI 2016, Part I, LNAI 9875, pp. 121–130, 2016.
DOI: 10.1007/978-3-319-45243-2_11

There are two possible approaches to this problem. The first one is improving the quality of the texts, in order to facilitate their automatic process. This text normalization task deals with several problems like the use of slang, word shortening, letter omissions and bad spelling [23]. The application of these techniques cleans the texts and improves the performance of the lexical analysis over them. The other approach is to improve the sentiment analysis process using the structured information [5] in addition to the tweeted text. Several researches have studied [2, 20] how the external information can improve the sentiment analysis task. The obtained results show that the external information is a reliable source of knowledge about sentiment and opinion of texts.

In this paper we expand the knowledge about the structured information used in opinion mining field, and its incorporation to the text analysis classical techniques. We study their application to the polarity detection in Spanish language and specifically in the subjective detection task. The rest of the paper is organized as follows: In Sect. 2, the context of this work is presented. In Sect. 3, a Structured and Unstructured information review is presented. Section 4 covers the experimental procedures, and conclusions are introduced in Sect. 5.

2   Related Work

Our study is focused on two aspects of opinion mining: the application of the contextual information and the Spanish polarity classification. In this section, we will review the papers which our work is based on.

2.1   Contextual Applications in Sentiment Analysis

The primary objective of the Sentiment Analysis field is the automatic retrieval of subjectivity and opinion polarity. However, determining their scope is a very complex task and their areas of application are extensive. There are several surveys that summarize the main applications and the most common techniques in sentiment analysis [1, 16, 21].

In the field of the application of contextual information, there are several researchers who use the additional information available in social networks in the classification tasks. In 2011, Pennacchiotti and Popescu [22] presented a generic model for user classification in social media that combines linguistic features and explicit social network features. They also emphasize one of the main problems of contextual information, the difficulty of collecting the social network features of a dataset. Mislove et al. [19] analysed data on a set of Twitter users in order to compare them with the U.S. population. To this end, they developed several techniques to enrich the information available of each user, detecting the gender, the ethnicity and geographic distribution of the users. This was one of the first studies that addressed the idea of the sampling bias and the study of the dataset population as an approach to improve predictions or measurements. Bermingham and Smeaton [4] modeled the political sentiment in order to predict electoral results in Twitter, including sociolinguistic features and unconventional punctuation. In the psychiatry sphere, De Choudhury et al. [7] developed a SVM classifier that can predict the likelihood of an individual to be depressed using Twitter. This work demonstrates the potential of the social networks as a tool for measuring and predicting emotional states of the users and gives new insights about the feature measures. Some of these features, used in their research, are the diurnal trends of the users, the volume of replies and the ego networks. Jiang et al. [13] present a target-depend sentiment classifier using the relations between tweets.

2.2   Spanish Polarity Classification

During the last years, research groups have published a large amount of approaches and methods in the sentiment analysis sphere, and have generated lexicons and polarity dictionaries that facilitate the tasks. Nevertheless, these tools are language dependent. Usually these are generated in one language and, at times, are translated to some other languages. This, combined with the difficulty of establish standard linguistic rules between languages, causes a performance decrease when adapting the tools to other languages [10].

A lot of papers have been published on the field of sentiment analysis in social media, specifically focused on the Spanish language. Vilares and Alonso [1], reviewed a large quantity of bibliographic references in the Spanish scope. Also the TASS workshop [26], a satellite event of the SEPLN Conference[1], presents a huge amount of algorithms and techniques based on opinion extraction in Twitter.

3   Structured and Unstructured Information

Twitter contains a large amount of information about each tweet in addition to the tweeted text. Hashtags, retweets, replies, mentions, followers and many other relations bring us a considerable volume of information about the user network and all its components. This can be a knowledge source about users and their opinions, as we have seen in previous researches, and bring an improvement to the sentiment analysis tasks. Assuming that the use of structured information in sentiment analysis tasks has been proved, our aim is to check their efficiency in the subdomain of automatic sentiment analysis at global level. This subdomain consist of performing an automatic sentiment analysis to determine the global polarity of each message about any topic, without any previous topic discrimination. According to our research proposal, we pretend to study the new possibilities of the structured information in the global level of sentiment analysis, adapting features used in concrete domains as politics [5] or psychiatry [10], and other features not used yet.

Currently, does not exist a unequivocal terminology to refer to the contextual information of the tweets. Barbosa and Feng [2] name them as

---

[1] http://www.sepln.org/.

"Tweet Syntax Features" and Liu [14] refears to them as "Twitter specific clues". In this paper we use the terminology of Structured and Unstructured information, described in Cotelo et al. [5]. In order to achieve the proposed task, it's necessary a full understanding of the Twitter structure and of how their components (Users, Texts, Communities) relate among them. The Fig. 1 represents the most frequent components of the social network and their relations, considered relevants to this work. Also, as part of the study of the data, the structural information of the social network has been classified according to its origin, that is, the component where the information origins.

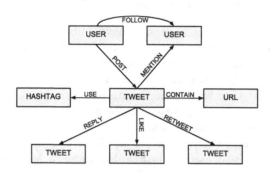

Fig. 1. Twitter structure representation.

As part of our research we have classified the available tweet data in four categories:

Text attributes. This category comprehends the attributes that appear in the text, but that does not depend on the words representation. These attributes emphasize on how the tweets are written, including the characteristic properties of Twitter that are part of the text, but not encompass semantic or sintactic analysis. Technically, this category could be considered as a subcategory of Tweet attributes, as long as the text is a part of each tweet, even so we consider this features clearly difference from the Tweet attributes. Examples include: *hathtags, links, emoticons, punctuation, retweet, used language.*

Tweet attributes. This category comprehends all the characteristics that define a only tweet but not are part of the text. These usually refer to the living process of the tweet within the network, the relations of the users with them or the way it had been posted. Examples include: *quantity of retweets, quantity of favourites, creation date/time, the application that sent the tweet, associated place.*

User attributes. This category comprehends the attributes relative to the authors of the tweets. These attributes represent several aspects of the users that may be relevant to understand the context of a tweet. Occasionally, these aspects are compiled to generate úser profilesthat simplify the user representation in the system. Examples include: *location, political affiliation, post habits.*

Topographic attributes. This category comprehends all the measures about the network topology. The topographic attributes often require some complex calculations and help us to know the role of a user or tweet in the network. Examples include: *Modularity class of user, In-degree and Out-degree of users, Network Communities of the mentioned users.*

4   Experiments

Our purpose is to predict the subjectivity of tweets using the structured components. The main characteristics of the subjectivity classification task are reviewed below. In order to detail the realized experiments, we also describe the corpus chosen, the features studied and the classification algorithms.

4.1   Subjectivity Detection

Liu [14] defines the subjectivity classification as follows: *"Subjectivity classification classifies sentences into two classes: subjective and objective. An objective sentence expresses some factual information, while a subjective sentence usually gives personal views and opinions."*. The subjectivity detection problem has been studied for several years in different areas, especially the approaches based on supervised learning. Since the beginning of the sentiment classification researches, subjectivity has also been explored as part of the global sentiment classification area. The sentiment classification can be expressed as a classification problem if three or more classes: Positive opinion, Negative opinion and no opinion, although these three classes are expanded in several cases. Furthermore, the problem can be divided into two classification subproblems; the opinion detection task first and the distribution between positive and negative opinions later on.

In this paper we address this first problem that is usually named Subjectivity Detection problem. In the Twitter research area, several authors have worked on the sentiment classification in which the subjectivity detection plays an important role. Barbosa and Feng [2] use some twitter features to implement a subjectivity classification. Davidov et al. [6] propose a sentiment classifier that uses punctuation-based features in posted texts. Due to their short length, each tweet is considered as a single sentence and accordingly, each tweet has only a single sentiment polarity.

4.2   Selected Corpus

The scope of our study is focused on the detection of subjectivity in Tweet texts in Spanish language. At present, only a few number of datasets that satisfy these conditions can be found in the state of art. Only the multi-language dataset presented in Volkova et al. [27] and the TASS dataset [26] includes texts with no sentiment in Spanish language.

Finally we decided to use the datasets of the sentiment analysis task at TASS'15[2] workshop [26]. This is an evaluation workshop for sentiment analysis focused on Spanish language, organized as a satellite event of the annual conference of the Spanish Society for Natural Language Processing (SEPLN). This paper is focused on the first task of the workshop, that consists in performing an automatic sentiment analysis to determine the global polarity of each message in the provided corpus. Tweets are divided into six different polarity labels: strong positive (P+), positive (P), neutral (NEU), negative (N), strong negative (N+) and one additional no sentiment tag (NONE). Tweets are also divided into two sets, the training dataset with 7.219 (11 %) items and the test dataset with 60.798 (89 %) items. Additionally, the task includes a 1.000 items dataset, a balanced and hand-labeled subset of the test dataset, that we use as evaluation of the performance of our systems.

The dataset contains a 20.54 % (Train) and 12,30 % (Test) of tweets tagged as NONE, that are considered as True results by our classifier, whereas the rest of labels are considered as False results. As our aim is the detection of objective and subjective texts as a first step of the polarity detection task, the use of a balanced or binary corpus was discarded. The significance of obtaining not only a high accuracy but also a high F1 measures have been taken into account and are explained in the Experimental Results section.

4.3 Features

Several features have been used to address the subjectivity classification problem [14], the vast majority based in the text: bag-of-words, vectorial representations of words, n-grams, etc. In the structural features area, some authors have studied the use of several features in the subjectivity classification of English tweets. Barbosa and Feng [2] exploit the use of retweets, hashtags, replies, links, punctuation, emoticons and the number of upper cases in a subjectivity classifier. Davidov et al. [6] use the length of the words and the punctuation signs as features.

Based on the approach to structured and unstructured information presented in by Cotelo et al. [5], we implement a single sentiment classifier for subjectivity detection, combining two classifiers, each one trained with a different type of information. As shown above, the corpus contains six different labels and as consequence it is not balanced in respect to the objectivity-subjectivity axis. This has been taken into account during the training and evaluation process. In order to train the classifier based on structural information, we have composed a feature list:

URL, Exclamation marks, Emoticons and Uppercase words.
According to the work of Barbosa and Feng [2], this features has been used in our work.

---

[2] Workshop on Sentiment Analysis at SEPLN Conference.

Uppercase Percent. In addition to the number of uppercase words, we used the percentage of uppercase characters of the total characters. This technique differs from the feature proposed by Barbosa and Feng [2], not only counting the words that starts with upper case, but counting all the characters. This ensures that the tweets with all capitals texts, tipically used for emphasis or "shouting", are taking into account.

Favorites. Twitter includes an option called "favorite", that allows the users to like individual tweets. Our study detected a relation between the average of "favorites" by tweet and their sentiment polarity.

Modularity Class. In Twitter, users may subscribe to other users tweets. This is known as "following" and establishes a directed graph of relations between users. During the conduct of our study, we proposed to extract the relations between the users (authors of the tweets) through their "following" relations and genetared their relation graph. Their modularity class revealed the existence of only three communities. A preliminary research shows that this communities are formed by associated individuals related to the left/right political parties or ideologies, and a third group of neutral celebrities. Used as feature, the modularity of each user generates a increase of accuracy and f1-measure in the classifier.

Graph Degrees. Some other attributes of the relation graph have been proposed and tested as features. In detail, the In-Degree and Out-Degree punctuations of the authors have proven to be useful to classification task.

RT. Twitter includes an option called "Retweet", that allows to share a message from another user. This boolean feature expresses if the analysed tweet is a "retweet" of an original tweet.

Ellipsis. During our study, it was noted that some objective tweets includes an ellipsis. In Twitter, ellipsis is often used to make observations about external information, like headlines, urls, or quotes from other users.

The second classifier was based on unstructural information. The selected model to represent the texts was the commonly used the bag-of-words [11]. This model represents each tweet as a matrix of token counts of its words.

4.4   Experimental Results

Considering that the dataset is unbalanced, as shown in Sect. 4.2, we decided to extract three different measures to evaluate the performance of the tested systems; Accuracy, Macro-F1 and NONE-F1. The accuracy is the proportion of true results among the total of the dataset, however, when the prior probabilities of the classes are very different, this metric can be misleading [12]. The macroaveraged F1-measure considers precision and recall, and provides information of how the system performs overall across the dataset. The so-called NONE-F1 are the micro-F1 measure of the NONE (or True) labels of the system. The F1-measure is considered as a relevant score for evaluating the accuracy of a test with a unbalanced dataset. Also we consider relevant to evaluate the specific F1-measure of the NONE labels, in order to rank the contribution of the classifier to a polarity detection task. Obtained results are reviewed in the Table 1.

The first approach to the task consists on a single classifier with the structural features described in Sect. 4.3. Multiple classifier models were tested, like LogisticRegression, Naive Bayes, and Random Forest, obtaining the best results with a GradientBoosting classifier [9], reaching a 70.8 % in Macro-F1 and a 43.0 % in NONE-F1. Then, a new test has been realized using the bag-of-words model, in order to contrast the performance of structured and unstructured approaches. The best results have been obtained with a LogisticRegression model with a balanced mode to automatically adjust weights, obtaining a 70.4 % in Macro-F1 and a 46.8 % in NONE-F1. The similarity of the results shows that both approaches have a relevance for the task, although the poor results involve that the task is complex.

We also investigated the chance of combining both approaches in order to improve the results of the classification task. To do this, we selected two different approaches; using both feature lists in a single classifier and a stacked generalization [28]. In the first case, the structural features of each tweet were added to their matrix representation, generating a new features list. This is a very simple way to merge both models and checks if both feature lists are directly complementary or need more complex techniques to improve the results. The best results has been realized with a GradientBoosting classifier and obtained a 69.2 % in Macro-F1 and a 43.5 % in NONE-F1. This technique does not improve the results, and in several cases decreases them, proving that is necessary the use of other techniques to merge the heterogeneous features effectively.

At least, we realized a stacked generalization work for combining both models. For the level-0 generalizer we use five different classifiers; Logistic Regression, GradientBoosting, Multinomial Naive Bayes, Random Forest and Calibrated with Isotonic Regression. Each of the classifiers were trained with both models, generating a ten classifiers array that formed the level-0 models. Then we used a Logistic Regression model for the level-1 classification model. We found that the use of regression models obtain best results, according to the presented by Ting and Witten in [25]. This approach obtained a 90.22 % in Macro-F1 and a 55.66 % in NONE-F1, being the best obtained results. This improvement implies that a complex technique, like the stacking, benefits from the heterogeneous features in relation to the other approaches.

Table 1. Results for subjectivity detection.

System	Accuracy	Macro F1	NONE-F1
Meta-Information	89.5 %	70.8 %	43.2 %
Bag-of-Words	79.3 %	70.4 %	46.8 %
MI+BoW	88.3 %	69.2 %	43.5 %
Stacking MI+BoW	89.8 %	90.20 %	55.65 %

5   Conclusions and Future Work

Our objective in this study was to learn about the contextual information, their uses at the subjectivity detection task and their application improving the text based models. Exist several previous approaches to the contextual data and we have attempted to adapt these knowledge to the global polarity detection task and to the spanish language. Our study has verified a hypothesis already applied in other social areas and expanded the knowledge relative to the contextual information, adding new ways to use the contextual information to the previous approaches of the state of art. Also we presented a contextual data classification for a better understanding of their nature and characteristics. We presented a first interaction of a subjectivity detection approach which uses some contextual elements to build its features. This approach overtakes the basic classifiers and achieves to combine the structured and unstructured information, establishing a method to complement the standard classification techniques. Although the accuracy and f1-measure are around 90 %, the poor values in the micro average reveal that exists an huge margin for improvement in the task. As future work, we want to connect our work with a complete polarity detection task, applying the extracted knowledge in other sentiment categories, exploring new contextual features. We want to perform a more extensive analysis to check more Twitter components and their relation with the different polarities, considering that distinct features could be related with only a particular sentiment category. Also, we seek to apply the contextual features with more complex models that include lexicons of polarity and semantic resources to really see the impact of them.

References

1. Alonso, M.A., Vilares, D.: A review on political analysis and social media. Procesamiento del Lenguaje Nat. **56**, 13–24 (2016)
2. Barbosa, L., Feng, J.: Robust sentiment detection on twitter from biased and noisy data. In: Proceedings of the 23rd International Conference on Computational Linguistics: Posters, pp. 36–44 (2010)
3. Belkaroui, R., Faiz, R.: Towards events tweet contextualization using social influence model and users conversations. In: Proceedings of the 5th International Conference on Web Intelligence, Mining and Semantics, p. 3. ACM (2015)
4. Bermingham, A., Smeaton, A.F.: On using Twitter to monitor political sentiment and predict election results (2011)
5. Cotelo, J.M., Cruz, F., Ortega, F.J., Troyano, J.A.: Explorando Twitter mediante la integracin de informacin estructurada y no estructurada. Procesamiento del Lenguaje Nat. **55**, 75–82 (2015)
6. Davidov, D., Tsur, O., Rappoport, A.: Enhanced sentiment learning using Twitter hashtags and smileys. In: Proceedings of the 23rd International Conference on Computational Linguistics: Posters (2010)
7. De Choudhury, M., Gamon, M., Counts, S., Horvitz, E.: Predicting depression via social media. In: ICWSM, p. 2 (2013)
8. Esparza, S.G., OMahony, M.P., Smyth, B.: Mining the real-time web: a novel approach to product recommendation. Knowl. Based Syst. **29**, 3–11 (2012)

9. Friedman, J.H.: Greedy function approximation: a gradient boosting machine. Ann. Stat. **29**, 1189–1232 (2001)
10. Han, B., Cook, P., Baldwin, T.: unimelb: Spanish text normalisation. In: Tweet-Norm@ SEPLN, pp. 32–36 (2013)
11. Harris, Z.S.: Distributional structure. Word **10**(2–3), 146–162 (1954)
12. Jeni, L.A., Cohn, J.F., De La Torre, F.: Facing imbalanced data-recommendations for the use of performance metrics. In: 2013 Humaine Association Conference on Affective Computing and Intelligent Interaction (ACII), pp. 245–251. IEEE (2013)
13. Jiang, L., Yu, M., Zhou, M., Liu, X., Zhao, T.: Target-dependent Twitter sentiment classification. In: Proceedings of the 49th Annual Meeting of the Association for Computational Linguistics: Human Language Technologies, vol. 1, pp. 151–160 (2011)
14. Liu, B.: Sentiment analysis and opinion mining. Synth. Lect. Hum. Lang. Technol. **5**(1), 1–167 (2012)
15. Liu, B., Zhang, L.: A survey of opinion mining and sentiment analysis. In: Aggarwal, C.C., Zhai, C.X. (eds.) Mining Text Data, pp. 415–463. Springer, New York (2012)
16. Martínez-Cámara, E., Martín-Valdivia, M.T., Ureña-López, L.A., Montejo-Ráez, A.R.: Sentiment analysis in Twitter. Nat. Lang. Eng. **20**(01), 1–28 (2014)
17. Medhat, W., Hassan, A., Korashy, H.: Sentiment analysis algorithms and applications: a survey. Ain Shams Eng. J. **5**(4), 1093–1113 (2014)
18. Mejova, Y., Srinivasan, P., Boynton, B.: GOP primary season on Twitter: popular political sentiment in social media. In: Proceedings of the Sixth ACM International Conference on Web Search and Data Mining. ACM (2013)
19. Mislove, A., Lehmann, S., Ahn, Y.Y., Onnela, J.P., Rosenquist, J.N.: Understanding the demographics of Twitter users. ICWSM **11**, 5 (2011)
20. Monti, C., Rozza, A., Zapella, G., Zignani, M., Arvidsson, A., Colleoni, E.: Modelling political disaffection from Twitter data. In: Proceedings of the Second International Workshop on Issues of Sentiment Discovery and Opinion Mining (WISDOM 2013) (2013)
21. Pang, B., Lee, L.: Opinion mining and sentiment analysis. Found. Trends Inf. Retrieval **2**(1–2), 1–135 (2008)
22. Pennacchiotti, M., Popescu, A.M.: A machine learning approach to Twitter user classification. ICWSM **11**(1), 281–288 (2011)
23. Porta, J., Sancho, J.L.: Word normalization in Twitter using finite-state transducers. In: Tweet-Norm@ SEPLN, vol. 1086, pp. 49–53 (2013)
24. Smith, C.: DMR Twitter Statistic Report. Last modified 26 Feb 2016. http://expandedramblings.com/index.php/downloads/twitter-statistic-report/. Accessed 28 Mar 2016
25. Ting, K.M., Witten, I.H.: Issues in stacked generalization. J. Artif. Intell. Res. (JAIR) **10**, 271–289 (1999)
26. Villena-Román, J., García-Morera, J., García-Cumbreras, M.A., Martínez-Cámara, E., Martín-Valdivia, M.T., Ureã-López, L.A.: Overview of TASS 2015. In: Proceedings of TASS 2015: Workshop on Sentiment Analysis at SEPLN, vol. 1397. CEUR-WS.org (2015)
27. Volkova, S., Wilson, T., Yarowsky, D.: Exploring demographic language variations to improve multilingual sentiment analysis in social media. In: EMNLP, pp. 1815–1827 (2013)
28. Wolpert, D.H.: Stacked generalization. Neural Netw. **5**(2), 241–259 (1992)

WSD-TIC: Word Sense Disambiguation Using Taxonomic Information Content

Mohamed Ben Aouicha, Mohamed Ali Hadj Taieb[(⊠)],
and Hania Ibn Marai

Multimedia Information System and Advanced Computing Laboratory,
Sfax University, Sfax, Tunisia
mohamedali.hadjtaieb@gmail.com

Abstract. Word sense disambiguation (WSD) is the ability to identify the meaning of words in context in a computational manner. WSD is considered as an AI-complete problem, that is, a task whose solution is at least as hard as the most difficult problems in artificial intelligence. This is basically used in application like information retrieval, machine translation, information extraction because of its semantics understanding. This paper describes the proposed approach (WSD-TIC) which is based on the words surrounding the polysemous word in a context. Each meaning of these words is represented by a vector composed of weighted nouns using taxonomic information content. The main emphasis of this paper is feature selection for disambiguation purpose. The assessment of WSD systems is discussed in the context of the Senseval campaign, aiming at the objective evaluation of our proposal to the systems participating in several different disambiguation tasks.

Keywords: Word Sense Disambiguation · Information content · Gloss · WordNet

1 Introduction

One of the first problems encountered in most natural language processing systems is ambiguity of polysemous words. This concept was introduced for the first time by Warren in 1949 [1]. He stressed that the ambiguity of words must be solved to enable automatic translation between languages. This ambiguity may be syntactic or semantic. The first was largely solved by using the markers parts of speech predicting the grammatical category of words and with very high accuracy as the approach of Erik Brill [2]. The problem is that words often have more than one meaning. The identification of the correct meaning of a word is called Word Sense Disambiguation (WSD). The objective, therefore, of a WSD system is to choose the right sense for a word in a context, usually a sentence, if that word has more than one meaning.

WSD existing approaches are usually based on various principles. They can be divided into three categories: the corpus-based, the knowledge-based and hybrid approaches which are based on the combination of both resources (corpus and knowledge resource). One of the problems with the conventional methods of WSD is the need to use an annotated corpus. In this respect, our goal was to develop a new

© Springer International Publishing Switzerland 2016
N.T. Nguyen et al. (Eds.): ICCCI 2016, Part I, LNAI 9875, pp. 131–142, 2016.
DOI: 10.1007/978-3-319-45243-2_12

approach to WSD without needing to an annotated corpus, based on the words surrounding the target word and a weighting mechanism applied on the nouns using taxonomic information content method through the WordNet [3] « is a » taxonomy.

The rest of the paper is organized as follows: Sect. 2 provides an overview about the WSD approaches. Section 3 presents our proposal WSD-TIC including the weighting mechanism and the disambiguation process based on co-occurring words. Section 4 reports on the evaluation and comparison of our approach through different benchmarks. The final section is devoted to presenting our conclusions and future work.

2 Related Works

Several approaches have been developed as part of the WSD. This section contains a description of a number of existing approaches in the field of WSD and the characteristics of each of them. We can distinguish three categories of approaches: corpora-based, knowledge-based and hybrid.

2.1 Corpus-Based Approaches

The corpus-based approaches exploit the textual content of the corpus. This approach seeks to find the correct meaning of an ambiguous word based on the context formed by the neighboring words (co-occurrence). In what follows, we present a study of some works belonging to such approaches.

Véronis [4] proposed HyperLex, an approach to automatically determine the various uses of a word in a corpus. The correct meaning of an ambiguous word is determined using the concept of co-occurrence graphs. A co-occurrence graph is constructed from the context where the word appears (nouns and adjectives). For a polysemous word, a corpus is created through the search of the word in the web using the meta-search engine Copernic Agent via both singular and plural forms of the word. They select the component receiving the highest weight as disambiguating.

Nameh et al. approach [5] is intended to affect the right sense in an ambiguous word by comparing the context in which it appears with the existing texts in the corpus annotated by the senses. This algorithm is divided into two main steps; the first is the extraction of feature vectors. In fact, for the ambiguous word, the vector of characteristics is represented by the set of words that appear in context. Each component of the vector is represented by the number of times that a word appears in the context of ambiguous word. Then, they compare the vector of ambiguous word with any other vector already built using the cosine measure.

2.2 Knowledge-Based Approaches

The rise of electronic dictionaries and lexical database offered another direction in the development of automatic disambiguation field. This new perspective has materialized through the methods based on the knowledge that try to extract from these resources automatically the information necessary for disambiguation. These approaches use an external resource (typically WordNet) to extract useful information for disambiguation.

Lesk [6] calculates the number of words in common between two definitions from a dictionary. Applied to the disambiguation, Lesk inputs a context (sentence) and the word to disambiguate; it compares the gloss of each sense of the target word with the glosses of any other word in the sentence. The correct meaning is the one assigned to the definition that shares the greatest number of words with the definitions of other words.

Banerjee and Pedersen [7] proposed a new approach derived from Lesk improving the definitions of words through their enrichment by the definitions of concepts which are linked to a given hierarchized concept (hyperonymy, metonymy, holonomy, etc.).

Sinha and Mihalcea [8] have proposed an approach based on the graphs to represent the meaning of ambiguous words. Initially, they created the graph representing the words, their meaning and the dependencies between them. Thereafter, they used a combination of three semantic similarity measures (Jiang and Conrath [9], Leacock et Chodorow [10] and Lesk [6]) for calculating the proximity between the different grammatical categories.

2.3 Hybrid Approaches

To assign the correct meaning of words, hybrid approaches combine several types of resources (word frequency, context, definition and relationships).

Hessami et al. proposed an approach [11] which uses the notion of trees and graphs of co-occurrence. They use WordNet as an external resource to extract the meaning of words. The algorithm is applied to a text corpus by treating its sentences. It allows the disambiguation of all words of context (phrase). They create a graph $G(V, E)$ with V is the set of vertices and E is the set of edges. Then, they create trees for each sense in G from the relationships « is a » of WordNet; then they search the tree node that exists in the graph: If it exists, an edge is created in the graph between the root of the tree and this node.

Basile et al. [12] inspired from the Lesk algorithm. Their basic idea was to improve the Lesk algorithm by replacing the notion of overlap between the glosses by measuring cosine similarity. They use Babelnet [13] as inventory senses. Having w_i (target word), the first step is to find its way from Babelnet. Creating the context: the words right and left words of w_i in context. Then, the glosses are expanded with linked concepts to a particular sense.

3 WSD-TIC: Approach of Word Sense Disambiguation Based on Taxonomic Information Content

Our basic idea was to introduce the concept of IC in WSD process which is divided into two main modules. The first module concerns the pre-processing including the extraction of grammatical categories, simple and compound nouns and their transformation into their canonical forms. The second module focuses on the steps of the disambiguation process. It begins by representing each target word by a feature vector. A feature vector is formed by the weighted nouns co-occurring with the target word. They are created using the names of glosses (extracted from WordNet) and synset glosses that are related to the target concept. The weight is computed using an IC-computing method. Compared to the related works, our proposal is considered a hybrid approach because it exploits the textual information (glosses) and the knowledge structure through the WordNet « is a » taxonomy for calculating the information content.

The basic idea is the computing of semantic relatedness degree between all senses of the target word and any sense of any word belonging to its context using the cosine measure. Each combination is represented by the sum of the values of semantic similarities between a sense s_{kj} and a set of senses each from a word (pertaining to the context). The combination having the highest sum will be chosen as representative of the senses s_{kj}. The last step concerns the choice of the sense with the highest value as the correct sense of the word that will be annotated. The corresponding sense to an ambiguous word is done using WordNet 3.0. However, before embarking on the disambiguation process, a pretreatment phase composed of two steps is necessary.

Extraction of simple and compound nouns: it is used to browse the sentence and seek all simple and compound nouns. This is important because it affects the results of the disambiguation process due to the fact that the meaning of a compound noun differs from the meaning of its composed words. For example, the words *"computer"*, *"science"* and *"computer science"* have different senses.

Determining part of speech: In this step, we considered that the words with grammatical category noun, verb, adjective or adverb. Every other word is eliminated. Grammatical categories of words are extracted using the Stanford parser because it treats the sentence as a whole and determines grammatical categories of words through the complete structure of the sentence. Researching the canonical forms of words is necessary to gain access to WordNet.

The disambiguation process takes as entry the target word and its context. Let C be the context constituted from n words: $C=\{w_1, w_2, w_3, .., w_n\}$ with w_k is the word to disambiguate ($1 \leq k \leq n$). The first step is to check if the target word has one or more senses. If it has only one sense, it will be the disambiguating sense. Else, it will follow the disambiguation process. The steps of this process are described by the following Algorithm:

	Algorithm : WSD-TIC
	Input
1	w: word to be annotated,
2	V_s: vector formed by the words composing the sentence s. It contains the pairs $<w_i, POS_{w_i}>$ where w_i is a word and POS_{w_i} is its part of speech.
	Output
3	Idx: the index of the selected sense of the word w
	Begin
4	Senses$\leftarrow$ sensesInWordNet(w) // the senses of w in WordNet 3.0
5	**If** ($|$Senses$|$ = = 1) **Then**
6	choose the singleton $s_1 \in$ *Senses* as the target sense of w
7	Idx$\leftarrow$1
8	**Else**
9	$C_w\leftarrow$ extractContext (s, w, *pos*, n) // *pos* is the position of w in the sentence s //and n the size of the context
10	**For** each $w_i \in C_w$ **do**
11	SF $\leftarrow$ extractSenseFeatures(w_i) //SF is a set of vectors representing the senses //of the word w_i and composed from the extracted nouns
12	**For** each $s \in$ SF **do**
13	computeNounsWeights (s) // compute the weight of each noun in vector s
14	**End For**
15	**End For**
16	Idx$\leftarrow$ findCorrectSense (SF, w) // the index of the selected sense of the //word w
17	**End If**
18	**End**

The method "*extractSenseFeatures(w_i)*" is intended to represent every meaning of a word present in the context of the word w by a feature vector. Nouns are generally the least ambiguous words with respect to verbs, hence our choice of nouns is only to represent the feature vectors of the meaning of a word. For a given sense of the word, its feature vector consists of all words extracted from glosses and the glosses of synsets that are connected to it. Since the synset glosses are generally short, we have chosen to enrich them by synset glosses that are related to them.

This method aims to extract for each sense assigned to a word in WordNet. The same way is applied to different grammatical categories of words by changing only the relations according to each category. For a given synset, its gloss is divided into two parts: the definition and the example. Each part follows a parsing step to determine the part of speech of each word. Thereafter, only the nouns that are extracted from glosses of the concerned synset and its related are grouped in a single vector. Any word belonging to this vector will be weighted using a mechanism based on the information content (IC).

3.1 Weighting Mechanism Based on IC

The weighted mechanism is based on noun gloss overlap quantified using nouns' weights computed with our IC computing method. All the nouns pertaining to the set of glosses of the ancestors' subgraph assigned to target sense are extracted. In the next paragraphs, we detail the steps of the computing process in order to provide the weight of the word w_i.

3.1.1 Computing Noun Weight

In this section, we describe the method used to compute the weight assigned to each noun pertaining to the set of glosses representing a specific concept. A concept is represented by a set of nouns extracted from the glosses of the concepts pertaining to its ancestors' subgraph. The weight attributed to each word w is based on the IC computing method proposed by Hadj Taieb et al. [14] by using the set of synsets in WordNet $Syn(w)$ ascribed to the target word w as follows:

$$weight_W(w) = \sum_{c \in Syn(w)} IC(c) \tag{1}$$

Next, this function is exploited to quantify the weight assigned to a concept $weight_C(c_i)$. The function $weight_C(c_i)$ is expressed as follows:

$$weight_C(c) = \sum_{w \in Nouns(c)} weight_W(w) \tag{2}$$

Where $Nouns(c)$ is the set of nouns extracted from the glosses representing the concept c in WordNet. In fact, from WordNet, we use the glosses of the concepts pertaining to $Hyper(c)$ representing the hypernyms of c including itself.

$weight_C(c_i)$: a value assigned to concept c_i computed based on the nouns of the glosses of the ancestors' subgraph of c_i.

3.1.2 Computing Information Content

Hadj Taieb et al. [14] proposed a method that quantifies the concept's IC based on the subgraph of ancestors. The ancestors' subgraph extracted from the taxonomy "*is a*" represents the propagation of features from a parent to the descendant by adding some specificities. The IC of a concept is computed based mainly on the contribution of each ancestor according to its depth and descendants' number. The IC value of a concept C is computed as follows:

$$IC(C) = \sum_{c \in Hyper(C)} Score(c) \tag{3}$$

where $Score(c)$ indicates the contribution of each ancestor pertaining to the set $Hyper(C)$. This score is computed as follows:

$$Score(c) = \left(\sum_{c' \in DirectHyper(c)} \frac{Depth(c')}{|Hypo(c')|} \right) \times |Hypo(c)| \qquad (4)$$

where c and c' are concepts, $DirectHyper(c)$ is the set of direct parents of c, and $Hypo(c)$ is the set of direct and indirect descendants, including the concept c.

3.2 Disambiguation Process

Figure 1 shows that the target word w_k is surrounded by the words pertaining to its context. Each word w_i is represented by a set of m_i senses in WordNet: $\{s_{i1}, s_{i2}, s_{i3}, \ldots, s_{im_i}\}$ where $i \in [1, n]$, and $i \neq k$ with each sense is defined by a vector of weighted nouns.

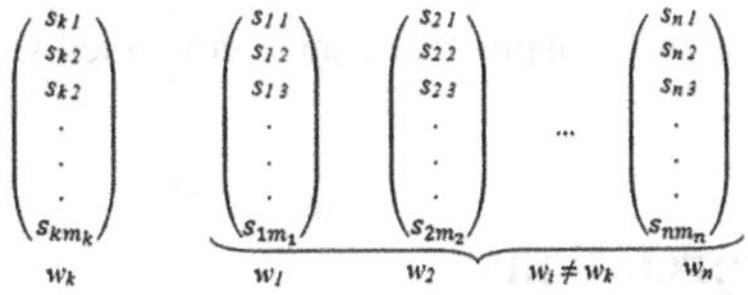

Fig. 1. Representation of the words in the context of w_k with their meanings in WordNet

$SR(s_{kj}, s_{il})$ represents the semantic relatedness between the senses s_{kj} and s_{il} where $1 \leq j \leq m_k$, $1 \leq i \leq n$, $i \neq k$ and $1 \leq 1 \leq m_i$.

$SR(s_{kj}, s_{il})$ is a function exploited in computing semantic relatedness between two senses s_{kj} and $s_{i\,l}$. The value is calculated using the cosine measure between the vectors representing each sense (A sense is represented by the weighted words using the IC).

The first step is to calculate, for each sense of the target word, the values of its semantic relatedness with any sense of any word belonging to the context of the word. Figure 2 illustrates all proximity values to calculate. Each vector represents all the values of proximity between a sense s_{kj} of the target word and the meaning of a word belonging to its context.

The next step is to generate all possible combinations between semantic relatedness values obtained in the previous step as it is illustrated in Fig. 2. For each of the obtained combinations, the semantic relatedness values are summed into one value. Each sense of the word w_k is represented by a value. Among the meanings of the target word, the sense chosen as the disambiguating sense is the one assigned to the highest value.

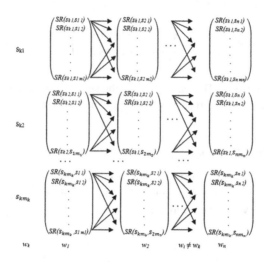

Fig. 2. Representation of different semantic relatedness computed values using the words of the context and the target word w_k

4   Results and Interpretations

In this section, we introduce all the experiments conducted to evaluate our approach. We have used the JWNL[1] package (Java WordNet Library) for interfacing WordNet. JSemcor[2] is used for interfacing the *SemCor* and *Senseval* data. Two different datasets are used: Senseval2 and Senseval3. At the beginning, we present the exploited datasets. Subsequently, we present the evaluation metrics. At the end, we detail the results and their interpretations.

4.1   Datasets and Evaluation Metrics

Several datasets, named as benchmarks are used in the evaluation of systems disambiguation of words.

Senseval2 is a dataset interested in evaluating word sense disambiguation of the meaning of words under the guidance of both associations ACL and SIGLEX. Senseval2 is a set of manually annotated data by experts. This set is annotated with WordNet 1.7 that has been standardized in WordNet 3.0. It consists of three files with 2518 words composed from 475 adjectives, 299 adverbs, 1163 nouns and 581 verbs.

Senseval3 has been annotated by experts using the WordNet. It consists of three files containing 2081 words. These 2081 words are divided into 364 adjectives, 15 adverbs, 951 nouns and 751 verbs.

---

[1] JWNL WordNet Java API: http://sourceforge.net/projects/jwordnet.

[2] http://projects.csail.mit.edu/jsemcor/.

Navigli [15] stated that a WSD approach can be assessed using the metrics recall, precision and F-measure as follows:

Recall is used to return the percentage of words correctly annotated by the approach relative to the total number of words (of all dataset) to annotate (to disambiguate).

$$Recall = \frac{number\ of\ words\ coorectly\ annotated\ by\ the\ proposed\ apporach}{number\ of\ total\ words\ annotated\ in\ dataset} \tag{5}$$

Precision refers to the measure of the effectiveness of an approach. It is defined as the number of words correctly annotated relative to the total number of words annotated by the approach.

$$Precision = \frac{number\ of\ words\ correctly\ annotated\ by\ proposed\ approach}{number\ of\ words\ annotated\ by\ the\ proposed\ approach} \tag{6}$$

F-measure is based on the combination of both precision and recall measures. It is defined as follows:

$$F - measure = \frac{2 * recall * precision}{recall + precision} \tag{7}$$

4.2   Results

In the following section, we introduce a comparison of our approach with some other unsupervised approaches that have been evaluated on the same benchmarks that we used. The results are used from the published ones according to the Senseval2 and Senseval3 events. We got the best results for a size of context window (n = 6).

Table 1 focuses on the benchmark Senseval2.[3] It shows a comparison of our approach with some works that have been evaluated using the task *"all words"* [16]. The values are sorted in ascending order according to the recall metric.

Table 1 shows that our system has resulted in a higher recall than many other unsupervised systems. The maximum values are obtained by supervised systems. These values are explained by their use of annotated corpus. But the development of this type of corpus is expensive. This represents a disadvantage for supervised approaches. Also, we can note that some approaches have high precision values and in against part their recall values are low such as *IIT* and *Sussex-sel*.

Senseval3 [17] is considered more difficult to disambiguate by [11] due to the high number of verbs which are fine-refined in WordNet. Table 2 displays the recall, precision and F-measure values obtained by our system and a number of other systems that have been evaluated on this benchmark. The values are sorted in ascending order of the recall.

---

[3] www.hipposmond.com/senseval2/Results/all_graphs.xls.

Table 1. Comparison of results of all data senseval2 *"all words"*

System	Precison	Recall	F-measure
IIT 1	0.287	0.033	0.059
IIT 3	0.294	0.034	0.060
IIT 2	0.328	0.038	0.068
Sussex - sel	0.598	0.14	0.226
Sussex - sel-ospd-ana	0.545	0.169	0.257
Sussex - sel-ospd	0.566	0.169	0.260
Universiti Sains Malaysia 2	0.36	0.36	0.36
CL Research-DIMAP	0.451	0.451	0.451
WSD-TIC	**0.471**	**0.450**	**0.460**
UCLA - gchao	0.5	0.449	0.473
UNED-AW-U2	0.556	0.55	0.552

Table 2. Comparison of results of all data senseval3 *"all words"*

System	Recall	Precision	F-measure
DLSI-UA-all-Nous	0.275	0.343	0.305
merl.system2	0.352	0.48	0.406
autoPSNVs	0.354	0.563	0.434
autoPS	0.433	0.49	0.459
IRST-DDD-09	0.441	0.729	0.549
upv-unige-CIAOSENSO2-eaw	0.451	0.608	0.517
WSD-TIC	**0.454**	**0.471**	**0.462**
KUNLP-Eng-ALL	0.496	0.51	0.502
IRST-DDD-LSI	0.496	0.661	0.566
DFA-Unsup-AW	0.546	0.557	0.551
IRST-DDD-00	0.582	0.583	0.582

The problem here is the fine granularity of WordNet which can be resolved through the use of the most frequent senses. Moreover, our proposal shows high accuracy with compound nouns.

5   Conclusion and Future Work

This paper presents the WSD-TIC as an approach for word sense disambiguation. The proposal is exploited for determining the correct meaning of an ambiguous word using the words surrounding it in the context of the target word and all relationships provided by the hierarchical structure of WordNet. To determine the meaning of a word to disambiguate, we proposed to use a semantic relatedness measurement which is based on the Information Content (IC) of concepts. Each sense of a polysemous word is represented by a feature vector, each of which composed of weighted nouns through an IC computing method. These nouns are taken from the definition of a synset in

WordNet (gloss). These glosses are extended by those in relationship with them (e.g. hyponymy, hypernymy, holonymy, pertainym, etc.). The cosine measure is then used to calculate the distance between the meaning of the target word and the meaning of words in the same context. WSD-TIC is assessed using the benchmarks Senseval2 and Senseval3 and compared to the unsupervised approaches. The results are encouraging. Considering the promising results yielded this work, further studies are needed to the use of other semantic relatedness measures.

References

1. Weaver, W.: Translation. In: Locke, W.N., Boothe, A.D. (eds.) Machine Translation of Languages, pp. 15–23. MIT Press, Cambridge (1955)
2. Brill, E.: Transformation-based error-driven learning and natural language processing: a case study in part-of-speech tagging. Comput. Linguist. **21**, 543–565 (1995)
3. Fellbaum, C. (ed.): WordNet: An Electronic Lexical Database (Language, Speech, and Communication). The MIT Press, Cambridge (1998). Illustrated edition
4. VÃ©ronis, J.: HyperLex: lexical cartography for information retrieval. Comput. Speech Lang. **18**, 223–252 (2004)
5. Nameh, M., Fakhrahmad, S.M., Jahromi, M.Z.: A new approach to word sense disambiguation based on context similarity. In: Proceedings of the Workshop of the World Congress on Engineering (2011)
6. Lesk, M.: Automatic sense disambiguation using machine readable dictionaries: how to tell a pine cone from an ice cream cone. In: Proceedings of the 5th Annual International Conference on Systems Documentation, pp. 24–26. ACM, Toronto (1986)
7. Banerjee, S., Pedersen, T.: Extended gloss overlaps as a measure of semantic relatedness. In: Proceedings of the 18th International Joint Conference on Artificial Intelligence, pp. 805–810. Morgan Kaufmann Publishers Inc., Acapulco (2003)
8. Sinha, R., Mihalcea, R.: Unsupervised graph-basedword sense disambiguation using measures of word semantic similarity. In: Proceedings of the International Conference on Semantic Computing, pp. 363–369. IEEE Computer Society, Washington (2007)
9. Jiang, J.J., Conrath, D.W.: Semantic similarity based on corpus statistics and lexical taxonomy. CoRR cmp-lg/9709008 (1997)
10. Leacock, C., Chodorow, M.: Combining local context and WordNet similarity for word sense identification. WordNet Electron. Lexical Database **49**(2), 265–283 (1998). Fellfaum, C. (ed.) MIT Press, Cambridge
11. Hessami, F.E., Mahmoudi, Jadidinejad, A.: Unsupervised graph-based word sense disambiguation using lexical relation of WordNet. Int. J. Comput. Sci. Issues (IJCSI) (2011)
12. Basile, P., Caputo, A., Semeraro, G.: An enhanced lesk word sense disambiguation algorithm through a distributional semantic model. In: 25th International Conference on Computational Linguistics, Proceedings of the Conference: COLING 2014, Technical papers, 23–29 August 2014, pp. 1591–1600. Dublin, Ireland (2014)
13. Navigli, R., Ponzetto, S.P.: BabelNet: the automatic construction, evaluation and application of a wide-coverage multilingual semantic network. Artif. Intell. **193**, 217–250 (2012)
14. Hadj Taieb, M.A., Ben Aouicha, M., Ben Hamadou, A.: A new semantic relatedness measurement using WordNet features. Knowl. Inf. Syst. **41**, 467–497 (2014)
15. Navigli, R.: Word sense disambiguation: a survey. ACM Comput. Surv. **41**, 10:1–10:69 (2009)

What Makes Your Writing Style Unique? Significant Differences Between Two Famous Romanian Orators

Mihai Dascalu[1(✉)], Daniela Gîfu[2], and Stefan Trausan-Matu[1]

[1] Computer Science Department, University Politehnica of Bucharest, 313 Splaiul Independenţei, 060042, Bucureşti, Romania
{mihai.dascalu, stefan.trausan}@cs.pub.ro
[2] Faculty of Computer Science, "Alexandru Ioan Cuza" University, 16 General Berthelot, 700483 Iaşi, Romania
daniela.gifu@info.uaic.ro

Abstract. This paper introduces a novel, in-depth approach of analyzing the differences in writing style between two famous Romanian orators, based on automated textual complexity indices for Romanian language. The considered authors are: (a) Mihai Eminescu, Romania's national poet and a remarkable journalist of his time, and (b) Ion C. Brătianu, one of the most important Romanian politicians from the middle of the 18th century. Both orators have a common journalistic interest consisting in their desire to spread the word about political issues in Romania via the printing press, the most important public voice at that time. In addition, both authors exhibit writing style particularities, and our aim is to explore these differences through our *ReaderBench* framework that computes a wide range of lexical and semantic textual complexity indices for Romanian and other languages. The used corpus contains two collections of speeches for each orator that cover the period 1857–1880. The results of this study highlight the lexical and cohesive textual complexity indices that reflect very well the differences in writing style, measures relying on Latent Semantic Analysis (LSA) and Latent Dirichlet Allocation (LDA) semantic models.

Keywords: Writing style · Textual complexity for Romanian language · Comparable corpora · Famous orators

1 Introduction

Automated evaluation of writing styles represents a challenge among linguistics and emphasizes the importance of technology in order to facilitate research on language. In addition, quantifying differences between speeches in different languages and between authors has become a trending topic in the field of Natural Language Processing (NLP). The equivalent manual analysis is an extremely time consuming process that requires highly skilled annotators, especially in linguistics. Prior research has proposed methods for creating sets of comparable corpora [1–5] that contain similar texts across multiple languages, genres and authors, which can later on be used to assess linguistic differences.

© Springer International Publishing Switzerland 2016
N.T. Nguyen et al. (Eds.): ICCCI 2016, Part I, LNAI 9875, pp. 143–152, 2016.
DOI: 10.1007/978-3-319-45243-2_13

The novelty of this study is reflected in its focus to compare emblematic writing styles of two Romanian authors who marked our society and its transition towards a more transparent system. Vianu, a famous Romanian literary critic, highlighted the problem of theoretical stylistics, namely the dual intention within language: to communicate and to reflect [6]. The expressiveness found in different variations of writing styles makes its presence felt in the reflective dimension of communication. According to Coteanu [7], a word's expressiveness is latent and can be inferred from context, thus emphasizing the importance of cohesion in terms of discourse representation. This is the case of the particularly expressive journalistic texts pertaining to the two selected authors: *Mihai Eminescu* and *Ion C. Brătianu*, whose speeches are representative for Romanian oratory.

Mihai Eminescu, known as Romania's national poet, was instructed in Vienna and is one of the most important journalistic voices of his time. In fact, we speak of three different journalistic stages [8]: (a) the first stage, until his entry in the *Junimea* society in 1876; (b) the *Iasi period* that includes his work for the *Curierul de Iaşi* publication; (c) his activity for the *Timpul* newspaper between 1877 and 1883. During the previous stages, there are no concrete differences of structure or writing style, but rather tones, thickenings or blurs in his political ideas. *Ion C. Brătianu* [9, 10] was an important Romanian politician, instructed in Paris, whose main concern was to draw the attention of political circles in France to support the cause of Romanians and their national aspirations. The considered collection of journalistic texts coincides with the period when he returned from exile, after nine years, while being involved as the Minister of several Ministries (Finances, Internal Affairs, or War) within the Romanian United Principalities (Moldavia and Wallachia). Furthermore, in 1875 he laid the foundations of the Liberal National Party. Brătianu's speech could be easily recognized due to his preeminent ideas of national consciousness and democratic values, such as individual freedom and social equality.

In order to explore the oratorical styles of both Romanian personalities, we rely on a previously validated textual complexity model, integrated in our *ReaderBench* framework [11–14], adapted for Romanian language [12], that addresses multiple facets of text difficulty and comprehension [11]: *text features* (e.g., length, structure or use of punctuation) [15], *textual formality* (e.g., vocabulary, slang, phrasal verbs, use of idiomatic language, and so on) [16], and *textual styles* (e.g., simple/complex sentences, stylistic markers, cohesion, etc.) [17]. The selected textual complexity indices, presented in detail later on, are reflective of each author's writing style and address different layers of discourse analysis, namely lexical structure and semantics, with emphasis on cohesion. Analyses of writing styles in Romanian language are not singular as they became constituent parts in the current trends of interpreting language facts [18–21], but this study represent a first automated in-depth comparison of famous Romanian speeches.

This paper is structured as follows: section two provides details on the used corpus and of the automated method employed through the *ReaderBench* framework. Section three presents results and corresponding discussions, while the last section highlights conclusions and future work.

2 Automated Assessment of Writing Style

In this section we present the analyzed corpus structured into 2 collections of texts by Eminescu and Brătianu, as well as the *textual complexity* indices from *ReaderBench* framework used to characterize each author's writing style.

2.1 Corpus Selection

Our linguistic analysis is focused on exploring the differences in writing styles between the two Romanian personalities from the 19th century (more specifically 1857–1880). This was a period in which Romania was becoming a well-defined nation in the European political context; thus, the speeches had a strong nationalist tone and the shared emotional load had a high impact on the population. Many of the texts were preceded by corresponding public speeches as oratory in public spaces was the best communication channel at the time. Our corpus was built starting from newspaper articles and contains around 139,000 lexical tokens (see Table 1). The articles were converted from PDF format into plain text using Optical Character Recognition software, followed by manual corrections on the raw texts.

Table 1. General corpus statistics.

Orators	Period	N docs	N words	Newspaper sources
M. Eminescu	1877–1880	65	80,193	Pressa, România liberă, Românul(u), Timpul
I. C. Brătianu	1857–1875	45	58,237	Românul, Monitorul Oficial
Total		*110*	*138,430*	

2.2 Indices of Writing Style

Three main categories of textual complexity indices computed by the *ReaderBench* framework were adapted for Romanian and are used to reflect specific traits of writing style for each orator [12]. First, at *surface analysis*, *ReaderBench* makes use of the proxes (i.e., computer approximations of text difficulty) initially developed by Page [22, 23]. Our model integrates the most representative and commonly used proxes in automated essay grading systems [23, 24], for example: average word/phrase/ paragraph length in characters, average unique/content words (dictionary forms that are not stopwords) per phrase or paragraph, average number of commas per sentence or paragraph. Entropy, derived from Shannon's Information Theory [25, 26], is also a relevant metric for quantifying word or character diversity. While word entropy reflects a more varied vocabulary and is related to an increased working memory as more concepts are introduced to the reader, character entropy is a language specific characteristic [27].

Second, *semantic analysis* is centered on cohesion and represents the core of our model. According to McNamara et al. [28], textual complexity is strongly related to cohesion in terms of comprehension, due to the fact that the reader must create a coherent mental representation of the underlying information (i.e., the situation model [29]). Thus, the lack of cohesion flow can increase the difficulty of a text [30] as readers can easily loose interest by finding text segments too unrelated one to another. In order to evaluate local and global cohesion, our model uses Cohesion Network Analysis (CNA) [31] to compute cohesion as the average semantic similarity [32, 33] at the following levels: intra-paragraph (between sentences of each paragraph), inter-paragraph (between any pair of paragraphs), or adjacency/transition from one paragraph or sentence to the next one. Cohesion between any two text segments is estimated as the average value of the cosine similarity in Latent Semantic Analysis (LSA) vector spaces [34, 35] and the inverse of the Jensen Shannon dissimilarity (JSD) [36] between Latent Dirichlet Allocation (LDA) topic distributions [37, 38]. Both models are based on the bag-of-words approach and reflect co-occurrence patterns from an initial training text corpora. For this study, LSA and LDA semantic models were trained on a Romanian corpus of more than 2 million content words covering journalistic texts, literature, politics, science and religion.

LSA uses a sparse term-document matrix that contains for each word a normalized number of its occurrences within a given document (for example, log-entropy, term frequency-inverse document frequency). The dimensionality of this matrix is reduced by projecting the resulting matrices from the Singular Value Decomposition (SVD) [39] on the most important k dimensions. Words and documents are compared using a cosine distance between their vector representations in the projected semantic space. LDA is a generative probabilistic model based on topic distributions. A topic is a Dirichlet distribution [40] over the vocabulary in which thematically related concepts are grouped together based on co-occurrence patterns in the training text corpora. CNA also provides a scoring mechanism for quantifying the importance of each analysis element (sentence, paragraph or entire document) based on the relevance of the underlying content words [41]. This is useful for evaluating the impact of individual sentences in relation to the whole document. In addition to the cohesion-centered discourse representation, specific discourse connectors and conjuncts for Romanian language are also identified using cue phrases in order to evaluate the degree of discourse elaboration, based on the following categories: coordinating connectives; logical connectors; semi and quasi coordinators; conjunctions; disjunctions; simple and complex subordinators; addition, contrasts, sentence linking, order, reference, reason and purpose constructs.

Third, *word complexity* is focused on evaluating each word's difficulty from multiple perspectives of discourse analysis: (a) distance in characters between the word stem, the lemma and the inflected form, (in general, multiple prefixes and suffixes increase the difficulty a certain word), (b) distinguishability approximated as the inverse document frequency from the Romanian text corpora, and (c) the word polysemy count from the Romanian WordNet [42] (words with multiple senses tend to be more difficult to comprehend).

3 Results and Discussions

Statistical analyses were performed to investigate the differences in the writing styles of journalistic texts produced by the two famous Romanian orators. As mentioned in the previous section, our analyses were focused on lexical and semantic properties of the journalistic texts. First, all variable indices reported by *ReaderBench* were checked for normality and those that demonstrated non-normality were removed. Multicollinearity was then assessed as pair-wise correlations ($r > .70$); if writing style properties demonstrated multicollinearity, the index that demonstrated the strongest effect in the model was retained for the final analysis (see Table 2 for final list of indices and their descriptive statistics). As it was expected, character entropy is a language feature and there are no significant differences between authors.

Table 2. General statistics.

Index	M (SD) M. Eminescu ($N = 65$)	M (SD) I.C. Brătianu ($N = 45$)	M (SD) Corpus ($N = 110$)
Average word length	3.88 (0.23)	3.68 (0.2)	3.8 (0.24)
Standard deviation in word letters	2.67 (0.18)	2.56 (0.13)	2.63 (0.17)
Average words per sentence	31.20 (8.54)	33.87 (10.11)	32.29 (9.26)
Standard deviation in unique words per sentence	5.53 (1.22)	5.50 (1.54)	5.52 (1.35)
Word entropy	5.41 (0.22)	5.29 (0.28)	5.36 (0.25)
Character entropy	2.71 (0.02)	2.71 (0.02)	2.71 (0.02)
Average difference between word and stem	1.32 (0.17)	1.33 (0.20)	1.32 (0.18)
Average word polysemy count	5.66 (0.75)	6.24 (0.89)	5.89 (0.86)
Average sentence score	0.55 (0.27)	0.75 (0.27)	0.63 (0.28)
Average sentence-paragraph cohesion (LSA)	0.70 (0.09)	0.65 (0.10)	0.68 (0.10)
Average sentence-paragraph cohesion (LDA)	0.82 (0.10)	0.73 (0.11)	0.79 (0.12)
Average intra-paragraph cohesion (LSA)	0.16 (0.07)	0.22 (0.07)	0.19 (0.08)
Average intra-paragraph cohesion (LDA)	0.41 (0.07)	0.46 (0.05)	0.43 (0.07)

Afterwards, a multivariate analysis of variance (MANOVA) [43, 44] was conducted to examine whether the lexical and semantic properties of the journalistic texts differed between the two famous Romanian orators. Box's M test (104.308) of equality of covariance matrices was not significant, $p(.017) > \alpha(.001)$, indicating that there are no significant differences between the covariance matrices. For all the variables presented in Table 3, Levene's test of equality of error variances is not significant

($p > .05$); therefore, the MANOVA assumption that the variances of each variable are equal across the groups is met. There was a significant difference among the two authors, Wilks' $\lambda = .0.512$, $F(11,98) = 8.498$, $p < .001$ and partial $\eta^2 = .488$. The textual complexity indices from Table 3 presented in descending order of effect size denote the variables that were significantly different between the two orators. Sentence-paragraph cohesion in both LSA and LDA semantic models capture the average resemblance between each constituent phrase and its corresponding paragraph (i.e., local cohesion with the main idea of the paragraph), whereas intra-paragraph cohesion measures the cohesion between each pair of phrases of the same paragraph (i.e., local cohesion in-between phrases). Corroborated with Fig. 1, we can observe that Eminescu uses in general more elaborated words (higher length), fewer, but more diverse words per sentence, as well as more self-contained and cohesive paragraphs.

Table 3. Tests of between-subjects effects for significantly different indices.

Index	df	Mean square	F	p	Partial Eta squared
Average word length	1	1.096	23.705	<.001	.180
Average sentence-paragraph cohesion (LDA)	1	0.239	20.500	<.001	.160
Average intra-paragraph cohesion (LSA)	1	0.094	19.455	<.001	.153
Average sentence score	1	1.064	14.958	<.001	.122
Standard deviation in word letters	1	0.363	14.174	<.001	.116
Average word polysemy count	1	8.935	13.551	<.001	.111
Average sentence-paragraph cohesion (LSA)	1	0.063	7.004	.009	.061

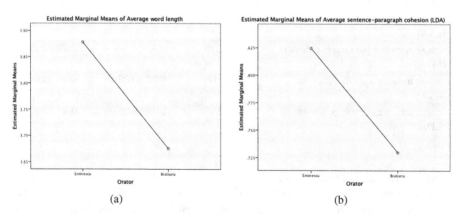

(a)   (b)

Fig. 1. Comparative views of writing styles reflected in textual complexity indices applied on the journalistic texts of both orators.

A stepwise Discriminant Function Analysis (DFA) was performed to predict the author of a given text based on the underlying writing style properties. The DFA retained five variables as significant predictors (*Average sentence-paragraph cohesion - LDA, Average words per sentence, Average intra-paragraph cohesion - LSA, Average word polysemy count, Average word length*) and removed the remaining variables as non-significant predictors.

The results prove that the DFA using these five indices significantly differentiated the texts pertaining to the two authors, Wilks' λ = .609, χ^2(df = 5) = 52.353 p < .001. The DFA correctly allocated 90 (50 + 40) of the 110 documents from the total set, resulting in an accuracy of 81.82 % (the chance level for this analysis is 50 %). For the leave-one-out cross-validation (LOOCV), the discriminant analysis allocated 89 (50 + 39) of the 110 texts for an accuracy of 80.90 % (see the confusion matrix reported in Table 4 for detailed results). The measure of agreement between the actual author and that assigned by the model produced a weighted Cohen's Kappa of 0.636, demonstrating substantial agreement.

Table 4. Confusion matrix for DFA classifying texts pertaining to different orators based on writing style properties.

		M. Eminescu	I.C. Bratianu	Total
Whole set	M. Eminescu	50	15	65
	I.C. Bratianu	5	40	45

		M. Eminescu	I.C. Bratianu	Total
Cross-validated	M. Eminescu	50	15	65
	I.C. Bratianu	6	39	45

4 Conclusions and Future Work

This research presents an in-depth study conducted to compare the work of two Romanian orators in terms of specificities of their writing style. The results reveal significant and interesting differences with regards to the degree of word elaboration (length and polysemy count), word diversity, as well as local cohesion reflected in the intra-paragraph and sentence-paragraph semantic similarity measures. Mihai Eminescu, probably due to the fact that he was also a great poet (considered the most important poet in Romania's literature), used more elaborated, lengthier words. Sentences contained fewer, but more diverse words, and paragraphs were more self-contained and cohesive that in the case of I.C. Brătianu. The journalistic texts of both orators are very complex (more than 30 words per sentence) and the selected features were successfully used to predict the author of a given text based on the underlying writing style properties, thus highlighting a clear demarcation between their work.

As extension of this study, we envision the inclusion of texts pertaining to other representative Romanian authors from different time periods and, potentially, other genres (for example, novels and essays) in order to identify additional individual traces of their writing style. This will also enable us to model trends in the time evolution of the Romanian language.

Acknowledgments. This work has been partially funded by the 2008-212578 LTfLL FP7 project, as well as the EC H2020 project RAGE (Realising and Applied Gaming Eco-System); http://www.rageproject.eu/ No. 644187.

References

1. de Saussure, F.: Cours de Linguistique Générale. Payot, Paris (1999)
2. Bo, L., Gaussier, E., Morin, E., Hazem, A.: Degré de comparabilité, extraction lexicale bilingue et recherche d'information interlingue. In: Conf´erence sur le Traitement Automatique des Langues Naturelles, vol. 1, pp. 211–222. LIRMM Montpellier, Montpellier (2011)
3. Morin, E., Daille, B.: Comparabilité de corpus et fouille terminologique multilingue. Traitement Automatique des Langues **47**(1), 113–136 (2006)
4. Gîfu, D.: Contrastive diachronic study on romanian language. In: FOI 2015, pp. 296–310. Institute of Mathematics and Computer Science, Academy of Sciences of Moldova (2015)
5. Aijmer, K., Altenberg, B., Johansson, M.: Languages in contrast: papers from a symposium on text-based cross-linguistic studies, Lund 4–5 March 1994, vol. 88. Lund studies in English (1996)
6. Vianu, T.: Arta prozatorilor români. Ed. Contemporană, Bucharest (1941)
7. Coteanu, I.: Stilistica Funcțională a Limbii Române, vol. 81. Editura Academiei, Bucharest (1993)
8. Ibrăileanu, G.: Spiritul Critic în Cultura Românească. Tipografia Moldova, Iași (2001)
9. Brătianu, I.C.: Memoire sur l'Empire d'Autriche dans la question d'Orient, Paris, France (1855)
10. Brătianu, I.C.: Memoire sur la situation de la Moldo–Valachie depuis la Traite de Paris, Paris, France (1857)
11. Dascalu, M.: Analyzing Discourse and Text Complexity for Learning and Collaborating. SCI, vol. 534. Springer, Cham (2014)
12. Dascalu, M., Gifu, D.: Evaluating the complexity of online Romanian press. In: 11th International Conference "Linguistic Resources and Tools for Processing the Romanian Language", Iasi, Romania, pp. 149–162 (2015)
13. Dascalu, M., Dessus, P., Bianco, M., Trausan-Matu, S., Nardy, A.: Mining texts, learner productions and strategies with ReaderBench. In: Peña-Ayala, A. (ed.) Educational Data Mining. SCI, vol. 524, pp. 335–377. Springer, Cham (2014)
14. Dascalu, M., Stavarache, L.L., Dessus, P., Trausan-Matu, S., McNamara, D.S., Bianco, M.: ReaderBench: an integrated cohesion-centered framework. In: Conole, G., Klobucar, T., Rensing, C., Konert, J., Lavoué, E. (eds.) EC-TEL 2015. LNCS, vol. 9307, pp. 505–508. Springer, Heidelberg (2015). doi:10.1007/978-3-319-24258-3_47
15. National Governors Association Center for Best Practices & Council of Chief State School Officers: Common Core State Standards. Authors, Washington D.C. (2010)

16. Eggins, S., Martin, J.R.: Genres and register of discourse. In: van Dijk, T.A. (ed.) Discourse as Structure and Process (Discourse Studies – A Multidisciplinary Introduction), vol. 1, pp. 231–232. Sage Publications, London (1997)

17. Biber, D.: A textual comparison of British and American Writing. Am. Speech **62**, 99–119 (1987)

18. Rosetti, A., Cazacu, B., Onu, L.: Istoria limbii române literare. Editura Minerva, București (1971)

19. Iordan, I.: Stilistica Limbii Române. Editura Științifică, București (1975)

20. Sala, M.: De la latină la română. Limba română, vol. 1. Editura Univers Enciclopedic & Academia Română, București (1998)

21. Guțu-Romalo, V.: Aspecte ale evoluției limbii române, Vol. Repere. Editura Humanitas Educațional, București (2005)

22. Slotnick, H.: Toward a theory of computer essay grading. J. Educ. Meas. **9**(4), 253–263 (1972)

23. Wresch, W.: The imminence of grading essays by computer—25 years later. Comput. Compos. **10**(2), 45–58 (1993)

24. Nelson, J., Perfetti, C., Liben, D., Liben, M.: Measures of text difficulty: Testing their predictive value for grade levels and student performance. Council of Chief State School Officers, Washington, DC (2012)

25. Shannon, C.E.: Prediction and entropy of printed English. Bell Syst. Tech. J. **30**, 50–64 (1951)

26. Shannon, C.E.: A mathematical theory of communication. Bell Syst. Tech. J., **27**, 379–423 & 623–656 (1948)

27. Gervasi, V., Ambriola, V.: Quantitative assessment of textual complexity. In: Barbaresi, M.L. (ed.) Complexity in Language and Text, pp. 197–228. Plus, Pisa (2002)

28. McNamara, D.S., Graesser, A.C., Louwerse, M.M.: Sources of text difficulty: Across the ages and genres. In: Sabatini, J.P., Albro, E., O'Reilly, T. (eds.) Measuring up: Advances in how we assess reading ability, pp. 89–116. R&L Education, Lanham (2012)

29. van Dijk, T.A., Kintsch, W.: Strategies of Discourse Comprehension. Academic Press, New York (1983)

30. Crossley, S.A., Dascalu, M., Trausan-Matu, S., Allen, L., McNamara, D.S.: Document Cohesion Flow: Striving towards Coherence. In: 38th Annual Meeting of the Cognitive Science Society. Cognitive Science Society, Philadelphia (in press)

31. Dascalu, M., Trausan-Matu, S., McNamara, D.S., Dessus, P.: ReaderBench – automated evaluation of collaboration based on cohesion and dialogism. Int. J. Comput.-Support. Collaborative Learn. **10**(4), 395–423 (2015)

32. Dascalu, M., Dessus, P., Trausan-Matu, Ş., Bianco, M., Nardy, A.: *ReaderBench*, an environment for analyzing text complexity and reading strategies. In: Lane, H.C., Yacef, K., Mostow, J., Pavlik, P. (eds.) AIED 2013. LNCS, vol. 7926, pp. 379–388. Springer, Heidelberg (2013)

33. Trausan-Matu, S., Dascalu, M., Dessus, P.: Textual complexity and discourse structure in computer-supported collaborative learning. In: Cerri, S.A., Clancey, W.J., Papadourakis, G., Panourgia, K. (eds.) ITS 2012. LNCS, vol. 7315, pp. 352–357. Springer, Heidelberg (2012)

34. Foltz, P.W., Kintsch, W., Landauer, T.K.: An analysis of textual coherence using latent semantic indexing. In: 3rd Annual Conference of the Society for Text and Discourse, Boulder, CO (1993)

35. Landauer, T.K., Dumais, S.T.: A solution to Plato's problem: the Latent semantic analysis theory of acquisition, induction and representation of knowledge. Psychol. Rev. **104**(2), 211–240 (1997)

36. Manning, C.D., Schütze, H.: Foundations of Statistical Natural Language Processing. MIT Press, Cambridge (1999)
37. Blei, D.M., Ng, A.Y., Jordan, M.I.: Latent Dirichlet allocation. J. Mach. Learn. Res. 3(4–5), 993–1022 (2003)
38. Blei, D.M., Lafferty, J.: Topic models. In: Srivastava, A., Sahami, M. (eds.) Text Mining: Classification, Clustering, and Applications, pp. 71–93. Chapman & Hall/CRC, London (2009)
39. Golub, G.H., Kahan, W.: Calculating the singular values and pseudo-inverse of a matrix. J. Soc. Ind. Appl. Math.: Ser. B, Numer. Anal. 2(2), 205–224 (1965)
40. Kotz, S., Balakrishnan, N., Johnson, N.L.: Dirichlet and Inverted Dirichlet Distributions. Continuous Multivariate Distributions, vol. 1, Models and Applications, pp. 485–527. Wiley, New York (2000)
41. Dascalu, M., Trausan-Matu, S., Dessus, P., McNamara, D.S.: Discourse cohesion: a signature of collaboration. In: 5th International Learning Analytics & Knowledge Conference (LAK 2015), pp. 350–354. ACM, Poughkeepsie (2015)
42. Tufiş, D., Barbu Mititelu, V., Bozianu, L., Mihăilă, C.: Romanian wordnet: new developments and applications. In: 3rd Global Wordnet Conference 2006 (GWC 2006), Jeju Island, Korea, pp. 337–344 (2006)
43. Stevens, J.P.: Applied multivariate statistics for the social sciences. Lawrence Erblaum, Mahwah (2002)
44. Garson, G.D.: Multivariate GLM, MANOVA, and MANCOVA. Statistical Associates Publishing, Asheboro (2015)

A Novel Approach to Identify Factor Posing Pronunciation Disorders

Naim Terbeh[✉] and Mounir Zrigui

LaTICE Laboratory-Monastir Unit, 5000 Monastir, Tunisia
naim.terbeh@gmail.com, mounir.zrigui@fsm.rnu.tn

Abstract. Literature seems rich with approaches which are based on the features contained in the speech signal and natural language processing techniques to detect vocal pathologies in human speeches. From the literature, we can mention also that several factors (vocal pathology, non-native speaker, psychological state, age …) can pose pronunciation disorders [10]. But to our knowledge, no work has treated pathological speech to identify factor posing pronunciation disorders. The current work consists in introducing an original approach based on the forced alignment score [8] to identify the factor posing mispronunciations contained in the Arabic speech. We distinguish two main factors: the pronunciation disorders can be from native speakers with vocal pathology or from non-native speakers who do not master Arabic-phoneme pronunciation. The results are encouraging; we attain an identification rate of 95 %. Biologists and computer scientists can benefit from our proposed approach to design high performance systems of vocal pathology diagnostic.

Keywords: Pronunciation disorders · Forced alignment score · Vocal pathology · Non-native speakers

1 Introduction

Voice commands are practically used in different areas (voice services, quality control, avionics, assistance to people with disabilities, etc.) [19]. However, there are a lot of factors posing pronunciation disorders that reduce the quality of communication and falsify the vocal command to transmit. Accordingly, the intended results can be erroneous.

We distinguish four principal types of pronunciation disorders [18]:

- Suppression: This type of pronunciation defects can be expressed by the removal of one or many sounds from the correct pronunciation. In this case, the difficulty of comprehension of produced speech will increase if the suppression exceeds one sound.
- Substitution: In this type of pronunciation disorders, speakers substitute the intended sound by another. For example, to produce the sound "ت" instead of the sound "ك". This type of pronunciation disorders is more known in children speeches than in adult speeches. In this type of pronunciation defects, the comprehension of a produced speech becomes more difficult in the case of frequent substitution.

© Springer International Publishing Switzerland 2016
N.T. Nguyen et al. (Eds.): ICCCI 2016, Part I, LNAI 9875, pp. 153–162, 2016.
DOI: 10.1007/978-3-319-45243-2_14

- Distortion: This type of phonemic disorders appears when the intended sound is badly produced but the new sound is similar to the desired one. For example, the pronunciation of the sound "ت" instead of the sound "ط". This type of mispronunciation is more known in adult speeches than in children speeches. Despite the fact that this type of pronunciation defects occurs, the new pronounced sound is comprehensible in human-human communications, but it poses comprehension problems in human-machine communications.
- Addition: It is the least known type of pronunciation disorders. This type of pronunciation defects appears when the speaker adds one or more sounds to the intended pronunciation.

As speakers who suffer from pronunciation disorders are not sheltered from human-machine communication, biologists and computer scientists try to introduce new techniques to rectify mispronunciations contained in a speech signal. To apply the adequate rectifying treatment, a diagnostic step of speech, classified as pathological, is necessary. In this work, we introduce a novel methodology to identify the factor posing pronunciation disorders in the Arabic continuous speech.

This paper will be structured as follows. First, Sect. 2 is consecrated to cite some works addressing the theme of speech diagnostic from the literature. Then, Sect. 3 is dedicated to detail our proposed methodology. After that, we presented the experimental results in the Sect. 4. Finally, the conclusion and future works are drown in Sect. 5.

2   State of the Art

In the literature, there are several studies based on characteristics contained in the speech signal to detect pronunciation disorders. There are also several approaches which are based on Natural Language Processing (NLP) techniques addressing speech classification. Among these works, we cite:

- Vahid et al. proposed in [7] an original approach based on artificial neural networks [11] to detect pronunciation disorders and distinguish between healthy and pathological speeches.
- In [13], Kukharchik et al. combined between the change of wavelet characteristics [14] and support vector machines [12] to detect vocal pathologies in a speech signal.
- Plante et al. used signal processing techniques to detect phonatory disorders in the children speeches. It was also the objective of the work in [15].
- Based on the hidden Markov model [6] and the LBG algorithm [5], Vahid et al. suggested in [9] a new approach to detect vocal pathologies in a speech signal and classify speech into healthy or pathological.
- Terbeh et al. put forward in [1] an original methodology based on phonetic modeling and phonetic distance to detect pronunciation disorders and classify Arabic continuous speech into healthy or pathological.

The literature seems rich with studies addressing speech classification into healthy or pathological. But to our knowledge, no work addresses the pathological speech to identify the factor that poses this phonological disorder.

Our contribution consists in introducing an original approach to identify the factor causing the pronunciation disorders. Based on the forced alignment score, we distinguish two main factors: vocal pathologies and non-native speech. Our tests' base is formed by Arabic continuous speech classified into a pathological one.

3 Methodology

Statistics show that the number of speakers with phonemic disabilities is in a successive increase [17]. Each pronunciation defect case is different from another depending on the factor that poses this disability. To help concerned speakers to practice the human-machine communication, a speech diagnostic step is necessary to identify the factor causing this vocal problem.

Our objective is to identify the factor posing pronunciation defects contained in the Arabic continuous speech. The principal idea consists in comparing, for each non-problematic phoneme, between the forced alignment score referring to healthy native speakers and speakers who suffer from pronunciation disorders. Based on this comparison, we distinguish two main factors: a native speaker with vocal pathology or a non-native speaker who learns spoken Arabic language as L2.

Our proposed methodology can be summarized in three steps:

- Generation of a numerical model of Arabic speech
- Speech classification
- Identification of factors posing pronunciation disorders

These are going to be detailed in the following subsections.

3.1 Forced Alignment

This treatment consists in aligning the speech signals with the corresponding transcriptions and affects for each phoneme a score according to the distance that separates them from the norm, which is the acoustic model. The current treatment was been realized based on the sphinx_align tool.

3.2 Numerical Arabic-Speech Model Generation

This first task is consecrated to calculate the numerical model of the spoken Arabic language. To generate this model, we must at first calculate the forced alignment score for all Arabic phonemes by different native speakers. For this objective we use a speech base containing 1750 Arabic words recorded by five healthy native speakers (for each phoneme, we use 50 words selected by an Arabic linguistic expert). In the following procedure, we express step by step the calculation of the referenced Numerical Arabic-Speech Model (NASM):

Algorithm 1.   Calculation algorithm of NASM

For each Arabic phoneme P_i $(1 \leq i \leq 35)$

1. n words containing this phoneme are recorded by m healthy native speakers (we use 50 words recorded by five native speakers).
2. Calculate the forced alignment score (average of forced alignment scores calculated for different speakers).
3. Calculate the standard deviation between forced alignment scores of different speakers.

Two sets are generated, S and δ, which contain respectively the forced alignment scores of all Arabic phonemes and the standard deviation between the forced alignment scores. These two sets form the referenced NASM. Figure 1 presents the general form of this referenced NASM.

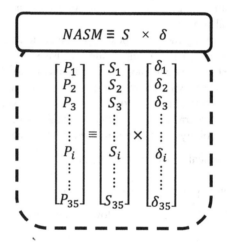

Fig. 1. General form of the referenced NASM

For each new speaker who suffers from pronunciation defects, we compare between their forced alignment score and that of the referenced NASM to determinate the factor posing the vocal disorders.

3.3   Speech Classification

As the test base is formed by pathological speeches, the current section is dedicated to Arabic speech classification into healthy or pathological. We use a speech base of 118 speech sequences, as explained in Table 1:

Table 1. Summary of corpus used in speech classification task

Speech bases	Speakers	Classification results
100 speech sequences	Native speakers (healthy and pathological)	60 are pathological
		40 are healthy
18 speech sequences	Non-native speakers	18 are pathological

In the current treatment, we utilize a speech classification system elaborated by Terbeh et al. [1]. Speech sequences classified as pathological form the test base of our proposed approach in identifying the factor posing pronunciation disorders.

3.4   Identification of Factors Posing Pronunciation Disorders

The present section is dedicated to identify whether the pronunciation disorders are caused by vocal pathology or non-native speech. For this task, we follow the next procedure:

For each new speaker who suffers from pronunciation disorders:

---

1. We generate the set $B=\{P_1, P_2, ..., P_m\}(m<35)$ of non-problematic phonemes (phonemes does not pose pronunciation disorders).
2. We suppose k the number of Arabic non-problematic phonemes P_i such as their forced alignment scores Sd_i verifies:

$$Sd_i \geq S_i\text{-}\delta_i \qquad (1)$$

We distinguish two cases:
* If $k \geq \frac{m}{2}$, then the speaker is **native** and suffers from **vocal pathology**.
* Else ($k < \frac{m}{2}$), the speaker is **non-native** who learns Arabic spoken language as L2.

Where S_i is the forced alignment score reference obtained by healthy native speakers for the phoneme P_i and δ_i is the standard deviation between the forced alignment scores of different healthy native speakers for the phoneme P_i.

3.5   General Overview of the Proposed Approach

To recapitulate, our proposed methodology in identifying the factor posing pronunciation disorders in the Arabic continuous speech can be summarized in the following algorithm:

Algorithm 2. Identification of factors posing pronunciation disorders in Arabic speech

Begin

1. Use 36 Arabic phonemes $\{P_1, P_2, ..., P_i, ..., P_{36}\}$ (35 Arabic phonemes and the phoneme SIL for the silence periods) and n native and healthy speakers.

2. Consider two sets $S=E=\emptyset$

3. For each phoneme P_i ($i=1,35$):

* Consider the set $A_i=\{S_{i1}, S_{i2}, ..., S_{ij}, ..., S_{in}\}$ containing the forced alignment scores of phoneme P_i for all n speakers (S_{ij} is the forced alignment score of phoneme P_i for the speaker n°j)

* S_i=Average(A_i)

* δ_i=Standard-deviation(A_i)

* $S=S \cup S_i$

* $E=E \cup \delta_i$

End for

4. Calculate for each new speaker with pronunciation disorders:

*B=$\{P_1, P_2, ..., P_i, ..., P_m\}$: set of non-problematic phonemes ($m<35$).

*D=$\{Sd_1, Sd_2, ..., Sd_i, ..., Sd_m\}$: set of forced alignment scores for the m non-problematic phonemes.

5. Consider k as a counter ($k=0$)

6. For each Arabic non-problematic phoneme P_i, compare between its scores S_i ($1<=i<=35$) in the set S (forced alignment scores of native speakers) and Sd_i in the set D (forced alignment scores proper to speaker with pronunciation disorders):

* if $Sd_i \geq S_i-\delta_i$ then $k=k+1$

End for

7. Distinguish two cases:

* If $\left|{}^k\!/_m\right|$=1 then the speaker is native and suffers from vocal pathology

* Else, the speaker is non-native who learns Arabic spoken language as L2

End

---

4   Tests and Experimental Results

4.1   Test Conditions

Tests are realized in the following conditions:

- We use 36 Arabic phonemes as noted in [2] (including the phoneme SIL; for silence periods.
- For each Arabic phoneme, 50 words were selected by an Arabic linguistic expert and recorded by five different healthy native speakers. This base was used to calculate the reference forced alignment scores of Arabic phonemes.
- For each Arabic phoneme:
  - * Selected words were recorded by five different healthy native speakers and we generated a forced alignment score for everyone.
  - * The reference forced alignment score is the average of these five scores.

- The reference model (NASM) is the combination of two vectors arranging respectively the forced alignment scores of all 35 Arabic phonemes and the standard deviations between the forced alignment scores calculate by different speakers.
- In the speech classification task (healthy or pathological speech), we use the system elaborated by Terbeh et al. [1].
- The test base contains 78 Arabic speech sequences classified as pathological using the classification system realized by Terbeh et al. [1] (see Table 1):

  * 60 were recorded by native speakers who suffer from vocal pathologies.
  * 18 were recorded by non-native speakers who learn the spoken Arabic language as L2 (in different learning levels).

4.2 Experimental Results

The forced alignment score reflects the mastery level of the phonemes' pronunciation. In Fig. 2, we present an extract from the forced alignment scores obtained for Arabic phonemes by native and healthy speakers.

The task of identification of factors posing pronunciation disorders can be summarized in Table 2:

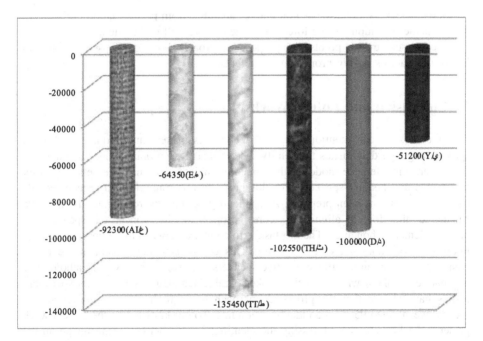

Fig. 2. An extract from the forced alignment scores obtained for Arabic phonemes: forced alignment score was calculated by five healthy native speakers

Table 2. Identification results of factors posing pronunciation disorders

Speech bases	Results	Identification rate	
60 speech sequences recorded by native speakers with vocal pathologies	-57 native speakers with vocal pathology	95 %	**94.87 %**
	-3 non-native speakers		
18 speech sequences recorded by non-native speakers	-one native speaker with vocal pathology	94.44 %	
	-17 non-native speakers		

4.3   Discussion

The results in the second table show that one among non-native speakers was identified as a native speaker who suffers from vocal pathology. This misidentification can be explained by the fact that:

- This speaker resides for a long-time in an Arabic country, which has allowed him to master most of the Arabic phonemes.
- He is a learner of spoken Arabic language in an advanced level and he masters most of the Arabic phonemes.

Native speakers identified as non-native can suffer from pronunciation problems in the phonemes duration, so the forced alignment score will be falsified.

Except the previous points, our proposed approach shows a high performance in identifying factors posing pronunciation defects.

5   Conclusion and Future Works

To conclude, we can mention that NLP techniques present an adequate method in spoken language diagnostics to identify factors that pose pronunciation disorders. In this paper, our reference model of the forced alignment score has been elaborated by five native and healthy speakers. For this purpose, a corpus of 1750 speech records (isolated words) has been prepared. After identifying the mispronunciations, we have calculated the forced alignment scores proper to the concerned speaker for non-problematic phonemes. The last task is to compare between these previous scores: reference score and that of the speaker. This comparison has generated the factor posing pronunciation disorders: a native speakers who suffer from vocal pathology or non-native speakers who learn the spoken Arabic language as L2. The experiment results have shown that the proposed approach has a high identification accuracy. Indeed, we attained 95 % as an identification rate. To our knowledge, the current work presents the first attempt addressing the identification of factors posing mispronunciations in the Arabic continuous speech. Thanks to these encouraging results, our suggested approach can be a reference for future works focalizing on pathological speech diagnostic.

As future works, we can mention that these results can be used as a phonetic imprint in security applications, especially to identify the speaker nationality (native or non-native) in phone communications.

It may be possible to benefit from this work to elaborate an automatic speech correction system for people suffering from pronunciation disorders [3, 4] and assist non-native speakers to master the spoken Arabic language [16].

We can also profit from our proposed methodology to elaborate applications of automatic speaker and dialect identification [20].

Acknowledgments. We would like to benefit from this opportunity to express my deepest regards to all members of the evaluation research committee in the ICCCI scientific conference. We would like also to extend our advance thanks to Mr. Mounir ZRIGUI for his valuable advices and encouragement.

References

1. Terbeh, N., Maraoui, M., Zrigui, M.: Probabilistic approach for detection of vocal pathologies in the Arabic speech. In: Gelbukh, A. (ed.). LNCS, vol. 9042, pp. 606–616. Springer, Heidelberg (2015)
2. Alghamdi, M., Almuhtasib, H., Elshafei, M.: Arabic phonological rules. King Saud Univ. J. Comput. Sci. Inf. **16**, 1–25 (2004)
3. Terbeh, N., Labidi, M., Zrigui, M.: Automatic speech correction: a step to speech recognition for people with disabilities. In: ICTA 2013, Hammamet-Tunisia, 23–26 October 2013 (2013)
4. Terbeh, N., Zrigui, M.: Vers la Correction Automatique de la Parole Arabe. In: Citala 2014, Oujda-Morocco, 26–27 November 2014 (2014)
5. Patane, G., Russo, M.: The enhanced LBG algorithm. Neural Netw. **14**(9), 1219–1237 (2001)
6. Bréhilin, L., Gascuel, O.: Modèles de Markov caches et apprentissage de sequences
7. Majidnezhad, V., Kheidorov, I.: An ANN-based method for detecting vocal fold pathology. Int. J. Comput. Appl. **62**(7), 1–4 (2013)
8. Jurafsky, D., Ward, W., Zhang, B., Herold, K., Yu, X., Zhang, S.: What kind of pronunciation variation is hard for triphones to model? In: ICASSP 2001, Salt Lake City, UT, 7–11 May 2001
9. Majidnezhad, V., Kheidorov, I.: A HMM-based method for vocal fold pathology diagnosis. IJCSI Int. J. Comput. Sci. Issues **9**(6), 135–138 (2012). No. 2
10. Kim, J., Kumar, N., Tsiartas, A., Li, M., Narayanan, S.: Intelligibility classification of pathological speech using fusion of multiple subsystems. In: Proceedings of Interspeech, Portland, Oregon, USA, pp. 534–537 (2012)
11. Paquet, P.: L'utilisation des réseaux de neurones artificiels en finance. Document de recherche n° 1997-1 (1997)
12. Archaux, C., Laanaya, H., Martin, A., Khenchaf, A.: An SVM based churn detector in prepaid mobile telephony (2004)
13. Kukharchik, P., Martynov, D., Kheidorov, I., Kotov, O.: Vocal fold pathology detection using modified wavelet-like features and support vector machines. In: 15th European Signal Processing Conference (EUSIPCO 2007), Poznan, Poland, 3–7 September 2007

14. Damerval, C.: Ondelettes pour la détection de caractéristiques en traitement d'images. Doctoral thesis, Mai 2008
15. Plante, F., Christian, B.-V.: Détection acoustique des pathologies phonatoires chez l'enfant. Doctoral thesis (1993)
16. Terbeh, N., Zrigui, M.: Vocal pathologies detection and mispronounced phonemes identification: case of Arabic continuous speech. In: LREC 2016, Portorož-Slovenia, 23–28 May 2016 (2016)
17. http://www.un.org/french/disabilities/default.asp?navid=35&pid=833, [consulted 6 April 2016]
18. http://kenanaonline.com/users/dkkhaledelnagar/photos/1238136361, [consulted 24 April 2016]
19. Blanc-Brude, T.: Intégration de commandes vocales dans un environnement d'apprentissage par l'action: enjeux ergonomiques. Doctoral dissertation, Grenoble 1 (2004)
20. Biadsy, F., Hirschberg, J., Habash, N.: Spoken Arabic dialect identification using phonotactic modeling. In: Proceedings of the EACL 2009 Workshop on Computational Approaches to Semitic Languages, pp. 53–61. Association for Computational Linguistics (2009)

Towards an Automatic Intention Recognition from Client Request

Noura Labidi[1](✉), Tarak Chaari[2], and Rafik Bouaziz[1]

[1] MIR@CL Laboratory, FSEGS, B.P. 1088, 3018 Sfax, Tunisia
noura.labidi@yahoo.fr, rafik.bouaziz@usf.tn
[2] ReDCAD Laboratory, University of Sfax, B.P. 1173, 3038 Sfax, Tunisia
tarak.chaari@redcad.org

Abstract. Nowadays, the relentless growth of the IT (Information Technology) market and the evolution of Service-oriented architectures (SOA) make the establishment of Service Level Agreements (SLA) between providers and clients a complex task. In fact, clients find many IT offers with complex terms especially if they do not share the same technical knowledge with providers. These latter have to well understand clients' requirements in order to be able to properly address their needs. In this context, ontologies can help in bridging the gap between provider's offers and client's needs. In this paper, we define firstly an ontology structure that models clients' intentions. Furthermore, we propose an approach for intention recognition from textual request written in English to automatically populate the intention ontology structure. An illustrative case is finally presented to prove the accurate performance of our proposed approach.

Keywords: Text analysis · Intentional structure · Ontology population · Term extraction · Knowledge modeling

1 Introduction

With regard to the complexity of Information Systems (IS), nowadays, SOA prove the optimal solution for a multitude of companies. However, the guarantee of performance in the SOA is essential to protect customers' rights against their suppliers. SLA are the solution supposed to define these guarantees [4]. Nevertheless, to reach such an agreement, several problems must be resolved. In fact, SLA has become increasingly complex with the large increase in the number of providers and their offers. Then, understanding these contracts is very difficult for customers who may not understand the providers' offers. So they may use their own language and own knowledge to express freely their request which is different from the provider's technical language. In that case, providers may not understand the client's intention. But, such an understanding is essential to provide better products and services to consumers.

Overall, facilitating the understanding between providers and customers requires a knowledge representation of client's intentions. Thus, there is a great

© Springer International Publishing Switzerland 2016
N.T. Nguyen et al. (Eds.): ICCCI 2016, Part I, LNAI 9875, pp. 163–172, 2016.
DOI: 10.1007/978-3-319-45243-2_15

need to an intelligent system able to represent, understand and interpret the client's intentions from its textual request targeting a specific provider product. However, the development of such a system is a long task which often requires linguistics knowledge and the use of automatic natural language processing techniques.

To the best of our knowledge, there is still no reported study of this problem. In this paper, we define a structure of client intention ontology and we present an approach that analyses and understands automatically the client request. Indeed, the proposed approach extracts the client's intention and builds an intention in accordance with the intention ontology structure. The proposed recognition approach is based on a set of linguistic rules. These rules facilitate the building of intentional structure which is then loaded as instance of client intention ontology. The originality of the proposed approach is, on the one hand, that the entire process of the extraction and population approach is made in an automatic and a semantic way. On the other hand, it can be applied both on a short and a long text.

The remainder of this paper is structured as follows. We begin first of all, by the related works. Section 3 presents the proposed ontology structure to model clients' intentions. Then, Sect. 4 exposes the basics of our approach for an automatic intention recognition from client textual request and gives details on the corresponding modules. Section 5 presents an illustrative case. Finally, the last section concludes the paper and deals with future works.

2   Related Works

Writing is an intentional action. This is why the reader must understand the transmitted information or knowledge by the writer. Intention recognition is the process which tackles this issue effectively. The majority of works tackling the intention recognition aims to improve online services for users of web service, but are differentiated by techniques used in their approaches of intention extraction.

In [11,12], the authors proposed a goal-driven intention extraction approach which extracts user intention from terms in web service query. These works aim to satisfy users intention by knowing which goal should be achieved. For that, authors use heterogeneous information as inputs such as user profile as background knowledge, domain ontology and goal structure in a multi agent architecture named *BDIAgent* (Believe-Desire-Intention). Furthermore, in the context of Web service, authors in [10] propose a quality and context awareness intention web service ontology. Some related methods are also proposed in [7,13]. They also treat the user request but otherwise; they classify a query submitted to a search engine to determine the semantic intent of Web queries. The works cited above are different from our issue because their input is user-submitted keyword queries which do not exceed 3 or 4 words. So, the extraction process is obvious.

In the same issue of intention classification but of social media posts, an approach based on intent classification from a paragraph written by a user of the forum is proposed in [2]. The authors focus on identifying intention from posts in online discussion forums; they propose a new transfer learning method (Co-class). The classifier used is based on labeled data in some domains and applies it to a new domain without labeling any training data. Authors in that work are interested only in the intention to buy, which restricts the extraction process. On the other hand, the labeling process is a manual process which is so hard, time consuming and expensive.

In order to improve the performance of an information research system, the authors in [8] use learning techniques and statistical methods to identify the intentions of the authors of scientific publications in the field of computer science. The authors built a recognition system called $RICAD$ [6] based on a semi-automatic method to extract intentional information from a domain specific corpus. The output of this method intends to enrich a knowledge base by new detected intentions. The final result of this approach is ontology of intentions. The authors of this work consider the intention composed of goal, mean and reason but they do not explicit the intention' components in the ontology proposed. In a more advanced technology, the authors in [1] built an *Intention Insider* system architecture which is intended to be a cloud service. The tool used natural language processing (NLP) and text mining techniques [3] to extract intention from online forums to discover people's intentions in the social channel. The authors of this work were based on intention templates for extracting intentions of members of the forums without defining a specific intentional structure.

Our analysis of the state of the art led us to conclude that each work cited has both advantages and weaknesses: in some contributions such as in [1,9] many manual interventions are needed which is laborious, tedious and long. Therefore, an automation of intention recognition process is needed. *BDI Agent* is an automatic system for detecting and extracting intentions. However, it needs to know the user's profile to detect its intention. Such information may not be available if a client simply expresses his needs by writing a free text without leaving traces on its own characteristics. In addition, in the solutions proposed in [1,11,12], the authors are interested only in one component of the intention (the goal or the object) and they do not define a structure for the intention which enables a more understanding for the client' request. Contrariwise, the author in [9] defines a structure for intention and specifies its components. But, the proposed system ($RICAD$) is not able to provide the components of the intention extracted explicitly which requires an additional effort and time.

3   The Proposed Ontology Structure to Model Clients' Intentions

In our context of work, we hypothesized that the client expresses his needs freely with an informal text. Our objective is to extract the client's intention from his request by transforming automatically the client's expression in an intentional structure. To deal with this issue, we started by building a structure for the intention. We defined an ontology for client intention which describes the client intention model. We aimed to load the information extracted from the client's request as instance in the intention ontology.

To identify intention components, the authors of [6] define an intention as I (A, G*, M*, R*) where: I represents the intention carried out by an action A; this action A expresses what the author of the intention wants to do, G represents the goal to achieve by performing the action, M represents the means to express how the action is accomplished and R represents the reason why the author chooses this action. The symbol * indicates that the number of acts composing the intention can be in the interval [0, N]. This intentional model has been enriched by the authors in [5] by adding specific elements for service level agreements negotiation. They built the *ClientOnto* ontology. In the same perspective, we consider that to improve understanding what the client meant, we should divide the intention into fragments and each part has its own meaning. In order to achieve this, we created a corpus of hundreds of textual consumers' requests extracted from forums. On the basis of this corpus, we created an ontology structure to model the clients' intentions. The proposed ontology structure illustrated as a class diagram according to the UML standard[1], is presented in Fig. 1. We consider that an intention is composed of an action, a subject, goals and constraints. Indeed, an action concerns one subject. A subject can be under constraints. Each constraint is composed of a property, an operator and a threshold. The action performed is to achieve 0 or many goals.

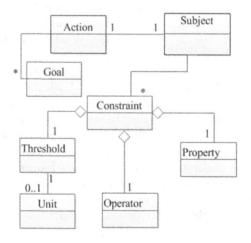

Fig. 1. Intention ontology structure

---

[1] http://www.omg.org/spec/UML/.

4   An Overview of the Approach for Automatic Intention Recognition from Client' Request

In this section, we present in detail the overall architecture of our approach which is based on the automated process described in Fig. 2. Our input is a client request represented as an informal text. The process of recognition and extraction of intention from this text involves three modules. The first module is an initialization one. It consists of the extraction of semantic components from the input text. The second module consists in searching the corresponding concepts (Action, Subject, Goal, Property, Operator, Threshold or Unit) in the intention ontology for each component derived from the first step. In the last module, we automatically add relationships between the concepts' instances in order to populate the ontology with the new instance of intention created.

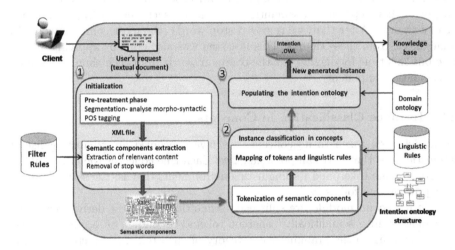

Fig. 2. The architecture of our intention recognition approach

In the following sub sections, we give more details about the three modules of our approach.

4.1   Initialization

Our goal in this module consists in extracting from the input text any relevant content helping to understand the intention of the client. The steps are described below:

Pre-treatment phase: We start with a pre-processing data which sets up the input text to the other sub-sequent operations. Each text introduced is then segmented in order to do a morphological and syntactic analysis (morpho-syntactic analysis). We use for that purpose a natural language processing method which is *Parts-of-Speech* tagging (PoS) (each word is labeled with its PoS such as noun, verb, adverb, etc.). The resulting output is a segmented text by word and by chunk (a chunk is a technical term used to refer to a piece of text) tagged by the mean of the analyzer *Treetagger Chunker*. The latter generates as output an XML file that contains the result of segmentation and tagging step.

Semantic components extraction: On the basis of POS tagging, we defined few filter rules to extract the relevant content from the client's request. This step is achieved by developing a shallow parser. The latter parses the tagged output from the first step chunk by chunk and word by word, and applies filter rules to extract the relevant content according to its chunk tag and the grammatical category (verb, singular noun, etc.) of the words composing the chunk. We maintain just Noun Chunks (NC), Verb Complex (VC), specifically those composed of infinitive verb, ADJeCtive chunks (ADJC) and Prepositional Chunk (PC) (starting by the preposition IN followed by a Proper Noun NP). We obtained as output of this step a list of semantic components; each one is composed of one or more words. After that, we removed stop words (the most common words in a language) from these chunks. This last step was also done by applying a filter rule but this time by digging out the common words from each chunk conserved as semantic component.

4.2   Instance Classification in Concepts

To attach each token to its concept, we have conceived an exhausting list of linguistic rules that cover all possible cases. Each rule concerns the grammatical category of the token and its linguistic environment, that is, the series of units that precede and follow it.

In this second module, we split each semantic component derived from the first module into semantically coherent tokens. Secondly, we apply linguistic rules according to the grammatical category of each token (extracted from the POS attribute). We use in these rules a set of tags relative to the *Treetagger Tag Set*[2]. By applying the suitable rule for the token, we can associate it the corresponding concept from the intention ontology. We present in Table 1 some of these rules. Additional other cases are treated and a list of other linguistic rules is identified similarly, however we omit them in this paper for space limits.

4.3   Populating the Intention Ontology

We intend, in this step, complete the population of the ontology by the new intention instance. As input of this step, we have instances of concepts. In order to maintain the coherence of the ontology structure defined previously, it remains to add some other instances and the relationships between instances of concepts involved in the ontology. We start by creating a relationship between the Action instance and the Subject instance. Then, we create instances of associations

---

[2] https://courses.washington.edu/hypertxt/csar-v02/penntable.html.

Table 1. An extract of linguistic rules for each grammatical category

POS	Rule	Cases	Description	Comments
To+VV	R1	To+VV	If the token t is a verb in the base form (VV) then t is an instance of the concept ACTION	As an action is usually expressed by a verb.
	R2	VV NN/ VV DT NN/ VV DT JJ NN/ VV DT JJ JJ NN	If the token t is a singular or masculine noun (NN), and is preceded at the position p-i by the action verb (i ∈ {1, 2, 3, 4}) and is not followed by a singular (NN) or plural noun (NNS) then t is an instance of the concept SUBJECT	The first occurrence of a noun which follows the action of the intention represents the subject of the intention.
NN	R3	PPS NN/NN IN	If (!R2) and if t is preceded by a possessive pronoun (PPS) or followed by one of these prepositions (IN) "for, around" then t is an instance of the concept PROPERTY	A noun that is not the subject of the intention is one of the properties representing constraints on the subject.
	R4	NN NN	If (!R3) and if t is followed by a singular noun then t concatenated with this noun forms an instance of the concept PROPERTY	A noun followed by another noun constitutes the same instance for the concept PROPERTY.
JJ	R5	JJ	If the token t is an adjective (JJ), then t is an instance of the concept THRESHOLD, except in rare cases	An adjective represents an approximate value for a property which is considered as a threshold.
CD	R6	CD	If the token t is a cardinal number (CD), then t is an instance of the concept THRESHOLD	A threshold is commonly represented by a cardinal number.
RB	R7	RB ~~JJ~~	If t is an adverb (RB) and is not followed by an adjective (e.g. very good) then t is an instance of the concept OPERATOR	An adverb (different from not) ensures the relationship between a property and a threshold.
NNS	R8	CD NNS	If the token t is a plural noun (NNS) and preceded by a cardinal number, then t is an instance of the concept UNIT	A unit usually follows the amount

between Action instance and Goal instances, as many as there are instances in Goals. In the other hand, each Unit instance is linked with its nearest Threshold instance.

Given that in a client request a number of constraints are expressed implicitly; we create a new Constraint instance to each specific combination of three instances of the concepts Property, Operator and Threshold. Then, we should link each created Constraint instance with each instance of its components. Here, we choose the more nearest combination of instances: we create a relationship between a Constraint instance and an Operator instance and its nearest

Threshold and Property instances (in the input text). At the end, each Constraint instance is linked to the Subject instance.

In many cases, we could found lack of information in the client text; the latter may not contain enough information to create a constraint. In this case, we add default values by acceding to a domain ontology to find a suitable values corresponding to other instances of the constraint components.

5   Illustrative Case

All along the present study, many experiments have been fulfilled to evaluate the applicability and the feasibility of our proposed approach of client intention recognition. In this section, we consider an illustrative case which belongs to a client's request.

In the remainder of this section, we will show the results of applying the different steps of our approach to the selected example.

Assume that the client wants to buy a cell and inputs the following request expression: *"I want to buy a new cell. I want it with a good screen resolution. Budget around 25000 dollars."*

Considering the intention ontology structure, the terms that will form the instance of intention are those outlined. An object diagram for this example is given in Fig. 3 to show clearly what we should have as a result after processing the different modules of the approach.

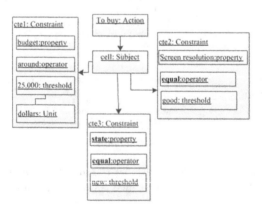

Fig. 3. Object diagram relative to the client request example

In the first step of the first module, after analyzing the input text with the *Treetagger-chunker*, the analyzer generates an XML file that contains the results of segmentation accompanied by POS tags and chunk tags. The second step applies filter rules. Thus, we obtain the following chunks: *I, to buy, a new cell,*

I, it, a good screen resolution, budget, around, 25.000 dollars. After removing stop words, we obtain: *to buy, new cell, good screen resolution, budget, around, 25.000 dollars.*

In the second phase of our proposed approach, we do the tokenization for the chunks and then we apply linguistic rules to match each token with the corresponding concept of the ontology. Based on the PoS of each token, we choose the category of linguistic rules to be applied. The result of this step is presented in Table 2. Then, we populate the intention ontology with the resulting instances.

Table 2. Matching tokens to ontology concepts

Token	POS	Applied rule	Adequate concept
to buy	To+VV	(R1)	ACTION
new	JJ	(R5)	THRESHOLD
cell	NN	(R2)	SUBJECT
good	JJ	(R5)	THRESHOLD
screen	NN	(R4)	PROPERTY
resolution	NN		
budget	NN	(R3)	PROPERTY
around	RB	(R7)	OPERATOR
25.000	CD	(R6)	THRESHOLD
dollars	NNS	(R8)	UNIT

The last step consists in creating relationships between instances of concepts (in conformity with the ontology structure) and creating constraint instances. In the case of a lack of information to create some constraints, we add some concept instances (e.g. adding the property "state" and the operator "equal" as suitable values for the threshold "new" for the constraint cte3). We obtain, at the end, a full instance for the intention ontology in conformity with Fig. 3.

6   Conclusion

In this paper, we introduced an automatic approach for client intention extraction. We believe that having a structure for the intention makes the understanding and the representation of client' requirements and needs clear and easy. In this context, we have defined an approach that picks out relevant information from an informal text which is input by a client after analyzing the text using *Treetagger Chunker* and matching each term to its suitable concept in the intention ontology. We have developed a Java prototype that implements all the steps of the approach including the linguistic rules. Currently, we continue testing our tool on other case studies and we are working on a deep experimentation by analyzing many clients' requests extracted from online forums (e.g. *Tom's Guide*[3]) to evaluate the prototype. As future perspectives to our work, we plan to resolve any ambiguities derived from each step and causing conflicts in other steps of the approach and to use richer formal languages to represent intention ontologies.

References

1. Castellanos, M., Hsu, M., Dayal, U., Ghosh, R., Dekhil, M., Ceja, C., Puchi, M., Ruiz, P.: Intention insider: discovering people's intentions in the social channel. In: Proceedings of the 15th International Conference on Extending Database Technology, pp. 614–617. ACM (2012)
2. Chen, Z., Liu, B., Hsu, M., Castellanos, M., Ghosh, R.: Identifying intention posts in discussion forums. In: HLT-NAACL, pp. 1041–1050 (2013)
3. Ding, X., Liu, B., Zhang, L.: Entity discovery and assignment for opinion mining applications. In: Proceedings of the 15th ACM SIGKDD International Conference on Knowledge Discovery and Data Mining, pp. 1125–1134. ACM (2009)
4. Fakhfakh, K.: Approche sémantique basée sur les intentions pour la modélisation, la négociation et la surveillance des contrats de qualité de service. Ph.D. thesis, Université des Sciences Sociales-Toulouse I (2011)
5. Fakhfakh, K., Tazi, S., Drira, K., Chaari, T., Jmaiel, M.: Enhancing client intentions analysis for service level agreements establishment assistance. In: 2010 Third International Conference on Communication Theory, Reliability, and Quality of Service (CTRQ), pp. 237–242. IEEE (2010)
6. Hassan, K., Ali, E., Chantal, S.d., Said, T.: Ontointention: an ontology for documents intentions. In: Second International Conference on Research Challenges in Information Science, RCIS 2008, pp. 301–306. IEEE (2008)
7. Hu, D.H., Shen, D., Sun, J.-T., Yang, Q., Chen, Z.: Context-aware online commercial intention detection. In: Zhou, Z.-H., Washio, T. (eds.) ACML 2009. LNCS, vol. 5828, pp. 135–149. Springer, Heidelberg (2009)
8. Kanso, H., Elhore, A., Soule-Dupuy, C., Tazi, S.: Recognition and extraction of intentions based on ontology. In: 3rd International Conference on Information and Communication Technologies: From Theory to Applications, ICTTA 2008, pp. 1–5. IEEE (2008)
9. Kanso, H.: Vers la reconnaissance des intentions de communication: application au contenu de publications scientifiques. Ph.D. thesis, Toulouse (2009)
10. Khanfir, E., Ben Djmeaa, R., Amous, I.: Quality and context awareness intention web service ontology. In: 2015 IEEE World Congress on Services (SERVICES), pp. 121–125. IEEE (2015)
11. Lee, C.H., Cheng, Y.R., Liu, A.: Aiding user intention satisfaction with case-based reasoning in ATIS applications. In: IEEE International Conference on Systems, Man and Cybernetics, SMC 2006, vol. 1, pp. 440–445. IEEE (2006)
12. Lee, C.H.L.: Toward intention-aware services provision. In: 2007 IEEE Region 10 Conference, TENCON 2007, pp. 1–4. IEEE (2007)
13. Li, X.: Understanding the semantic structure of noun phrase queries. In: Proceedings of the 48th Annual Meeting of the Association for Computational Linguistics, pp. 1337–1345. Association for Computational Linguistics (2010)

---

[3] http://www.tomshardware.fr/forum/.

Data Mining Methods and Applications

Fuzzy Cognitive Maps for Long-Term Prognosis of the Evolution of Atmospheric Pollution, Based on Climate Change Scenarios: The Case of Athens

Vardis-Dimitris Anezakis[1], Konstantinos Dermetzis[1(✉)],
Lazaros Iliadis[1], and Stefanos Spartalis[2]

[1] Lab of Forest-Environmental Informatics and Computational Intelligence,
Department of Forestry and Management of the Environment
and Natural Resources, Democritus University of Thrace,
193 Pandazidou Street, 68200 N Orestiada, Greece
{danezaki, kdemertz, liliadis}@fmenr.duth.gr
[2] Laboratory of Computational Mathematics, Department of Production
and Management Engineering, School of Engineering, Democritus University
of Thrace, V. Sofias 12, Prokat, Building A1, 67100 Xanthi, Greece
sspart@pme.duth.gr

Abstract. Air pollution is related to the concentration of harmful substances in the lower layers of the atmosphere and it is one of the most serious problems threatening the modern way of life. Determination of the conditions that cause maximization of the problem and assessment of the catalytic effect of relative humidity and temperature are important research subjects in the evaluation of environmental risk. This research effort describes an innovative model towards the forecasting of both primary and secondary air pollutants in the center of Athens, by employing Soft Computing Techniques. More specifically, Fuzzy Cognitive Maps are used to analyze the conditions and to correlate the factors contributing to air pollution. According to the climate change scenarios till 2100, there is going to be a serious fluctuation of the average temperature and rainfall in a global scale. This modeling effort aims in forecasting the evolution of the air pollutants concentrations in Athens as a consequence of the upcoming climate change.

Keywords: Fuzzy Cognitive Maps · Air pollutants · Climate change models · Soft Computing Techniques

1 Introduction

Air pollution is the condition in which air is contaminated by foreign substances, radiation or other forms of energy, in quantities or duration that can have harmful or poisonous effects in the health of living organisms. Moreover, they can upset the ecological balance in large or small geographical scale. These pollutants are either emitted directly by various human activities, like energy production from solid or liquid fuels, transport, industry and heating and they are known as primary ones

© Springer International Publishing Switzerland 2016
N.T. Nguyen et al. (Eds.): ICCCI 2016, Part I, LNAI 9875, pp. 175–186, 2016.
DOI: 10.1007/978-3-319-45243-2_16

(e.g. CO, NO, NO_2, SO_2) or they are formed in the atmosphere under proper conditions and they are known as secondary ones (e.g. O_3). The assessment of air pollution consequences requires a comprehensive spatiotemporal analysis of the favoring conditions and the search for interrelations between pollutants and meteorological plus photochemical factors affecting it [2–4]. This research effort proposes an innovative analytical Soft Computing model towards long term estimation of the pollutants concentrations. More specifically it carries out a descriptive representation of complex correlations between atmospheric pollutants with the method of Fuzzy Cognitive Maps (FCM), based on historical data of the Athens center for the period 2000–2012. Additionally, it proposes a sophisticated method of predicting the values of pollutants in relation to the fluctuation of temperature and rainfall values, as reflected by the number of projections of climate model GFDL_CM2.0 for the period 2020–2099.

1.1 Related Literature - Data

Important research efforts have been carried out on the short term, towards forecasting of the air quality in medium cities or major urban centers like Athens, using statistical methods and without taking into serious consideration the effect of the fluctuation of temperature and precipitation in the 21st century due to climate change [15].

Gordaliza et al. [6] developed coherent storylines about ordinary people living under diverse scenarios of low/high CO_2. Luiz et al. [12] constructed FCM in order to understand the viability of Clean Development Mechanism (CDM) projects in South Africa and how they would influence greenhouse gas (GHG) emissions. Zhang et al. [20] explored the application of fuzzy cognitive maps on getting stakeholders' perspectives and they employed graph theory indices on quantifying them. Pathinathan and Ponnivalavan [16] analyzed the hazards of plastic pollution using Induced Fuzzy Cognitive Maps (IFCMs). IFCMs are a fuzzy-graph modeling approach based on expert's opinion. Amer et al. [1] developed three future scenarios using FCM for the national wind energy sector of a developing country. Mesa-Frias et al. [13] developed a novel method based on FCM to quantify the framing assumptions in the assessment of health impact. Fons et al. [5] proposed a model of an eco-industrial park and used FCM to analyze the impacts of this model in terms of pollution and waste disposal.

The motivation for this research was the development of a rational system, capable of analyzing effectively the actual conditions during incidents of high air pollution in urban areas and also to model the long term evolution of emissions due to climate change. FCM were developed based exclusively on measurable factors resulting from the correlation analysis of the actual data and not on the opinion of experts as usually. The above perspectives add validity and reliability in the overall inference process. Moreover, the development and use of a fuzzy system to forecast future values of air pollutants based on climate change scenarios is an interesting innovation that significantly improves the quality and value of the proposed model. The design, development and testing of this system are described herein. The topography of Athens prevents the diffusion of pollutants. The pollution data come from the "Patissia" area in the center of Athens. Relativity data analysis with FCM was performed for the air pollution measuring station of "Patissia" for the period 2000 to 2012, in order to obtain a symbolic representation of existing complex correlations between the atmospheric pollutants. The above measuring station (who is distinguished for its consistency and reliability as a few missing values were observed) is storing hourly values of CO (in mg/m^3), (NO, NO$_2$, O$_3$ and SO$_2$ in μg/m^3). During a day the station of Patissia is full of traffic and this is the reason of high values of air pollution concentrations. Additionally, records related to six meteorological factors namely: air temperature (Temp), relative humidity RH), air pressure(PR), solar radiation (SR), wind speed (WS) and wind direction (WD) were obtained from the station of "Thiseion" 9 km far from the sea.

During data pre-processing all the records with missing values for one or more parameters were removed from the dataset. Outliers are very important as they are always considered for the activation of the civil protection mechanisms. For this reason, they were not removed from the datasets in order to obtain representative training samples offering potential generalization in future forecasting models. Finally in order to tackle the problem of features with different range, in which the higher values most affect the cost function with respect to the characteristics of the smaller ones, without being more important, a normalization process was performed in the interval [−1, +1].

2   Theoretical Framework and Methodology

2.1   Correlation Analysis

In order to test the level of linear relationship between meteorological parameters and air pollutants, the typical relativity analysis was performed, using the parametric correlation coefficient of Pearson (r). The Pearson linear correlation coefficient between two parameters X and Y is defined based on a sample of n pairs of observations (x_i, y_i) $i = 1, 2, \ldots, n$, and it is denoted as $r(X, Y)$ or more briefly as r. The variables $\bar{x}$ and $\bar{y}$ are the averages of (xi, yi). The r is the covariance (CovX,Y) of the two variables divided by the product of their standard deviations (sx,sy). It is given by the following Eq. 1:

$$r = \frac{s_{xy}}{s_x s_y} = \frac{\sum_{i=1}^{v}(x_i - \bar{x})(y_i - \bar{y})}{\sqrt{\sum_{i=1}^{v}(x_i - \bar{x})^2}\sqrt{\sum_{i=1}^{v}(y_i - \bar{y})^2}} = \frac{\sum_{i=1}^{v} x_i y_i - v\bar{x}\bar{y}}{\sqrt{\sum_{i=1}^{v} x_i^2 - v\bar{x}^2}\sqrt{\sum_{i=1}^{v} y_i^2 - v\bar{y}^2}} \quad (1)$$

The correlation coefficient is a pure number in the interval [−1, 1]. More specifically, when $0 < r \leq 1$, then X, Y are linearly positively correlated and when $-1 < r < 0$, then X, Y are negatively correlated. When $r = 0$ or close to zero there is no correlation between them.

2.2   Fuzzy Cognitive Maps

FCM are fuzzy-graph structures. In the model of a fuzzy cognitive map, the nodes are linked together by edges and each edge connecting two nodes describes the change in the activation value. The direction of the edge implies which node affects the other.

The sign of the causality relationship is positive if there is a direct influence, negative if there is an inverse influence relation and zero if the two nodes are uncorrelated. The causal relationships are described by the use of fuzzy linguistics and they are fuzzified by using membership functions taking values in the closed interval [−1, 1] [14, 17, 18].

Unlike the majority of complex dynamic systems, characterized by nonlinearity and high uncertainty, the fuzzy cognitive maps use advanced learning techniques in order to choose appropriate weights for the causal connections between the examined variables. This is done in order to reflect the examined problem with absolute realism.

Combining the theoretical background of fuzzy logic, FCM cover the comparison and characterization purpose of the reference sets, towards modeling and solving complex problems for which there is no structured mathematical model.

2.3   GFDL_CM 2.0

Flato et al. (2013) [21] finds robust relationship between the ability of the GFDL_CM2.0 model to represent interannual variability of near-surface air temperature and the amplitude of future warming. Thus, the GFDL_CM2.0 climate change model has been chosen as the most suitable for this research area. Also this model generally provides a smooth gradual temperature rise up to $2.9°$ C plus reduced rainfall of −245.96 mm up to 2099. The GFDL-CM2.0 is a coupled Atmospheric - Ocean general circulation model, developed at NOAA's Geophysical Fluid Dynamics Laboratory (GFDL). It is divided into four modules: the atmosphere model (AM2P13), the ocean (OM3P4), the dry (LM2) and the ice cover (SIS). The horizontal resolution of the atmospheric and land model is $2°$ latitude $\times$ $2.5°$ longitude whereas for the oceanic it is $1° \times 1°$. The atmospheric model has twenty-four vertical planes whereas the land model has eighteen for the heat storage and the oceanic fifty. The atmospheric model uses a B-grid dynamical core, a k-profile planetary boundary layer scheme [11] and a simple local parameterization of the vertical momentum transport by cumulus convection. The land-cover-type distribution is a combination of a potential natural vegetation of type one and a historical land use distribution dataset. GFDL-CM2.0 uses explicit fresh water fluxes to simulate the exchange of water across the air-sea interface, rather than virtual salt fluxes [7]. Subgrid-scale parameterizations of ocean model OM3.0 include K-profile parameterization (KPP) vertical mixing, neutral physics [8, 9] and a spatially dependent anisotropic viscosity [10]. Air-sea fluxes are computed on the ocean model time step, which is 1 h in OM3.0. SIS (sea ice model) is a dynamical model [19] with three-vertical layer thermodynamics (two ice, one snow), and a scheme for prognosing five different ice thickness categories and open water at each grid point. GFDL-CM2.0 model make use of the Flexible Modeling System (FMS) coupler for calculating and passing fluxes between its atmosphere, land and ocean components.

3   Description of the Proposed Model

The air pollution evolution modeling system comprises of the following four distinct algorithmic stages: Modeling, Grid, Scenarios and Forecasting. In the first stage all of the associated parameters are added and named and then they are interconnected by synapses to create the causal positive or negative correlations. The fuzzification of the correlations, i.e. the description of each interface in verbal common terms was accomplished by selecting six Linguistics namely: Three positive scales (low positive (+), middle positive (++), high positive (+++)). Three negative scales (low negative (−), middle negative (−−), high negative (−−−)) corresponding to fuzzy weights (Table 1) (Fig. 1).

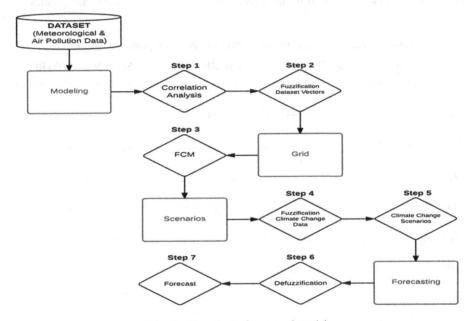

Fig. 1. Flowchart of proposed model

Table 1. Effect and value of six linguistics which corresponding to fuzzy weights

Effect	Value
high positive (+++)	1
middle positive (++)	0.5
low positive (+)	0.25
low negative (−)	−0.25
middle negative (−−)	−0.5
high negative (−−−)	−1

The description of the algorithmic steps is done in the next paragraph:

Step 1 (Modeling): Application for the calculation of the degree of correlation between the variables under consideration: Carbon monoxide (CO), nitrogen monoxide (NO), nitrogen dioxide (NO_2), ozone (O_3), sulfur dioxide (SO_2), temperature air (AirTemp), humidity (RH), atmospheric pressure (PR), solar radiation (SR), sunshine (SUN), wind speed (WS), wind direction (WD) and rainfall (RF).

Step 2: Partitioning of the variables with negative correlation from the ones with positive correlation, with the use of the assigned Linguistics over the initial crisp values. Three successive and overlapping triangular membership functions were employed in order to classify the correlations to the corresponding fuzzy sets (Linguistics) "Low", "Medium" and "High". The following Table 2 presents clearly the fuzzification of the correlation results (assignment of the corresponding Linguistics).

Table 2. Fuzzification of the correlation analysis with proper Linguistics

	CO	NO	NO₂	O₃	SO₂	AirTemp	RH	PR	SR	SUN	WS	WD	RF
CO	1	+++	++	---	++	–	+	+	–	+	---	+	---
NO	+++	1	++	---	++	---	++	+	–	+	---	+	---
NO₂	+++	+++	1	---	++	+	–	–	+	+	---	+	---
O₃	---	---	---	1	---	++	---	–	+	+	+++	–	---
SO₂	+++	+++	++	---	1	---	+	++	–	–	---	+	---
AirTemp	–	–	++	+++	---	1	---	–	+++	+++	+	+	---
RH	++	+++	–	---	+	---	1	+	---	---	---	–	+++
PR	+	+	–	–	+++	---	+	1	–	–	–	---	---
SR	–	–	+	+	–	++	---	–	1	+++	++	+	---
SUN	+	+	++	++	–	++	---	–	+++	1	++	+	---
WS	---	---	---	+++	---	+	---	–	++	++	1	–	+
WD	++	++	+++	---	+	+	–	---	++	++	---	1	+
RF	---	---	---	---	---	---	+++	---	---	---	+	+	1

Step 3 (Grid): It involves the design of the FCM following the input and the interconnection of all correlated variables, based on the Linguistics that emerged after the fuzzyfication of the crisp numerical values.

The algorithm simulating the interactions between two nodes of the FCM was implemented by performing a repetitive calculation of the new link value corresponding to each node. This value depends on the weight of the node from which an edge begins and also on the weight of the edge joining the two nodes. The transfer function estimates the new value of each node and the weight of each connection. The negative type of influence is depicted with an orange color and the positive with a blue color. The degree of influence depends on the thickness of each line. The higher the influence the thicker the line, as you can see in the Fig. 2 (Table 3).

Table 3. The degree of influence

	CO	NO	NO$_2$	O$_3$	SO$_2$	AirTemp	RF
AirTemp	−	−	++	+++	−−	1	−−
RF	−−	−−−	−−−	−−−	−−	−−	1
	−	−	−	−	−		

The degree of influence between some variables depicted in the Fig. 2.

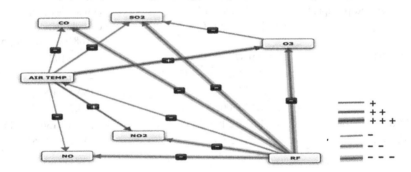

Fig. 2. A FCM between air temperature, rainfall and air pollutants. (Color figure online)

3.1 Scenarios and Forecasting

In the third stage, the changes in the values of temperature and precipitation due to climate change (CC) for the period 2020 till 2099 are fuzzified in order to obtain the corresponding linguistics. The whole process is based on the GFDL_CM2.0 CC model. This phase also includes extended testing of various scenarios.

Step 4 (Scenarios): Partitioning of the scenarios variables, based on the changes in temperature and precipitation, according to the global climate model GFDL_CM2.0.

Finally, the obtained crisp numerical values are fuzzified, with the use of four triangular fuzzy membership functions (FMF) and eight semi-triangular fuzzy membership functions (S-FMF). Two of FMF and four of S-FMF are related to temperature changes in the closed interval [0, + 2.86] °C, which is the highest temperature increase for the area under study in the specified time interval. The first S-FMF, the FMF and the next S-FMF refer to the interval [0, 1.43 (median)] °C. These FMF correspond to the fuzzy sets: low negative (−), middle negative (−−), high negative (−−−), whereas the linguistic high negative (−−−) contains values close to the smallest estimated change. The next S-FMF, the FMF and the last S-FMF refer to the interval [1.43 (median), 2.86 (the highest fluctuation based on the model)] °C. These FMF correspond to the low positive (+), middle positive (++), high positive (+++), with the high positive (+++) being close to the maximum temperature change. In the same way, two FMF and four S-FMF were developed for the precipitation with crisp values in the interval [−108.48, −219.98] mm. Due to the fact that precipitation appears reduced in the future values, the first S-FMF, the FMF and the last S-FMF cover the precipitation reduction in the interval [−164.23, −219.98] mm which corresponds to high precipitation reduction. The next S-FMF, the FMF and the last S-FMF were used for the smaller changes [−108.48, −164.23 (median)] mm which correspond to the fuzzy sets low positive (+), middle positive (++), high positive (+++), with high positive (+++) declaring the lowest rainfall reduction (−108,48 mm) (Table 4).

Table 4. FMF and S-FMF boundaries (Temperature, rainfall)

Fuzzy sets corresponding to temperature and precipitation changes	FMF and S-FMF boundaries in the closed interval [0, + 2.86] °C	FMF and S-FMF boundaries in the closed interval [−108.48, −219.98] mm
−−− (S-FMF)	0 0.572	−220 −197.7
−− (FMF)	0.143 0.715 1.287	−214.4 −192.1 −169.8
− (S-FMF)	0.858 1.43	−186.5 −164.2
+ (S-FMF)	1.43 2.002	−164.2 −141.9
++ (FMF)	1.573 2.145 2.717	−158.6 −136.3 −114
+++ (S-FMF)	2.288 2.86	−130.7 −108.5

Step 5 (Scenarios): It includes extended testing of various scenarios based on the potential changes in the temperature and precipitation and moreover its influence in the air quality of Athens. The fuzzy Linguistics produced by the use of climate change scenarios are defuzzified in order to obtain the forecast of the potential future crisp values of the air pollutants. In this way we perform a projection in the distant future for the problem of environmental degradation due to air pollution.

Step 6 (Forecasting): For the defuzzification the centroid function was used which estimates the center of gravity of the fuzzy set distribution.

$$x = \frac{\int x \cdot \mu(x)dx}{\int \mu(x)dx} \tag{6}$$

Step 7: The index of the magnitude of change in the pollutants concentration, is calculated based on the amount of relative change of each parameters value.

$$RelativeChange = \frac{FutureValue - InitialValue}{InitialValue} \tag{7}$$

4   Results and Discussion

After applying various CC scenarios (including various potential changes in temperature and precipitation) under the GFDL_CM2.0 model for the center of Athens, the forecasted Relative Changes (RC) in the concentration of pollutants were obtained. Finally, 36 scenarios were developed by the combination of temperature and the rainfall. Table 5 presents the most important of the forecasted values.

Table 5. Relative changes of air pollutants based of the climate change scenarios

ID	AirTemp (AT)	RF	O_3	SO_2	CO	NO	NO_2
1	high negative (+0 °C)	high negative (−219.98 mm)	0	0.2	0.06	0.05	0.09
2	high negative (+0 °C)	high positive (−108.48 mm)	−0.02	0.03	−0.02	−0.01	−0.33
3	low negative (+1.43 °C)	high negative (−219.98 mm)	0.03	0.18	0.06	0.05	0.12
4	low negative (+1.43 °C)	high positive (−108.48 mm)	−0.01	−0.06	−0.05	−0.02	−0.21
5	high positive (+2.86 °C)	high negative (−219.98 mm)	0.19	0.14	0.05	0.03	0.14
6	high positive (+2.86 °C)	high positive (−108.48 mm)	0.11	−0.38	−0.29	−0.28	−0.11

Attempting a thorough presentation of the most important RC observed, it is obvious that a zero increase in temperature (0 °C), combined with a large reduction of rainfall (−219.98 mm) contributes significantly to the growth of all primary pollutants while it does not affect the secondary pollutants such as O_3 (ID1). On the other hand, a very slight decrease in rainfall (−108.48 mm) helps to increase the SO_2, while helping to reduce NO_2, NO, O_3 and CO (ID2). Increasing the temperature to the value of the median (+1.43 °C) in combination with a minimum reduction of precipitation (−108.48 mm) allows the reduction of both primary and secondary pollutants (ID4), whereas absolutely the opposite (ID3) is observed when the rainfall is reduced significantly (-219.98 mm). A large increase in the temperature higher than the median ($1.43\ °C$) and as high as the upper limit of the CC model ($2.86\ °C$) combined with a high decrease in the millimeters of rainfall (-219.98 mm) contributes to the increase of all pollutants, especially to the increase of O_3 (ID5). Finally, a very small reduction in the millimeters of rainfall (-108.48 mm) leads to a significant reduction of all pollutants concentration except for the values of O_3 which rises significantly (ID6). An AT increase from $0\ °C$ to 2.86 regardless the rainfall reduction results to an increase of O_3 and reduces the values of SO_2 and NO, CO. The highest values for O_3 and NO_2 appear in the extreme AT scenario (ID5) whereas the lowest air pollutants values appear in the ID2 (zero AT increase) and the highest values for SO_2, CO, NO are observed in ID1 and ID3 (stable or low increase of AT). It is important to mention that we observe the highest values for all pollutants with the highest decrease of rainfall ($-219{,}98$ mm) in ID1 and ID5 and their lowest in ID2 and ID6 scenario with the least rainfall reduction ($-108, 48$ mm). At Table 6 the average values of historical period of 2000–2012 used as initial values for the estimation of forecasted values. The forecasted Relative Changes (RC) determine the values of each pollutant. The last two decades of the 21st century (2080–2099) are very interesting, due to the extreme scenarios observed (ID5). According to this case the temperature will increase by 15 % ($2.86\ °C$) which represents the Linguistic High Positive change (+++) and the precipitation will be reduced by 49 % (-219.98 mm) corresponding to the Linguistic High Negative ($---$) change. This combination will cause significant increase of all primary and secondary pollutants (presented in the following Table 6).

Table 6. Air pollutants forecasted values based on the extreme scenario

Period	O_3	SO_2	CO	NO	NO_2
2000–2012	21.64	20.53	2.47	115.79	87.11
2080–2099	25.75	23.40	2.59	119.26	99.30
Absolute increase value	+4.11	+2.87	+0.12	+3.47	+12.19

A potential verification of the extreme scenario will cause a significant increase in the concentrations of O_3 and NOx which will imply the increase of the cardiovascular and respiratory diseases and will be accompanied by frequent presence of thick summer photochemical smog. Also the high values of SO_2 and CO, will increase the frequency of smog during the winter months, while serious problems may occur on the surfaces of monuments and historic buildings from acid rain. Since there are no actual similar future projections and forecasting efforts for Athens we cannot check the accuracy of the obtained results. However, both the methodology employed and the produced scenarios with the forecasted evolution of air pollution till 2099 are very useful as they will motivate researchers towards more flexible modeling attempts beyond the typical statistical ones.

5   Conclusions

In this paper we have proposed the use of an innovative method for analysis and modeling of air quality. We have also developed and applied an air pollution and forecasting system which is based on CC models and uses Soft Computing techniques. More specifically, Linguistic representation of the correlations between atmospheric pollutants is performed by employing Fuzzy Cognitive Maps. Based on the above approach we have obtained relative changes of air pollutants values for the center of Athens for the period 2000–2012. Additionally, a projection in the future is performed regarding the evolution of the pollutants concentrations, based on the fluctuation of temperature and precipitation till 2100, as reflected by the projections of climate model GFDL_CM2.0. The workings of the model were tested in various scenarios and presented important information about the hazards of air quality and the effects of air pollution. From this point of view, though the forecasted results cannot be checked for accuracy, the fact that this paper introduces a flexible Soft Computing approach to produce long term projection of air quality based on scenarios of CC opens innovative horizons to researchers introducing an alternative algorithm. In the future we will try to improve the model by employing optimization methods (e.g. genetic algorithms, swarm intelligence) or hybrid approaches.

References

1. Amer, M., Jetter, A.J., Daim, T.U.: Scenario planning for the national wind energy sector through fuzzy cognitive maps. In: Technology Management in the IT-Driven Services (PICMET) Proceedings of PICMET 2013, pp. 2153–2162 (2013)
2. Bougoudis, I., Demertzis, K., Iliadis, L.: HISYCOL a hybrid computational intelligence system for combined machine learning: the case of air pollution modeling in Athens. Neural Comput. Appl. **27**, 1191–1206 (2015). doi:10.1007/s00521-015-1927-7. Springer
3. Bougoudis, I., Demertzis, K., Iliadis, L.: Fast and low cost prediction of extreme air pollution values with hybrid unsupervised learning. In: Integrated Computer-Aided Engineering, Vol. Preprint. NO. Preprint, pp. 1–13. IOS Press (2015). doi:10.3233/ICA-150505
4. Bougoudis, I., Iliadis, L., Papaleonidas, A.: Fuzzy inference ANN ensembles for air pollutants modeling in a major urban area: the case of Athens. Eng. Appl. Neural Netw. Commun. Comput. Inf. Sci. **459**, 1–14 (2014)
5. Fons, S., Achari, G., Ross, T.: A fuzzy cognitive mapping analysis of the impacts of an eco-industrial park. J. Intell. Fuzzy Syst. **15**(2), 75–88 (2004)
6. Gordaliza, J.A., Florez, R.E.V.: Using fuzzy cognitive maps to support complex environmental issues learning. In: Proceedings of New Perspectives in Science Education Conference, 2nd edn. (2013)
7. Griffies, S.M.: Fundamentals of Ocean Models, p. 496. Princeton University Press, Princeton (2004)
8. Griffies, S.M., Gnanadesikan, A., Pacanowski, R., Larichev, V., Dukowicz, J.K., Smith, R. D.: Isopycnal mixing in a z-coordinate ocean model. J. Phys. Oceanogr. **28**, 805–830 (1998)
9. Griffies, S.M.: Gent–McWilliams skew flux. J. Phys. Oceanogr. **28**, 831–841 (1998)

10. Large, W., Danasbogulu, G., McWilliams, J., Gent, P., Bryan, F.O.: Equatorial circulation of a global ocean climate model with anisotropic viscosity. J. Phys. Oceanogr. **31**, 518–536 (2001)

11. Lock, P., Brown, R., Bush, R., Martin, M., Smith, B.: A new boundary layer mixing scheme. Scheme description and single-column model tests. Mon. Weather Rev. **128**, 3187–3199 (2000)

12. Luiz, J., Muller, E.: Greenhouse gas emission reduction under the kyoto protocol: the South African example. Int. Bus. Econ. Res. J. **7**, 75–92 (2008)

13. Marco, F., Chalabi, Z., Foss, M.: Assessing framing assumptions in quantitative health impact assessments: a housing intervention example. Environ. Int. **59**, 133–140 (2013)

14. Papageorgiou, E.I., Salmeron, J.L.: A review of fuzzy cognitive maps research during the last decade. IEEE Trans. Fuzzy Syst. **21**(1), 66–79 (2013)

15. Paschalidou, A.: University of Ioannina, Ph.d. thesis development of box model for the air pollution forecasting in medium size cities (2007). (in Greek)

16. Pathinathan, T., Ponnivalavan, K.: The study of hazards of plastic pollution using induced fuzzy cognitive maps (IFCMS). J. Comput. Algorithm **3**, 671–674 (2014)

17. Salmeron, J.L., Froelich, W.: Dynamic optimization of fuzzy cognitive maps for time series forecasting. Knowl. Based Syst. **105**, 29–37 (2016). Forthcoming

18. Vidal, R., Salmeron, J.L., Mena, A., Chulvi, V.: Fuzzy cognitive map-based selection of TRIZ trends for eco-innovation of ceramic industry products. J. Cleaner Prod. **107**, 202–214 (2015)

19. Winton, M.: A reformulated three-layer sea ice model. J. Atmos. Oceanic Technol. **17**, 525–531 (2000)

20. Zhang, H., Song, J., Su, C., He, M.: Human attitudes in environmental management: fuzzy cognitive maps and policy option simulations analysis for a coal-mine ecosystem in China. J. Environ. Manag. **115**, 227–234 (2013)

21. http://www.climatechange2013.org/images/report/WG1AR5_Chapter09_FINAL.pdf

The Results of a Complex Analysis of the Modified Pratt-Yaskorskiy Performance Metrics Based on the Two-Dimensional Markov-Renewal-Process

Viktor Geringer[1]($\boxtimes$), Dmitry Dubinin[2], and Alexander Kochegurov[3]

[1] Faculty of Engineering, Baden-Wuerttemberg Cooperative State University, Friedrichshafen, Germany
geringer@dhbw-ravensburg.de
[2] Tomsk State University of Control Systems and Radioelectronics, Lenin Avenue 40, Tomsk, Russian Federation 634050
dima@info.tusur.ru
[3] National Research Tomsk Polytechnic University, Lenin Avenue 30, Tomsk, Russian Federation 634050
kaicc@tpu.ru

Abstract. The paper presents the results of a quantitative estimation of the edge detection quality using modified Pratt-Yaskorskiy criterion, as well as generalization and adaptation of both approaches based on the generalized quality criterion as part of «CS sF» stochastic simulation software package. The reference images are approximated by the two-dimensional high rise renewal stream offering the stationarity properties with no aftereffects and ordinariness. The efficiency of the proposed metrics is considered for three edging algorithms (Marr-Hildreth, ISEF and Canny) at different levels of the additive normal noise. The estimated errors of the first and second kind are given, which allow referring to the efficiency of the proposed generalized quality criterion.

Keywords: Stochastic computer simulation · Research on models · Reference image · Edge detection · Quality metrics · Performance evaluation · Comparison of algorithms

1 Introduction

At present the solution of a great scope of scientific and technical tasks is related to the processing and analysis of information in such format as images. The images appear as result and object of research in aerospace, navigation, surface remote sounding and in many other areas of human activity. The image of the outline drawing is one of the basic stages of the image processing, as well as its analysis and identification.

This paper presents the results of stochastic simulation to assess the quality of the outline drawing image on the basis of Pratt criterion [1] and its modification proposed by Yaskorskiy [2] and the generalized quality criterion subject to three main types of errors (Fig. 1) discussed in [3–5].

© Springer International Publishing Switzerland 2016
N.T. Nguyen et al. (Eds.): ICCCI 2016, Part I, LNAI 9875, pp. 187–196, 2016.
DOI: 10.1007/978-3-319-45243-2_17

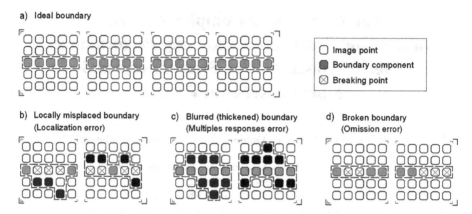

Fig. 1. The main types of errors introduced by search and localization operators

The reference images (RIs) are produced with the two-dimensional high rise renewal stream [6]. In the course of simulation and further analysis, the RIs have been exposed to the additive noise with a normal law of distribution [7]. Analysis of the efficiency of the proposed quality metrics of the outline drawing image was performed for three classical edging algorithms (Marr-Hildreth [8], ISEF [9] and Canny [10]).

2 Objective of the Work

The objective of this work is to produce the modification of the Pratt metrics proposed by Yaskorskiy on condition of the metrics generalization of all the main types of errors introduced by the search and localization operators; the adaptation of the generalized metrics as part of «**CS sF**» [11] for the detailed analysis of the algorithms under study and addressing the synthesis of new algorithms focused on a certain object domain.

3 Stages of Simulation and Research Methods

Pursuing the objective the following procedures were undertaken:

(a) generation of the reference bitmap images with specific morphological and statistical properties using «**CS sF**» package,
(b) image compositing of the noise component with a normal law of distribution and a specified peak signal-to-noise ratio (PSNR),
(c) edge detection with the various edge detectors,
(d) integration of the modified Pratt-Yaskorskiy criterion as part of «**CS sF**», evaluation and comparison of the edge detection quality by various localization operators on the basis of the proposed criterion.

4 Mathematical Formalization of the Task

Very often, the classical Pratt metrics is used to compare the derived results. Its standard form is known as [1]:

$$FoM = \frac{1}{\max\{I_I,\ I_A\}} \sum_{i=1}^{I_A} \frac{1}{1 + \alpha \cdot d(i)^2} \tag{1}$$

where I_I and I_A – the number of elements in the ideal and actual edge points [1]; $d(i)$ – offset value of i-th element of the found line normally to the ideal outline; α – scale factor, which takes into account the error value due to the dislocated edge (see Fig. 1b).

The preparation specificity of the reference bitmaps images as part of «CS sF» was taken into account in adaptation of the original metrics recording (1). The reference images are formed on the basis of the vector-mode description of the outline drawing [3, 11]. Perception of the "a priori" outline location in the reference image allows presentation of the Pratt metrics general format as follows:

$$FoM_{LE} = \frac{1}{I_I - CN} \sum_{i=1}^{I_I} \frac{1}{1 + \alpha \cdot d(i)^2} \tag{2}$$

where CN – the number of points in the real outline drawing being out of metrics calculation (offset value of i–th point of the found line normally to the ideal outline is in excess of the domain under analysis (5 x 5), where $d_{max} = 2$).

In accounting of the edge thickening and smear, one more element is also to be considered:

$$FoM_{MRE} = \frac{1}{I_I - CNC} \sum_{i=1}^{I_I} \frac{1}{1 + \beta \cdot b(i)^2} \tag{3}$$

where CNC – the number of elements in the real outline drawing being out of metrics calculation (e.g. central point of element 170 of «A» morphology character set, which image is the intersection of the vertical and horizontal lines in the center of the working area, that does not allow the central point accounting); β – scale factor, which takes into account the error value due to the thickened (smeared) edge (Fig. 1(c); $b(i)$ – the number of the smeared points of the ideal outline in i-th domain.

Thus, the Pratt metrics allows accounting of the dislocated (localization error), smeared and thickened edges (multiples responses error), see. Figs. 2 and 3. The disadvantages of both representations (2) and (3) appear to be poor response of the metrics to omissions of the edge elements and the length of discontinuity (the sequence of the edge elements related to each other). Thereby there is conservative value of the metrics in evaluation of the bitmap outline field generated at the low s/n ratio (PSNR: Peak signal-to-noise ratio). The Yaskorskiy paper [2] presents one of the possible solutions to the problem. Let us present one of the Yaskorskiy proposals by modified Pratt criterion:

$$FoM_{OE} = \frac{1}{N} \sum_{j=1}^{N} \frac{1}{1 + \chi \cdot n(j)^2} \qquad (4)$$

where N – the number of the domains under analysis (in the present paper 5×5); χ – scale factor, which takes into account the error value due to the number of the omitted points in the given domain (Fig. 4); $n(j)$ – the total number of the omitted edge points in j-th domain under analysis.

In generalization of the results, it is proposed to combine three components (2), (3) and (4) based on the multiplicative quality criterion and to present it as follows:

$$MPFoM = FoM_{LE} \bullet FoM_{MRE} \bullet FoM_{OB} \qquad (5)$$

where $MPFoM$ – Modification Pratt's Figure of Merit.

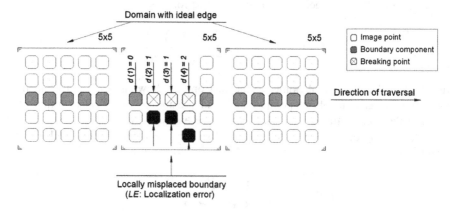

Fig. 2. Dislocated edge (error type: LE – Localization error)

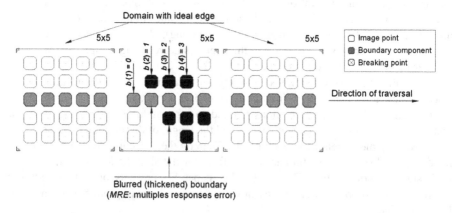

Fig. 3. Smeared or thickened edge (error type: MRE– multiple responses error)

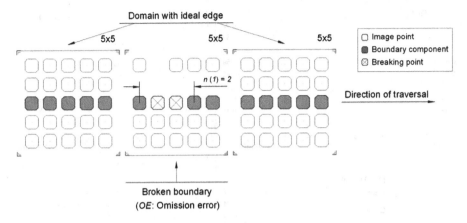

Fig. 4. Broken edge (error type: *OE*– Omission error)

5   Results

The proposed methodology was tested with «CS sF». The reference images of «*A*» morphology were generated using the two-dimensional high rise renewal stream [12–14]. The studies were performed at different values of contrast [15] of the black and white RI. Figure 5 shows the averaged values of luminance (L_A: average luminance) and RIs contrast (C_{RMS}: root mean square contrast) used for the stochastic simulation. The model of ideal step (Step edge) [16] was used as the outline edge. The "Delt" value described the minimum luminance differential of the image segments.

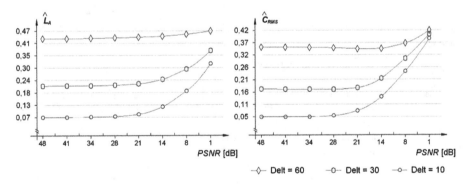

Fig. 5. L_A (estimate of the average brightness), C_{RMS} (contrast) for the RIs of «*A*» morphology

Figure 6 presents the dependence of the components of the generalized quality criterion and the signal-noise ratio.

Figure 7 shows the results of the edge detection quality. Presented there are the dependency graphs of the errors of the first and the second kind [5, 17]. By comparison of the results shown in Fig. 6 and the results shown in Fig. 7, one should bear in mind that better detection accords with higher values FoM_{LE}, FoM_{MRE}, FoM_{OE} and lower error values P_{nd} and P_{fa}.

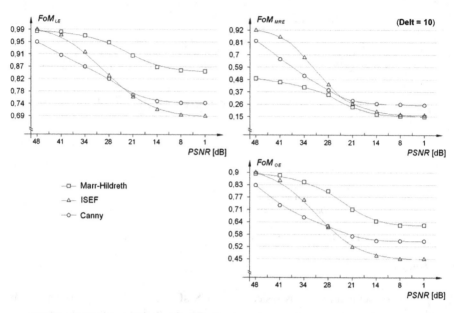

Fig. 6. *FoM*_{MRE}, *FoM*_{OE} and *FoM*_{LE} for the reference images of «*A*» morphology

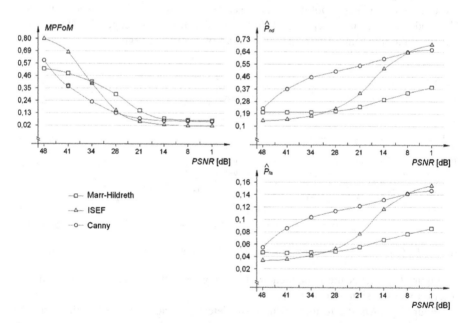

Fig. 7. *MPFoM*, errors P_{nd} (of the first kind) and P_{fa} (of the second kind) for the RIs of «*A*» morphology

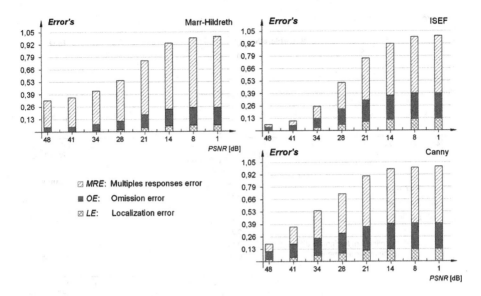

Fig. 8. The weight comparison of the main types of errors (Delt = 10)

Three dependencies were drawn up (Figs. 8 and 9) for the side-by-side comparison of the major errors of the search and localization operators to be used for comparison of the error introduced errors *MRE*, *OE* and *LE* for different operators of the image edge detection.

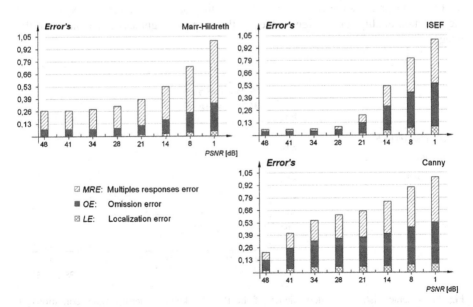

Fig. 9. The weight comparison of the main types of errors (Delt = 60)

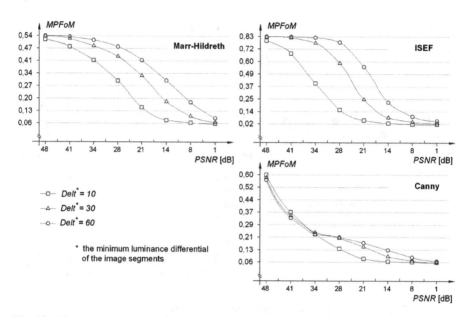

Fig. 10. The output of the search and localization operators under conditions of various brightness and contrast

In the course of investigation of the localization operators' behavior there were obtained the dependencies of the performance quality (modified Pratt-Yaskorskiy metrics) under conditions of different values of the average brightness L_A and contrast C_{RMS} of the reference images (Fig. 10).

Figure 11 presents the calculation results of the characteristic value dependency of the detection quality for different values of the average brightness and contrast of the images.

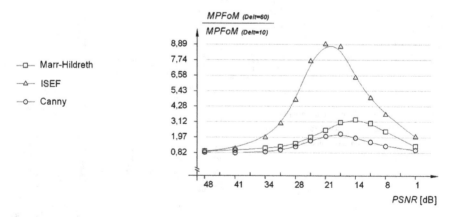

Fig. 11. Characteristic value dependency of the Pratt-Yaskorskiy metrics under conditions of various brightness and contrast

6 Conclusion

The findings of the numerical study using the «CS sF» software and algorithmic package showed that the modified Pratt-Yaskorskiy quality criterion (5) can be used for the unbiased quality estimation of the search and edge detection algorithms. The studies performed by the example of three known detection algorithms (Marr-Hildreth, ISEF and Canny) according to this method allowed identification of the preferred domains for each algorithm depending on the s/n ratio and the type of morphology underlying the imaging.

Acknowledgment. We express our sincere gratitude and appreciation to our families for their delicacy, support and understanding. We express special gratitude to technologist Helene Geringer for her assistance in refining the style of the paper, as well as for preparation of the illustrative material.

References

1. Abdou, I.A., Pratt, W.: Quantitative design and evaluation of enhancement/thresholding edge detectors. Proc. IEEE **67**(5), 753–763 (1979). (ISSN 0018–9219)
2. Yaskorsky, A.V.: Kriterii otsenki kachestva raboty detektorov konturov. [Criteria for the performance evaluation of the edge detectors]. Optoelectron. Instrum. Data Process. **3**, 127–128 (1987). (In Russ.)
3. Dubinin, D., Geringer, V., Kochegurov, A., Reif, K.: An efficient method to evaluate the performance of edge detection techniques by a two-dimensional semi-Markov model. In: CICA 2014, pp. 1–7. IEEE Press, Orlando (2014). doi:10.1109/CICA.2014.7013248
4. Nguyen, T.B., Zhou, D.: Contextual and non-contextual performance evaluation of edge detectors. Pattern Recogn. Lett. **21**(9), 805–816 (2000)
5. Boaventura, I. Gonzaga, A.: Method to evaluate the performance of edge detector, In: Proceedings of SIBGRAPI 2009, XXIInd Brazilian Symposium on Computer Graphics and Image Processing, pp. 1–3 (2009). Rio de Janeiro (ISSN: 2176–0853)
6. Dubinin, D., Geringer, V., Kochegurov, A., Reif, K.: Ein stochastischer algorithmus zur bildgenerierung durch einen zweidimensionalen markoff-erneuerungsprozess Oldenbourg Wissenschaftsverlag. at- Automatisierungstechnik **62**(1), 57–64 (2014)
7. Dubinin, D., Geringer, V., Kochegurov, A.: A particular method of modelling stochastic intensity fields by isotropic, one-step Markov chains, tm – Technisches Messen, no. 5, pp. 271–276. Scientific Publisher, Oldenbourg (2012)
8. Marr, D., Hildreth, E.: Theory of edge detection. Proc. Roy. Soc. Lond. B **207**, 187–217 (1980)
9. Shen, J., Castan, S.: An optimal linear operator for edge detection. In: IEEE Proceedings of the Conference on Vision and Pattern Recognition, pp. 109–114 (1986)
10. Canny, J.: A computational approach to edge detection. IEEE Trans. Pattern Anal. Mach. Intell. **8**(6), 679–698 (1986)
11. Dubinin, D., Geringer, V., Kochegurov, A., Reif, K.: Bundled software for simulation modeling. In: Proceedings of the International Symposium on Signals, Circuits and Systems (ISSCS 2013), pp. 1–4. ISSCS Press, Romania, (IEEE Catalog Number: CFP13816-CDR) (2013)

12. Dubinin, D., Kochegurov, A., Laevski, V.: Metodika modelirovaniya sluchaynykh yarkostnykh poley, approksimirovannykh odnorodnymi odnourovnevymi markovskimi tsepyami. [A particular method of generating stochastic intensity fields by two-dimensional semi-Markov model]. J. Inf. Sci. **4**(12), 35–40 (2011). ICM & MG SB RAS Publisher, Novosibirsk. (In Russ.)

13. Laevski, V.: Algorithm of one-level Markovian fields construction. Bull. Tomsk Polytech. Univ. **309**(8), 32–36 (2006). (ISSN 1684-8519)

14. Buimov, A.: Korrelyacionno- ekstremal'naya obrabotka izobrazhenij. [Correlation-Extremal Processing of Images]. In: Tarasenko, V. (ed.), p. 132. TSU Publisher, Tomsk (1987). (In Russ.)

15. Peli, E.: Contrast in complex images. J. Opt. Soc. Am. **7**(10), 2032–2040 (1990). doi:10. 1364/JOSAA.7.002032

16. Chidiac, H., Ziou, D.: Classification of image edges. In: Vision Interface 1999, pp. 17–24. Troise-Rivieres, Canada (1999)

17. Denisov, V., Dubinin, D., Kochegurov, A., Geringer, V.: Rezul'taty issledovaniya kompleks-nogo metoda otsenki kachestva okonturivaniya na osnove dvukhmernogo tochechnogo potoka vosstanovleniya. [The results of the investigation of the integrated performance evaluation method of edge detection based on the two-dimensional renewal process]. In: Vestnik SibGAU, no. 2, pp. 300–309 (2015). (In Russ.)

Generic Ensemble-Based Representation of Global Cardiovascular Dynamics for Personalized Treatment Discovery and Optimization

Olga Senyukova[1]([⊠]), Valeriy Gavrishchaka[2]([⊠]), Maria Sasonko[3],
Yuri Gurfinkel[3], Svetlana Gorokhova[3], and Nikolay Antsygin[4]

[1] Faculty CMC, Lomonosov Moscow State University, GSP-1, Leninskie Gory,
Moscow 119991, Russian Federation
osenyukova@graphics.cs.msu.ru
[2] Department of Physics, West Virginia University, Morgantown, WV 26506, USA
gavrishchaka@gmail.com
[3] Research Clinical Center of JSC Russian Railways, Chasovaya Str., 20,
Moscow 125315, Russian Federation
{msasonko,yugurf}@yandex.ru, cafedra2004@mail.ru
[4] Children's City Hospital #1, Avangardnaya Str., 14, St. Petersburg 198205,
Russian Federation
anciginnick@gmail.com

Abstract. Accurate and timely diagnostics does not warranty successful treatment outcome due to subtle personal differences, especially in the
case of complex or rare cardiac abnormalities. A proper representation of
global cardio dynamics could be used for quick and objective matching of
the current patient to former cases with known treatment plans and outcomes. Previously we have proposed the approach for heart rate variability (HRV) analysis based on ensembles of different measures discovered
by boosting algorithms. Unlike original HRV techniques, ensemble-based
metrics could be much more accurate in early detection of short-lived or
emerging abnormal regimes and slow changes in long-range dynamic patterns. Here we demonstrate that the same metrics applied to long HRV
time series, collected by Holter monitors or other means, could provide
effective characterization of global cardiovascular dynamics for decision
support in discovery and optimization of personalized treatments.

Keywords: Computer-aided diagnostics · Heart rate variability
analysis · Ensemble learning · Personalized medicine

1 Introduction

In express diagnostics, preventive monitoring and personalization of medical
treatment, it is important to find and correctly interpret quantitative measures capable of detecting emerging and transient abnormalities and other subtle
regime changes. A proper representation of global cardio dynamics may offer fast

© Springer International Publishing Switzerland 2016
N.T. Nguyen et al. (Eds.): ICCCI 2016, Part I, LNAI 9875, pp. 197–207, 2016.
DOI: 10.1007/978-3-319-45243-2_18

and objective matching of the current patient to former cases with known treatment plans and outcomes. This could be used for decision support in discovery and optimization of personalized treatment of cardiovascular pathology.

Cardiac diagnostics based on ECG combines several desirable features and is widely used by medical practitioners and researchers. Heart rate variability (HRV) analysis, as opposed to analysis of electrocardiogram (ECG) waveform, offers a set of measures that are sensitive to subtle changes in heart rate dynamics and can provide complementary insight in cardiac diagnostics [11,14]. HRV time series is also referred to as RR time series, where RR or interbeat interval is the interval between two successive peaks (R points) in ECG waveform. Most of HRV analysis tools currently used in practice are based on time- and frequency-domain linear indicators [11]. Methods from nonlinear dynamics (NLD) provide more natural modeling framework for adaptive biological systems with multiple feedback loops [5,14]. Compared to linear indicators, many NLD-based measures are less sensitive to data artifacts, e.g. due to patient motion and other activities, and to overall nonstationarity. However, many linear and NLD indicators can be used only for long time series [14]. It means that they are not applicable for early detection of both short-lived precursors of emerging physiological regimes and abnormalities with transient patterns that appear on shorter data segments.

Previously we have illustrated that these challenges could be overcome by combining existing HRV measures into boosting-like ensemble learning classification framework [3]. Boosting is better than most other combination techniques because it is capable of discovering an ensemble of complementary models with significantly lower bias (higher accuracy) and lower variance (better stability) compared to each individual model [2]. Potentially more flexible data-driven models, like neural networks, require much more training samples for stable calculation. Moreover, they work like "black box" and lack interpretability. In contrast, boosting-based classifiers combine stability and interpretability since they are constructed from the well-understood low-complexity base models [3].

Additionally, individual base classifiers from the ensemble together implicitly encode many different physiological regimes. In our recent publications [3,9] we referred to this utilization of ensemble internal structure as ensemble decomposition learning (EDL) and described benefits of EDL-based approach for quantitative description and detection of complex and rare states that are hard to quantify by other means.

In this paper, we demonstrate that a collection of multi-dimensional EDL vectors, computed on subsequent segments of long ECG time series, provides robust characterization of global multi-state cardio dynamics. We propose to compute the distance between two subjects as the length of Minimum Spanning Tree (MST) built on the distance matrix, obtained by calculating Euclidean distances between each EDL vector of one subject with each EDL vector of the other subject. On real data we show that this new metrics could provide true personalized description of the global physiological state and is capable of fine differentiation even within the same abnormality type which makes it usable for effective discovery and optimization of personalized treatment in cardiology and in a wider scope of personalized medicine applications.

2 Boosting-Based Framework for the Discovery and Optimization of Ensemble-Based Physiological Indicators and Metrics

Although our framework is generic and suitable for analysis of different physiological time series (e.g. electrocardiogram (ECG), electroencephalogram (EEG), electromyogram (EMG) and gait), we use HRV measures computed from ECG time series in all examples due to our current focus on cardiological applications.

2.1 Ensemble-Based Indicators

The well-known NLD indicators applicable for HRV analysis are based on detrended fluctuation analysis (DFA) [8], multi-scale entropy (MSE) [1], and multi-fractal analysis (MFA) including MFA extension of DFA [6]. The comparable performance is also demonstrated by advanced linear indicators based on power spectrum analysis of the RR time series [11]. One of the widely used indicators of this type is a power spectrum ratio of the low-frequency band (0.04–0.15 Hz) to the high frequency band (0.4–0.15 Hz). Results presented in this paper are based on indicators derived from the described families of HRV measures. However, our framework is open to any other HRV metric that can offer complementary value in cardiac state differentiation.

In general, HRV measures require long time series for stable calculation [14]. However, HRV indicators have to be computed on short segments in order to capture early signs of developing and/or intermittent abnormalities or to detect subtle initial effects of treatment procedures. Otherwise, indicator computed on a long time series will average out these short-lived effects and will fail to detect them. Unlike traditional HRV measures, the proposed ensemble-based indicators are suitable for short RR time series [3].

A typical boosting algorithm, such as AdaBoost [2], for the two-class classification problem, works as follows. Firstly, equal and normalized weights are assigned to all training samples. At each iteration t the optimal base classifier $h_t(x)$ is chosen using a weighted error function. Here x is an input vector. Samples, misclassified by the best model at the current iteration, are penalized by increasing their weights for the next iteration. Therefore, on each iteration, the algorithm focuses on the most hard to classify samples. The final ensemble of base classifiers, given below, classifies the unknown sample as class $+1$ when $H_T(x) > 0$ and as -1 otherwise:

$$H_T(x) = \sum_{t=1}^{T} \alpha_t h_t(x), \tag{1}$$

where α_t are combination coefficients obtained on each iteration, and T is the total number of iterations. Several variations of boosting algorithm, including regime adjustments and important regularization procedures, exist.

A natural choice of base classifiers could be low-complexity HRV indicators $h(\beta[p], \gamma)$. β may correspond, for example, to either a DFA scaling exponent (or its MFA extension), a slope of MSE curve, or a power spectrum ratio, γ is a threshold level (decision boundary) and p is a vector of adjustable parameters of the chosen measure. On each iteration, choosing the best indicator β and optimizing over (p, γ), we obtain an ensemble of classifiers (1).

2.2 Ensemble Decomposition Learning for Quantification of Complex and Rare States

The best classifiers for each boosting iteration, $h_t(x)$, together provide good global performance of the final ensemble. Indeed, $h_t(x)$ are local experts for different implicit regimes or domains of a whole feature space. Therefore, partial information of a wide variety of dynamical regimes becomes implicitly encoded in the obtained ensemble of classifiers. However, this information is not available in the aggregated form, $H_T(x)$. Extraction of this underutilized knowledge could be formalized in terms of ensemble decomposition learning (EDL) [9]. Ensemble decomposition feature vector can be introduced as follows:

$$D(x) = [\alpha_1 h_1(x), \alpha_2 h_2(x), \ldots, \alpha_T h_T(x)]. \tag{2}$$

Each sample can be represented by this vector after applying an ensemble classifier $H_T(x)$ to it, which provides detailed and informative state representation of the considered system. Also, such representation could be difficult or impossible to obtain by other means. Indeed, our procedure requires just two (e.g. normal and abnormal) or a few well-populated classes. Representation for rare classes/states is generated implicitly by solving two-class classification problem. Obtaining direct representation for such states would be very challenging or practically impossible due to severe data limitation.

2.3 Ensemble-Based Representation of Global Cardiovascular Dynamics and Its Utility for Personalized Treatment Discovery and Optimization

Relative comparison of $H(x)$ values computed from short RR segments is the most direct usage of boosting-based meta-indicator for express diagnostics and for early detection of emerging pathologies [3,10]. However, in the early stages of the developing abnormality or in cases of its intermittent nature, a large fraction of short RR segments may be normal, which increases probability of missing such abnormalities. Similarly, false alarms caused by artifacts (e.g. due to patient's motion or short-term emotional variations) will also decrease robustness of diagnostics based on short RR segments. Therefore, stability and accuracy can be significantly increased by analyzing global ECG dynamics characterized by the full distribution of aggregated values, $H(x)$, or EDL vectors calculated on a large number of consecutive RR segments [10]. Such RR data are readily available from standard long-term Holter recordings.

Previously, we have already suggested using statistical moments of distribution of $H(x)$ values calculated on consecutive RR segments [10]. Given availability of long ECG time series, such an approach could significantly reduce noise and increase diagnostics accuracy, while preserving essential ability to detect short-term or emerging cardiac events. However, much more detailed representation of the global ECG dynamics, offered by collection of EDL vectors computed on consecutive RR segments, could go beyond just abnormality detection and classification.

A proper representation of global cardio dynamics could be used for quick and objective matching of the current patient to former cases with known treatment plans and outcomes. Direct comparison of full ECG (RR) time series from two individuals is not effective due to high level of noise and natural long-term variations of the ECG time series. Collection of consecutive EDL vectors provides effectively filtered representation using natural discretization (classification framework) that removes unimportant variations but preserves differentiation among key micro-states. By calculating Euclidean distances between each EDL vector of one subject with each EDL vector of another subject, distance matrix is obtained. However, we need single-number distance measure between two subjects that aggregates comparisons between all of these micro-states represented by EDL vectors. Large distance matrix could be noisy by itself and usage of simple averages or medians from all cross-EDL distances is far from optimal as shown in the next section on real data.

The described challenge of handling distance matrix is similar to that encountered in financial applications dealing with quantification of the market state using large and noisy correlation matrices of thousands of stocks. Recently, it was shown that graph-based approaches such as Minimum Spanning Tree (MST) could offer significant advantages over the more traditional approaches based on random matrix theory [7,13]. MST representation is motivated by the human perception which organizes information with the most economical encoding. A spanning tree is a connected graph containing all vertices of the original graph without loops [12]. The spanning tree length is defined as the sum of the weights of its edges. MST is a spanning tree with minimal length among all spanning trees connecting the nodes of the graph. MST of the graph can be derived with Prim's or Kruskal's algorithm [12].

In the context of stock correlation matrix, individual stocks correspond to MST vertices and weight of the edge (i, j) is the distance, $d_{i,j} = sqrt(2 \cdot (1 - rho_{i,j}))$, between stocks (vertices) i and j, where $rho_{i,j}$ is a correlation between those stocks [7]. MST provides robust low-dimensional representation of the original correlation matrix. The aggregated MST measure is its length defined as

$$L = \frac{1}{N-1} \sum_{d_{ij} \in T} d_{ij}, \qquad (3)$$

where $(N-1)$ is the number of edges present in MST. MST length is an important aggregated measure that is sensitive to market downturns when correlation between stocks increases [7].

Similarly, MST approach can be used to measure similarity or "distance" between global physiological states of two patients represented by the collection of EDL vectors (2). If the length of RR time series permits computation of N EDL vectors from N consecutive segments for each subject, we can create cross-subject distance (proximity) matrix, where distances between all EDL vectors of one subject and all EDL vectors of another subject are computed. Information from $N(N-1)/2$ numbers of distance matrix d_{ij} will be represented with $(N-1)$ edges of MST. The distances between EDL vectors i and j, given by (2), are defined as $l1$ or $l2$ norm in T-dimensional space. The "distance" between global physiological states of two subjects will be given by the length (3) of the obtained MST. Such distance measure should be much more stable and informative than simple averaging of all distances between EDL vector pairs.

Effectiveness of the described approach was already demonstrated in the context of biometric application based on the analysis of gait time series [4]. However, that result was not fully surprising, since certain gait features are known to be quite individual and effective as biometric modalities. On the other hand, ECG patterns are only analyzed to establish existence of abnormality, its type and severity. To the best of our knowledge, there were no studies demonstrating that ECG patterns can be used for much finer state differentiation, including robust personalization within the same abnormality. Here we apply described MST-based metric for cross-subject comparison of long ECG time series for the first time. Our preliminary results are encouraging.

3   Results and Discussion

Analysis presented in this section is mostly based on real-patient ECG data from http://www.physionet.org. We used RR data from 52 subjects with normal sinus rhythm, 27 subjects with congestive heart failure (CHF), 84 subjects with long-term atrial fibrillation (LTAF), and 48 subjects with different types of arrhythmia. Up to 24 h of RR data for each normal, CHF, and LTAF subjects are available. In addition, up to 30 min of RR data are available for each subject with arrhythmia. We have also added 78 intervals (each of 30 min) from patients with supraventricular arrhythmias to expand the arrhythmia data set. It should be noted that, while various cardiac abnormalities can be accompanied by arrhythmia, a separate arrhythmia sample, considered here, represents arrhythmia-only condition.

In all calculations presented in this section, the full data sets described above are used. The training dataset for ensemble learning algorithms includes no more than 50 % of normal, CHF, and arrhythmia data combined. LTAF data have not being used in the training phase. Since base classifiers are low-complexity with small number of adjustable parameters, we have not observed any significant differences between in-sample and out-of-sample results. Number of boosting

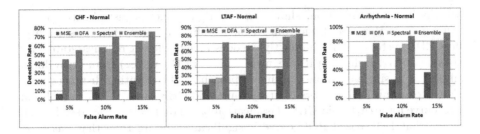

Fig. 1. Detection rates of CHF, LTAF and arrhythmia for different false alarm rates.

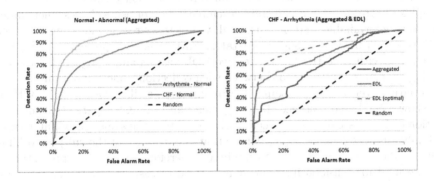

Fig. 2. Left: ROC curves of the aggregated ensemble-based indicator. Right: ROC curves based on EDL metrics of the same indicator using full ensemble (green) and MSE-only subset (dashed green). (Color figure online)

iterations in the considered examples is 30. Typically, the best-on-average single HRV indicators, which are also picked up at the 1-st boosting iteration, are DFA and power spectrum ratio, while MSE is very important complementary component in the final ensemble and could be important in differentiation of abnormality types.

Performance comparison of typical ensemble classifier with each of single HRV measures in the context of normal/abnormal classification is demonstrated in Fig. 1. Since all measures are computed on short RR segments of 256 beats, this analysis relates to express test when only short ECG time series is available. Ensemble-based indicator shows significant improvement in all three cases. More detailed discussion of similar results is available in our previous publications [3,9,10].

Even though boosting-based approach could significantly improve accuracy of the normal-abnormal classifier, without re-training, such indicator may not be able to differentiate between abnormality types [9] as illustrated in Fig. 2. Receiver operating characteristic (ROC) curves of the ensemble indicator, shown on the left panel, demonstrate good normal-abnormal classification for both CHF and arrhythmia. On the other hand, ROC curve of the same indicator for CHF vs arrhythmia classification, shown on the right panel is just slightly better than random. However, besides aggregated value (1), ensemble classifier also offers EDL vector (2) that implicitly encodes many different regimes/states. By choosing a single reference RR segment (and its EDL vector) from arrhythmia sample (as in single-example learning approach) and computing distances from this vector to all EDL vectors of arrhythmia and CHF samples, one can obtain ROC curve for arrhythmia-CHF classification based on EDL metrics (see Fig. 2). We see that EDL-based ROC curve is significantly better than that based on the aggregated value. By choosing certain sub-components of the ensemble (e.g. only MSE), one can further improve differentiation based on EDL metrics.

Thus, even in the absence of data for direct classifier estimation for certain complex or rare states, one can train ensemble indicator using classes where large data sets are available, and then use EDL vector to represent wide range of other important cardiological conditions/states. Furthermore, representing long RR time series with collection of consecutive EDL vectors and using MST for computing aggregated distance (3) between such collections, one can obtain fine differentiation even within the same abnormality type as illustrated in Fig. 3. Here we show MST-based distances between each pair of 20 CHF patients using collection of 50 consecutive EDL vectors obtained from the same normal-abnormal ensemble indicator as in Fig. 1. We see that the distance of the subject to himself is either minimal or close to minimal. Therefore, since our representation can be used for self-identification, it is natural to assume that other subjects, close to the considered patient in terms of our metrics, have very similar cardiac conditions and responses to personalized treatments.

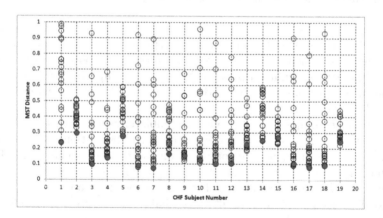

Fig. 3. MST-based distances between each pair of 20 CHF patients using collection of 50 consecutive EDL vectors computed on 256-beat RR segments. For each subject, distances to all other subjects is represented by black circles, while distance to his own portion of RR time series, not overlapping with the original one, is shown by a red circle. (Color figure online)

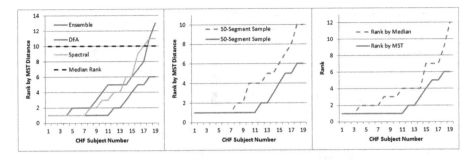

Fig. 4. Rank of the MST-based distance of each CHF subject to himself against distances to other subjects. Rank 1 correspond to minimal distance. Left: ensemble vs single indicator. Middle: 10-segment vs 50-segment representation. Right: MST-based distance vs simple median of all segment-to-segment distances.

Importance of (1) using ensemble measures, (2) MST for handling cross-subject distance metrics, and (3) long ECG time series, is demonstrated in Fig. 4. Here we rank distance of each CHF subject to himself against distances to other subjects. Minimal distance corresponds to rank 1, next after minimal to rank 2, etc. In the case of ideal self-identification, all ranking numbers would be 1. Unless otherwise specified, all presented results are for 50 consecutive RR segments. Left panel shows that ensemble-based measure is significantly better than any individual HRV measures. Middle panel demonstrates improvement of self-identification accuracy when RR time series is increased from 10 to 50 256-beat segments. Finally, right panel shows superiority of MST over simple median of all segment-to-segment distances.

While presented illustrations support utility of our metrics for fine-grain characterization of the personal cardiac state, the ultimate test of our metrics would consist of identification of subjects with close cardiac states before treatment and establishing whether the same specialized treatments produce similar outcomes. While we are still in the process of collecting such detailed test data, results of preliminary application of our overall framework to real clinical data are very encouraging.

It is important that our ensemble-based indicators remain applicable to new data sets, that are collected using different equipment and/or protocols, without re-calibration. Indeed, new data sources and various clinical recordings containing complex and rare cases, where application of our measures would be important, may have limited amount of reference data for re-training. It was already mentioned that our ensemble indicators demonstrate very good out-of-sample stability. However, they were tested on samples from the same source (physionet.org) as training data.

Here we applied our ensemble-based indicator, calibrated on physionet.org data, to long-term Holter recordings obtained at Research Clinical Center of JSC Russian Railways (Russian Federation). As illustrated in Fig. 5, conclusions inferred from our ensemble-based HRV measures are in line with cardiologists'

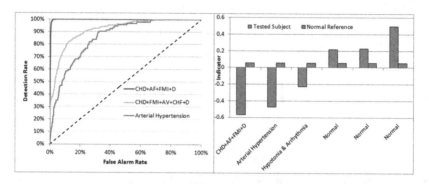

Fig. 5. Left: ROC curves of the ensemble-based indicator for normal-abnormal classification for the patient with arterial hypertension (green line) and two patients with combination of coronary heart disease (CHD), fibrosis after myocardial infarction (FMI), atrial fibrillation (AF), type 2 diabetes (D), congestive heart failure (CHF), and atrioventricular block of 2nd degree (AV). Right: 10-th percentiles of the distribution of ensemble indicator values computed on consecutive RR segments of the full Holter recordings. (Color figure online)

diagnostics. Here, for demonstration purposes, we used long Holter recordings from 11 normal (healthy) subjects and several patients with different combinations of abnormality types and severity. ROC curves for normal-abnormal classification based on 256-beat RR segments, shown in the left panel of Fig. 5, indicate good differentiation capabilities of our indicator even for express diagnostics from single short RR segments.

However, for emerging abnormalities or conditions with intermittent heart rate dynamics, analysis of short time series may not be able to detect abnormality, which will be observed as deteriorating ROC curves based on short RR segments. Nevertheless, using tail measures (e.g. 10-th percentile) of the distribution of indicator values, $H(x)$, computed on consecutive RR segments of long Holter recording, one can still detect abnormal behavior and correctly quantify severity of the abnormality. This is illustrated in the right panel of Fig. 5. Finally, preliminary analysis of ECG recordings from Children's City Hospital #1 of St. Petersburg (Russian Federation) also confirmed applicability of our ensemble measures without re-training. Further comprehensive testing of the described ensemble-based measures, including distance metrics between global cardiological states of different patients, on real clinical data is warranted.

References

1. Costa, M., Goldberger, A.L., Peng, C.-K.: Multiscale entropy analysis of physiologic time series. Phys. Rev. Lett. E **71**, 021906 (2005)
2. Freund, Y., Schapire, R.E.: A decision-theoretic generalization of on-line learning and an application to boosting. J. Comput. Syst. Sci. **55**(1), 119–139 (1997)

3. Gavrishchaka, V.V., Senyukova, O.V.: Robust algorithmic detection of cardiac pathologies from short periods of RR data. In: Pham, T., Jain, L.C. (eds.) Knowledge-Based Systems in Biomedicine. SCI, vol. 450, pp. 137–153. Springer, Heidelberg (2013)
4. Gavrishchaka, V., Senyukova, O., Davis, K.: Multi-complexity ensemble measures for gait time series analysis: application to diagnostics, monitoring and biometrics. In: Signal and Image Analysis for Biomedical and Life Sciences. Advances in Experimental Medicine and Biology, vol. 823, pp. 107–126. Springer International Publishing, Cham (2015)
5. Kantz, H., Schreiber, T.: Nonlinear Time Series Analysis. Cambridge University Press, Cambridge (1997)
6. Makowiec, D., Dudkowska, A., Zwierz, M., et al.: Scale invariant properties in heart rate signals. Acta Phys. Pol. B 37(5), 1627–1639 (2006)
7. Onnela, J.-P., Chakraborti, A., Kaski, K., Kertesz, J., Kanto, A.: Dynamics of market correlations: taxonomy and portfolio analysis. Phys. Rev. E 68, 056110 (2003)
8. Peng, C.-K., Havlin, S., Stanley, E.H., Goldberger, A.L.: Quantification of scaling exponents and crossover phenomena in nonstationary heartbeat time series. Chaos 5, 82–87 (1995)
9. Senyukova, O., Gavrishchaka, V.: Ensemble decomposition learning for optimal utilization of implicitly encoded knowledge in biomedical applications. In: IASTED International Conference on Computational Intelligence and Bioinformatics, pp. 69–73. ACTA Press, Calgary (2011)
10. Senyukova, O., Gavrishchaka, V., Koepke, M.: Universal multi-complexity measures for physiological state quantification in intelligent diagnostics and monitoring systems. In: Pham, T.D., Ichikawa, K., Oyama-Higa, M., Coomans, D., Jiang, X. (eds.) ACBIT 2013. CCIS, vol. 404, pp. 76–90. Springer, Heidelberg (2014)
11. Task Force of the European Society of Cardiology, the North American Society of Pacing and Electrophysiology: Heart rate variability: standards of measurement, physiological interpretation, and clinical use. Circulation 93, 1043–1065 (1996)
12. Theodoridis, S., Koutroumbas, K.: Pattern Recognition. Academic Press, San Diego (1998)
13. Tumminello, M., Lillo, F., Mantegna, R.N.: Correlation, hierarchies, and networks in financial markets. J. Econ. Behav. Organ. 75, 40–58 (2010)
14. Voss, A., Schulz, S., Schroederet, R., et al.: Methods derived from nonlinear dynamics for analysing heart rate variability. Philosophical Trans. Royal Soci. A 367, 277–296 (2008)

Avoiding the Curse of Dimensionality in Local Binary Patterns

Karel Petranek[(⊠)], Jan Vanek, and Eva Milkova

University of Hradec Kralove,
Rokitanskeho 62, 50002 Hradec Kralove, Czech Republic
{karel.petranek, eva.milkova}@uhk.cz,
vanek.conf@gmail.com

Abstract. Local Binary Patterns is a popular grayscale texture operator used in computer vision for classifying textures. The output of the operator is a bit string of a defined length, usually 8, 16 or 24 bits, describing local texture features. We focus on the problem of succinctly representing the patterns using alternative means and compressing them to reduce the number of dimensions. These reductions lead to simpler connections of Local Binary Patterns with machine learning algorithms such as neural networks or support vector machines, improve computation speed and simplify information retrieval from images. We study the distribution of Local Binary Patterns in 100000 natural images and show the advantages of our reduction technique by comparing it to existing algorithms developed by Ojala et al. We have also confirmed Ojala's findings about the uniform LBP proportions.

Keywords: Dimensionality reduction · Local binary patterns · Image analysis

1 Introduction

Texture analysis is one of the basic tasks in image recognition and analysis tasks. Textures can contain interesting structural information about objects in the scene and to a degree they are invariant to light and viewpoint transformations. Texture information can be used as the input for segmentation [1, 2], scene reconstruction [3] or machine learning algorithms [4–6].

Traditionally, textures have been processed by transforming the 2D image signal into the frequency domain using Fourier or more generally wavelet transforms and performing analysis in the frequency domain [7, 8]. There is a computational disadvantage to this approach because the transform can be costly and may need to be computed twice – into the frequency domain and back, depending on the application.

Local Binary Patterns (LBP) operator [9] is a texture operator that works directly on the raster information. There is no need to transform the image into the frequency domain. The output of the LBP operator is a bit string representing texture class at the given image pixel. This bit string is then directly used for further analysis in the subsequent higher-level algorithms. Other advantages of the LBP operator include an easy compensation for image rotation, scale, brightness and contrast transformations [10].

© Springer International Publishing Switzerland 2016
N.T. Nguyen et al. (Eds.): ICCCI 2016, Part I, LNAI 9875, pp. 208–217, 2016.
DOI: 10.1007/978-3-319-45243-2_19

Our focus in this paper is to address one disadvantage of LBP – their high dimensionality. We show that the traditional LBP representation as a high-dimensional bit string can be reduced to a 2-component real vector while keeping most of the relevant information.

We first describe existing algorithms for computing LBPs on a given grayscale image. In the second part of the paper, we describe the dimensionality reduction and its trade-offs compared to standard LBP.

2 LBP Extraction

The LBP operator assigns a bit string B to every pixel of the input image. B can also be re-interpreted as an integer representing texture class at the given point in the image. The individual bits of the LBP are derived from intensity values around the given pixel (see Fig. 1). First, a circular neighbourhood of a pixel $P_{x,y}$ with the radius r is analysed and intensities of every point in this neighbourhood are compared in a clockwise order to the intensity of $P_{x,y}$. Points with a lower intensity than $P_{x,y}$ produce a 0 in the resulting bit string and points with a higher intensity produce 1. The number of points p in the neighbourhood is a parameter of the operator and is usually set to 8, 16 or 24 to match integer sizes in common CPUs to allow for reasonably big lookup tables. More formally, the LBP value at the given image point $P_{x,y}$ can be defined as:

$$\text{LBP}_{p,r}(P_{x,y}) = \sum_{i=0}^{p-1} 2^i 1_{[0,\infty)}(P_{x_i,y_i} - P_{x,y})$$

where $\text{LBP}_{p,r}(P_{x,y})$ is the LBP value represented as a p-bit integer at $P_{x,y}$, p the number of points in the circular neighbourhood, r radius of the neighbourhood, P_{x_i,y_i} image

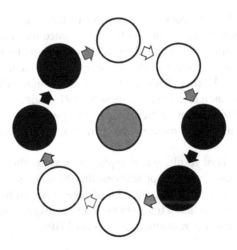

Fig. 1. A circular 8-neighbourhood representing a Local Binary Pattern. Black points have a lower intensity than the central point, white points have a higher intensity.

intensity of the ith point in the neighbourhood and $1_{[0,\infty)}(v)$ the non-negative indicator function returning 1 for non-negative v and 0 for negative v.

Changing the radius r leads to a bigger or smaller neighbourhood. It is possible to achieve scale invariance by computing LBP for multiple values of r. Usually, r is increased exponentially, leading to a logarithmic number of LBP calculations per pixel in proportion to the image size.

The LBP operator output is invariant to monotonic changes in brightness and contrast as neither adding, nor multiplying the image intensity by a constant factor changes the sign of the difference between $P_{x,y}$ and the points in its circular neighbourhood.

2.1 Rotation Invariance

Input image rotation results in a bit shift of the LBP bit string as the starting bit of the LBP gets rotated which in turn results in a different texture class being recognized. When recognizing textures, this is an unwanted side effect. The aim of most computer vision systems is to detect textures or objects under any rotation.

Pietikäinen [11] presents a simple method to make LBP values independent of the image rotation. The bit string is first rotated to its base position which is a cyclic rotation of the bit string where most zero bits are at the beginning of the string – the 8 bit pattern 11000011 gets normalized to 00001111. The base position can be found using a lookup table for small bit strings up to about 24 bits. For longer bit strings or for the lookup table computation a fast processor instruction can be used to repeatedly rotate the pattern. The base position is then found as the minimum of all integers produced by these bit string rotations.

2.2 Uniform LBP

There are exponentially many rotationally invariant LBP values, $O(2^p)$ where p is the pattern size. This results in an exponentially small chance that some patterns occur in the input data. Ojala [9] observed that about 80 % of patterns found in the images are concentrated in 8 bins and noticed that these patterns contain at most 2 changes between 0 and 1 in the bit string. This has led to the definition of uniform LBP which specifies rotation invariant patterns with at most two 0/1 changes in the bit string. Other patterns are grouped together into a "remaining" category. This gives a total of $p + 2$ categories which significantly reduces the LBP space and allows for a simpler manipulation.

LBP provides excellent results when applied to texture classification [10]. However, for many machine learning applications the 20 % remaining values result in suboptimal classification results because a lot of potentially useful information is lost [12]. The aim is therefore to find a trade-off between the exponential nature of rotation invariant LBPs and the linear restriction of uniform LBPs.

3   Reducing LBP Dimensionality

When LBPs are used in conjunction with machine learning algorithms such as neural networks or SVMs that work with real-valued vectors as inputs, the size of the resulting bit string leads to big input vectors. There are p inputs for each pixel; therefore even for 8 bit LBPs the input vector size is 8 times the image size. Uniform LBPs do not help in this case as their output is a categorical value which has to be split to $p + 2$ individual inputs – even worse than vanilla rotation-invariant LBPs.

Our solution avoids using the bit stream as the final representation. We transform the bit stream into two variables – the number of transitions t between 0 and 1 and The number of zero bits z. Because these variables are counts, we can use them directly as inputs to machine learning algorithms, needing only two inputs per image pixel. Using these two variables, we are able to express all the uniform patterns regardless of their neighbourhood size p. The combinations of t and z also allow expressing more patterns than only the uniform patterns, leading to a smaller information loss. In general, there is a quadratic amount of $t \times z$ combinations with respect to p but only a linear amount of uniform patterns.

Note that using only the z and t numbers to represent a pattern leads to a lossy representation. There is a quadratic number of patterns that are theoretically expressible using z and t but the total number of patterns is 2^p. This also holds for the number of rotation invariant LBPs which also grow exponentially with p.

Even though the reduction loses some information, it maintains information about relative lighting in the neighbourhood (the number of zeros) and gradient stability (the number of transitions). The numbers z and t are also naturally rotation invariant which speeds up the computation as there is no need for lookup tables or bitwise rotations to find the appropriate rotation invariant LBP.

It is beneficial to normalize the values of z and t into the interval $[0; 1]$ when they are used as inputs to machine learning algorithms. As both the number of zeros and the number of transitions are limited by p, the normalization can be computed by dividing z and t by p.

4   Results

We analyzed the occurrences of various LBPs on images from the ILSVRC 2014 test set which contains 100 k images selected from the ImageNet dataset [13]. First, we verified that Ojala's assertion of uniform patterns being 80 % of all patterns [10] also holds for natural images present in the ImageNet dataset. Figure 2 shows the distribution of 8 bit patterns with radius 1 in the test set. Uniform patterns take 84.6 % which corresponds to the number reported by Ojala. Similar results were achieved for 16 bit patterns with radius 4 where the uniform patterns make 54.6 % of all patterns, corresponding roughly to the 57.6 % reported by Ojala.

Because the proposed z, t reduction of LBP is a lossy reduction, we analysed the information loss that occurs during the reduction and compared it to 8 bit and 16 bit uniform patterns. The histograms in Fig. 3 show how many rotation invariant

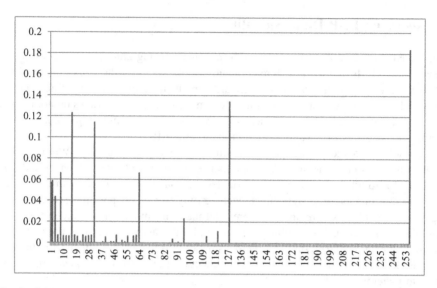

Fig. 2. 8-bit rotation invariant Local Binary Pattern relative frequencies in the ILSVRC 2014 test set which contains 100 k natural images [13].

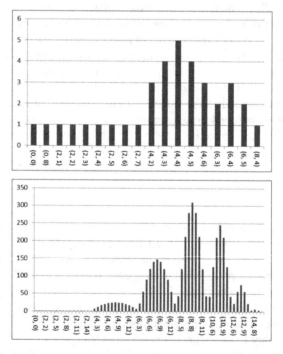

Fig. 3. Number of rotation invariant patterns for the given t, z combinations for 8 bit patterns (top) and 16 bit patterns (bottom). The information loss occurs mainly in complex patterns with many transitions that are less likely to appear in images. Some column labels in the 16 bit version were omitted for clarity.

LBPs are aliased for the given t, z combination. If the reduction were lossless, each bin would contain only one element.

Note that the bins corresponding to uniform patterns $(t \leq 2)$ contain only one element, therefore the z, t reduction is able to represent all uniform patterns without any information loss. Our solution also treats efficiently the non-uniform patterns as the more frequently occurring patterns with fewer transitions are aliased significantly less often than patterns with more transitions, especially for larger LBPs (see Fig. 3 bottom and Fig. 4).

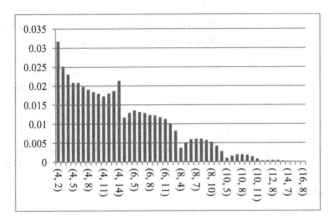

Fig. 4. The frequencies of t, z combinations in the ILSVRC 2014 test set for 16 bit patterns. Patterns with fewer transitions occur significantly more frequently than patterns with more transitions.

4.1 Accuracy and Training Performance

We applied LBP, rotation invariant LBP, uniform LBP and our proposed method to two popular image data sets, MNIST [14] and CIFAR-10 [15] (grayscale version). The converted datasets were used as inputs to state-of-the-art neural networks for classifying the MNIST and CIFAR-10 images. We compare the classification accuracy and training times in tables Tables 2, 3, 4, 5. The Keras deep learning framework [16] with the TensorFlow backend was chosen for the experiments because of its simplicity and predefined standard neural network architectures for both the MNIST and CIFAR-10 datasets.

In total, 4 experiments with 3 different network architectures were conducted. We used a simpler, feed forward network with 2 hidden layers for both MNIST and CIFAR-10, a convolutional network with 4 hidden layers for MNIST and a deeper convolutional network with 6 hidden layers for CIFAR-10. These networks are based on the built-in Keras models and use the same training parameters and architecture [17] (mnist_mlp, mnist_cnn and cifar10_cnn). The network architectures and parameters are summarized in Table 1.

Table 1. Network architectures and training parameters used for comparing the proposed method with existing approaches.

	Simple	CNN-MNIST	CNN-CIFAR-10
Architecture[a]	IL, FC(512), DO (20 %), FC (512), DO (20 %), FC(10)	IL, CL(3 × 3), CL (3 × 3), MP(2 × 2), DO(25 %), FC(128), DO(50 %), FC(10)	IL, CL(3 × 3), MP(2 × 2), DO(25 %), CL(3 × 3), CL (3 × 3), MP(2 × 2), DO (25 %), FC(512), DO (50 %), FC(10)
Hidden Layer Activations	Rectified Linear Unit	Rectified Linear Unit	Rectified Linear Unit
Output and Loss functions	Softmax +Categorical Cross Entropy	Softmax+Categorical Cross Entropy	Softmax+Categorical Cross Entropy
Optimizer	RMSProp	AdaDelta	Nesterov's SGD[b] with Momentum, learning rate 0.01, momentum 0.9 and weight decay of 10^{-6}
Epochs	15	15	15

[a]The model is always a sequence of layers denoted from left to right (IL – Input Layer, FC – Fully Connected Layer, DO – Drop-Out, MP – Max-Pooling Layer; see Keras [16] for more details).
[b]Stochastic Gradient Descent.

The networks were trained on a laptop with an nVidia GTX 660 M GPU with 2 GB of VRAM and an i7-3610QM processor; GPU accelerated training was enabled.

The results confirm the hypothesis that training will take less time while retaining or surpassing the accuracy of the existing binary LBP approaches. The time savings are especially large (up to 50 %) for simpler models where the input dimensionality significantly affects the complexity of the whole model. Most LBP applications such as texture classification or face recognition fall into this category where LBPs are used as the feature extractor and only a simple classifier is used to decide the final class [4, 18].

The proposed method is on-par or exceeds the accuracy of existing LBPs with the only exception being the MNIST dataset with a high-precision convolutional neural network where even the small information loss of our method already slightly affects the accuracy compared to vanilla LBPs (our method still outperforms uniform LBPs).

The results in Table 4 appear to favor the uniform LBP implementation in terms of accuracy. However, the network is too simple to actually learn anything for vanilla LBP, rotation invariant LBP and uniform LBP. The test accuracy for uniform LBP went from 33.33 % in the first epoch to 31.54 % in the final epoch, meaning the network actually diverged. Inputs from our method were the only inputs that showed a steady learning progress (from 9.98 % in the first epoch to 18.67 % in the final epoch).

Table 2. Comparison of training time and accuracy on the Simple network and the MNIST data set. The proposed method has similar accuracy to the other LBP variants while training much faster.

MNIST+Simple	Test set accuracy	Training time (seconds)
LBP	98.27 %	238
LBP (rot. invariant)	98.07 %	236
LBP (uniform)	97.25 %	256
Proposed method	97.92 %	148

Table 3. Comparison of training time and accuracy on the CNN-MNIST network and the MNIST data set. The proposed method has similar accuracy to the uniform LBP but lower accuracy than vanilla LBP due to smaller information content. Training time is dominated by complex hidden layers but the proposed method still trains faster than other approaches.

MNIST+CNN-MNIST	Test set accuracy	Training time (Seconds)
LBP	98.17 %	1570
LBP (rot. invariant)	98.07 %	1569
LBP (uniform)	96.91 %	1583
Proposed method	97.06 %	1524

Table 4. Comparison of training time and accuracy on the Simple network and the CIFAR-10 data set. The proposed method was the only method that allowed the network to learn (from 9.98 % to 18.67 % accuracy, see discussion below) while also training about 2 times faster.

CIFAR-10+Simple	Test set accuracy	Training time (Seconds)
LBP	10.00 %	229
LBP (rot. invariant)	10.00 %	231
LBP (uniform)	31.54 %	274
Proposed method	18.67 %	129

Table 5. Comparison of training time and accuracy on the CNN-CIFAR-10 network and the CIFAR-10 data set. The proposed method has higher accuracy than other LBP variants because the rich structure of CIFAR-10 images and the simplicity of our method makes up for the information loss. Training time is dominated by complex hidden layers but the proposed method still trains faster than other approaches.

CIFAR-10+CNN-CIFAR-10	Test set accuracy	Training time (Seconds)
LBP	30.4 %	2850
LBP (rot. invariant)	28.62 %	2850
LBP (uniform)	29.68 %	2866
Proposed method	38.38 %	2805

5 Conclusion

We demonstrate a simple technique for reducing the dimensionality of Local Binary Patterns and thus improving their utility for machine learning tasks and information retrieval from images. Training and recognizing LBP transformed images can be 4–12 times faster using the proposed technique depending on LBP size. We achieve this by reducing the LBP bit vector to two numerical values. While this reduction loses information compared to basic rotation invariant LBPs, it keeps more information than the most commonly used uniform LBPs while remaining rotation invariant and is therefore preferable for machine learning tasks. The reduced values can also be calculated without first computing the LBPs itself, leading to higher computational efficiency. Further performance improvements are possible by exploiting the locality of LBPs and computing them in parallel for each pixel of the input image.

We demonstrated the impact of our reduction technique by analysing LBPs on a large real-world image dataset with 100 k natural images and compared it to the commonly used uniform patterns. As a side result, we also confirmed the conclusions of Ojala et al. about LBP distribution in images. We also showed that our approach leads to significant computational savings when used in conjunction with existing machine learning algorithms while retaining accuracy.

While the described algorithms assume a grayscale input image, we expect it is possible to extend them to work on multi-channel colour images. We will explore these possibilities in our future work.

Acknowledgement. This paper was supported by the research project SPEV, University of Hradec Kralove, Faculty of Informatics and Management, 2016.

References

1. Ojala, T., Pietikäinen, M.: Unsupervised texture segmentation using feature distributions. Pattern Recognit. **32**, 477–486 (1999)
2. Qing, X., Jie, Y., Siyi, D.: Texture segmentation using LBP embedded region competition. Electron. Lett. Comput. Vis. Image Anal. **5**, 41–47 (2005)
3. Rara, H., Farag, A., Elhabian, S., Ali, A., Miller, W., Starr, T., Davis, T.: Face recognition at-a-distance using texture and sparse-stereo reconstruction. In: 2010 Fourth IEEE International Conference on Biometrics: Theory Applications and Systems (BTAS), pp. 1–6. IEEE (2010)
4. Shan, C.: Learning local binary patterns for gender classification on real-world face images. Pattern Recognit. Lett. **33**, 431–437 (2012)
5. Wang, X., Han, T.X., Yan, S.: An HOG-LBP human detector with partial occlusion handling. In: 2009 IEEE 12th International Conference on Computer Vision, pp. 32–39. IEEE (2009)
6. Trefný, J., Matas, J.: Extended set of local binary patterns for rapid object detection. In: Proceedings of the Computer Vision Winter Workshop (2010)
7. Chang, T., Kuo, C.-C.: Texture analysis and classification with tree-structured wavelet transform. IEEE Trans. Image Process. **2**, 429–441 (1993)

8. Livens, S., Scheunders, P., Van de Wouwer, G., Van Dyck, D.: Wavelets for texture analysis, an overview. In: Sixth International Conference on Image Processing and Its Applications, 1997, pp. 581–585. IET (1997)
9. Ojala, T., Pietikäinen, M., Mäenpää, T.: Gray scale and rotation invariant texture classification with local binary patterns. In: Vernon, D. (ed.) ECCV 2000. LNCS, vol. 1842, pp. 404–420. Springer, Heidelberg (2000)
10. Ojala, T., Pietikainen, M., Maenpaa, T.: Multiresolution gray-scale and rotation invariant texture classification with local binary patterns. IEEE Trans. Pattern Anal. Mach. Intell. 24, 971–987 (2002)
11. Pietikäinen, M., Ojala, T., Xu, Z.: Rotation-invariant texture classification using feature distributions. Pattern Recognit. 33, 43–52 (2000)
12. Guo, Z., Zhang, D., Mou, X.: Hierarchical multiscale LBP for face and palmprint recognition. In: 2010 17th IEEE International Conference on Image Processing (ICIP), pp. 4521–4524. IEEE (2010)
13. Russakovsky, O., Deng, J., Su, H., Krause, J., Satheesh, S., Ma, S., Huang, Z., Karpathy, A., Khosla, A., Bernstein, M., Berg, A.C., Fei-Fei, L.: ImageNet Large Scale Visual Recognition Challenge. ArXiv Prepr. arXiv:14090575. (2014)
14. Lecun, Y., Cortes, C.: The MNIST database of handwritten digits. http://yann.lecun.com/exdb/mnist/
15. Krizhevsky, A., Hinton, G.: Learning multiple layers of features from tiny images. Comput. Sci. Dep. Univ. Tor. Technical Report 1, 7 (2009)
16. Chollet, F.: Keras Deep Learning Framework. GitHub (2015)
17. Chollet, F.: Keras Examples. https://github.com/fchollet/keras/blob/master/examples/
18. Liu, L., Zhao, L., Long, Y., Kuang, G., Fieguth, P.: Extended local binary patterns for texture classification. Image Vis. Comput. 30, 86–99 (2012)

On the Foundations of *Multinomial* Sequence Based Estimation

B. John Oommen[1(✉)] and Sang-Woon Kim[2]

[1] School of Computer Science, Carleton University, Ottawa K1S 5B6, Canada
oommen@scs.carleton.ca
[2] Department of Computer Engineering, Myongji University,
Yongin 17058, South Korea
kimsw@mju.ac.kr

Abstract. This paper deals with the relatively new field of sequence-based estimation which involves utilizing both the information in the observations and in their sequence of appearance. Our intention is to obtain Maximum Likelihood estimates by "extracting" the information contained in the observations when perceived as a *sequence* rather than as a *set*. The results of [15] introduced the concepts of Sequence Based Estimation (SBE) for the Binomial distribution. This current paper generalizes these results for the multinomial "two-at-a-time" scenario. We invoke a novel phenomenon called "Occlusion" that can be described as follows: By "concealing" certain observations, we map the estimation problem onto a lower-dimensional binomial space. Once these occluded SBEs have been computed, we demonstrate how the overall Multinomial SBE (MSBE) can be obtained by mapping several lower-dimensional estimates onto the original higher-dimensional space. We formally prove and experimentally demonstrate the convergence of the corresponding estimates.

Keywords: Estimation using sequential information · Sequence Based Estimation · Estimation of multinomials · Fused estimation methods · Sequential information

1 Introduction

Estimation is the central aspect associated with the training phase of classification and Machine Learning. Since the *sequence*-based paradigm for supervised

---

The first author is a *Fellow: IEEE* and *Fellow: IAPR*. The work was done while he was visiting at Myongji University, Yongin, Korea. He also holds an *Adjunct Professorship* with the Department of Information and Communication Technology, University of Agder, Grimstad, Norway. The work was partially supported by NSERC, the Natural Sciences and Engineering Research Council of Canada and a grant from the National Research Foundation of Korea. This work was also generously supported by the National Research Foundation of Korea funded by the Korean Government (NRF-2012R1A1A2041661).

© Springer International Publishing Switzerland 2016
N.T. Nguyen et al. (Eds.): ICCCI 2016, Part I, LNAI 9875, pp. 218–229, 2016.
DOI: 10.1007/978-3-319-45243-2_20

learning that is explored in this paper is relatively new, we shall first motivate its perspective. Estimation methods generally fall into various categories, including the Maximum Likelihood Estimates (MLEs) and the Bayesian family of estimates [1,3,4,7,20] which are well-known for having good computational and statistical properties. Consider the strategy used for developing the MLE of the parameter of a distribution, $f_X(\theta)$, whose parameter to be estimated is θ. The input to the estimation process is the set of points/observations $\mathcal{X} = \{x_1, x_2, \ldots, x_N\}$, whose elements are assumed to be generated independently and identically as per the distribution, $f_X(\theta)$. The process for computing the Maximum Likelihood (ML) estimate involves deriving the likelihood function, i.e., the likelihood of the distribution, $f_X(\theta)$, generating the sample points/observations $\mathcal{X}$ given θ, which is then maximized (by traditional optimization or calculus methods) to yield the estimate, $\widehat{\theta}_{MLE}$. The general characteristic sought for is that the estimate $\widehat{\theta}_{MLE}$ converges to the true (unknown) θ with probability one, or in a mean square sense. The Bayesian schemes work with a similar goal, except that rather than them using Likelihood functions, they compute the posterior distributions assuming that θ itself is a random variable with a known distributional form. Bayesian and ML estimates generally possess desirable convergence properties. Indeed, the theory of estimation has been studied for hundreds of years [1,3,10,17–19], and it has been the backbone for the learning (training) phase of statistical pattern recognition systems [4,7,9,20,21].

Traditionally, the ML and Bayesian estimation paradigms work within the model that the data, from which the parameters are to be estimated, is known, and that it is treated as a *set*. The position that we respectfully submit is that traditional ML and Bayesian methods ignore and discard[1] valuable *sequence*-based information. The goal of this paper is to "extract" and "utilize" the information contained in the observations when they are perceived *both as a set* and in *their sequence of appearance*. Put in a nutshell, this paper deals with the relatively new field of sequence-based estimation in which the goal is to estimate the parameters of a distribution by maximally "squeezing" out the *set*-based and *sequence*-based information latent in the observations.

The consequences of solving this problem are potentially many. Estimation, as researchers in almost all fields of science and engineering will agree, is a fundamental issue, in which the practitioner is given a set of observations involving the random variable, and his task is to estimate the parameters which govern the generation of these observations. Since, by definition, the problem involves random variables, decisions, predictions, regressions and classification related to the problem are, in some way, dependent on the practitioner obtaining reliable estimates of the parameters that characterize the underlying random variables.

More specifically, suppose that the user received $\mathcal{X}$ as a sequence of data points as in a typical real-life (or real-time) application such as those obtained in a data-mining domain involving sequences, or in data involving radio or

---

[1] This information is, of course, traditionally used when we want to consider *dependence* information, as in the case of Markov models and *n*-gram statistics.

television news items. The question that we have investigated is the following: "Is there any information in the fact that in $\mathcal{X}$, x_i specifically precedes x_{i+1}?". Or in a more general case, "Is there any information in the fact that in $\mathcal{X}$, the **sequence** $x_i x_{i+1} \ldots x_{i+j}$ occurs $n_{i,i+1,\ldots i+j}$ times?". Our position, which we proved in [15] for binomial random variables, is that even though $\mathcal{X}$ is generated by an i.i.d. process, there is information in these pieces of sequential data which can be "maximally" utilized to yield the so-called family of Sequence Based Estimators (SBEs). The problem was initially studied in [15], but only for the case of binomial random variables.

As far as we know, apart from the results in [15], there are no other reported results which utilize sequential information in obtaining such estimates. Also, as highlighted in [15], unlike the use of sequence information in syntactic pattern recognition, grammatical inference and in modeling channels using Hidden Markov Models (which involve estimating the bigram and n-gram probabilities of *dependent* streams of data [2,4,6]), in our case, we assume that the elements in the stream of data, $\mathcal{X}$, occur *independently*, and yet have information not utilized by traditional MLE schemes.

The contributions of this paper can be catalogued as follows:

1. This paper lists the first reported results for obtaining the MLEs of the parameters (i.e., the vector of probabilities responsible for the generation) of a multinomial distribution, when the data is processed both as a *set* of observations and as a *sequence* in which the samples occur in the set. These estimates are called the Multinomial Sequence Based Estimates (MSBEs).
2. The paper pioneers the concept of obtaining MSBEs by invoking the phenomenon of "Occlusion" in which certain observations are hidden or concealed to first yield binomial SBEs, and these are subsequently fused to yield the MSBE.
3. The paper contains the formal results[2] for the MSBE schemes when the sequence is processed in pairs. They have all been experimentally verified.

To the best of our knowledge, apart from our previous results of [15], all of these are novel to the field of estimation, learning and classification. Also, in the interest of space and brevity, the proofs of the theoretical results presented here are omitted. They are found in [16].

2   On Obtaining MSBEs Using Occluded SBEs

Before we proceed with the theoretical and experimental results, it is necessary for us to formalize the notation that will be used[3].

Notation 1: To be consistent, we introduce the following notation.

---

[2] The paper lists numerous theorems whose proofs are found in [16]. The results for longer subsequences (i.e., three-at-a-time, four-at-a-time etc.) are also found in [16].

[3] We apologize for this cumbersome notation, but this is unavoidable considering the complexity of the problem and the ensuing analysis.

- X is a multinomially distributed random variable, obeying the distribution S.
- $\mathcal{X} = \{x_1, x_2, \ldots, x_J\}$ is a realization of a sequence of occurrences of X, where each $x_i \in \mathcal{D}$.
- An index $a \in \mathcal{D}$ is said to be the unconstrained variable in any computation if all the other estimates $\{s_i\}$ are specified in terms of s_a, where $i \neq a$. It will soon be clear that in any computation there can only be *a single* unconstrained variable. The other variables are defined in terms of it.
- $\mathcal{X}^{ab} = \{x_1, x_2, \ldots, x_{N_{ab}}\}$ is called the *Occluded* sequence of $\mathcal{X}$ (with N_{ab} items) with respect to a and b, if it is obtained from $\mathcal{X}$ by deleting the occurrences of all the elements except a and b. Whenever we refer to the sequence $\mathcal{X}^{ab} = \{x_1, x_2, \ldots, x_{N_{ab}}\}$, we always imply that the first variable (in this case a) is the unconstrained variable.
- Let $< j_1 j_2 \ldots, j_k >$ be the subsequence[4] examined in the *Occluded* sequence $\mathcal{X}^{ab}$, where each $j_m, (1 \leq m \leq k)$, is either a or b. Then[5]:
    - The BSBE, for s_a obtained by examining in $\mathcal{X}^{ab}$ the subsequence $< j_1 j_2 \ldots, j_k >$ will be given by $\left. \widehat{q}_a \right|^{ab}_{<j_1 j_2 \ldots, j_k>}$, where, as before, the first variable (in this case a) is the unconstrained variable.
    - Similarly, the BSBE, for s_b obtained by examining in $\mathcal{X}^{ab}$ the subsequence $< j_1 j_2 \ldots, j_k >$ will be given by $\left. \widehat{q}_b \right|^{ab}_{<j_1 j_2 \ldots, j_k>}$, where the first variable (in this case a) is the unconstrained variable.
- Consider the sequence $\mathcal{X}$ in which the index a is the unconstrained variable. Let $< j_1 j_2 \ldots, j_k >$ be the subsequence examined in the sequence $\mathcal{X}$, where each $j_m, (1 \leq m \leq k)$, is either a or '*', where each '*' is the *same* variable, say $c \in (\mathcal{D} - \{a\})$. Then:
    - The MSBE for s_a (where a is the unconstrained variable) obtained by examining in $\mathcal{X}$ the sequence $< j_1 j_2 \ldots, j_k >$ will be given by $\left. \widehat{s}_a \right|^{a}_{<j_1 j_2 \ldots, j_k>}$ where each j_i that is not a is replaced by a '*', and where each '*' is the *same* variable, say $c \in (\mathcal{D} - \{a\})$.
    - For any constrained variable b, the MSBE for s_b obtained by examining in $\mathcal{X}$ the sequence $< j_1 j_2 \ldots, j_k >$ will be given by $\left. \widehat{s}_b \right|^{ab}_{<j_1 j_2 \ldots, j_k>}$, where a is the unconstrained variable.
- Trivially, for all a and b: $\left. \sum_{b \neq a} \widehat{s}_b \right|^{ab}_{<j_1 j_2 \ldots, j_k>} = 1 - \left. \widehat{s}_a \right|^{a}_{<j_1 j_2 \ldots, j_k>}$. □

Example of Notation 1: Let $\mathcal{D} = \{1, 2, 3, 4\}$, and $\mathcal{X} = 134211232341122$. Then, the *Occluded* sequence $\mathcal{X}^{12}$ is obtained by erasing from $\mathcal{X}$ all occurrences of 3 and 4, and has the form $\mathcal{X}^{12} = 1211221122$. Observe that N_{12} is 10. Then:

---

[4] For the present, we consider non-overlapping subsequences. We shall later extend this to overlapping sequences when we report the experimental results.

[5] The reader must take pains to differentiate between the q's and the s's, because the former refer to the BSBEs and the latter to the MSBEs.

- If 1 is the unconstrained variable, the BSBE of s_1 obtained by examining $\mathcal{X}^{12}$ for all occurrences of the sequence $< 121 >$ will be given by $\widehat{q}_1 \Big|_{<121>}^{12}$.
- If 2 is the unconstrained variable, the BSBE of s_4 obtained by examining all occurrences of the sequence $< 224 >$ will be given by $\widehat{q}_4 \Big|_{<224>}^{24}$.
- If in any specific computation, 4 is the unconstrained variable, the MSBE of s_4 obtained by examining all occurrences of the sequence $< **4 >$ will be given by $\widehat{s}_4 \Big|_{<**4>}^{4}$, and will be obtained by normalizing using the quantities $\widehat{s}_1 \Big|_{<114>}^{41}$, $\widehat{s}_2 \Big|_{<224>}^{42}$ and $\widehat{s}_3 \Big|_{<334>}^{43}$. $\qquad\square$

2.1 The Fundamental Theorem of Fusing Occluded Estimates

Our first task is to formulate how we can compute the MSBEs by utilizing information gleaned by the *Binomial* SBEs (BSBEs) obtained from the set of occluded sequences. Consider an occluded sequence, $\mathcal{X}^{ab}$, extracted from the original sequence, $\mathcal{X}$, by removing all the variables except a and b. In the sequence being examined, we choose one variable, say a to be the unconstrained variable. We shall first attempt to obtain BSBEs of the relative proportions of s_a and s_b, the quantities to be estimated, from $\mathcal{X}^{ab}$. Thereafter, we utilize the set of these relative proportions to compute the MSBEs of all the variables. We formalize this in what we call the *Fundamental Theorem of Fusing Occluded Estimates*.

Theorem 1. *For every pair of indices, a and b, let $\mathcal{X}^{ab}$ be the Occluded sequence, extracted from the original sequence, $\mathcal{X}$, by removing all the variables except a and b. If we consider a to be the unconstrained variable, we define*
$q_a = \frac{s_a}{s_a+s_b}$ *and* $q_b = \frac{s_b}{s_a+s_b}$, *where* $q_a + q_b = 1$. *Now let* $\widehat{q}_a \Big|_{\pi(a,b)}^{ab} \neq 0$ *and*
$\widehat{q}_b \Big|_{\pi(a,b)}^{ab} = 1 - \widehat{q}_a \Big|_{\pi(a,b)}^{ab}$ *be the BSBEs of q_a and q_b respectively based on the occurrence[6] of any specific subsequence $\pi(a,b)$. Then, if c is a dummy variable[7] representing any of the variables, the MSBEs of s_a and s_b obtained by examining the occurrences[8] of $\pi(a,b)$ in every $\mathcal{X}^{ab}$ are:*

$$\widehat{s}_a \Big|_{\pi(a,b)}^{a} = \frac{1}{\sum_{\forall c} \rho_c}, \quad and \quad \widehat{s}_b \Big|_{\pi(a,b)}^{ab} = \frac{\widehat{q}_b \Big|_{\pi(a,b)}^{ab}}{\sum_{\forall c} \rho_c}, \tag{1}$$

where $\rho_a = 1$ and $\forall c \neq a$, $\rho_c = \dfrac{\widehat{q}_c \Big|_{\pi(a,c)}^{ac}}{\widehat{q}_a \Big|_{\pi(a,c)}^{ac}}$.

---

[6] How BSBEs are obtained for specific instantiations of $\pi(a,b)$ is discussed later.
[7] The fact that c is a dummy variable will not be repeated in future invocations.
[8] This, of course, makes sense only if $\forall c, \widehat{q}_a \Big|_{\pi(a,c)}^{ac} \neq 0$.

Proof. This is the central theorem of this paper. With a being unconstrained, let the BSBE of q_a based on the occurrence of any specific subsequence $\pi(a,b)$ be $\widehat{q_a}\Big|_{\pi(a,b)}^{ab}$. Clearly, $\widehat{q_b}\Big|_{\pi(a,b)}^{ab} = 1 - \widehat{q_a}\Big|_{\pi(a,b)}^{ab}$. The MSBE is then obtained by resorting to the Weak Law of Large Numbers which guarantees that if the sequence examined is "large enough", the ratio between the various probabilities is also the ratio between their estimates, thus providing a mechanism to normalize the corresponding estimates.

The proof of the result is omitted due to space considerations. It is in [16]. $\square$

3   MSBEs Using Pair-Wise Sequential Information

3.1   Theoretical Results

The following results for MSBEs are true when the sequential information is processed in pairs.

Theorem 2. *Let* $q_a = \frac{s_a}{s_a + s_b}$ *and* $q_b = \frac{s_b}{s_a + s_b}$, *where* $q_a + q_b = 1$. *Then,* $\widehat{q_a}\Big|_{<aa>}^{ab}$ *and* $\widehat{q_b}\Big|_{<aa>}^{ab}$, *the BSBEs of* q_a *and* q_b *obtained by examining the occurrences of* $<aa>$ *in* $\mathcal{X}^{ab}$ *are:*

$$\widehat{q_a}\Big|_{<aa>}^{ab} = \sqrt{\frac{n_{aa}}{N_{ab}/2}}, \quad and \quad \widehat{q_b}\Big|_{<aa>}^{ab} = 1 - \sqrt{\frac{n_{aa}}{N_{ab}/2}}, \tag{2}$$

where n_{aa} *is the number of occurrences of* $<aa>$ *from among the* $\frac{N_{ab}}{2}$ *non-overlapping subsequences*[9] *of length 2 in* $\mathcal{X}^{ab}$. *Consequently,*

$$\widehat{s_a}\Big|_{<aa>}^{a} = \frac{1}{\sum_{\forall c} \rho_c}, \quad and \quad \widehat{s_b}\Big|_{<aa>}^{ab} = \frac{\widehat{q_b}\Big|_{<aa>}^{ab}}{\sum_{\forall c} \rho_c}, \tag{3}$$

where $\rho_a = 1$ *and* $\forall c \neq a, \rho_c = \frac{1 - \sqrt{\frac{n_{aa}}{N_{ac}/2}}}{\sqrt{\frac{n_{aa}}{N_{ac}/2}}}$.

Proof. The proof of the theorem is found in [16]. $\square$

The following example will help clarify the concepts of how the BSBEs are computed and how the MSBE is obtained from the BSBEs.

Example I: Let us suppose that:

$$\mathcal{X} = \{2, 2, 3, 3, 1, 1, 2, 1, 1, 2, 3, 2, 3, 1, 1, 2, 1, 1, 2, 2, 2, 1, 3\}.$$

---

[9] Observe that it would be statistically advantageous (since the number of occurrences obtained would be almost doubled) if all the overlapping $N_{ab} - 1$ subsequences of length 2 were considered. The computational consequences of this are given in [16].

We shall consider the MSBEs for the case when the variable 1 is unconstrained. This will highlight why our present results are *far more complex* than
the corresponding binomial results derived in [15]. Indeed, the extension of the
binomial to the multinomial case depends *on the identity of the unconstrained
variable.*

Estimation of the MSBE when 1 is the Unconstrained Variable

First of all, $\mathcal{X}^{12} = \{2, 2, 1, 1, 2, 1, 1, 2, 2, 1, 1, 2, 1, 1, 2, 2, 2, 1\}$,
and, $\mathcal{X}^{13} = \{3, 3, 1, 1, 1, 1, 3, 3, 1, 1, 1, 1, 1, 3\}$.

From the set $\mathcal{X}^{12}$, we see that $N_{12} = 18$, and in $\mathcal{X}^{12}$, $n_{11} = 4$, and so, as per
Eq. (2):

$$\widehat{q}_1\Big|_{<11>}^{12} = \sqrt{\tfrac{4}{9}} = \tfrac{2}{3}, \text{ and}$$

$$\widehat{q}_2\Big|_{<11>}^{12} = 1 - \sqrt{\tfrac{4}{9}} = \tfrac{1}{3}.$$

Thus, $\widehat{q}_2\Big|_{<11>}^{12} = (0.5) \cdot \widehat{q}_1\Big|_{<11>}^{12}$.

Again, from the set $\mathcal{X}^{13}$, we see that $N_{13} = 14$, and in $\mathcal{X}^{13}$, $n_{11} = 4$, and so,
as per Eq. (2):

$$\widehat{q}_1\Big|_{<11>}^{14} = \sqrt{\tfrac{4}{7}} = 0.7559, \text{ and}$$

$$\widehat{q}_3\Big|_{<11>}^{13} = 1 - \sqrt{\tfrac{4}{7}} = 0.2441.$$

Thus, $\widehat{q}_3\Big|_{<11>}^{13} = (0.323) \cdot \widehat{q}_1\Big|_{<11>}^{13}$.

Normalizing the above with regard to the relative proportions to vari-
able 1 as per Theorem 1, implies normalizing $[\theta \quad 0.5\theta \quad 0.323\theta]^T$. This yields
the MSBE of $[s_1 \quad s_2 \quad s_3]^T$, with 1 being the unconstrained variable, to be
$[0.5485 \quad 0.2743 \quad 0.1772]^T$.   □

The corresponding results for $\widehat{s}_a\Big|_{<bb>}^{a}$, $\widehat{s}_a\Big|_{<ab>}^{a}$ and $\widehat{s}_a\Big|_{<ba>}^{a}$ etc. follow.

Theorem 3. *Let* $q_a = \frac{s_a}{s_a + s_b}$ *and* $q_b = \frac{s_b}{s_a + s_b}$, *where* $q_a + q_b = 1$. *Then,* $\widehat{q}_a\Big|_{<bb>}^{ab}$
and $\widehat{q}_b\Big|_{<bb>}^{ab}$, *the BSBEs of* q_a *and* q_b *obtained by examining the occurrences of*
$< bb >$ *in* $\mathcal{X}^{ab}$ *are:*

$$\widehat{q}_a\Big|_{<bb>}^{ab} = 1 - \sqrt{\frac{n_{bb}}{N_{ab}/2}}, \quad \text{and} \quad \widehat{q}_b\Big|_{<bb>}^{ab} = \sqrt{\frac{n_{bb}}{N_{ab}/2}}. \tag{4}$$

where n_{bb} *is the number of occurrences of* $< bb >$ *from among the* $\frac{N_{ab}}{2}$ *non-
overlapping subsequences of length 2 in* $\mathcal{X}^{ab}$. *Consequently,*

$$\widehat{s}_a\Big|_{<bb>}^{a} = \frac{1}{\sum_{\forall c} \rho_c}, \quad \text{and} \quad \widehat{s}_b\Big|_{<bb>}^{ab} = \frac{\widehat{q}_b\Big|_{<bb>}^{ab}}{\sum_{\forall c} \rho_c}, \tag{5}$$

where $\rho_a = 1$ *and* $\forall c \neq a, \rho_c = \dfrac{\sqrt{\frac{n_{cc}}{N_{ac}/2}}}{1 - \sqrt{\frac{n_{cc}}{N_{ac}/2}}}.$

Proof. The proof is similar to that of Theorem 2. The details are omitted.   □

Theorem 4. *Let* $q_a = \frac{s_a}{s_a+s_b}$ *and* $q_b = \frac{s_b}{s_a+s_b}$, *where* $q_a + q_b = 1$. *Then,* $\left.\widehat{q_a}\right|^{ab}_{<ab>}$, *the BSBE of* q_a *obtained by examining the occurrences of* $< ab >$ *in* $\mathcal{X}^{ab}$, *can be obtained if and only if the roots of the quadratic equation* $\lambda^2 - \lambda + \frac{n_{ab}}{N_{ab}/2} = 0$ *are real (where* n_{ab} *is the number of occurrences of* $< ab >$ *from among the* $\frac{N_{ab}}{2}$ *non-overlapping subsequences of length 2 in* $\mathcal{X}^{ab}$). *Its value,* λ_a, *is the root whose value is closest to* $\widehat{q_a}$. *Further, in such a case,* $\left.\widehat{q_b}\right|^{ab}_{<ab>} = 1 - \left.\widehat{q_a}\right|^{ab}_{<ab>} = \lambda_b$. *Finally,*

$$\left.\widehat{s_a}\right|^{a}_{<ab>} = \frac{1}{\sum_{\forall c}\rho_c}, \quad and \quad \left.\widehat{s_b}\right|^{ab}_{<ab>} = \frac{\left.\widehat{q_b}\right|^{ab}_{<ab>}}{\sum_{\forall c}\rho_c}, \tag{6}$$

where $\rho_a = 1$ *and* $\forall c \neq a, \rho_c = \frac{\lambda_c}{\lambda_a}$.

Proof. The proof of this theorem is also included in [16].   □

The final theorem about the MSBE computed using the occurrences of $< ba >$ in $\mathcal{X}^{ab}$ is given below. Its proof is identical to the one above.

Theorem 5. *Let* $q_a = \frac{s_a}{s_a+s_b}$ *and* $q_b = \frac{s_b}{s_a+s_b}$, *where* $q_a + q_b = 1$. *Then,* $\left.\widehat{q_a}\right|^{ab}_{<ba>}$, *the BSBEs of* q_a *obtained by examining the occurrences of* $< ba >$ *in* $\mathcal{X}^{ab}$, *can be obtained if and only if the roots of the quadratic equation* $\lambda^2 - \lambda + \frac{n_{ba}}{N_{ab}/2} = 0$ *are real (where* n_{ba} *is the number of occurrences of* $< ba >$ *from among the* $\frac{N_{ab}}{2}$ *non-overlapping subsequences of length 2 in* $\mathcal{X}^{ab}$). *Its value,* λ_a, *is the root whose value is closest to* $\widehat{q_a}$. *Further, in such a case,* $\left.\widehat{q_b}\right|^{ab}_{<ba>} = 1 - \left.\widehat{q_a}\right|^{ab}_{<ba>} = \lambda_b$. *Finally,*

$$\left.\widehat{s_a}\right|^{a}_{<ba>} = \frac{1}{\sum_{\forall c}\rho_c}, \quad and \quad \left.\widehat{s_b}\right|^{ab}_{<ba>} = \frac{\left.\widehat{q_b}\right|^{ab}_{<ba>}}{\sum_{\forall c}\rho_c}, \tag{7}$$

where $\rho_a = 1$ *and* $\forall c \neq a, \rho_c = \frac{\lambda_c}{\lambda_a}$.   □

Notice that the four estimates $\left.\widehat{s_a}\right|^{a}_{<aa>}$, $\left.\widehat{s_a}\right|^{a}_{<ab>}$, $\left.\widehat{s_a}\right|^{a}_{<ba>}$ and $\left.\widehat{s_a}\right|^{a}_{<bb>}$ are not linearly independent. Indeed, this is true because: $n_{aa} + n_{ab} + n_{ba} + n_{bb} = \frac{N_{ab}}{2}$.

3.2   Experimental Results: Sequences of Pairs

In this section, we present the results of our simulations[10] on synthetic data for the case when the sequence is processed in pairs. In every case, we have

---

[10] In the tables, values of *unity/zero* represent the cases when the roots are complex or when the number of occurrences of the event concerned are zero.

considered the $N_{ab} - 1$ *overlapping* subsequences of length 2 for the occluded sequence $\mathcal{X}^{ab}$. Thus, for all $b \neq a$, we have used the following expressions to obtain *computational* approximations of the true corresponding estimates derived in Theorems 2 to 5 respectively:

$$\widehat{q}_a \Big|_{<aa>}^{ab} = \sqrt{\frac{n_{aa}}{N_{ab}-1}},$$

The roots of $\lambda^2 - \lambda + \frac{n_{ab}}{N_{ab}-1} = 0$,

The roots of $\lambda^2 - \lambda + \frac{n_{ba}}{N_{ab}-1} = 0$, and

$$\widehat{q}_a \Big|_{<bb>}^{ab} = 1 - \sqrt{\frac{n_{bb}}{N_{ab}-1}},$$

where in each case, we have used $(N_{ab} - 1)$ instead of $\frac{N_{ab}}{2}$.

In every case examined, the multinomial distribution was S, where $S = [s_1, s_2 \ldots s_d]^T$, with d taking values 3, 4 and 5.

The MSBE process for the estimation of S was extensively tested for numerous distributions and for different dimensionalities, but in the interest of brevity, we merely cite a single specific example for a given value of d. In each case, the estimation algorithms were presented with random occurrences of the variables for $N = 390625$ (i.e., 5^8) time instances. Each table reports the results of the estimation for the specific value of d, and in each table, the respective actual value of S used has been specified. To render the comparison meaningful, we have also used the identical data stream to follow the "traditional" MLE computation, i.e., the one that does not utilize the sequential information.

To compare the value of S to its estimate, we have also computed the Euclidean distance between S and its estimates, $\hat{S}$, namely $E_{MLE} = ||S - \widehat{S_{MLE}}||$ and $E_{MSBE} = ||S - \widehat{S_{MSBE}}||$, where $\widehat{S_{MLE}}$ was the ML estimate, and $\widehat{S_{MSBE}}$ was evaluated using the corresponding result depending on the pair of symbols examined in the occluded sequence. The results are tabulated in [16] and respectively, and when the pairs examined in every $\mathcal{X}^{ab}$ were aa, ab, ba and bb. To demonstrate the true convergence properties of the estimates, we have also reported the values of the ensemble averages of the estimates in Tables 1 and 2 respectively, taken over an ensemble of 100 experiments. The convergence of every single estimate is remarkable.

To be more specific, for the case when $d = 3$ and $S = [0.6 \quad 0.25 \quad 0.15]^T$ and when the pair examined in every $\mathcal{X}^{ab}$ was aa, the E_{MSBE} had the ensemble average of 0.1263 when only $N = 25$ symbols were processed (please see Table 1). This value decreased to 0.1247 when $N = 125$ symbols were processed. This error was marginally lower (due to the sampling variance) than the asymptotic error at $N = 5^8$ of 0.1272. The reader should also observe the manner in which the E_{MSBE} closely followed the E_{MLE}.

By way of comparison, when the pair examined in every $\mathcal{X}^{ab}$ was ab, (again for the case when $d = 3$) the value E_{MSBE} had the ensemble average of 0.1740 when only $N = 25$ symbols were processed. The progressive decrease of the error was again observed. It became 0.1412 when $N = 125$ symbols were processed, and became very close to the steady-state value when even as few as 625 samples were examined. Due to the sampling error caused by the random sequences, the

Table 1. A table of the *ensemble* averages (taken over 100 experiments) of E_{MLE}, the error of the MLE, and the error of the MSBE, E_{MSBE}, at time N, for $d = 3$, where the latter MSBEs were estimated by using the formal expressions of Theorems 2 to 5 approximated using the issues discussed in the beginning of this section. Here $d = 3$ and $S = [0.6 \quad 0.25 \quad 0.15]^T$. In the case of the MSBE, in each column, we mention the pair being examined, i.e., whether it is $< aa >$, $< ab >$, $< ba >$ or $< bb >$.

N	E_{MLE}	$E_{MSBE}\mid_{<aa>}$	$E_{MSBE}\mid_{<ab>}$	$E_{MSBE}\mid_{<ba>}$	$E_{MSBE}\mid_{<bb>}$
5^2 (25)	0.1091	0.1263	0.1740	0.1712	0.1725
5^3 (125)	0.1221	0.1247	0.1412	0.1036	0.1122
5^4 (625)	0.1252	0.1258	0.1292	0.1248	0.1269
5^5 (3,125)	0.1270	0.1272	0.1277	0.1284	0.1281
5^6 (15,625)	0.1273	0.1274	0.1273	0.1281	0.1280
5^7 (78,125)	0.1272	0.1272	0.1270	0.1274	0.1273
5^8 (390,625)	0.1272	0.1272	0.1272	0.1273	0.1273

Table 2. A table of the *ensemble* averages (taken over 100 experiments) of E_{MLE}, the error of the MLE, and the error of the MSBE, E_{MSBE}, at time N, for $d = 5$, where the latter MSBEs were estimated by using the formal expressions of Theorems 2 to 5 approximated using the issues discussed in the beginning of this section Here $d = 5$ and $S = [0.33 \quad 0.25 \quad 0.18 \quad 0.14 \quad 0.10]^T$. In the case of the MSBE, in each column, we mention the pair being examined.

N	E_{MLE}	$E_{MSBE}\mid_{<aa>}$	$E_{MSBE}\mid_{<ab>}$	$E_{MSBE}\mid_{<ba>}$	$E_{MSBE}\mid_{<bb>}$
5^2 (25)	0.1363	NaN	0.2217	0.2242	0.2120
5^3 (125)	0.1696	0.1746	0.1952	0.1751	0.1581
5^4 (625)	0.1864	0.1861	0.1932	0.1516	0.1541
5^5 (3,125)	0.1862	0.1856	0.1906	0.1466	0.1468
5^6 (15,625)	0.1882	0.1883	0.1888	0.1495	0.1497
5^7 (78,125)	0.1879	0.1879	0.1881	0.1769	0.1770
5^8 (390,625)	0.1880	0.1879	0.1880	0.1882	0.1882

MLE and MSBEs taken for a *single* experiment don't follow such a regular pattern, especially for small values of N.

Due to space limitations, the theoretical and experimental results for the cases when the subsequences are of lengths 3 and 4 are found in [16].

4   Conclusions

In this paper, we have investigated the relatively new field of sequence-based estimation. The pioneering work in this area [15] introduced the concepts of Sequence Based Estimation (SBE) for Binomial distributions. This paper has generalized the latter results for multinomial distributions. The rationale motivating the development of SBEs and MSBEs is that traditional ML and Bayesian estimation ignore/discard valuable *sequence*-based information. SBEs "extract" the information contained in the observations when perceived as a *sequence*. In this paper, we have generalized the results of [15] for the multinomial case. Our strategy involves a novel and previously-unreported phenomenon called "Occlusion" where by hiding (or concealing) certain observations, we map the original estimation problem onto a lower-dimensional binomial space. We have also shown how these consequent occluded SBEs can be fused to yield overall Multinomial SBE (MSBE). This is achieved by mapping several lower-dimensional estimates, that are all bound by rigid probability constraints, onto the original higher-dimensional space. The theoretical and experimental results for the cases when the subsequences are of lengths 3 and 4 are found in [16].

References

1. Bickel, P., Doksum, K.: Mathematical Statistics: Basic Ideas and Selected Topics, vol. 1, 2nd edn. Prentice Hall, Upper Saddle River (2000)
2. Bunke, H.: Structural and syntactic pattern recognition. In: Chen, C.H., Pau, L.F., Wang, P.S.P. (eds.) Handbook of Pattern Recognition and Computer Vision, pp. 163–209. World Scientific-25, River Edge (1993)
3. Casella, G., Berger, R.: Statistical Inference, 2nd edn. Brooks/Cole Publisher Company, Pacific Grove (2001)
4. Duda, R., Hart, P., Stork, D.: Pattern Classification, 2nd edn. John Wiley and Sons, Inc., New York (2000)
5. El-Gendy, M.A., Bose, A., Shin, K.G.: Evolution of the internet QoS and support for soft real-time applications. Proc. IEEE **91**, 1086–1104 (2003)
6. Friedman, M., Kandel, A.: Introduction to Pattern Recognition - Statistical, Structural, Neural and Fuzzy Logic Approaches. World Scientific, New Jersey (1999)
7. Fukunaga, K.: Introduction to Statistical Pattern Recognition. Academic Press, San Diego (1990)
8. Goldberg, S.: Probability: An Introduction. Prentice-Hall, Englewood Cliffs (1960)
9. Herbrich, R.: Learning Kernel Classifiers: Theory and Algorithms. MIT Press, Cambridge (2001)
10. Jones, B., Garthwaite, P., Jolliffe, I.: Statistical Inference, 2nd edn. Oxford University Press, New York (2002)
11. Kittler, J., Hatef, M., Duin, R.P.W., Matas, J.: On combining classifiers. IEEE Trans. Pattern Anal. Mach. Intell. **PAMI–20**, 226–239 (1998)
12. Kreyszig, E.: Advanced Engineering Mathematics, 8th edn. John Wiley & Sons, New York (1999)
13. Kuncheva, L.I., Bezdek, J.C., Duin, R.P.W.: Decision templates for multiple classifier fusion: an experimental comparison. Pattern Recogn. **34**, 299–414 (2001)
14. Kuncheva, L.I.: A theoretical study on six classifier fusion strategies. IEEE Trans. Pattern Anal. Mach. Intell. **PAMI–24**, 281–286 (2002)
15. Oommen, B.J., Kim, S.-W., Horn, G.: On the estimation of independent binomial random variables using occurrence and sequential information. Pattern Recogn. **40**(11), 3263–3276 (2007)

16. Oommen, B.J., Kim, S-W.: Occlusion-based estimation of independent multinomial random variables using occurrence and sequential information. To be submitted for Publication
17. Ross, S.: Introduction to Probability Models, 2nd edn. Academic Press, Orlando (2002)
18. Shao, J.: Mathematical Statistics, 2nd edn. Springer, Heidelberg (2003)
19. Sprinthall, R.: Basic Statistical Analysis. Allyn and Bacon, Boston (2002)
20. van der Heijden, F., Duin, R.P.W., de Ridder, D., Tax, D.M.J.: Classification, Parameter Estimation and State Estimation: An Engineering Approach using MATLAB. John Wiley and Sons Ltd, England (2004)
21. Webb, A.: Statistical Pattern Recognition, 2nd edn. John Wiley & Sons, New York (2002)

Global Solar Radiation Prediction Using Backward Propagation Artificial Neural Network for the City of Addis Ababa, Ethiopia

Younas Worki[1], Eshetie Berhan[1], and Ondrej Krejcar[2(✉)]

[1] School of Mechanical and Industrial Engineering, Addis Ababa Institute
of Technology, Addis Ababa University, Addis Ababa, Ethiopia
yonaswork2006@gmail.com, berhan.eshetie@gmail.com
[2] Faculty of Informatics and Management,
Center for Basic and Applied Research, University of Hradec Kralove,
Rokitanskeho 62, 50003 Hradec Kralove, Czech Republic
Ondrej@Krejcar.org

Abstract. Ethiopia is located close to the equatorial belt that receives abundant solar energy. For Ethiopia, to achieve the optimum utilization of solar energy, it is necessary to evaluate the incident solar radiation over the countries of interest. Though, sophisticated and costly equipment are available but they are very limited for developing countries' like Ethiopia. This paper is therefore tries to explore the use of artificial neural network method for predicting the daily global solar radiation in the horizontal surface using secondary data in the city of Addis Ababa. For this purpose, the meteorological data of 1195 days from one station in Addis Ababa along the years 1985–1987 were used for training testing and validating the model All independent variables (Min and Max Temperature, humidity, sunshine hour and wind speed were normalized and added to the model. Then, Back propagation (BP) Artificial Neural Network (ANN) method was applied for prediction and training respectively to determine the most suitable independent (input) variables. The results obtained by the ANN model were validated with the actual data and error values were found within acceptable limits. The findings of the study show that the Root Mean Square Error (RMSE) is found to be 0.11 and correlation coefficient (R) value was obtained 0.901 during prediction.

Keywords: Global solar radiation · Artificial Neural Network · Backpropagation · Climatological parameters · Multi-layer feedforward neural network

1 Introduction

Due to geographical advantages, it has recently been realized that renewable energy sources can have a beneficial impact, compared with other energy sources, with regard technical, environmental, and political vision in the country Ethiopia [1–3]. Solar energy is the most ancient source, and; it is the root material for almost all fossil and renewable types. Solar energy is freely available and could be easily harnessed to reduce our

© Springer International Publishing Switzerland 2016
N.T. Nguyen et al. (Eds.): ICCCI 2016, Part I, LNAI 9875, pp. 230–238, 2016.
DOI: 10.1007/978-3-319-45243-2_21

reliance on hydrocarbon-based energy by both, passive and active designs. Precise solar-radiation estimation tools are critical in the design of solar systems [4, 13, 14].

Information about the availability of solar energy on flat surface is crucial for the target design and study of solar energy systems [14]. For a country like Ethiopia, the efficient application of solar energy seems unavoidable because of plentiful sunshine available during the year [3]. In many countries the traditional way of knowing the amount of Global Solar Radiation (GSR) in a particular region is to install Pyranometers as many locations as possible thus requiring daily maintenance, data recording, and consequently increasing cost of GSR data collection [5, 12]. Companies and institutions in Ethiopia commonly uses annual or monthly average GSR values which are available from different sources via websites and solar system sizing simulation software for solar system sizing. However, since such data were developed based on climatic and geographical analogy of other neighbouring areas with large uncertainties it leads to inefficient and costly system sizing [12, 14].

Therefore, it is rather more economical to develop methods to estimate the GSR using climatological parameters. For example, using artificial neural network as an estimation tool has proved its efficiency in predicting different parameters. Various studies on Artificial Neural Networks (ANN) for prediction of GSR with different approaches have been conducted in various parts of the world [11–18]. Climatological and meteorological parameters are important parameters in indicating the amount of solar radiation in a selected region. So applying ANN can be valuable in determining the effects of meteorological parameters and finally prediction of solar radiation [3, 4]. This paper is therefore, ANN model that will take five weather variables as an input for predicting solar radiation in the city of Addis Ababa, Ethiopia using Backward Propagation ANN.

2 Problem Definitions

In order to build reliable solar energy systems, meteorological GSR weather information should be available for the region where the systems are planning to build. Many countries have developed their own GSR prediction models as a valuable long-term energy-planning tool. The Ethiopian Metrological institute still lacks models for the GSR of the country and in particular for the city of Addis Ababa, where various power systems are designed and implemented. This work is the first attempt to generate a weather model for the city of Addis Ababa and which will later be extended to other Ethiopian cities.

Addis Ababa has a sub-tropical highland climate. The city has a complex mix of highland climate zones with temperature of up to 10 C depending on the elevation and prevailing wind pattern. As the rest of other cities in Ethiopia, Addis Ababa, has also four main seasons. The four seasons are Summer (Kiremt or Meher in Ethiopian Language) from June to August with heavy rainfall in the three months, Autum (Belg) from September to November spring season and sometimes known as the harvest season, Winter (Bega) from December to February with frost in the morning specially in January and the Spring (Tseday) from March to May with occasional shower and may is the hottest season.

3   Literature Review

An Artificial Neural Networks (ANN) is a type of artificial intelligence (computer system) that attempts to mimic the way the human brain processes and stores information. It works by creating connections between mathematical processing elements called neurons. In Artificial Neural Network, knowledge is encoded into the network through the strength of the connections between different neurons, called weights, and by creating groups, or layers, of neurons that work in parallel [5, 12, 18, 19]. The system learns through a process of determining the number of neurons or nodes and adjusting the weights for the connections based on training data. Artificial neural networks (ANNs) is developed in the 1980s. They have contributed to the advancement of science and engineering in the field of scientific computing.

The most popular algorithm for ANNs is the back-propagation neural network (BPNN), owing to its advantages in terms of a simple structure and ease of implementation for solving problems such as function approximation, time series forecasting, pattern recognition, and process control [19].

The BPNN training processes are commonly divided into two phase. The first one isforward propagation of information and the second one is backward propagation of errors. These are the most widely used approaches in the neural network model. The Back-propagation (BP) neural networks commonly have better advantage than the other in terms of their nonlinear characteristic, self-learning, self-organizing, and adaptive capacity. BP sare widely applied in function approximation and data fitting. Since 1986, the use of BP neural networks have resulted in tremendous achievements in the study of spatial distribution. These networks have achieved better performance than other methods [8, 11, 19]. A number of scholars have also carried out research on sunshine distribution using BPNN method. Benghanem et al. [6] used a BP neural network model to estimate and model daily global solar radiation sing data from 1998 to 2002 at the National Renewable Energy Laboratory. Jiang [9, 10] employed BP neural network models to estimate monthly mean daily global solar radiation in eight typical cities in China, while Cao and Cao [7] developed a new method combining an artificial neural network and wavelet analysis to forecast solar irradiance.

Furthermore, Ali, [5] presented a BP neural network to estimate global solar radiation (GSR) as a function of air temperature data in a semi-arid environment. The models were then trained to estimate GSR as a function of the maximum and minimum air temperature and extra-terrestrial radiation. BPNN can simple represented as shown in Fig. 1 with the numbers of neurons in the input, hidden, and output layers are I, H, and O, respectively.

Let define $X = \{x_1, x_2, ..., x_I\}$ as an arbitrary input of the sample vector in the model, $Z = \{z_1, z_2, ..., z_O\}$ be the actual output vector, $D = \{d_1, d_2, ..., d_O\}$ be the expected output vector, $w_{ij}(1 \leq i \leq I, 1 \leq j \leq H)$ be the weight that connects the values between the input vector and hidden layers vectors. Further; let $v_{jk}(1 \leq j \leq H, 1 \leq k \leq O)$ be the expected values between the hidden layers and the output layer, $\theta_j(1 \leq j \leq H)$ be the output threshold of each neuron in the hidden layers, $r_k(1 \leq k \leq O)$ be the threshold of each neuron in the output layer, and let $f(x)$ be an activation function that usually employs the sigmoid function, and given by $f(x) = \frac{1}{1+e^{-x}}$. The output of a

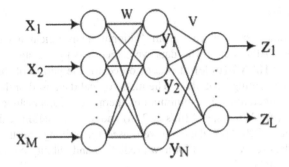

Fig. 1. Structure of a typical Three Layer BPNN [12]

hidden layers neuron are calculated by $y_{j=f\left(\sum_{k=1}^{I} w_{ij \cdot x_i - \theta_i}\right)}$, the output of an output layer neuron is calculated as $z_{k=f\left(\sum_{k=1}^{O} v_{jk \cdot y_j - r_k}\right)}$ and the error of an output layer neuron is calculated by using $E = \frac{1}{2}\sum_{k=1}^{O}(d_k - z_k)^2$ which is exactly equals to $\frac{1}{2}\sum_{k=1}^{O}(d_k - f\left(\sum_{k=1}^{O} v_{jk \cdot y_j - r_k}\right))^2$.

Since BPNN is following a gradient descent algorithm, then both the weight and threshold values are adjusted and calculated by the following formulas. $\Delta w_{ij} = -\rho \frac{\partial E}{\partial w_{ij}}$, $\Delta v_{jk} = -\rho \frac{\partial E}{\partial v_{jk}}$, $\Delta \theta_j = -\rho \frac{\partial E}{\partial \theta_j}$, $\Delta r_k = -\rho \frac{\partial E}{\partial r_k}$, where ρ the learning rate set by the user. The training does not terminate until the total error is smaller than the set value. The computed parameters such as the connecting weight between the input and the hidden layers, the value between the hidden and the output layer, the output threshold value of each neuron in the hidden layers and the threshold of each neuron in the output layer are used for predicting and fitting the model.

4 Research Methodology

4.1 Study Area

The city of Addis Ababa, capital city of Ethiopia, is located between longitude 9° 1′ N and latitude 38° 44′ E with an elevation of 2356 m height and covers an area of 527 km². The mean annual temperature in Addis Ababa, Ethiopia is very mild at 15.9 °C. The variation of average monthly temperatures is 3 °C which is an extremely low range. In the winter time records showed temperatures by day reach 23.3 °C (74 °F) on average dropping to 7.3 °C (45.2 °F) overnight. In spring time temperatures increase reaching 24.7 °C (76.4 °F) mostly in the afternoon with the overnight peaks of 10.7 °C (51.2 °F). During summer average high temperatures are 21 °C (69.8 °F) and average low temperatures are 10 °C (50 °F).

4.2   Data Sources

Meteorological data were collected from National metrological agency (NMA) of Ethiopia for Addis Ababa city. The data were recorded by the agency at a specific station name called HAAB (Bole). It was recorded for the period of three years from 1985 to 1987 GC, covering 1195 days. The metrological data used in this research are: daily maximum temperature (°C), minimum temperature (°C), sunshine (hours), mean relative humidity (%), Wind speed (m/s at 2 m) and solar radiation (cal/cm^2/day. The first two years' data (720 days' data set) were used for training and the next one-year data (375 days' data set) were used for new prediction and validity on the best network as an independent data.

However, almost all of these metrological data has some missed values. The missed values were proportionally ranging from 12 % to 25 % at different days between the years 1985 to 1987. IBM SPSS statistics 22 software was used to statistically replace those missed input/output values using data imputation techniques before the data is being normalized. A portion of the normalized input output data is presented in Table 1.

Table 1. Normalized input/output parameters and sample dataset

Input/output parameters		Daily values						
		1	2	3	4	5	...	1095
Input parameters	Daily average relative humidity (%)	0.13	0.03	−0.09	−0.12	−0.05	...	0.15
	Daily max. Temperature (°C)	0.45	0.50	0.55	0.41	0.39	...	0.65
	Daily min. Temperature (°C)	−0.21	−0.27	−0.31	−0.29	−0.33	...	0.52
	Daily sunshine (h)	0.52	0.61	0.61	0.67	0.67	...	−0.02
	Daily wind speed (m/s at 2 m)	−0.61	−0.38	−0.25	0.19	−0.31	...	−0.63
Output/Target Parameters	Radiation (kwh/m^2/day)	0.19	0.12	0.33	0.39	−0.21	...	0.00

5   Results and Discussions

A statistical analysis including Root Mean Square error (RMSE), and deterministic coefficient (R2) is showed in order to evaluate the performance accuracy of the developed BPANN models. Further, to verify whether there is any causal performance trend in the ANN models under study, RMSE and correlation coefficient was compared under each approximation. RMSE specifies information on the temporary performance and measures the difference of predicted values in the area of the actual measured data. The lower the RMSE, the more accurate is the estimation whereas the higher is the R value is better on the model. The comparison between predicted ANN models and the measured GSR data for all the nine approximations are plotted on Fig. 2. This is the number of times, that the validation was run. The Root Mean Square Error (RMSE) value increases from the fifth to the seventh approximation, but drastically decrease and become stable afterwards and attains its minimum value, which is 0.11 at this point. Similar to the ninth approximation, the correlation coefficient R also recorded its maximum value. The correlation coefficient R also increased from six to the ninth approximation. The BPANN achieves its best performance on the ninth approximation, which is R = 0.901.

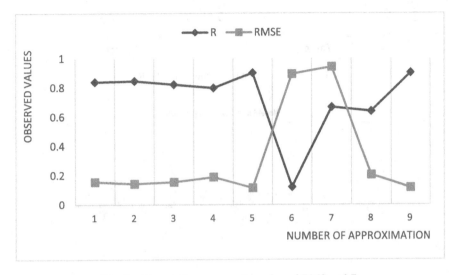

Fig. 2. The relations among the value of RMS and R

Moreover, based on the RMSE and R value, the research is also further validated and fitted to the regression model and its finding is shown at Fig. 3. The predicted results of the regression model are closed to 45° lines, which indicates that the difference between the predicted and the measured values are relatively smaller. The statistical characteristics of RMSE = 0.013 and R = 0.9204 were obtained on the training phase and RMSE = 0.020 and R = 0.80064 in the testing phase. As a result, the proposed model deemed an efficient to predict the GSR for the Addis Ababa for practical purpose such as solar power system [16, 18].

For further illustration the real amount of GSR data were scaled and presented to the model. This illustration is done using measured daily GSR data for the years 1985–1987. The result is demonstrated on Fig. 4. The result depicted the difference between the predicted and measured values. The error values of the model are concentrated around the zero line, which indicates the prediction were accurate. The predicted ANN models provide a very good prediction of Daily GSR behavior for the city of Addis Ababa with minimum error and RMSE value of 0.11 and correlation coefficient of R equals to 0.901.

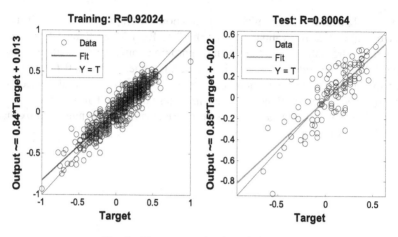

Fig. 3. The measured value of the GSR

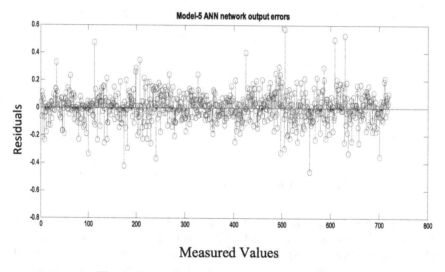

Fig. 4. The predicted and residual during validation

6   Conclusion

The study's purposes incorporated finding a simple model for forecasting solar energy for the city of Addis Ababa through use of ANN approximation using Backward Propagation techniques. A three years' data that were collected daily were used for the of training purpose and testing purpose. During validation, the model resulted in a best fit after the ninth approximation with correlation coefficients of 0.901, with the lowest error values RMSE: 0.1140. From the findings it can be concluded that BPANN can be used to predict the GSR of Addis Ababa at a minimum cost and reasonable accuracy. The researchers recommended the Ethiopian metrology agency to use the developed

BPANN model for prediction of solar radiation 24 h in advance that would help efficiently optimize energy distribution of new and already installed solar power systems. This finding is in line with the findings of [16, 18, 20].

Acknowledgment. This work and the contribution were supported by project "Smart Solutions for Ubiquitous Computing Environments" FIM, University of Hradec Kralove, Czech Republic (under ID: UHK-FIM-SP-2016-2102).

References

1. Mekonnen, S.A.: Solar energy assessment in Ethiopia: modeling and measurement. Addis Ababa University Department of Environmental Science, July 2007
2. Kassahun, G.S.: Predictive modelling of Kaliti wastewater treatment plant performance using Artificial Neural Networks, Chemical Engineering with Specialization in Environmental Engineering, February 2012
3. Ethio Resource Group with Partners, Solar and Wind Energy Utilization and Project Development Scenarios, Ethiopian Rural Energy Development and Promotion Center, October 2007
4. Azadeh, A., Maghsoudi, A., Sohrabkhani, S.: An integrated artificial neural networks approach for predicting global radiation. Energy Convers. Manage. **50**(6), 1497–1505 (2009)
5. Ali, R.: Estimating global solar radiation using artificial neural network and air temperature data in a semi-arid environment. Renew. Energy **35**(9), 2131–2135 (2010)
6. Benghanem, M., Mellit, A., Alarm, S.N.: ANN-based modelling and estimation of daily global solar radiation data: a case study. Energy Convers. Manage. **50**(7), 1644–1655 (2009)
7. Cao, J.C., Cao, S.H.: Study of forecasting solar irradiance using neural networks with preprocessing sample data by wavelet analysis. Energy **31**(15), 3435–3445 (2006)
8. Chang, F.-J., Kao, L., Kuo, Y.-M., et al.: Artificial neural networks for estimating regional arsenic concentrations in a black foot disease area in Taiwan. J. Hydrol. **388**, 65–76 (2010)
9. Jiang, Y.: Computation of monthly mean daily global solar radiation in China using artificial neural networks and comparison with other empirical models. Energy **34**(9), 1276–1283 (2009)
10. Singh, K.P., Gupta, S.: Artificial intelligence based modeling for predicting the disinfection by-products in water. Chemom. Intell. Lab. Syst. **114**, 122–131 (2012)
11. Vakili, M., Sabbagh-Yazdi, S.-R., Kalhor, K., Khosrojerdi, S.: Using Artificial Neural Networks for prediction of global solar radiation in Tehran considering particulate matter air pollution. Energy Procedia **74**, 1205–1212 (2015)
12. Olatomiwa, L., Mekhilef, S., Shamshirband, S., Petković, D.: Adaptive neuro-fuzzy approach for solar radiation prediction in Nigeria. Renew. Sustain. Energy Rev. **51**, 1784–1791 (2015)
13. Jiang, H., Dong, Y., Wang, J., Li, Y.: Intelligent optimization models based on hard-ridge penalty and RBF for forecasting global solar radiation. Energy Convers. Manage. **95**(1), 42–58 (2015)
14. Mohammadi, K., Shamshirband, S., Petković, D., Khorasanizadeh, H.: Determining the most important variables for diffuse solar radiation prediction using adaptive neuro-fuzzy methodology; case study: City of Kerman. Iran, Renew. Sustain. Energy Rev. **53**, 1570–1579 (2016)

15. Akarslan, E., Hocaoglu, F.O.: A novel adaptive approach for hourly solar radiation forecasting. Renew. Energy **87**(1), 628–633 (2016)
16. Kumar, R., Aggarwal, R.K., Sharma, J.D.: Comparison of regression and artificial neural network models for estimation of global solar radiations. Renew. Sustain. Energy Rev. **52**, 1294–1299 (2015)
17. Çelik, Ö., Teke, A., Yıldırım, H.B.: The optimized artificial neural network model with Levenberg–Marquardt algorithm for global solar radiation estimation in Eastern Mediterranean Region of Turkey. J. Cleaner Prod. **116**(10), 1–12 (2016)
18. Eriko, I., Kenji, O., Yoichi, M.I., Makoto, A.: A neural network approach to simple prediction of soil nitrification potential: a case study in Japanese temperate forests. Ecol. Model. **219**(1–2), 200–211 (2008)
19. Hu, J., Zhou, G., Xu, X.: Using an improved back propagation neural network to study spatial distribution of sunshine illumination from sensor network data. Ecol. Model. **26**, 86–96 (2013)
20. Krejcar, O., Mahdal, M.: Optimized solar energy power supply for remote wireless sensors based on IEEE 802.15.4 standard. Int. J. Photoenergy **2012**, 9 (2012). doi:10.1155/2012/305102. Article ID: 305102

Determining the Criteria for Setting Input Parameters of the Fuzzy Inference Model of P&R Car Parks Locating

Michał Lower$^{(\boxtimes)}$ and Anna Lower

Wroclaw University of Science and Technology,
ul. Wyb. Wyspianskiego 27, 50-370 Wroclaw, Poland
{michal.lower,anna.lower}@pwr.edu.pl

Abstract. The paper presents the fuzzy inference model of evaluation of P&R facilities location. In such a system there are car parks where travellers can change from car to public transport. Recognising proper locations of the P&R facilities is a key aspect for the system. On the basis of previous studies it was found that developing algorithms supporting the determination of input parameters to the model is necessary. The authors created recurrent algorithm for determining the parameter of the road quality. For the parameters of quantity of public transport means and the distance to the city centre the algorithm based on the exponential function was made. For determining the parameter describing the connection to the main road the fuzzy inference model was built. Thanks to the developed methodology the model of P&R locating has become more universal and possible to use for a broader group of experts.

Keywords: Fuzzy logic inference · Park and ride system · P&R facilities · P&R location

1 Introduction

The problem of urban congestion is currently an important issue considered in spacial planning. Due to environmental factors and the widely understood quality of life of cities' inhabitants, the aim is to reduce car traffic in cities with a special focus on central zones. Best practice is to introduce significant limitations in the movement of individual vehicles and replace it with a well-functioning public transport system with the ability to use park and ride (P&R) complex interchange nodes. Park and ride system has been strongly developed in Europe in the last 20 years. It takes different forms depending on local circumstances and expectations. The most common form is such a kind where there are car parks located close to the intermodal hubs. These are places where travelers can change from their cars to public transport - rail and bus to get to their destination. [2]. The consequence of the choice of public transport is the reduction of number of cars in the city centre. This provides benefits in terms of lower congestion and the expected reduction of air pollution. Recognising proper locations of the

© Springer International Publishing Switzerland 2016
N.T. Nguyen et al. (Eds.): ICCCI 2016, Part I, LNAI 9875, pp. 239–248, 2016.
DOI: 10.1007/978-3-319-45243-2_22

P&R facilities is a key aspect for the system. Practice and study [1] indicates that P&R facilities are not used if they are uncomfortable located for potential users, even if a level of congestion is high. The criteria for the quality assessment of the selected location are formulated in descriptive and general terms. Research outsourced to specialists is expensive and time consuming. Most focus is given to the examination of a few pre-selected places [12]. Practice has proved that the intuitive choice of the location of these places, without a detailed analysis of all the circumstances often produces adverse effects. As a result, the existing facilities are not used as intended [7]. The choice of the place of P&R facilities is a research issue carried out by scientists, e.g. the authors of the paper [4] define locations in a very precise manner based on the input data. Possible applications of such a model are very limited because obtaining such detailed and precise data requires additional funds and effort. Built mathematical models often do not treat the problem comprehensively, e.g. it is assumed that a city has linear organisation and has been developed along one important communications corridor [14] and private cars of inhabitants of settlements in the outskirts of the city are the main source of morning traffic. The authors propose to consider also a model with a number of transport corridors. However, they state that it would be especially difficult because the catchment areas of P&R objects would have complex shapes, e.g. parabolic boundaries.

There is also a number of studies based on GIS platform. The authors use statistical research [3,5] but the study is not a simple way to indicate the location of the parking lot. Data collected by GIS platform can only provide a basis for analysis for a group of experts. Moreover, the current algorithms used in GIS platform do not analyse a number of important parameters, especially those whose data are ambiguous and imprecise [6].

In our previous studies we applied expert knowledge to fuzzy inference model of P&R car parks locating [6,7,10]. With such a model of location assessment even a less experienced person can use it, e.g. urban planners, officials. The model can help obtaining fast results, so a large number of proposals can be examined in a short time. The use of inference based on fuzzy logic allowed to take into account factors that are difficult to be clearly estimated. This allowed to use the knowledge and judgement of an expert effectively. Such an approach greatly facilitated and accelerated the analysis of the problem.

Fuzzy logic, which is one of the newest areas of mathematics, is widely used for inference in such cases. It is particularly useful in situations where the input parameters for evaluation are difficult to determine and are based on intuitive expert knowledge. The examples of such applications can be found in publications [8,9,11].

2   Research of P&R Facilities Locating with the Use of Fuzzy Logic

Specific features of P&R system determine the usefulness of the selected site to place there P&R facilities. Appropriate placement of a P&R facility in the

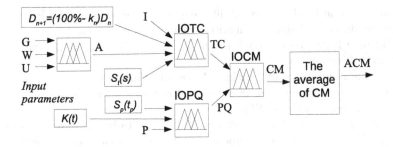

Fig. 1. The scheme of the fuzzy inference model of P&R car parks locating.

communication system makes it more popular, and thus, increases the capacity of the major roads. The P&R facilities should offer convenient access to down-town which is usually the most attractive area [1,7,13]. All researchers are consistent on three issues pertaining to the location of a P&R facility. The proximity of major roads is important, P&R should be visible from the road, moreover: a location on the verge of congestion seems ideal as it allows the users to use their cars on the less crowded part of the road [7].

The value of the site for locating the P&R facility can be assessed on the basis of a number of conditions and parameters. They can be categorized according to the territorial conditionality and to the public transport [10]. Public transport is to take over passenger traffic in order to reduce car traffic. On the basis of this expert classification the inference has been divided into two local models of inference. The final result is the effect of fuzzy inference based on the results of inference of local models. The territorial conditions indicator is designated by IOTC model. The public transport quality indicator is determined by IOPQ model. On the basis of the results of computing of IOTC and IOPQ models the final score is designated by IOCM model. Additionally, the parameter I can be given as a range of values from minimum to maximum so, ultimately, the final result is averaged. During testing we decided to add another fuzzy local model which is to determine a parameter A. All four local models are Mamdani type. The scheme of the complete inference is presented in Fig. 1.

The triangular membership functions for each value of linguistic variable and the centre of mass method are used in all fuzzy models. The range of values of input and output parameters is 0–100% [10].

The following parameters have been adopted as linguistic variables of the fuzzy inference model IOTC:

I – traffic intensity – the number of cars aiming to the city every day (the city as a destination).

D – quality of the road connecting traffic sources with the P&R facility (intersections, obstacles, etc.).

A – connection between the main road and the parking place (G - ease of access from the main road, W – visibility of the parking facility from the main road, U – the level of accessibility, e.g. P&R location on the left or right side, readability of signs). Linguistic variable values: *good, bad.*

S_t – distance from the car park to the centre of the city, in the category of territorial indicators.

TC – territorial indicator of location, the result of inference of the IOTC model. The following parameters have been adopted as linguistic variables of the fuzzy inference model IOPQ:

K – quantity of different means of city transport (railway, subway, bus, tram, etc.), the number of communication lines within each type, frequency of running.

P – the distance from the car park to the nearest public transport stop.

S_p – the distance from the car park to the centre of the city, in the category of public transport indicators. The parameter is dependent on the time required to get to the city centre by public transport (t_p) in comparison with the time required for driving there by car (t_w).

PQ – public transport quality indicator, the result of inference of the IOPQ model [10].

3   Algorithms Defining the Input Parameters

After conducting a series of studies of cities in terms of location P&R facilities [6,7,10] the need to create algorithms that help to assess the value of the input parameters was noticed. The value of parameters will be determined on the basis of the experience gained in these studies.

3.1   The Parameters I and P

It turned out that the inflow of vehicles can be found directly from the available traffic information. Therefore, at this stage of studies, an algorithm defining the parameter I on the basis of other factors has not been developed. Similarly, the parameter P has been decided to be determined directly by an expert. If it turns out that there are problems with obtaining necessary information then an additional algorithm will be developed.

3.2   The Parameter D

When assessing the value of the parameter D a location at the beginning of agglomeration was adopted as 100 %. The quality of access to the parking lot is reduced depending on changes in road class in conjunction with the traffic and the amount of road obstacles according to formula (1).

$$D_{n+1} = (100\% - k_n)D_n \qquad (1)$$

where $D_0 = 100\%$ and it means there are no obstacles.

n – the number of the next obstacle,

k_n – the percentage weight of the next obstacle,
D_n – indicator of the road quality including the obstacles from 0 to n.

Parameter k_n was determined on the basis of changes in the quality of the road, the size of traffic jam at the intersections or other obstacles (e.g. level crossings, restrictions, etc.). The colours refer to a 4-point scale reading of Google maps with traffic intensity taken into account:

– four–leg intersection, orange colour, $k_n = 4$
– four–leg intersection, red colour, $2 \times 4\% \ k_n = 8$
– four–leg intersection, brown colour – $4 \times 4\% \ k_n = 16$
– three–leg intersection or four–leg intersection without possibility to take left turn from minor road, changing colour to red $2 \times 2\% \ k_n = 4$
– three–leg intersection or four–leg intersection without possibility to take left–turn from minor road, changing colour to brown $4 \times 2\% \ k_n = 8$
– four–leg intersection with four–line road – $k_n = 16$
– transformation of the width of the road – from four-lane to two-lane road, $k_n = 20$

3.3   The Parameter A

Parameter A - access to the parking lot from the major road. The parameter is determined by the Mamdani fuzzy inference model, where the input parameters are defined as follows:

G – ease of access from the main road, the value is *very good* when the location of the facility is directly next to the road which is the main inlet corridor,
W – visibility of the parking facility from the main road,
U – the level of accessibility, e.g. P&R location on the left or right side, readability of signs.

Membership functions of all parameters are defined as triangular. The function vertices (the value of the membership function equals 1) for the parameter A has been adopted according to the following scale:

– 0 % -*insufficient*
– 25 % - *mediocre*
– 50 % -*sufficient*
– 75 % - *good*
– 100 % - *very good*

The following fuzzy inference rules has been defined:

1. **If** G is *good* **and** W is *good* **and** U is *small* **then** A is *very good*
2. **If** G is *good* **and** W is *good* **and** U is *big* **then** A is *good*
3. **If** G is *good* **and** W is *poor* **and** U is *small* **then** A is *very good*
4. **If** G is *good* **and** W is *poor* **and** U is *big* **then** A is *sufficient*

5. **If** G is *bad* **and** W is *good* **and** U is *small* **then** A is *sufficient*
6. **If** G is *bad* **and** W is *good* **and** U is *big* **then** A is *mediocre*
7. **If** G is *bad* **and** W is *poor* **and** U is *small* **then** A is *mediocre*
8. **If** G is *bad* **and** W is *poor* **and** U is *big* **then** A is *insufficient*

All membership functions meet the unity rule, i.e. the sum of all values of the membership functions at each point of the domain of function is equal to 1.

According to the fuzzy inference rules, the value of the parameter A is *very good* when the location of the facility is directly next to the major road, the object can be seen from the major road and from the moment one can see it the access is straightforward. The value of the parameter A is reduced when the object is not visible or when it is necessary to use an additional way of a lower class that supports entry. The parameter A has low value for an object located close to the road, but with a great difficulty of access, e.g. after seeing it, one must drive further approx. 700 m on the main road, and then pull off the road on the lower class road and pass through several intersections to get to it. Such a situation, experts also found uncomfortable.

3.4   The Parameter K

Parameter K - the waiting time for public transport (the combination between a number of modes of public transport and the waiting time for a single mode). The value of K is 100 when the waiting time is equal to 0.1 of the time that an access by car would take. The waiting time of 60 min. gives the value of parameter $K = 0$. In view of the fact that exceeding the waiting time of 30 min. gives extremely negative feeling of the traveller, the longer waiting does not deepen frustration in significant way. Thus, the calculation can not be linear. Therefore, to calculate the parameter K the exponential function (2) was used.

$$y = a \cdot b^{(c-t)} \tag{2}$$

where t – the value of the function, the waiting time (minutes).

It was concluded that the critical point of the waiting time is $t = 20$ min and it will respond to parameter K at 30 %. Thus, it is assumed that $a = 30$ and $c = 20$. Assuming that $t = 2$ min, K should be about 100. Therefore, for the parameter b the value was set at 1.07. Finally, the formula has adopted the following shape (3) and (4).

$$K = \begin{cases} y_k & \text{if } y_k < 100 \\ 100 & \text{if } y_k \geq 100 \end{cases} \tag{3}$$

where:

$$y_k = 30 \cdot 1.07^{(20-t)} \tag{4}$$

3.5   The Parameter S

Parameter S has been divided into two parameters - S_t and S_p. In the category of territorial indicators the value of parameter S_t was adopted as 100 % (it is a distance s_{max} to city centre) in a place situated on the border of the congestion area during the traffic peak hours. The closer to the city centre the lower parameter value. In the place where the parking lot may become a destination point the value of S_t is 0 % (it is a distance s_{min} to city centre). It is approx 5 min. pedestrian access to the city centre. To determine the parameters S_t in the program supporting the work of an expert it is assumed that the intermediate values of S_t are proportional to the distance according to the above described rule. Therefore, the formula computing the value of S_t is presented in (5).

$$S_t = \begin{cases} 0 & \text{for } s \leq s_{min} \\ \dfrac{100\,(s - s_{min})}{s_{max} - s_{min}} & \text{for } s_{min} < s < s_{max} \\ 100 & \text{for } s \geq s_{max} \end{cases} \quad (5)$$

where s is a distance to city centre measured in kilometres.

Parameter S_p is a distance to the city centre measured by the time required to reach it by means of public transport. The shorter the time, the higher the quality of the location. This means that in spite of the desired long distance (territorially), efficiently organized public transport (the shortest route, traffic priority, etc.) allows to comfortably achieve the destination point in a short time. In the category of public transport indicators, the parameter S_p was adopted as 100 % when a travel time to the city centre by means of public transport does not exceed 10 min. The value of parameter decreases with increasing of the time of drive. The relationship between time of travel by car and a commuting time by public transport is also very important. The location of P&R facility is attractive when a commuting time by public transport is shorter than time of travel by car. Therefore, the calculation of the real travel time is an important element that influences the evaluation of parameter S_p. For this purpose the average movement speed in the city (v_a) should be determined, eg. for medium-sized cities it may be 30 km/h. The average speed mainly depends on traffic congestion. At the next step the distance (s_s) from the examined P&R facility location to the city centre should be estimated. On the basis of these two values the travel time to the city centre (t_w) should be determined according to the formula (6).

$$t_w\,[min] = \frac{60 \cdot s_s\,[km]}{v_a\left[\dfrac{km}{h}\right]} \quad (6)$$

It should be assumed that the commuting time by public transport (t_p) equal to the calculated time t_w gives the value of the parameter $S_p = 50$. If $t_p < t_w$ the parameter S_p takes values greater than 50. If $t_p > t_w$ the parameter S_p takes values less than 50. The commuting time t_p equal to the time calculated for the fastest means of public transport for the distance to the destination point, measured in a straight line, gives the value of the parameter $S_p = 100$.

The value of parameter $S_p = 0$ must be expertly determined depending on the kind of city. It may occur in situation where:

$$t_p = 2 \cdot t_w \tag{7}$$

To determine the parameter S_p in the program supporting the work of an expert the formula (2) was used. It was assumed that $c = t_w$ and $t = t_p$. It was adopted that for S_p the critical point is the moment in which $t_p = t_w$ and then S_p should be 50 %, so for such a situation, a value $a = 50$. Finally S_p is shown in the formulas (8) and (9).

$$S_p = \begin{cases} y_s & \text{if } y_s < 100 \\ 100 & \text{if } y_s \geq 100 \end{cases} \tag{8}$$

where:

$$y_s = 50 \cdot b_s^{(t_w - t_p)} \tag{9}$$

The parameter b_s can be calculated from the Eq. (10) which was deduced on the base of (9) for $S_p = 100\,\%$. It should be noted that the value b_s determines the specificity of the city. To determine this value t_{wx} i t_{px} should be determined first. The value of t_{wx} is the time to drive through the stretch of road L_x by car while the value of t_{px} is the time it takes to overcome the stretch of road L_x by the fastest means of public transport. The value L_x is designated as the distance in a straight line to the city centre from the furthest possible location of the P&R parking lot. In medium-sized cities surveyed by our team the value t_{px} was about 10 min while in the big city - Warsaw 39 min. It is assumed that the P&R system is functioning properly when there are traffic amenities allowing public transport to get better average travel times than by private car.

$$100 = 50 \cdot b_s^{(t_{wx} - t_{px})} \tag{10}$$

4   The Tests Results

Input data prepared by experts for selected medium-sized and big cities, as described in [6,7] has been used to test the new methodology for determining the criteria for setting input parameters. For the selected big cities, such as Warsaw and Poznan tests were based on data obtained from the expert analysis with the use of our model and compared with the data calculated by the new procedures. The results obtained from the experts differed from data determined with the help of our new procedures, however, the proportion and hierarchy among locations were maintained. If there was a difference and it was greater than 10 % (e.g. 20–30%), this difference was almost the same for all parameters within one city. In the final result relativity of assessment between locations was preserved. Hierarchy of location quality turned out to be the same. Such a result should

Table 1. "METRO MLOCINY" location of the P&R facility in Warsaw

	I	D	G	W	U	A	K	P	S_t	b_s	S_p	TC	PQ	ACM
Previous model	80–90	70				60	95	70	80		80	78	89	90
New procedures	80–90	69	50	40	91	46	90	70	62	1.034	78	77	82	86

be considered as good as in research it is important to determine the relationship between quality locations in a city and not the absolute assessment of these locations. By assumption, in the fuzzy inference model the expert knowledge is devoid of absolute reference point.

The example of comparing of results of both methodologies for one location are shown in Table 1. It can be seen that the big difference is in the values of parameter A. This result was consulted with the expert once again. The expert concluded that after the re-analysis of the facility location a new value of the parameter A can be considered better. The expert stated that the P&R facility is quite away from the main road and in the earlier analysis it was taken into account insufficiently.

The big difference can be observed also in the values of the parameter S_t. In this case the proportion and hierarchy among locations in the same city were maintained.

5 Conclusions

The test results of new methodology for determining the criteria for setting input parameters are very significant. When formulating input parameter values by experts, without the help of these algorithms, expert knowledge combined with a good knowledge of the city is important. It was found that the expert who is not a resident of the city, has big problems with self-determining of input parameters of the model. The proposed procedures allow the determination of input parameters for experts who are not residents nor users of the examined city. Thanks to the developed methodology the fuzzy inference model of P&R locating has become more universal and possible to be used by a broader group of experts. It is important to add that experts who are users of questioned cities, took the presented proposals of solutions with satisfaction. They concluded that the proposed algorithms simplified analysis and the executed analysis became more relative. Experts expressed their views about the difference in the results determined by presented algorithms and parameters defined directly by themselves. These experts found that in their opinion parameters determined by the proposed algorithms were more credible than these determined in direct way.

References

1. Aros-Vera, F., Marianov, V., Mitchell, J.E.: p-Hub approach for the optimal park-and-ride facility location problem. Eur. J. Oper. Res. **226**(2), 277–285 (2013)

2. Clayton, W., Ben-Elia, E., Parkhurst, G., Ricci, M.: Where to park? a behavioural comparison of bus park and ride and city centre car park usage in Bath, UK. J. Trans. Geogr. **36**, 124–133 (2014)
3. Faghri, A., Lang, A., Hamad, K., Henck, H.: Integrated knowledge-based geographic information system for determining optimal location of park-and-ride facilities. J. Urban Planning Dev. **128**(1), 18–41 (2002)
4. Farhan, B., Murray, A.T.: Siting park-and-ride facilities using a multi-objective spatial optimization model. Comput. Oper. Res. **35**(2), 445–456 (2008). Part special issue: Location modeling dedicated to the memory of Charles S. ReVelle
5. Horner, M.W., Grubesic, T.H.: A GIS-based planning approach to locating urban rail terminals. Transportation **28**(1), 55–77 (2001)
6. Lower, A., Lower, M., Masztalski, R., Pach, P., Szumilas, A.: Locating P&R facilities by the fuzzy inference - case of medium - sized cities. In: SGEM Conference Proceedings ELSEVIER Products (2016)
7. Lower, A., Lower, M., Masztalski, R., Szumilas, A.: The location of park and ride facilities using the fuzzy inference model. Architect. Environ. Eng. **2**(10), 507 (2015)
8. Lower, M., Magott, J., Skorupski, J.: Analysis of air traffic incidents using event trees with fuzzy probabilities. Fuzzy Sets and Syst. **293**, 50–79 (2016). Theme: Applications of Fuzzy Sets
9. Lower, M.: Self-organizing fuzzy logic steering algorithm in complex air condition system. In: Håkansson, A., Nguyen, N.T., Hartung, R.L., Howlett, R.J., Jain, L.C. (eds.) KES-AMSTA 2009. LNCS, vol. 5559, pp. 440–449. Springer, Heidelberg (2009)
10. Lower, M., Lower, A.: Evaluation of the location of the P&R facilities using fuzzy logic rules. In: Zamojski, W., Mazurkiewicz, J., Sugier, J., Walkowiak, T., Kacprzyk, J. (eds.) Theory and Engineering of Complex Systems and Dependability: Proceedings of the Tenth International Conference on Dependability and Complex Systems DepCoS-RELCOMEX, 29 June–3 July 2015. Brunów, Poland, pp. 255–264. Springer, Cham (2015)
11. Lower, M., Magott, J., Skorupski, J.: Air traffic incidents analysis with the use of fuzzy sets. In: Rutkowski, L., Korytkowski, M., Scherer, R., Tadeusiewicz, R., Zadeh, L.A., Zurada, J.M. (eds.) ICAISC 2013, Part I. LNCS, vol. 7894, pp. 306–317. Springer, Heidelberg (2013)
12. Malasek, W., Seinke, J., Wagner, J.: Analiza mozliwosci lokalizacji parkingow P+R w rejonie glownych wlotow drogowych do Warszawy (Analysis of possibilities of location of P&R car parks in the area of the main inlet roads of Warsaw). In: Warsaw Development Planning Office J.S.C. (2009)
13. Meek, S., Ison, S., Enoch, M.: Role of busbased park and ride in the UK: a temporal and evaluative review. Transp. Rev. **28**(6), 781–803 (2008)
14. Wang, J.Y., Yang, H., Lindsey, R.: Locating and pricing park-and-ride facilities in a linear monocentric city with deterministic mode choice. Transp. Res. Part B Methodol. **38**(8), 709–731 (2004)

On Systematic Approach to Discovering Periodic Patterns in Event Logs

Marcin Zimniak[1(✉)] and Janusz R. Getta[2(✉)]

[1] Faculty of Computer Science, TU Chemnitz, Chemnitz, Germany
marcin.zimniak@cs.tu-chemnitz.de
[2] School of Computer Science and Software Engineering, University of Wollongong,
Wollongong, Australia
jrg@uow.edu.au

Abstract. Discovering periodic patterns from historical information is a computationally hard problem due to the large amounts of historical data to be analyzed and due to a high complexity of the patterns. This work shows how the derivations rules for periodic patterns can be applied to discover complex patterns in case of logs of events. The paper defines a concept of periodic pattern and its validation in a workload trace created from the logs of events. A system of derivations rules that transforms periodic patterns into the logically equivalent ones is proposed. The paper presents a systematic approach based on the system of derivation rules to discovery of periodic patterns in logs of events.

1 Introduction

Discovering periodic patterns is a time consuming process due to the extremely large amounts of workload data to be analyzed and due to a high level of complexity of the patterns. An important property of complex periodic patterns is similarity of their frequencies with the frequencies of their simple components. Similarity of frequencies makes possible discovery of simple periodic patterns first and creation of complex patterns later on. Such approach requires a system of *derivations rules* for periodic patterns and appropriate strategies for the logical inferences of patterns. The first objective of this work is to discover and to investigate a system of derivation rules that can be used to create the complex periodic patterns from the simple ones. The second objective is to invent the methods and algorithms that systematically apply the rules to discover the complex periodic patterns from information included in the logs of events.

In this work we adopt a model of historical information used for process mining [1] where a *log of events* is defined as a sequence of pairs ⟨*eventidentifier, timestamp*⟩. Timestamps associated with the events and predefined sequence of disjoint time units are used to create a workload trace that contains information about the total number of times each event has been processed within each time unit. Next, we define a system of derivation rules that transform periodic patterns in a way that increases their structural complexity. Afterwards these rules are used to create a sequence of algorithms that

© Springer International Publishing Switzerland 2016
N.T. Nguyen et al. (Eds.): ICCCI 2016, Part I, LNAI 9875, pp. 249–259, 2016.
DOI: 10.1007/978-3-319-45243-2_23

transforms a workload trace into elementary periodic patterns and later on into complex ones. The algorithms apply the derivation rules in a way that preserves equivalence of a transformed set of patterns with no loss of information from the original workload trace.

The paper is organized in the following way. The next section reviews the previous research works related to discovering periodic patterns in historical information. A Sect. 3 defines the concepts of multisets, time units and it shows how a log of events is transformed into a workload trace. A Sect. 4 defines a concept of periodic pattern and its validation in a workload trace. A system of derivation rules for periodic patterns is proposed in a Sect. 5. A Sect. 6 presents the algorithms that use the derivation rules to discover complex periodic patterns. Finally, a Sect. 7 concludes the paper.

2   Previous Work

The works on frequent itemsets or various approaches to association rules mining [2] or frequent episodes [3] inspired the works on cyclic patterns. A starting point to many research studies on cyclic patterns is work [4] that defines the principle concepts of cycle pruning, cycle skipping, cycle elimination heuristics.

Discovering periodic patterns in event logs appears to be quite similar to periodicity mining in time series [5] where the long sequences of elementary data items partitioned into a number of ranges and associated with the timestamps are analyzed to find the cyclic trends. However, due to the internal structures of complex data processing operations, like for example SQL statements, its analysis cannot be treated in the same way as analysis of sequences of atomic data items like numbers of characters.

The latest works on discovering periodic patterns address the concepts of full periodicity, partial periodicity, perfect and imperfect periodicity [6] and the most recently asynchronous periodicity [7].

The model of periodicity considered in this paper is a variation of the model introduced in [8].

3   Workload Trace

Let e be a unique identifier of an event, for example identifier of query processing plan in a database system, or an identifier of flight booking routine in a flight reservation system, etc. A *log of events* is a sequence of pairs $\langle e_1 : t_1 \rangle$, ..., $\langle e_n : t_n \rangle$ where each e_i is a unique identifier of an event, t_i is a timestamp when the processing of an event e_i has started, and $t_1 \leq \ldots \leq t_n$. Note, that information about event e_i started at the different moments in time is recorded many times in a log of events.

A log of events is transformed into a *workload trace* at a data preparation stage. To define a concept of *workload trace* we consider a period of time $\langle t_{start}, t_{end} \rangle$ over which a log of events is recorded. The period of time is divided into a contiguous sequence of disjoint and fixed size *elementary time units*

$\langle t_e^{(i)}, \tau_e \rangle$ where $t_e^{(i)}$ for $i = 1, \ldots, n$ is a timestamp when an elementary time unit starts and τ_e is a length of the unit. The period $< t_{start}, t_{end} >$ consists of elementary time units such that $t_{start} = t_e^{(1)}$ and $t_e^{(i+1)} = t_e^{(i)} + \tau_e$ and $t_e^{(n)} + \tau_e = t_{end}$.

A time unit $\langle t, \tau \rangle$ consists of one or more consecutive elementary time units. A nonempty sequence U of n disjoint time units $< t^{(i)}, \tau^{(i)} > i = 1, \ldots, n$ over $\langle t_{start}, t_{end} \rangle$ is any sequence of time units that satisfies the following properties: $t_{start} \leq t^{(1)}$ and $t^{(i)} + \tau^{(i)} \leq t^{(i+1)}$ and $t^{(n)} + \tau^{(n)} \leq t_{end}$. Note, that individual time units in a sequence may have different length.

As a simple example consider a log of events that starts on $t_{01:01:2007:0:00am}$ and ends on $t_{31:01:2007:12:00pm}$. Then, a sequence of disjoint time units called as *morning tea time* consists of the time units $\langle t_{01:01:2007:10:30am}, 30 \rangle$, $\langle t_{02:01:2007:10:30am}, 30 \rangle, \ldots, \langle t_{31:01:2007:10:30am}, 20 \rangle$ where a time unit on the last day of month is a bit shorter than the previous ones.

A definition of a workload trace is based on a concept of *multiset*. A *multiset* M is a pair $\langle S, f \rangle$ where S is a finite set and $f : S \to N^+$ is a function such that $\sum_{s \in S} f(s) < \infty$ which is called as a *cardinality* and it is denoted by $|M|$. The function determines multiplicity of each element in S. If $M' = \langle S, f' \rangle$ is another multiset on S, then we say that M' is a *submultiset* of M denoted by $M' \subseteq M$ if $f'(x) \leq f(x)$ for all $x \in S$. In the rest of this paper we shall denote a multiset $\langle \{e_1, \ldots, e_m\}, f \rangle$ where $f(e_i) = k_i$ for $i = 1, \ldots, m$ as $(e_1^{k_1}, \ldots, e_m^{k_m})$. We shall denote an empty multiset $\langle \emptyset, f \rangle$ as $\emptyset$. We shall abbreviate a single element multiset (e^k) as e^k.

Let n denotes the total number of time units in U and let $U[i]$ denotes the i-th time unit in U where i changes from 1 to n. A *workload trace of an event e* is a sequence W_e of n multisets of events such that $W_e[i] = (e^{k_i})$ or $W_e[i] = \emptyset$ for $i = 1, \ldots, n$ and $k_i \geq 1$ equal to the total number of times processing of an event e started in the i-th time unit $U[i]$. If processing of an event e starts in a time unit i then it may happen that entire period of time over the event is processed overlaps on one or more finite number of successive time units $i + 1, i + 2, \ldots$, k. In this work we consider only periodic processing of events determined by the time units where processing of each event has started and we ignore the overlaps of its processing period on the successive time units.

Let E be a set of all events whose occurrences are recorded in a log $L(E)$ over time units U and saved in a reduced event table. A *workload trace of a log $L(E)$* is denoted by W_L and $W_L[i] = \biguplus_{e \in E} W_e[i], \forall i = 1, \ldots, |U|$, i.e. it is a multiset union over the respective time units of workload traces of all events included in E.

4   Periodic Pattern

A class of CRP periodic patterns is defined as a triple $\langle C, R, P \rangle$ where

(1) a *carrier C* defines a structure of periodically repeated events, computations, queries, etc.,

(2) a *range R* determines a time scope of periodic repetitions of a *carrier* measured in time units, for example from time unit t_i to a time unit t_j,

(3) a *periodicity P* determines when the next periodic repetition of a *carrier* may happen, for example after p time units from the latest occurrence of a *carrier* with possible delay by k time units.

In this work we consider a subclass of *CRP* periodic patterns defined as a triple $\langle C, f : t, p \rangle$ where:

(1) a *carrier C* is a nonempty sequence of at least one nonempty multisets of events,

(2) a *range f:t* is a pair of natural numbers that determine a location of the *first* cycle and the *total* number of cycles in the pattern,

(3) a *periodicity p* is a natural number which determines a distance between every two adjacent cycles,

(4) and the values of $f{:}t$ and p must satisfy the conditions $f, t \geq 1$ and $p \geq 0$ and $f + (t-1) * p + |C| - 1 \leq |U|$ and if $t = 1$ then $p = 0$.

The following sequence of definitions leads to a concept of validation of periodic pattern in a workload trace. Let C be a sequence of multisets where $|C| \leq n$. A *trace of C* spanning over n multisets and starting at a time unit f where $f + |C| - 1 \leq n$ is denoted by $tr(C, f, n)$ and it is defined as sequence of $f - 1$ empty multisets followed by a sequence of multisets C and and followed by $n - (f + 1) - |C|$ empty multisets. For example, $trace(e_1 e_2^2, 3, 5)$ is a sequence of multisets $\emptyset \emptyset e_1 e_2^2 \emptyset$.

A *trace of a periodic pattern* $\langle C, f : t, p \rangle$ over n time units where $f + (t - 1) * p + |C| - 1 \leq n$ is denoted by $\text{TR}(\langle C, f : t, p \rangle, n)$ and it is defined as a union $tr(C, f, n) \uplus tr(C, f + p, n) \uplus \ldots \uplus tr(C, f + (t - 1) * p, n)$. In the other words, a trace of periodic pattern is a union of traces of its carrier over n multisets such that each trace starts at the time units $f, f + p, \ldots, f + (t-1) * p$. For example, a trace of periodic pattern $\langle e_1 e_2^2, 2 : 2, 1 \rangle$ over 5 time units is the following union of sequences of multisets $\emptyset e_1 e_2^2 \emptyset \emptyset \uplus \emptyset \emptyset e_1 e_2^2 \emptyset = \emptyset e_1 (e_1, e_2^2) e_2^2 \emptyset$.

In a special case when a value of $t = 1$, i.e. when a pattern consists of only one cycle, a value of parameter p must be equal to 0 for example, a trace of periodic pattern $\langle e_1 e_2^2, 2 : 1, 0 \rangle$ over 5 time units is equal to $\emptyset e_1 e_2^2 \emptyset \emptyset$.

In another special case when a value of parameter $p = 0$ and value of parameter $t > 1$, a trace of periodic pattern $\langle e_1 e_2^2, 2 : 2, 0 \rangle$ over 5 time units is equal to $\emptyset e_1^2 e_2^4 \emptyset \emptyset$.

A periodic pattern $\langle C, f : t, p \rangle$ *is valid in a workload histogram* W_L recorded over n time units if $TR(\langle C, f : t, p \rangle, n)[i] \subseteq W_L[i]$ for $i = 1, \ldots, n$. In the other words a periodic pattern is valid in a workload trace that spans over n time units if every element of trace of the pattern over n time units is included in the respective element of a workload W_L.

For example, a periodic pattern $\langle e_1 e_2^2, 2 : 2, 1 \rangle$ is valid in a workload trace $e_1^3 e_1 (e_1^2, e_2^2) e_2^2 \emptyset$ because every element of its trace $\emptyset e_1 (e_1, e_2^2) e_2^2 \emptyset$ is included in the respective element of the workload trace.

5   Derivation Rules

The derivation rules presented in this section use the periodic patterns valid in a workload trace W_L that spans over n time units to create new periodic patterns also valid in W_L.

Rule 0 (*Discovery*) Let C be a submultiset of events such that $C \subseteq W_L[f]$ for $f \in \{1, \ldots, n\}$. Then a periodic pattern $\langle C, f : 1, 0 \rangle$ is valid in W_L. A *discovery* rule means that it is always possible to create a single cycle periodic pattern valid in W_L from any non-empty submultiset of an element in a workload trace.

Rule 1 (*Normalization*) If a periodic pattern $\langle C, f : t, p \rangle$ is valid in a workload W_L then a periodic pattern $\langle C', f' : t, p \rangle$ such that C' is obtained from C through elimination of all i leading empty multisets and all trailing empty multisets and such that $f' = f + i$ is valid in W_L. *Normalization* rule allows for elimination of leading and/or trailing empty multisets from a carrier of a periodic pattern.

Rule 2 (*Split*) If a periodic pattern $\langle C, f : 1, p \rangle$ is valid in a workload W_L then the following four cases are possible.

(1) If $t = 2$ then the periodic patterns $\langle C, f : 1, 0 \rangle$ and $\langle C, f + p : 1, 0 \rangle$ are valid in a workload trace W_L.
(2) If $t > 2$ and then the periodic patterns $\langle C, f : t, 0 \rangle$ and $\langle C, f + p : t - 1, p \rangle$ are valid in a workload trace W_L or the periodic patterns $\langle C, f : t - 1, p \rangle$ and $\langle C, f + (t - 1) * p : 1, 0 \rangle$ are valid in a workload trace W_L.
(3) If $t > 3$ and $f_{split} = f + i * p$ for $3 \leq i \leq t - 1$ then the periodic patterns $\langle C, f : i - 1, p \rangle$ and $\langle C, f_{split} : t - i + 1, p \rangle$ are valid in a workload trace W_L.

A case (1) of *split* rule divides a pattern that consist of two cycles into two single cycle patterns. A case (2) "cuts of" a single cycle periodic pattern from either left or right side of a pattern that consist of more than two cycles. Finally, a case (3) splits a periodic pattern that has more than three cycles into two patterns with more than one cycle.

Rule 3 (*Synthesis*) If the periodic patterns $\langle C, f_i : t_i, p_i \rangle$ and $\langle C, f_j : t_j, p_j \rangle$ are valid in a workload trace W_L and $f_i < f_j$ then the following four cases are possible.

(1) If $t_i = t_j = 1$ and $f_i < f_j$ then a periodic pattern $\langle C, f_i : 2, f_j - f_i \rangle$ is valid in a workload trace W_L.
(2) If $t_i = 1$ and $t_j \neq 1$ and $f_j - f_i = p_j$ then a periodic pattern $\langle C, f_i : t_j + 1, p_j \rangle$ is valid in a workload trace W_L.
(3) If $t_j = 1$ and $t_i \neq 1$ and $f_j = f_i + t_i * p_i$ then a periodic pattern $\langle C, f_i : t_i + 1, p_i \rangle$ is valid in a workload trace W_L.
(4) If $t_j \neq 1$ and $t_i \neq 1$ and $p_i = p_j$ and $f_j = f_i + t_i * p_i$ then a periodic pattern $\langle C, f_i : t_i + t_j, p_i \rangle$ is valid in a workload trace W_L.

A case (1) of *synthesis* rule merges two single cycle pattern into one pattern. The cases (2) and (3) add single cycle patterns at the left/right end of a pattern. A case (4) concatenates two patterns such that both of them consist of more than one cycle.

The *split* and *synthesis* rules can be used to eliminate the cycles from a periodic pattern such that the result is still a valid periodic pattern.

Rule 4 (*Decomposition*) If a periodic pattern $\langle C, f : t, p \rangle$ is valid in a workload W_L then a periodic pattern $\langle C', f : t, p \rangle$ where a carrier C' is obtained from a carrier C by elimination of any multiset from any element of a carrier C is valid in W_L.

Rule 5 (*Composition*) If the periodic patterns $\langle C_i, f_i : t, p \rangle$ and $\langle C_j, f_j : t, p \rangle$ are valid in a workload W_L and $f_i \leq f_j$ then a periodic pattern $\langle C_k, f_i : t, p \rangle$ where $C_k = tr(C_i, 1, f_j - f_i + |C_j|) \uplus tr(C_j, f_j - f_i, f_j - f_i + |C_j|)$ is valid in W_L. For example, if the periodic patterns $\langle e_1 e_2^2, 1 : 3, 4 \rangle$ and periodic pattern $\langle e_1, 4 : 3, 4 \rangle$ are valid in a workload trace W_L then a periodic pattern $\langle e_1 e_2^2 \emptyset e_1, 1 : 3, 4 \rangle$ is valid in W_L.

The rules of *composition*, *split*, and *synthesis* can be used to group the adjacent cycles of a periodic pattern. Grouping of the cycles and *composition* derivation rules can be used to combine two periodic patterns with different carriers, ranges and periodicity.

6   Discovering Periodic Patterns

The algorithms for discovering the periodic patterns systematically apply the derivation rules presented in the previous section until no more periodic patterns that satisfy the pre-defined criteria can be derived. Derivation of periodic patterns must preserve the equivalance the original set of periodic patterns obtained from application of *discovery* rule to a workload trace W_L.

A process of data preparation starts from selection of n time units that span over a period of time when a log of events has been recorded. Then, a log of events $L(E)$ is filtered and it is converted into a workload trace W_L that consists of n elements one for each time unit.

6.1   Algorithm 0 (Vertical Slicing of Workload)

An input to the algorithm is a workload trace W_L. The algorithm does a vertical slicing of W_L and groups single event periodic patterns by individual events. On output we get the sets $E_1, \ldots, E_m$ such that each set E_i contain single event patterns in a form $\langle e_j^k, i : 1, 0 \rangle$. The algorithm is a sequence of the following steps.

(1) Mining periodic patterns starts from application of a *discovery* rule to the individual elements of a trace W_L to create the periodic patterns $\langle W_L[1], 1 : 1, 0 \rangle, \ldots, \langle W_L[n], n : 1, 0 \rangle$.

(2) In the next step, a *decomposition* rule is applied to each one of the patterns $\langle W_L[i], i : 1, 0 \rangle$ for $i = 1, \ldots, n$ to create the single event patterns consistent with a form $\langle e_j^k, i : 1, 0 \rangle$ for all events $e_j \in W_L[i]$ and for the largest possible value of k.

(3) Finally, the single event periodic patterns are grouped into the sets $E_1, \ldots, E_m$ such that the patterns with a carrier e_j^k are included into a set E_j for $j = 1, \ldots, m$ and for any value of k.

A periodic pattern $\langle e^k, f : t, p \rangle$ is called as an *elementary periodic pattern*. The sets of periodic patterns $E_1, \ldots, E_m$ obtained from Algorithm 0 are processed one by one by Algorithm 1 to create the elementary periodic patterns.

6.2 Algorithm 1 (Synthesis of Elementary Periodic Patterns)

An input to the algorithm is a set E of single cycle periodic patterns $\langle e^k, f : 1, 0 \rangle$ for any value of k. An output is a set of elementary periodic patterns such that for any two patterns in the set it is impossible to apply a *synthesis* rule to create a new pattern which is longer than two patterns used by the rule. The algorithm uses the following concept of *p-adjacency*. We say, that periodic patterns $\langle C, f_i : t_i, p_i \rangle$ and $\langle C, f_j : t_j, p_j \rangle$ are *p-adjacent* when it is possible to apply a *synthesis* rule to the patterns to get a pattern with a given value of parameter p. It simply means that the end of one periodic patterns is distant from the beginning of the other periodic pattern with the same carrier by p elements in a workload trace. For example, the periodic patterns $\langle e_1 e_2^2, 1 : 3, 3 \rangle$ and $\langle e_1 e_2^2, 10 : 1, 0 \rangle$ are *3-adjacent* because the application of a *synthesis* rule to the patterns yields a pattern $\langle e_1 e_2^2, 1 : 4, 3 \rangle$. The algorithm consists of the following steps.

(1) Set value of p to 1.
(2) Find in E a pair of *p-adjacent* periodic patterns $\langle e^{k_i}, f_i : t_i, p_i \rangle$ and $\langle e^{k_j}, f_j : t_j, p_j \rangle$ such that $f_i < f_j$. If a pair can be found then continue from step (3) else increase a value of p by one. If a new value of p is equal to n then quit the algorithm else repeat step (2).
(3) If $k_i = k_j$ then replace in E the pair $\langle e^{k_i}, f_i : t_i, p_i \rangle$ and $\langle e^{k_j}, f_j : t_j, p_j \rangle$ with a periodic pattern $\langle e^{k_i}, f_i : t_i + t_j, p \rangle$.
(4) If $k_i < k_j$ then replace in E the pair $\langle e^{k_i}, f_i : t_i, p_i \rangle$ and $\langle e^{k_j}, f_j : t_j, p_j \rangle$ with the periodic patterns $\langle e^{k_i}, f_i : t_i + t_j, p \rangle$ and $\langle e^{k_j - k_i}, f_j : t_j, p \rangle$.
(5) If $k_i > k_j$ then proceed analogously to (4) where both i and j are replaced.
(6) Return to step (2).

Algorithm 1 enforces no limitations on the minimal length of periodic patterns. Therefore, some of the synthesized patterns can be short with only few cycles. Such short periodic patterns can be combined into longer patterns through grouping of its carrier. Grouping of the short periodic patterns in the sets of periodic patterns $E_1, \ldots, E_m$ is performed by Algorithm 2.

6.3 Algorithm 2 (Grouping of Short Periodic Patterns)

An input to the algorithm is a set E of elementary periodic patterns $\langle e^k, f : t, p \rangle$ such that $p \leq p_{min}$ where p_{min} is the largest acceptable period of periodic patterns created by the algorithm. A threshold parameter p_{min} is needed to avoid accidental compositions of periodic patterns located too far from each other to be the real reflection of periodic processes occurring in the reality. For example, a set of patterns $\{\langle e, f : 2, 1\rangle, \langle e, f+3 : 2, 1\rangle, \langle e, f+6 : 2, 1\rangle\}$ can be combined into a pattern $\langle ee, f : 3, 3\rangle$ in the following way. First, application of *decomposition* rule creates 6 single cycle patterns $\langle e, f : 1, 0\rangle$, $\langle e, f+1 : 1, 0\rangle$, $\langle e, f+3 : 1, 0\rangle$, ... Then, *composition* rule is applied to create the patterns $\langle ee, f : 1, 0\rangle$, $\langle ee, f+3 : 1, 0\rangle$, $\langle ee, f+6 : 1, 0\rangle$. Finally, *composition* rule is used to get $\langle ee, f : 3, 3\rangle$. An output from the algorithm is a set of periodic patterns such that for any two patterns in the set it is impossible to apply a *split* and *composition* rule to create a new pattern which has more cycles than a given value of parameter t_{min}.

The algorithm consists of the following steps.

(1) Copy all periodic patterns $\langle e^k, f : t, p\rangle$ in a set E such that $t = 1$ or ($t > t_{min}$ and $p > p_{min}$) into a set of patterns E_{result} and remove such patterns from E.

(2) Use a *split* rule to transform each periodic pattern $\langle e^k, f : t, p\rangle$ left in E into the patterns $\langle e^k, f : 1, 0\rangle$, $\langle e^k, f+p : 1, 0\rangle$,...,$\langle e^k, f+(t-1)*p : 1, 0\rangle$ and next use a *composition* rule to transform the patterns into a pattern $\langle C, f : 1, 0\rangle$ where a carrier C is a sequence of e^k separated with $p-1$ empty sets and repeated $t-1$ times and ended with e^k. For example, pattern $\langle e^k, f : 3, 2\rangle$ is transformed into $\langle e^k \emptyset e^k \emptyset e^k, f : 1, 0\rangle$. For each new pattern obtained from the *split* and *composition* rules save in a rollback set information about the original form of a pattern.

(3) Set a value of p to 1.

(4) Find in E a pair of *p-adjacent* periodic patterns $\langle C, f_i : t_i, p_i\rangle$ and $\langle C, f_j : t_j, p_j\rangle$ such that $f_i < f_j$. If a pair can be found then continue from step (5) else increase a value of p by one. If a new value of p is equal to n then continue from a step (8) else repeat step (4).

(5) Use a *synthesis* rule to the pair of periodic patterns found in the previous step to create a new pattern $\langle C, f_i : t_i + t_j, p\rangle$.

(6) Replace in E the pair of patterns found in a step (4) with the new pattern created in a step (5) and return to a step (4).

(7) For all periodic patterns in E such that t is equal to 1 and such that the patterns have their respective information in a rollback set use associated rollback information saved in a step (2) to transform the patterns into their original form.

(8) Copy all periodic patterns in E into E_{result}.

Algorithm 2 is individually applied to each one of the sets of periodic patterns $E_1,\ldots,E_m$.

The algorithms presented so far are able to discover the periodic patterns whose carrier is a sequence of multisets of exactly the same event. However, the real world processes consists of many events and their periodic repetitions create periodic patterns whose carriers are sequences of multisets of many different events. For example, if an event e_1 is periodically processed and it always happens that an event e_2 follows in the nearest future an event e_1 then we expect to get a periodic pattern whose carrier looks like $e_1 \ldots e_2$. A *composition* rule can be used to merge the periodic patterns that have the same frequencies and such that their start points and total number of cycles do not differ a lot. Merging the patterns whose start points are far from each other is doubtful as such patterns do not reflect the same real world processes. To measure the similarity of the patterns we use a concept of (δ_f,δ_t)-*compatibility*. The periodic patterns $\langle C_i, f_i : t_i, p \rangle$ and $\langle C_j, f_j : t_j, p \rangle$ are (δ_f,δ_t)-*compatible* if $|f_i - f_j| \leq \delta_f$ and $|t_i - t_j| \leq \delta_t$.

6.4 Algorithm 3 (Composition of Complex Periodic Patterns)

Algorithm 3 applies a *composition* rule to create periodic patterns whose carrier includes many different events. Input to the algorithm is a unions of the sets of periodic patterns $E = E_1 \cup \ldots \cup E_n$ obtained from the previous algorithms. Algorithm 3 merges the periodic patterns that have different carriers, the same periodicity, and the differences between their start point and total number of cycles are smaller then the given values of parameters δ_f and δ_t. Output from the algorithm is a set E of periodic patterns such that any pair of periodic patterns in the set is not (δ_f,δ_t)-*compatible* for the assumed values of δ_f and δ_t. The algorithm consists of the following steps.

(1) Find in E a pair of (δ_f,δ_t)-*compatible* periodic patterns $\langle C_i, f_i : t_i, p \rangle$ and $\langle C_j, f_j : t_j, p \rangle$. If a pair cannot be found then quit the algorithm.

(2) Reorder the elements of pair such that $f_i \leq f_j$ and consider one of the following three case.

(3.1) If $t_i = t_j$ then apply a *composition* rule to the patterns included in a pair and create a periodic pattern $\langle C_k, f_i : t_i, p \rangle$ where $C_k = tr(C_i, 1, f_j - f_i + |C_j|) \uplus tr(C_j, f_j - f_i, f_j - f_i + |C_j|)$. Replace in E the periodic patterns included in the pair with the new periodic pattern.

(3.2) If $t_i < t_j$ then apply a *split* rule to a pattern $\langle C_j, f_j : t_j, p \rangle$ and create the patterns $\langle C_j, f_j : t_i, p \rangle$ and $\langle C_j, f_j + t_i * p : t_j - t_i, p \rangle$. Next, apply a *composition* rule to the patterns $\langle C_i, f_i : t_i, p \rangle$ and $\langle C_j, f_j : t_i, p \rangle$ and create a periodic pattern $\langle C_k, f_i : t_i, p \rangle$ where $C_k = tr(C_i, 1, f_j - f_i + |C_j|) \uplus tr(C_j, f_j - f_i, f_j - f_i + |C_j|)$. Replace in E the patterns $\langle C_i, f_i : t_i, p \rangle$ and $\langle C_j, f_j : t_i, p \rangle$ with the patterns $\langle C_k, f_i : t_i, p \rangle$ and $\langle C_j, f_j + t_i * p : t_j - t_i, p \rangle$.

(3.3) If $t_i > t_j$ then apply a *split* rule to a pattern $\langle C_i, f_i : t_i, p \rangle$ and proceed analogously to (3.2)

(4) Return to step (1).

To apply a *composition* rule in Algorithm 3 the periodic patterns must have identical values of parameter p. Such restriction seems to be justified when discovering complex patterns from their simple components. However, in a case when derivations are used to discover future performance bottlenecks periodicity must be adjusted with a *split* rule to create (δ_f,δ_t)-*compatible* patterns and to apply a *composition* rule later on.

7 Summary and Conclusions

This work shows how to discover the complex periodic patterns of events through systematic transformations of an initial set of elementary patterns obtained from a workload trace. A system of derivation rules proposed in the paper can be used to unify the parameters of periodic patterns and to create the complex patterns from the elementary ones. A sequence of algorithms applies the derivation rules transform the elementary periodic patterns into the complex periodic patterns.

A system of derivation rules for periodic patterns plays a central role in the proposed approach. Systematic application of the derivation rules whose correctness is based on a concept of validity in a workload trace assures correctness of the algorithms. An important property of the proposed sequence of algorithms is that all periodic patterns whose carriers do not have common elements or whose ranges do not overlap can be correctly discovered by the algorithms.

There are many important applications of periodic patterns discovering in many branches of computer science. Let us take for example some important problem in automated performance tuning which is anticipation of conflicting events. The derivation rules proposed in the paper can be used to predict the coincidences of any events in the future workloads. It is possible to find the coincidences of events through application of derivation rules to adjust the length and frequencies of patterns whose carriers include such events and through application of a composition rule.

A class of periodic patterns considered in the paper can be generalized into a more realistic class of *imperfect periodic patterns* by elimination of some components from a *trace* of periodic pattern. For example, such class of periodic patterns may allow for a single cycles of pattern to be missing and/or to be processed in a different time unit. The derivation rules presented in the paper can be quite easily adopted to represent the derivations of imperfect periodic patterns.

References

1. Van der Aalst, W.M.P.: Process Mining Discovery, Conformance and Enhancement of Business Processes. Springer, Heidelberg (2011)
2. Luna, J., Cano, A., Sakalauskas, V., Ventura, S.: Discovering useful patterns from multiple instance data. Inf. Sci. **357**, 23–38 (2016)
3. Mannila, H., Toivonen, H., Verkamo, A.I.: Discovery of frequent episodes in event sequences. Data Min. Knowl. Disc. **1**, 259–289 (1997)
4. Özden, B., Ramaswamy, S., Silberschatz, A.: Cyclic association rules. In: Proceedings of the Fourteenth International Conference on Data Engineering, pp. 412–421 (1998)
5. Rasheeed, F., Alshalalfa, M., Alhajj, R.: Efficient periodicity mining in time series databases using suffix trees. IEEE Trans. Knowl. Data Eng. **23**(1), 79–94 (2011)
6. Huang, K.Y., Chang, C.H.: SMCA: A general model for mining asynchronous periodic patterns in temporal databases. IEEE Trans. Knowl. Data Eng. **17**(6), 774–785 (2005)

7. Yeh, J.S., Lin, S.C., Hu, S.C.: Novel algorithms for asynchronous periodic pattern mining based on 2-d linked list. Int. J. Database Theory Appl. **5**(4), 33–43 (2012)
8. Getta, J., Zimniak, M., Benn, W.: Mining periodic patterns from nested event logs. In: The 14th IEEE International Conference on Computer and Information Technology, CIT 2014, pp. 160–167 (2014)

The Fuzzy Approach to Assessment
of ANOVA Results

Jacek Pietraszek[1](✉), Maciej Kołomycki[1], Agnieszka Szczotok[2],
and Renata Dwornicka[1]

[1] Department of Software Engineering and Applied Statistics, Cracow
University of Technology, Al. Jana Pawla II 37, 31-864 Kraków, Poland
pmpietra@gmail.com, mkolomycki@gmail.com,
renata.dwornicka@gmail.com
[2] Institute of Materials Science, Faculty of Materials Science and Metallurgy,
Silesian University of Technology, ul. Krasinskiego 8, 40-019 Katowice, Poland
agnieszka.szczotok@polsl.pl

Abstract. Typically, the analysis of variance (ANOVA) is used to compare
means in the subsets obtained through the division of a large numerical dataset
by assigning a categorical variable labels to dataset's values. The test criterion
for the decision on 'all equal' vs. 'not all equal' is a comparison of the significance level described by a well-known p-value and the a priori assigned
critical significance level, α, usually 0.05. This comparison is treated very
strictly basing on the crisp value; however, it should not be so, especially if
p-value is near α, because the certainty of the decision varies rather smoothly
from 'strongly not' to 'no opinion' to 'strongly yes'. It is very interesting to
analyze such results on the basis of the fuzzy arithmetic theory, using the
modified Buckley's fuzzy approach to the statistics combined with the bootstrap
approach, because it may be adopted to the cases where subjective assessments
are introduced as *quasi*-measurements.

Keywords: Analysis of variance · ANOVA · Alpha-cuts · Fuzzy numbers ·
Fuzzy statistics · Materials science · Bootstrap

1 Introduction

In general, uncertainty expresses our lack of knowledge about the future behavior and
about the states of an investigated phenomenon. The oldest solution, provided by
Pascal, uses a probabilistic approach based on the frequency of events; Kolmogorov
however formalized it, basing on an axiomatic approach and Borel's field of sets. Such
an approach assumes implicitly that there is a possibility, at least a potential one, to
replicate the test many times and experimentally determine the associated asymptotic
frequency. The investigations of Poincare [1] and Hadamard [2] provided at the end of
the 19th century and later in 1963 by Lorenz [3] revealed that a deterministic system
with large sensitivity may lead to chaos, practically indistinguishable from a probabilistic random system.

© Springer International Publishing Switzerland 2016
N.T. Nguyen et al. (Eds.): ICCCI 2016, Part I, LNAI 9875, pp. 260–268, 2016.
DOI: 10.1007/978-3-319-45243-2_24

A different approach to the uncertainty was presented in 1965 by Zadeh [4]. He proposed to introduce a specific extension of the set algebra: the membership function with a value varying from 0 to 1. Such an idea has been developed intensively since the 1960s, in particular, the fuzzy arithmetic defined for real fuzzy numbers by Dubois and Prade [5].

In 1968, Zadeh [6] already considered the relation of the probability and fuzzy descriptions of uncertainty, however, it was Buckley [7, 8] who offered in 2005 a consistent approach to fuzzy estimators related to random samples defined as crisp datasets. His concept of a fuzzy estimator is based on the mapping between a source pair: a significance level and its related confidence interval and the resulting alpha-cut. Such an approach requires rather difficult analytical transformations and it has appeared to be rather impractical in a more complicated analysis.

In 2006, Grzegorzewski [9] chose the theory of decision as a starting point and developed a more general classification that contained three elements: analyzed data, tested hypotheses and additional assumptions/conditions. The elements may be considered as fuzzy or non-fuzzy, which leads to many possible combinations. It is very interesting as a conceptual idea but there is lack of practical instructions for the evaluation of such tests.

Some elements of a fuzzy approach have been adopted by the design of experiments analysis [10], which however, led to a very difficult inference. A similar attempt to use the neural network approximation [11] revealed a large instability of results affected by a random procedure of neural network identification. This may also be caused by an unconsidered correlation between variables, which imposes a selection of the particular pair of triangular norms [12].

In a further analysis the authors adopted Buckley's approach but his very complicated analytical transformations were replaced with a bootstrap approach [13], which also led to the obtaining of empirical distributions. The distributions were used to construct alpha-cuts related to fuzzy estimators.

2 Methods

2.1 Buckley's Fuzzy Statistics

The fuzzy number is defined as a fuzzy subset of $\mathbf{R}$ [5] described by its membership function $\bar{A}$:

$$\bar{A} : \mathbb{R} \to [0, 1]. \tag{1}$$

Alpha-cut $\bar{A}[\alpha]$ is defined as a non-fuzzy subset of $\mathbf{R}$ where the membership function is greater than or equal to α i.e.:

$$\bar{A}[\alpha] = \{x \in \mathbb{R} : \bar{A}(x) \geq \alpha \wedge \alpha \in (0, 1]\}. \tag{2}$$

The value of alpha-cut for $\alpha = 0$ is specifically defined as a closed support of a membership function i.e.:

$$\bar{A}[0] = \overline{\{x \in \mathbb{R} : \bar{A}(x) \neq 0\}}. \tag{3}$$

Buckley restricted the use of fuzzy numbers to the subtype of 'triangular shaped fuzzy numbers', which means that a membership function is a combination of two monotonic functions: left – monotonically increasing and right – monotonically decreasing. Buckley used their inverse form to define the alpha-cut $\bar{Q}[\alpha]$ of the fuzzy number $\bar{Q}$ as a closed, bounded interval for $0 \leq \alpha \leq 1$ i.e.:

$$\bar{Q}[\alpha] = [q_1(\alpha), q_2(\alpha)], \tag{4}$$

where $q_1(\alpha)$ is an increasing function of α, $q_2(\alpha)$ is a decreasing function of α and $q_1(1) = q_2(1)$.

The key element in Buckley's approach is the identification of the fuzzy estimator alpha-cuts with the confidence intervals and the alpha with the significance level. The confidence interval is denoted as:

$$[\theta_1(\beta), \theta_2(\beta)] \tag{5}$$

where β is used as a symbol of the significance level because the traditional symbol α collides with the argument of the alpha-cut. The values of β vary from something very small but different from zero to less than 1 e.g. $0.01 \leq \beta < 1$. The value of 1 is treated in a special manner because it results in zero-length interval i.e. in a confidence interval of 0 confidence.

Thus the fuzzy estimator $\bar{\theta}$ of the statistics θ is defined inversely through its alpha-cut $\bar{\theta}[\alpha]$:

$$\bar{\theta}[\alpha] = [\theta_1(\alpha), \theta_2(\alpha)]. \tag{6}$$

2.2   ANOVA Analysis

The one-way analysis of variance (ANOVA) [14] compares means of subsets created from the large dataset by a categorical classification. If the subdivided data are sampled as i.i.d. (independent and identically distributed) normal random variables with common variance then the null hypothesis that the means are equal may be tested. The computational procedure is briefly described below according to the decomposition of the sum of squares as proposed by Fisher [15].

The raw dataset x_{ij}, $i = 1,\ldots, k; j = 1,\ldots, n_i$ is divided into k subgroups of n_j values each. The great mean $\bar{x}$ is evaluated

$$\bar{x} = \frac{1}{n}\sum_{i=1}^{k}\sum_{j}^{n_i} x_{ij}, \quad n = \sum_{i=1}^{k} n_i \tag{7}$$

and the subgroup also means

$$\bar{x}_i = \frac{1}{n_i} \sum_{j=1}^{n_i} x_{ij}. \tag{8}$$

Then the sum of square between the subgroups is evaluated

$$SS_{Factor} = \sum_{i=1}^{k} n_i(\bar{x}_i - \bar{x})^2 \tag{9}$$

as well as the sum of squares within the subgroups

$$SS_{Error} = \sum_{i=1}^{k} \sum_{j=1}^{n_i} (x_{ij} - \bar{x}_i)^2. \tag{10}$$

Next, the mean squares are evaluated

$$MS_{Factor} = \frac{SS_{Factor}}{k-1}, \tag{11}$$

$$MS_{Error} = \frac{SS_{Error}}{n-k}. \tag{12}$$

Finally, F statistics is evaluated

$$F = \frac{MS_{Factor}}{MS_{Error}}. \tag{13}$$

If the assumptions are met, the F statistics should follow Fisher's distribution with the degrees of freedom equal to $f_1 = k - 1$ and $f_2 = n - k$, respectively. The significance level *p-Value* is evaluated from an inverse cumulative distribution and the null hypothesis is rejected if *p-Value* is smaller than the critical significance level α assigned *a priori*.

2.3 Bootstrap of the Dataset

The idea of the bootstrap procedure is based on the iteratively conducted re-sampling from a raw dataset and generating a pool of alternative datasets [13]. However it should be noted that re-sampling bases on drawing from i.i.d. values, whereas the raw dataset does not contain such values. The key point is a linear model [16] constructing the background of ANOVA:

$$z_{ij} = \mu + a_i + \varepsilon_{ij}, \quad \varepsilon \sim N(0, \sigma^2), \tag{14}$$

where a_i is a subgroup effect deviating from a great mean μ.

This model should be identified prior to the bootstrap procedure giving μ, a_i, and residuals r_{ij} and then the bootstrap random draw should be taken from the pool of residuals r_{ij}. Iteratively, new bootstrapped datasets will be created based on Eq. 14 and re-sampled residuals and ANOVA procedure will be applied. The final results will constitute the large set of evaluated *p-Values*. This set will be a source for evaluation of confidence intervals and related alpha-cuts.

3   Materials

Nickel-based superalloys are used mainly in aircraft and power-generation turbines and typically produced by an investment casting process especially useful for making castings of complex and near-net shape geometry. Studies were performed on the IN 713C superalloy [17]. The polycrystalline castings of IN 713C were produced in the investment casting process conducted by the Laboratory for Aerospace Materials at Rzeszow University of Technology in Poland. Finally, castings were cut off. The cross-sections were included and prepared as metallographic samples from nickel-based superalloy. The microstructural investigations of the cross-sections of the casting were carried out by means of the Hitachi S-4200 scanning electron microscope. The recorded microphotographs were next subjected to a computer-aided image analysis by means of Met-Ilo program in order to estimate quantitatively the main parameters describing the ($\gamma + \gamma'$) eutectic islands that occurred in the investigated superalloy. The data obtained from the analysis of the GK casting were processed in the bootstrap analysis in this paper. The data obtained from the image analysis were transformed by the logarithmic formula (Eq. 15) because the eutectic area cannot achieve negative values while the ANOVA analysis requires the assumption of a normally distributed noise without bound limitations:

$$z = \ln(y) \tag{15}$$

where y denotes the obtained eutectic area and z denotes the transformed value.

The source data were divided into 6 groups with 31, 49, 80, 75, 64 and 61 values, respectively [17]. The technological aim was to check the homogeneity between the cross-sections.

4   Results

At the beginning, the classic analysis led to ANOVA table with crisp results (Table 1).

Table 1. ANOVA table for transformed data (source [17]).

Effect	SS	df	MS	F	p
Trace	3.318	5	0.664	0.411	0.841
Error	572.019	354	1.616	–	–
Total	575.338	359	1.603	–	–

Such results do not impose the rejection of a homogeneity hypothesis.

Next, the bootstrap procedure resulted in the dataset containing 10.000 records of bootstrapped F and p values. The dataset size (10.000) was selected for convenience: an easy selection of exact quantiles and – in reverse – an easy recalculation of the membership function values associated with confidence intervals.

The obtained values of F and p were sorted, associated with their $1/n$th of experimental probability (i.e. 1/10.000 for the mentioned case) and – through the inverse cumulative probability function – confidence intervals for different values of significance level were evaluated.

5   Analysis

The description statistics for the bootstrapped F and p values are presented in Table 2.

The distribution of bootstrapped F values is presented in Fig. 1. As mentioned by Szczotok et al. [17], this distribution significantly differs from theoretical F distribution. It may be caused both by a limited size of source raw data and by only an asymptotical equality of F distribution in Fisher's theorem.

Table 2. The description statistics for bootstrapped F and p values (the mode is related to the highest histogram class)

Value	mean	Median	mode	−95 %	+95 %
F_{boot}	1.414	1.246	0.75...1.00	0.247	3.531
p_{boot}	0.354	0.287	0.00...0.05	0.004	0.941

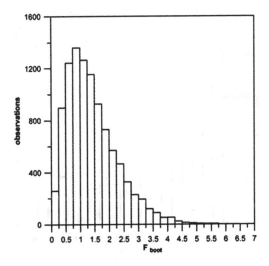

Fig. 1. Histogram of the bootstrapped F statistics

The fuzzy estimator of the bootstrapped F statistics – constructed on the basis of one-sided confidence interval – is presented in Fig. 2. The procedure of Buckley's recalculation of confidence intervals into the alpha membership function imposes that the maximum is set at zero value i.e. when all means are mutually equal without any random noise inside groups.

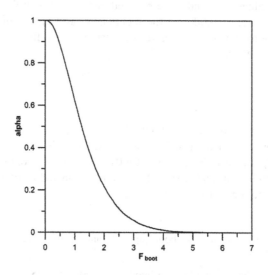

Fig. 2. Fuzzy estimator of the bootstrapped F statistics

The distribution of bootstrapped p values is presented in Fig. 3. The fuzzy estimator of the bootstrapped p values is presented in Fig. 4.

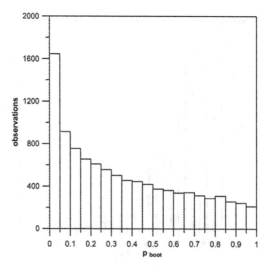

Fig. 3. Histogram of the bootstrapped p values

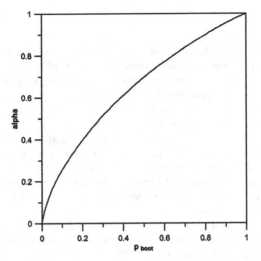

Fig. 4. Fuzzy estimator of the bootstrapped p values

The obtained fuzzy estimator may be used to reject or not the null hypothesis of ANOVA: simultaneous equality of all means. In the case mentioned here, this leads to the following fuzzy assessments:

- classic ANOVA $F = 0.411$ is related to the fuzzy assessment 0.922 (see Fig. 2),
- classic ANOVA $p = 0.841$ is related to the fuzzy assessment 0.922 (see Fig. 4).

Firstly, it is worth noting that both assessments are equal, which means that such fuzzy estimators are consistent in decision making. Secondly, such a relatively large value means that the hypothesis of means equality may be treated with a high degree of certainty. Such conclusions are a slightly different in comparison to the statistical orthodox: 'not reject', which does not mean 'accept'.

6 Conclusion

The authors have proposed a bootstrap approach to obtain fuzzy estimators of ANOVA key statistics: F value and p value. The aim was to achieve a support for decision making that would be more convenient than a typical *flip-flop* decision scheme in the classic Neyman-Pearson hypothesis testing based on a comparison of p value with critical significance level α assumed *a priori*. The classic decision scheme is especially difficult if p value is near α level.

The proposed scheme bases on Buckley's approach modified for one-sided confidence intervals. It appeared to be effective and consistent with the classic decision scheme.

Further activities will be oriented to develop consistent mathematical and algorithmic formalism of such fuzzy-probabilistic mixture and – finally – to adopt such formalism into DoE (the design of experiments) applications to analyze that cases where subjective assessments are introduced as *quasi*-measurements.

References

1. Poincaré, J.H.: Sur le problème des trois corps et les équations de la dynamique. Divergence des séries de M. Lindstedt. Acta Mathematica **13**, 1–270 (1890)
2. Hadamard, J.: Les surfaces à courbures opposées et leurs lignes géodesiques. Journal de Mathématiques Pures et Appliquées **4**, 27–73 (1898)
3. Lorenz, E.N.: Deterministic non-periodic flow. J. Atmos. Sci. **20**, 130–141 (1963)
4. Zadeh, L.A.: Fuzzy sets. Inf. Control **8**, 338–353 (1965)
5. Dubois, D., Prade, H.: Fuzzy real algebra: some results. Fuzzy Sets Syst. **2**, 327–348 (1979)
6. Zadeh, L.A.: Probability measures of fuzzy events. J. Math. Anal. Appl. **23**, 421–427 (1968)
7. Buckley, J.J.: Fuzzy statistics: hypothesis testing. Soft. Comput. **9**, 512–518 (2005)
8. Buckley, J.J.: Fuzzy Probability and Statistics. Springer Verlag, Heidelberg (2006)
9. Grzegorzewski, P.: Decision Support Under Uncertainty. Statistical Methods for Imprecise Data. EXIT, Warszawa (2006)
10. Pietraszek, J.: Fuzzy regression compared to classical experimental design in the case of flywheel assembly. In: Rutkowski, L., Korytkowski, M., Scherer, R., Tadeusiewicz, R., Zadeh, L.A., Zurada, J.M. (eds.) ICAISC 2012, Part I. LNCS, vol. 7267, pp. 310–317. Springer, Heidelberg (2012)
11. Pietraszek, J., Gadek-Moszczak, A.: The smooth bootstrap approach to the distribution of a shape in the ferritic stainless steel AISI 434L powders. Solid State Phenomen. **197**, 162–167 (2013)
12. Pietraszek, J.: The modified sequential-binary approach for fuzzy operations on correlated assessments. In: Rutkowski, L., Korytkowski, M., Scherer, R., Tadeusiewicz, R., Zadeh, L. A., Zurada, J.M. (eds.) ICAISC 2013, Part I. LNCS, vol. 7894, pp. 353–364. Springer, Heidelberg (2013)
13. Shao, J., Tu, D.: The Jackknife and Bootstrap. Springer, New York (1995)
14. Davies, L., Gather, U.: Robust Statistics. In: Gentle, J.E., Hardle, W.K., Mori, Y. (eds.) Handbook of Computational Statistics, vol. 2, pp. 711–749. Springer, Heidelberg (2012)
15. Fisher, R.A.: Statistical Methods for Research Workers. Oliver and Boyd Press, Edinburgh (1925)
16. Rutherford, A.: Introducing ANOVA and ANCOVA - A GLM Approach. SAGE Publications Ltd., London, Thousand Oaks, New Delhi (2001)
17. Szczotok, A., Nawrocki, J., Gądek-Moszczak, A., Kołomycki, M.: The bootstrap analysis of one-way ANOVA stability in the case of the ceramic shell mould of airfoil blade casting. Solid State Phenom. **235**, 24–30 (2015)

Virtual Road Condition Prediction Through License Plates in 3D Simulation

Orcan Alpar and Ondrej Krejcar[✉]

Faculty of Informatics and Management,
Center for Basic and Applied Research, University of Hradec Kralove,
Rokitanskeho 62, 50003 Hradec Kralove, Czech Republic
orcanalpar@hotmail.com, ondrej@krejcar.org

Abstract. Predicting the road conditions lie curves, slopes, hills, helps drivers react faster to avoid possible collisions in hypovigilance and besides, this kind of driver assistance system is more crucial for intelligent vehicles. Even though there are many radar, wifi, infrared systems and devices, what we propose in this paper is a monocular license plate segmentation to foresee the road ahead while cruising behind a blinding vehicle. License plates in the precalibrated images from 3D simulation are segmented and analyzed to identify the front car's angle of repose. Therefore the angles of the road are estimated frame by frame with calculated distances for prediction of the virtual road.

Keywords: Road condition · Driver assistance · License plates virtual road · 3d simulation · Monocular

1 Introduction

Considering the recent developments on intelligent transportation systems, cruise assistant systems and relevant technologies are the major branches and highly essential for avoiding possible traffic accidents. The prediction of the road conditions would be so useful to make the drivers prepare themselves for the curves and hills. However, while cruising on a rough road behind a blinding vehicle, it sometimes is so hard to foresee the conditions. Furthermore, the recognition of the front cars' position using a monocular camera is difficult, yet the license plate (LP) of the front car is simpler, as long as it is mounted on the vehicle.

The identification of road conditions through front cars' motion not only would assist the drivers but also could create a signal to warn the drivers, when necessary. Therefore in this research, we firstly focus on segmentation and continuous chasing of the front car's license plate frame through 3D simulation. The ease of localization of an LP is the main motivation of this paper, since the motion of the front car could simple be identified by skewness and distortion of the license plate in a frame as long as the plate is attached. On the contrary to general LP recognition systems, we don't need to localize and identify the digits and characters on the LP, so this step is omitted.

The main idea beneath this research topic is one of our previous article regarding a collision warning system by segmentation of LPs and calculation of the area in a serous game context [1]. We created a fuzzy warning system to create a warning signal when

© Springer International Publishing Switzerland 2016
N.T. Nguyen et al. (Eds.): ICCCI 2016, Part I, LNAI 9875, pp. 269–278, 2016.
DOI: 10.1007/978-3-319-45243-2_25

the front car is getting closer using LP segmentation and area calculation. The area estimated in each frame revealed the approximate distance and relative speed therefore a danger warning is generated once the front car is dangerously decelerating or cars are dramatically close. The localization of the LP is done by contrast identification as long as the LP contains darker digits on lighter background, which is also used in one of our previous research as well [2]. In a different point of view, we again calculated the contrast data however for corona segmentation and discrimination of brake light from rear lights. The main purpose of these papers are collision avoidance, when the cars are cruising significantly close.

The prediction of road condition without a vision of the driver while following a vehicle would help the driver to get ready for the upcoming curves and slopes, therefore the kernel of this paper could be considered as accident avoidance system. On the other hand, it would be a promising component for automatic driving assistant system of future intelligent vehicles, while following a big vehicle impeding the vision.

Given this fact, it is possible to find similar focus in the articles published in the past decade. The main aims of these articles are various though, such as driver warning systems [3–5], auto-steering systems [6–8] and auto-braking systems [9]. The sensors are various as well since we may find lasers [10], wireless systems [11], monocular cameras [12], GPS [13] and radar technologies [14].

Besides these papers, there are also a few eminent articles concerning the LPs and collision avoidance. Phelawan et al. [15] introduced a similar method to measure the distance between two vehicles. In their paper, they segmented the LPs as we did however they afterward compared with a database to estimate the distance. Likewise, Chan et al. [16] tried to reach the same outcome using the character heights of LP instead of the LP sizes. Considering the instruments used in the papers as research tool, simulation based papers are also remarkable such as [17, 18].

Briefly, we already have pre-calibrated LP images with Nikon D90 camera using 18–200 mm f/3.5-5.6G IF-ED AF-S VR DX lens which has sensor width of 23.6 and height of 15.8 with 24 mm crop factor. Therefore we already were able to estimate the distance of the plates on a frame after segmentation. We have a few assumptions that will rule over this research: firstly all algorithms in this paper are applicable for the LPs that have darker digits on lighter background, preferably black digit on white. Secondly we calibrated the LP images from European Union standards, all are 462×130 mm which correspond approximately 70.000 pixel2 in our images in 1 m distance with 24 mm focal length (36 mm full frame).

The mathematical foundations of these paper and the experiment conducted by the 3D model cars are presented in following sections. Initially, the models are introduced as well as the flowchart of our system. Moreover, the LP segmentation methodology is identified with the equations of contrast analysis. Finally the trapezoid analysis is introduced with the experiments. As an important notice: All models used in this paper are selected from DOSCH 3D add-on libraries of 3D Studio MAX and analyzed by MATLAB.

2 Cruising Behind a Blinding Big Vehicle

The main difficulty that the drivers encounter while cruising behind a big vehicle is unpredictability of the road ahead since it can be really hard to sense the road condition. We have already presented a system in [1] to warn the driver once the front car is decelerating, therefore in this paper we only focused on the structural condition of the road instead, while the cars are cruising in close speeds within a safe following distance.

In this paper we basically focused on four variables to predict the road conditions ahead: Distance, Hill, Lateral Slope and Curve which are used to predict the road for assisting the drivers. The distance is calculated by the area of the front car's license plate however the plate is normalized to be strictly perpendicular to the driver's sight which also means that the plates should be in perfect parallel. Therefore the major issue is to identify the position of the front car in 3D space from its license plate with grade, rotation and curve variables. The demonstration of these variables generated by 3D simulation with a model vehicle Volvo XC90 could be seen in Fig. 1.

Fig. 1. Several possible positions of the front car (on the left: curve to the left, in the middle: rotation to right, on the right: uphill)

In each example, the position of the license plates reveal the road on which the car is cruising. The front car could be parallel or any of these conditions therefore analyzing the trapezoid of the plate that the camera captured gives the relation of road condition vs estimated distance to predict the virtual road. Since the main information to be gathered is the inference from the trapezoid of the plate, we firstly processed the acquired images from the 3D model as in Fig. 2. This figure is the flowchart of our system as well as the graphical abstract of our paper.

Initially, the images are captured from the 3D simulation and grayscaled. The cumulative contrasts are calculated for rows and columns separately to find the region of interest (ROI). The ROI image is binarized and the trapezoid is drawn that covers the white area. In Fig. 2, the angle analysis is rather simple since the car has only rotated to its left and the blue boundary actually is a rectangle. The real case scenarios are more complicated and hard to analyze considering examination of various trapezoids. However as we know the height-width ratio of a single license plate, the angles could be identified by analytical geometry rules.

Considering the motion of a car on a rough surface, there can be three kind of effect on the images: distortion, skewness and rotation. We assume that the vision of a plate

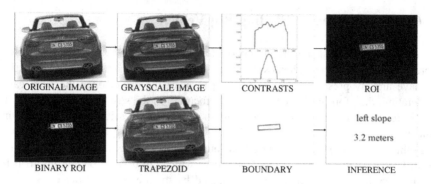

Fig. 2. Workflow of the proposed system.

will be distorted when turning left or right on the road, skewed when climbing hill and rotated when a car is on a lateral slope. Even though the distortion and skewness are similar in meaning, we divided into these subsets for better discrimination. Given these assumptions, distortion takes place in x-y, skewness in x-z and rotation in y-z planes. Finally the place of the trapezoid and the angles are analyzed to predict the curves, slopes and grades ahead.

3   Foundations of Segmentation

The system we propose starts with license plate segmentation without recognizing the letters. Although there are numerous methods in the literature for segmentation, we used contrast difference methodology with low-pass filtering and applying thresholds since we only focus on the plate itself.

For a red-green-blue (RGB) image taken from a camera mounted on a car or from a simulation like in this research, the image consists of numbers matrices in three layers. Therefore any pixel $p_{i,j,k}$ on this cube structure $I_{i,j,k}$ with three layers could be stated as:

$$p_{i,j,k} \in I_{i,j,k}(i = [1:w], j = [1:h], k = [1:3]) \tag{1}$$

where $I_{i,j,k}$ is the image, w is the width, h is the height and k is the color channel. With a standard grayscaling process that Matlab provides which is

$$\bar{p}_{i,j} \in G_{i,j} = 0.2989 p_{i,j,1} + 0.5870 p_{i,j,2} + 0.1140 p_{i,j,3} \tag{2}$$

the original image $I_{i,j,k}$ is turned into a grayscale image $G_{i,j}$ and the color channels are omitted. Once we obtain the grayscale image, there will be only one matrix containing numbers from 0 to 255. Since our first assumption is about the contrast of LPs that is black digits on white background, most of the pixel contrast are placed in LPs. Therefore, for every $G_{i,j}(i = [1:w], j = [1:h])$ absolute neighborhood differences over a threshold T are accumulated for each row and column by:

$$H_i = \left(\sum_{n=2}^{h} |G(i,n) - G(i,n-1)|, \; if \; |G(i,n) - G(i,n-1)| > T \right) \Big|_{i=1}^{w} \qquad (3)$$

$$V_j = \left(\sum_{m=2}^{w} |G(m,j) - G(m-1,j)|, \; if \; |G(m,j) - G(m-1,j)| > T \right) \Big|_{j=1}^{h} \qquad (4)$$

where G are grayscale images, H is horizontal and V is vertical totals, h is the height and w is the width of the image. For each single pixel, we applied a filter in which the moving averages are used namely,

$$AH_i = \frac{\left(\sum_{n=i-20}^{i+20} H_i \right) \Big|_{i=21}^{w}}{41} \qquad (5)$$

$$AV_j = \frac{\left(\sum_{m=j-20}^{j+20} V_j \right) \Big|_{j=21}^{h}}{41} \qquad (6)$$

Once the averages are calculated, largest averages are computed for every row and column to be used in dynamic thresholding by;

$$AH_{max} = \sum_{i=1}^{w} \max(AH_i) \qquad (7)$$

$$AV_{max} = \sum_{j=1}^{h} \max(AV_j) \qquad (8)$$

The averages are ultimately computed and the pixels less than the corresponding average are turned to zero and thus black.

$$AH_{ave} = \frac{AH_{max}}{w} \qquad (9)$$

$$AV_{ave} = \frac{AV_{max}}{h} \qquad (10)$$

$$AH_i = 0 \; if \; AH_i < AH_{ave} \big|_{i=1}^{w} \qquad (11)$$

$$AV_j = 0 \; if \; AV_j < AV_{ave} \big|_{i=1}^{h} \qquad (12)$$

The region of interest is found by these equations basically with applying dynamic and static thresholds and low-pass filtering. However the grayscale matrix $\bar{p}_{i,j} \in G_{i,j}$ is still consisting of mid values $0 \le \bar{p}_{i,j} \le 255$, therefore it is turned to binary image

$$\dot{p}_{i,j} \in B_{i,j}(i = [1:w], j = [1:h]) \text{ and } \dot{p}_{i,j} \in [0,1] \tag{13}$$

The image is manipulated and ready for trapezoid drawing and analysis.

4 Foundations of Trapezoid Analysis

The processed image now consists of binary values and the white pixel frame corresponds the LP with the dimensions of 462×130 mm. However in case of being not parallel to the ground, the angles should be estimated and the normalized LP should be found. Let P_f be the area of the license plate on the frame and P_r the area of the real license plate in one meter distance which is 70 k pixel^2. Since the closer LPs occupy larger place on the frames, the distance d in meters could simply be approximated by;

$$d = \sqrt{P_r / P_f} \tag{14}$$

From the position of the white pixels, the LP on the frame could be identified. If $x_{nw}, x_{ne}, x_{sw}, x_{se}$ are the horizontal and $y_{nw}, y_{ne}, y_{sw}, y_{se}$ are the vertical positions of the trapezoid like in Fig. 3.

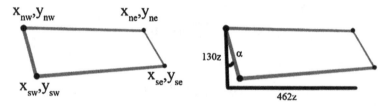

Fig. 3. The coordinates of the LP on a frame on the left. The normalized position on the right.

The ratio between the height and the width of the LP must be constant therefore the sides are represented by $130z$ and $452z$ where z is the ratio. If $x_{nw} = x_{sw} \wedge x_{ne} = x_{se} \wedge y_{ne} = y_{nw} \wedge y_{se} = y_{sw}$ the LP is perfectly parallel to the ground therefore the area P_f is estimated by

$$P_f = \left(\dot{x}_h - \dot{x}_l \right) \left(\dot{y}_h - \dot{y}_l \right) \tag{15}$$

where $\dot{x}_h = x_{ne} = x_{se}$, $\dot{x}_l = x_{nw} = x_{sw}$, $\dot{y}_h = y_{nw} = y_{ne}$, $\dot{y}_l = y_{sw} = y_{se}$. Additionally, in case of $x_{nw} < x_{sw} \wedge x_{ne} < x_{se}$ like in the example above in Fig. 3 the front car is turning left or there is a lateral slope to the left. Given this contradiction, the main discrimination point is the angles. If $x_{sw} - x_{nw} = x_{se} - x_{ne}$ then there should be only a lateral slope to the left. However, if $(y_{sw} - y_{nw})^2 + (x_{sw} - x_{nw})^2 > (y_{se} - y_{ne})^2 + (x_{se} - x_{ne})^2$ then we may consider the perspective effect and the car is also turning to left. In this situation the y-values should not match such as $y_{se} < y_{sw}$, and $y_{ne} < y_{nw}$, Turning right and right lateral slope analysis is the inverse of these equations above. To normalize the LP, the largest lateral edge is selected with max prompt and the corresponding x point is the reference point x_R of the normalized rectangle like in Fig. 3 right.

In Left Situation:

$$\left(\dot{x}_h - \dot{x}_l\right) = \sqrt{\left(y_{se} - y_{ne}\right)^2 + \left(x_{se} - x_{ne}\right)^2} \tag{16}$$

$$\left(\dot{y}_h - \dot{y}_l\right) = \sqrt{\left(y_{ne} - y_{nw}\right)^2 + \left(x_{ne} - x_{nw}\right)^2} \tag{17}$$

In Right Situation:

$$\left(\dot{x}_h - \dot{x}_l\right) = \sqrt{\left(y_{sw} - y_{nw}\right)^2 + \left(x_{sw} - x_{nw}\right)^2} \tag{18}$$

$$\left(\dot{y}_h - \dot{y}_l\right) = \sqrt{\left(y_{ne} - y_{nw}\right)^2 + \left(x_{ne} - x_{nw}\right)^2} \tag{19}$$

Finally in case of $x_{nw} < x_{sw} \wedge x_{ne} > x_{se}$ or inverse, there should be an uphill or downhill. Again in case of $x_{sw} - x_{nw} = x_{ne} - x_{se}$ the LP is still parallel to the ground, while if $x_{nw} < x_{sw} \wedge x_{ne} > x_{se}$ the reference point x_R will be uphill while if $x_{nw} > x_{sw} \wedge x_{ne} < x_{se}$ it will be the downhill. In each case, the distance is calculated using the reference points as mentioned above. The angle between the actual and the normalized plates is computed and the normalized plate is realized starting from the reference point. The area of the normalized plate is estimated by parameter z, namely:

$$P_f \cong 60000z^2 \tag{20}$$

Where

$$z = \frac{(\dot{x}_h - \dot{x}_l)\cos\alpha}{130} \tag{21}$$

for any reference point found on x-axis. Although the mathematical equations seem complicated, it could be clearer with the experiments conducted by 3D simulation.

5   Results and Experiments

The experiments conducted by manipulating a Range Rover in three axis and analyzing the frames by MATLAB using the software we wrote. The frames are 640×480 pixel^2 and the threshold T = 50 is used in all trials. In the simulation, a white environment is utilized which creates more contrast to identify the LP correctly yet the system could managed to recognize the plates. All experiment are done around 2 meters of distance between the cars since the first aim of this paper is enable to foresee the road ahead. Initially the parallel position of the front car is investigated.

As mentioned before, it is very unusual to find $y_{se} = y_{sw} \wedge y_{ne} = y_{nw}$ situation therefore, the system identifies a left curve even the car is perfectly parallel according to 3D Studio Max. As seen in Fig. 4, $y_{se} < y_{sw}$, and $y_{ne} < y_{nw}$, therefore the final inference is "turning to left".

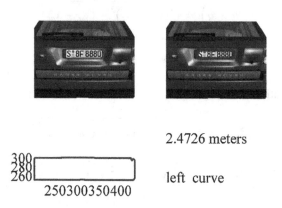

2.4726 meters

left curve

Fig. 4. Parallel Position

The second analysis is about the curves mostly encountered in urban traffic. Two types of frame is analyzed the results are presented in Figs. 5 and 6. In these trials, one lateral edge of LP is significantly smaller than the other.

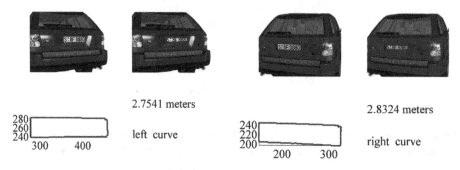

2.7541 meters

left curve

2.8324 meters

right curve

Fig. 5. Left curve (left) Right curve (right)

The final two experiments are about the lateral slope and the hills. The difference in hills situation is very small that can only be identified by $x_{nw} > x_{sw} \wedge x_{ne} < x_{se}$ inequalities for downhill. Nonetheless lateral slopes are easier to identify even if they are hard to encounter in urban roads.

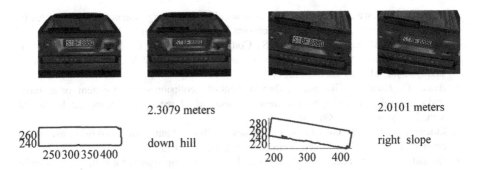

Fig. 6. Downhill (left) and lateral slope (right)

6 Conclusions and Discussions

The idea presented in this paper briefly is the prediction of the road condition while cruising a blinding huge vehicle, using the position of front car's license plate. 3D Simulation models are analyzed to segment the LP correctly prior to angle analysis. Afterwards, the boundaries of segmented LPs are analytically analyzed to decide on where the front car is cruising. Through the differences of the coordinates, the positions of the plates and thus the front vehicles are estimated.

Although the subject is so new, we achieved some encouraging results for the vehicles cruising close to each other. Yet, there still could be couple of enhancements as future research. Firstly, there wasn't any confidence intervals in if-then cycles of our algorithms, therefore the exactness is mandatory for most of the situations, especially for the vehicles straightly cruising that are also parallel to the ground. Secondly, achieving the exact value of the angles are not included in this research which could be implemented further.

Thirdly, we used 3D model cars which makes the manipulation of the angles in a simulation so simple while the results could shed light on real vehicles. Finally the analyses mentioned in this paper don't involve the combination of the road conditions which could be solved by an intelligent inference system. Implementing an inference system which will also solve the incorrect results by thresholding or filtering methods.

Acknowledgment. This work and the contribution were also supported by project "Smart Solutions for Ubiquitous Computing Environments" FIM, University of Hradec Kralove, Czech Republic (under ID: UHK-FIM-SP-2016-2102).

References

1. Alpar, O., Stojic, R.: Intelligent collision warning using license plate segmentation. J. Intell. Transp. Syst. (2015). doi:10.1080/15472450.2015.1120674
2. Alpar, O.: Corona segmentation for nighttime brake light detection. IET Intell. Transp. Syst. (2015). doi:10.1049/iet-its.2014.0281

3. Cabrera, A., Gowal, S., Martinoli, A.: A new collision warning system for lead vehicles in rear-end collisions. In: 2012 IEEE Intelligent Vehicles Symposium (IV) (2012)
4. Sengupta, R., Rezaei, S., Shladover, S., Cody, D., Dickey, S., Krishnan, H.: Cooperative collision warning systems: concept definition and experimental implementation. J. Intell. Transp. Syst. 11(3), 143–155 (2007)
5. Ararat, Ö., Kural, E., Güvenç, B.: Development of a collision warning system for adaptive cruise control vehicles using a comparison analysis of recent algorithms. In: Intelligent Vehicles Symposium 2006 (2006)
6. Eidehall, A., Pohl, J., Gustafsson, F., Ekmark, J.: Toward autonomous collision avoidance by steering. IEEE Trans. Intell. Transp. Syst. 8(1), 84–94 (2007)
7. Eskandarian, A., Soudbakhsh, D.: Enhanced active steering system for collision avoidance maneuvers. In: Proceedings of 11th International IEEE Conference on Intelligent Transportation Systems (ITSC 2008) (2008)
8. Kim, J., Hayakawa, S., Suzuki, T., Hayashi, K., Okuma, S., Tsuchida, N.: Modeling of driver's collision avoidance maneuver based on controller switching model. IEEE Trans. Syst. Man Cybern. Part B 35(6), 1131–1143 (2005)
9. Kaempchen, N., Schiele, B., Dietmayer, K.: Design of a dependable model vehicle for rear-end collision avoidance and its evaluation. In: 2010 IEEE Instrumentation and Measurement Technology Conference (I2MTC) (2010)
10. Kaempchen, N., Schiele, B., Dietmayer, K.: Situation assessment of an autonomous emergency brake for arbitrary vehicle-to-vehicle collision scenarios. IEEE Trans. Intell. Transp. Syst. 10(4), 678–687 (2009)
11. Kavitha, K.V.N., Bagubali, A., Shalini, L.: V2v wireless communication protocol for rear-end collision avoidance on highways with stringent propagation delay. In: Proceedings of International Conference on Advances in Recent Technologies in Communication and Computing (ARTCom 2009) (2009)
12. Ramirez, A., Ohn-Bar, E., Trivedi, M.: Integrating motion and appearance for overtaking vehicle detection. In: Proceedings of the Intelligent Vehicles Symposium. IEEE (2014)
13. Chang, B., Tsai, T., Young, C.: Intelligent data fusion system for predicting vehicle collision warning using vision/GPS sensing. Expert Syst. Appl. 37(3), 2439–2450 (2010)
14. Lee, K., Peng, H.: Evaluation of automotive forward collision warning and collision avoidance algorithms. Veh. Syst. Dyn. 43(10), 735–751 (2005)
15. Phelawan, J., Kittisut, P., Pornsuwancharoen, N.: A new technique for distance measurement of between vehicles to vehicles by plate car using image processing. Procedia Eng. 32, 348–353 (2012)
16. Chan, K., Ordys, A., Duran, O.: A system to measure gap distance between two vehicles using license plate character height. In: Bolc, L., Tadeusiewicz, R., Chmielewski, L.J., Wojciechowski, K. (eds.) ICCVG 2010, Part I. LNCS, vol. 6374, pp. 249–256. Springer, Heidelberg (2010)
17. Li, X., Shaobin, W., Li, F.: Fuzzy based collision avoidance control strategy considering crisis index in low speed urban area. In: 2014 IEEE Conference and Expo Transportation Electrification Asia-Pacific (ITEC Asia-Pacific), pp. 1–6 (2014)
18. Chang, C.Y., Chou, Y.R.: Development of fuzzy-based bus rear-end collision warning thresholds using a driving simulator. Intell. Transp. Syst. 10(2), 360–365 (2009)

Decision Support and Control Systems

Creating a Knowledge Base to Support the Concept of Lean Administration Using Expert System NEST

Radim Dolak[✉]

Department of Informatics and Mathematics, School of Business Administration
in Karviná, Silesian University in Opava, Karviná, Czech Republic
dolak@opf.slu.cz

Abstract. Concept of lean administration is a part of a comprehensive concept of lean company. The purpose of the concept of lean administration is to identify, classify and minimize waste in administrative processes. This paper discusses the following topics: lean administration principles, basic model of lean administration and creating lean administration knowledge base using IF-THEN rules. Case study focuses on process of creating a knowledge base that will support an implementation of lean administration concept. The main tasks of the knowledge base are mainly the following: identification of level of waste in administrative processes and recommendation of appropriate methods and tools of industrial engineering to reduce waste in administrative processes.

The knowledge base for lean administration was created in an expert system NEST and testing was done on the basis of fictitious data about administrative processes.

Keywords: Lean administration · Expert system · Knowledge base

1 Introduction

The concept of lean administration is implemented usually after implementation of the concepts of lean manufacturing and lean logistics. All mentioned concepts are part of the Lean Company concept which is a complex approach includes these lean concepts: manufacturing, logistics, administration and development. These concepts have originated in Japan in the famous automotive company Toyota.

Introduction of the concept of lean administration may be a way how to get a competitive advantage in the global competitive environment. The main principle of lean administration concept is the restriction of all unnecessary processes and activities without value to the customers and without profit for the company.

The main goal of this paper is describe possibility of using knowledge base for support lean administration concept implementation. There will be discussed general information about empty expert system called NEST such as structure of expert system and possibilities for knowledge representation. This expert system for diagnostic applications based on IF-THEN rules was used for creating and editing the knowledge base called "lean administration".

© Springer International Publishing Switzerland 2016
N.T. Nguyen et al. (Eds.): ICCCI 2016, Part I, LNAI 9875, pp. 281–291, 2016.
DOI: 10.1007/978-3-319-45243-2_26

The case study describes the following issues of knowledge engineering such as a process of creating a knowledge base, testing of knowledge base and finally identification of wasting in administration processes and recommendation of appropriate methods and tools of industrial engineering to reduce wasting in administration processes.

2 Related Work

Related work can be divided into area of lean administration and expert systems. There are several approaches for lean administration optimization. We can find a systematic approach with description of lean metrics in Tapping and Shuker [25]. Hobbs [11] is comparing differences between administrative processes and manufacturing processes and deals with administrative areas and non-value-added work. We can find a lot of successful implementation of lean administration in real case studies in a book called Lean administration: case studies in leadership and improvement [3]. One chapter of this book describes details about company which built a lean culture through the business on the foundation of a 5S program carefully and methodically implemented in all departments. Mann [21] is describing Lean principles in administrative in Chap. 6 and he confirms that lean is no longer just for manufacturing. He illustrates that the principles and many of the tools of lean are just as effective in administrative, technical, professional and also healthcare processes.

There are also many organizations and academies that provides complete consulting and educational services for Middle European industrial enterprises such as for example API [2], DMC Management Consulting [7] and MBtech [22]. These organizations and academies streamline administrative processes using a variety of methods and tools of industrial engineering.

Eriksen, Fischer and Mønsted [8] are defining three of the most effective methods of preventing errors in administration and service functions are target management of errors, sender control and error control. The most common methods that are used to standardize administration processes: introducing the principles of 5S, continuous elimination of waste, Value Stream Mapping, visual management, quality management, ergonomics, etc.

There are many sources and literature deals with topic of expert systems as for example [1, 6, 13] or [24] which provides overview about experts systems and process of creating a knowledge base of expert system. Russel and Norvig [24] published theory that expert systems that incorporate utility information have additional capabilities compared with pure inference systems because of possibility to make conclusions based on decision which questions to ask.

3 Lean Administration

Lean administration should be defined as a modern complex approach consists of a set of tools to eliminate waste in administration. Administrative processes are important part of enterprise structure and improving their productivity has a great impact on the productivity of other business processes. There are according Locher [20] four basic steps to the application of Lean: stabilize, standardize, visualize and continually improve.

There are many benefits from implementing lean administration concept in the company. Four Principles Lean Management Experts [19] deals with the following tangible improvements:

- revision of data management processes reduced errors by 25 %,
- optimizing an insurances claims process decreased errors by 98 % over the entire process,
- lean internal reporting optimization reduced unneeded content by 65 % and report generation lead time by 31 %.

Koenigsaecker [15] described another benefit: it was evaluated that the administrative standard work events and the average administrative standard work event resulted in an 80 % productivity gain.

3.1 Areas of Waste

Keyte and Locher [14] seeks to enable users to see administrative waste, identify its sources, and develop a future state that eliminates it so that scarce resources can then be focused on those activities truly necessary to create value. One chapter of their book also deals with identifying Office Waste.

Areas of waste in an office environment are detailed described in [19] and can be divided into 7 categories such as "Overproduction", "Movement", "Transportation & Handling", "Inventory", "Waiting", "Defects" and "Over-processing".

- Overproduction is represented for example by printing paperwork or processing an order before it is needed, too much information gathered, stored, and maintained. Overproduction can be eliminated according to [12] by balancing capacity and workload. There can be used some tools such as: SMED, Production levelling of Heijunka, One-piece flow cells.
- Movement is very typical type of wasting in many offices. It can be unnecessary movement such as walking between offices locations or walking to copier, printer, fax etc. Typical problem is also saving electronical files everywhere outside from central storage, looking for printed documents and other office items because they do not have a defined place.
- Transportation & Handling should be represented for example by following activities: excess movement of paperwork, multiple printed documents or multiple electronic data, complicated approval.
- Inventory is a type of wasting in administrative processes based on making and purchasing things before they are needed such as unnecessary documents, reports or email and all forms of batch processing that create inventory.
- Waiting in administrative is represented usually by slow computers, slow information system, waiting due to complicated system for approvals, waiting for customer information or waiting for clarification or correction of work received from upstream processes.

- Defects are besides wasting particularly serious problem of nonfunctional administrative processes that can cause problems and loss. Typical examples are errors during data entry (orders or invoice errors), engineering change orders, employee turnover and miscommunication.
- Over-processing is represented typically by re-entering data into multiple information systems, making extra copies, generating unused document (reports, printing and mailing, faxing), repetition of same information in different forms, use of different software in different departments.

Some authors such as for example [4] and [16] describes one more type of wasting that is generally called "not used creativity of employees". There will be also used this type of wasting in knowledge base of expert system for lean administration concept.

3.2 Methods and Tools

There are many methods and tools that provide the optimization in lean administrative processes. The most famous methods for eliminating waste in administrative processes are following: 5S, Kaizen, Value Stream Mapping, Kanban, Six Sigma, SMED, Production levelling of Heijunka, One-piece flow cells.

I will mention 5S methodology at first. According to Fabrizio and Tapping [10] are the activities at the heart of 5S for the Office (organizing, ordering, cleaning, standardizing, and sustaining all of these) completely logical. They are the basic rules for managing any effective workplace. The 5S system retains its fundamental power to change the workplace and involve everyone in improvement.

Kaizen is one of the systems of continuous improvement processes based on daily improvement in small steps. The basic philosophy of this approach is the statement that "nothing is so good that it could not be better." Office Kaizen is according Lareau [17] concerned primarily with creating a competitive advantage via the reduction of surface waste. Lareau [18] is defining Office Kaizen as the application of kaizen and its adjuncts to nonmanufacturing processes such as "typical office areas" including purchasing, logistics, finance, human resources, quality control, engineering, planning etc.

Value Stream Mapping represents according to Rother [23] all the actions (both value added and non-value added) currently required to bring a product through the main flows essential to every product. Taking a value stream perspective means working and improving complexly, not just optimizing the parts. Value Stream Mapping is according Erlach [9] especially connected with order processing.

Kanban is a scheduling system for lean production using just-in-time principles. Tautrim [26] wrote that Kanban is a readily recognizable signal that tells you that something is needed. Kanban is implemented in a simple case with cards.

Six Sigma methods integrate according to Carroll [5] principles of business, statistics, and engineering to achieve tangible results. The main purpose is doing business with a focus on eliminating defects through fundamental process knowledge.

4 Case Study: Creating a Knowledge Base to Support Lean Administration Concept

The case study deals with a process of creating a knowledge base in expert system NEST to support lean administration concept. The main goal is to create a knowledge base for analysis of administration processes, find wasting in administration processes and recommend appropriate methods and tools of industrial engineering for elimination of waste in administration processes. There are also sub-objectives including definition of basic model for lean administration concept and conversion of basic lean administration model in the form of rules for representing in the knowledge base of expert system NEST.

4.1 Expert System NEST

Expert system NEST is an empty expert system which includes inference mechanism. It was created during several years of research at the University of Economics in Prague, Czech Republic. More information about this expert system is published by Ivánek, Kempný and Laš [13]. Expert system NEST provides a graphical user interface for basic actions with knowledge base such as: creating and editing knowledge bases, integrity check, basic settings such as type of uncertainty, consultation process and finally also a recommendation statements with an explanation of the findings.

4.2 Acquisition of Knowledge for Building Knowledge Base About Lean Administration

Acquisition of knowledge for building knowledge base is the most important factor in knowledge engineering during process of creating a quality knowledge base which is expressed by the different types of rules. We need to establish very good cooperation between an expert in the relevant field and knowledge engineer. It is also possible to study relevant materials in the relevant field to acquire the necessary knowledge. We can find many sources and literature about lean administration processes and principles, methods and tools of industrial engineering or characteristics and benefits of this methods and tools as for example in [2, 4, 8, 10, 16, 18] or [26].

The basic model of lean administration knowledge base is based on two important relevant fields: criteria for lean administration concept and also methods and tools of industrial engineering to support lean administration concept.

4.3 Criteria for Lean Administration

We can name list of many basic criteria for lean administration. There is a general basic idea of lean administration: to reduce the basic types of waste in administrative processes. We can see different types of waste in areas of administrative processes in the following Fig. 1. It can be divided into these different categories: overproduction, movement, transportation, inventory, waiting, defects, over-processing and not used creativity of employees.

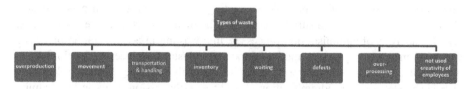

Fig. 1. The basic model of wasting in administration processes.

We can set the appropriate questions for each category of waste. It is necessary to study the basic constituents of individual types of waste for creating questions. Then it will be possible to identify the rate of waste for different categories of waste. We can find some key elements for lean administration processes in [2, 4] or [16].

It is necessary to study also important issue of methods and tools of industrial engineering because inference mechanism of expert system NEST will recommend as final statements appropriate methods and tools of industrial engineering for elimination of wasting in administrative processes such as: Standardized work, 5S, Process mapping, Value stream mapping, Kaizen, Planning, Workshops, Gemba, Kanban, Lean Six Sigma Office, One Piece Flow.

4.4 Building Knowledge Base in the Expert System NEST

Knowledge base of expert system NEST is based on using these elements for representing knowledge such as: attributes and propositions, rules, contexts and integrity constraints. Knowledge base is created by the NEST editor or by editing of XML file that are used for saving of knowledge base. When we are creating new knowledge base so we start with entering the global parameters such as the range of weights, the threshold of a global context and condition. We can also work with different types of inference mechanism such as standard logic, neural network or hybrid. Knowledge base in expert system NEST has the following hierarchical architecture:

- queries,
- intermediate statements,
- final statements.

We can see the structure of the knowledge base called "lean administration" in the following Fig. 2 as a cross-section of knowledge base for lean administration which is focused on overproduction area of waste, which is one of the eight types of wasting in administrative processes.

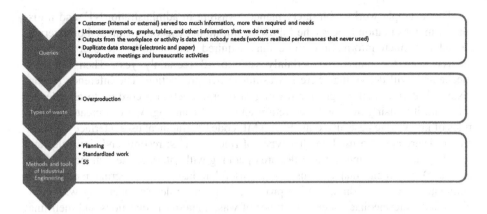

Fig. 2. Cross-section of knowledge base for overproduction in lean administration.

Expert system NEST supports typical IF-THEN rules. We can systematically classified the types of rules into the following groups: evaluation (basic) rules, recommending rules and specific rules.

Evaluation (basic) rules are used to derive the types of waste directly from the basic questions. We can see an example of evaluation (basic) rule in the following XML syntax from NEST editor:

```
<compositional_rule>
 <id>wasting in administration1-1</id>
 <condition>
 <conjunction>
  <literal>
   <id_attribute>wasting 1-1</id_attribute>
   <negation>0</negation>
  </literal>
 </conjunction>
 </condition>
 <conclusions>
  <conclusion>
   <id_attribute>wasting 1-overproduction
   </id_attribute>
   <negation>0</negation>
   <weight>1,750</weight>
  </conclusion>
 </conclusions>
</compositional_rule>
```

This example works with query named "wasting in administration1-1" and it displayed in consultation process the following question: "Customer (internal or external) served too much information, more than required and needs". The user will type answer from the range −3 (certainly yes) to +3 (certainly yes). Then inference mechanism will derive that there is wasting in overproduction. The inference mechanism will derive final weight for wasting in overproduction according to all relevant rules and it is using the combined return and direct chaining. We can create and edit rules of knowledge base also outside of XML code by graphical user interface of NEST editor. There are also used another types of rules such as: recommendation rules and specific rules. Recommendation rules are operating with intermediate statements (types of waste) and their final conclusions are useful industrial engineering methods for eliminating waste in administration processes. Specific rules are operating with combination of intermediate statements (types of waste) and basic questions and their final conclusions are useful industrial engineering methods for eliminating waste in administration processes.

We can see a partly structure of knowledge base in the form of graph like as output from NEST editor in the following Fig. 3. Rules are represented by edges and statements (queries, intermediate statements and final statements) are represented by nodes.

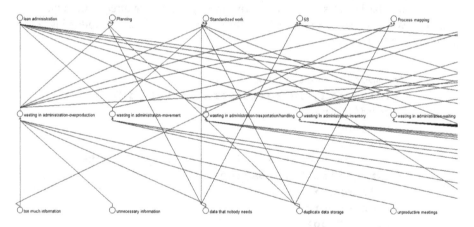

Fig. 3. Cross-section of the designed knowledge base lean administration.

As a result of consultation process with expert system NEST we will receive the following statements:

- overall information about lean administration status,
- recommendation of appropriate methods and tools of industrial engineering with weight of importance,
- summary of wasting in administration processes for different types of waste.

4.5   Process of Testing a Knowledge Base Using Fictitious Data

Process of testing a knowledge base of expert system is usually based on testing using fictitious or real data. We can start testing in the NEST expert system after the knowledge base is created. The knowledge base deals with lean administration was testing using fictitious data to verify basic functionality. Testing was based on consultation process with expert system. It was used random data for simulation different situation in lean administration to check if the results are in accordance with input data and knowledge base derive meaningful conclusions. Except of random data was also used specifically specified data in order to simulate the state of an ideal lean administration, and on the other hand the administration of a high degree of wastage.

We can see results of testing a knowledge base with the conclusions of waste in different areas of administration (intermediate statements) in Fig. 4.

Name	Min weight	Max weight	Status <	Type
lean administration	0,665	0,665	final	goal
wasting in administration-overproduction	2,529	2,529	final	intermediate
wasting in administration-movement	-1,333	-1,333	final	intermediate
wasting in administration-trasportation/handling	2,000	2,000	final	intermediate
wasting in administration-inventory	1,167	1,167	final	intermediate
wasting in administration-waiting	-2,997	-2,997	final	intermediate
wasting in administration-defects	-1,454	-1,454	final	intermediate
wasting in administration-overprocessing	-0,631	-0,631	final	intermediate
wasting in administration-not used HR creativity	-1,780	-1,780	final	intermediate

Fig. 4. Results of rule-based reasoning.

It was discovered after testing knowledge base using fictitious data that it will be necessary to perform further modifications in knowledge base. It seems important to add more direct and specific rules to improve more detailed decision-making skills of knowledge base in similar decision-making situations.

5   Conclusion

This paper describes lean administration issue and the possibility of using the knowledge base of expert system to support this concept. For a case study deals with creating a knowledge base of lean administration was selected expert system NEST.

Knowledge base has a three level architecture: queries (questions), intermediate statements (types of waste) and final statements (methods and tools of Industrial Engineering). The main tasks of created knowledge base are the following: identification of the wasting in administration processes and recommendation of appropriate methods and tools of industrial engineering to reduce wasting in administration processes. Knowledge base was tested by the using fictitious data.

As a future work, the knowledge base will be improved by adding more direct and specific rules and it will be also tested in using real data from administration processes in companies.

Acknowledgement. The work was supported by the IGS project number IGS/20/2016 called "Creating a knowledge base of expert system to support the concept of lean administration".

References

1. Akerkar, R., Sajja, P.: Knowledge-based systems. Jones and Bartlett Publishers, Sudbury (2010)
2. API - Academy of Productivity and innovations. http://www.e-api.cz/en/
3. Association for manufacturing excellence, Lean administration: case studies in leadership and improvement. Productivity Press, New York (2007)
4. Bejčková, J.: Štíhlá administrativa – základ prosperující společnosti (2. část). Úspěch: produktivita a inovace v souvislostech, vol. 1, pp. 10–11. API: Želevčice (2013)
5. Caroll, B.J.: Lean Performance ERP Project Management: Implementing the Virtual Lean Enterprise. CRC Press, Boca Raton (2007)
6. Coppin, B.: Artificial Intelligence Illuminated. Jones and Bartlett Publishers, Sudbury (2004)
7. DMC Management Consulting. http://www.dmc-cz.com/en
8. Eriksen, M., Fischer, T., Mønsted, L.: Lean Management in the Administration and Service Sector. Børsen Forlag, Copenhagen (2007)
9. Erlach, K.: Value Stream Design: The Way Towards a Lean Factory. Springer, Heidelberg (2013)
10. Fabrizio, T.A., Tapping, D.: 5S for the Office: Organizing the Workplace to Eliminate Waste. Productivity Press, New York (2006)
11. Hobbs, D.P.: Applied Lean Business Transformation: A Complete Project Management Approach. J. Ross Publishing, Fort Lauderdale (2011)
12. Chiarini, A.: Lean Organization: from the Tools of the Toyota Production System to Lean Office. Springer, Milan (2013)
13. Ivánek, J., Kempný, R., Laš, V.: Znalostní inženýrství. OPF Karviná, Karviná (2007)
14. Keyte, B., Locher, D.: The Complete Lean Enterprise: Value Stream Mapping for Administrative and Office Processes. Productivity Press, New York (2004)
15. Koenigsaecker, G.: Leading the Lean Enterprise Transformation. CRC Press, Boca Raton (2013)
16. Košturiak, J., Frolík, Z.: Štíhlý a inovativní podnik. Alpha Publishing, Praha (2006)
17. Lareau, W.: Office Kaizen: Transforming Office Operations into a Strategic Competitive Advantage. ASQ Quality Press, Milwaukee (2003)
18. Lareau, W.: Office Kaizen 2: Harnessing Leadership, Organizations, People, and Tools for Office Excellence. ASQ Quality Press, Milwaukee (2011)
19. Lean Administration/Four Principles. http://www.fourprinciples.ae/solutions/lean-administration#.VrX07lnnV_k
20. Locher, D.: Lean Office and Service Simplified: The Definitive How-To Guide. CRC Press, Boca Raton (2011)
21. Mann, D.: Creating a Lean Culture: Tools to Sustain Lean Conversions. CRC Press, Boca Raton (2015)
22. MBtech. https://www.mbtech-group.com

23. Rother, M., Shook, J.: Learning to See: Value Stream Mapping to Add Value and Eliminate Muda. The Lean Enterprise Institute, Cambridge (2003)
24. Russell, S., Norvig, P.: Artificial Intelligence: A Modern Approach. Pearson Education, New Jersey (2010)
25. Tapping, D., Shuker, T.: Value Stream Managwemet for the Lean Office. Productivity Press, New York (2003)
26. Tautrim, J.: Lean Administration: Taschenbuch/Beraterleitfaden: Wesentliche Konzepte und Werkzeuge für mehr Effizienz in der Verwaltung. epublu GmbH, Berlin (2014)

Fuzzy Bionic Hand Control in Real-Time Based on Electromyography Signal Analysis

Martin Tabakov[1(✉)], Krzysztof Fonal[1], Raed A. Abd-Alhameed[2],
and Rami Qahwaji[2]

[1] Department of Computational Intelligence,
Wroclaw University of Science and Technology, Wroclaw, Poland
martin.tabakow@pwr.edu.pl, 168132@student.pwr.edu.pl
[2] School of Electrical and Computer Science,
University of Bradford, Bradford, UK
{r.a.a.abd,r.qahwaji}@bradford.ac.uk

Abstract. In this paper a fuzzy model for control of bionic hand in real-time is proposed. The control process involves interpretation and analysis of surface electromyography signal (sEMG) acquired from patients with amputees. The work considers the use of force sensing resistor to achieve better control of the artificial hand. The classical type-1 Mamdani fuzzy control model is considered for this application. The conducted experiments show comparable results with respect to applied assumptions that give the confidence to implement the proposed concept into real-time control process.

Keywords: Fuzzy control · Mamdani model · Bionic limbs · Electromyography · Signal analysis

1 Introduction

The theoretical background and hardware implementations of bionic limbs has widely been expanded in recent years in the field of medical applications [5, 11, 15] and robotics domain [9] as a result of the available electronic components with sufficient computing power with a very low cost. The process of such applications applied fuzzy logic in the design of robotic systems, as it is capable to handle a non-linear complex systems [2, 8, 16]. Commonly, in the research concepts of bionic limbs engineering design, surface electromyography signals are the most often used control source for bionic limbs [5, 13, 18]. The research ideas have evolved so quickly that even neuro-controlled bionic arms, able to allow an amputee to move prosthetic arm as if it is a real limb directly by thinking, have been introduced [12, 14]. Nevertheless, most of the research areas consider the problem of bionic limbs control, as a classification problem, i.e. if a certain input conditions are recognized (sensor information and sEMG values) then implemented mechanism based on machine learning is applied [3]. Unfortunately, from patient perspective, there was a problem with such a concept, namely muscle contraction related to certain functionality should be entirely executed. Next, a classifier is used for functionality recognition. This concept does not give the patient the sensing of real control over the bionic hand. Therefore, in our interpretation of real-time signal

© Springer International Publishing Switzerland 2016
N.T. Nguyen et al. (Eds.): ICCCI 2016, Part I, LNAI 9875, pp. 292–302, 2016.
DOI: 10.1007/978-3-319-45243-2_27

analysis, we focused on a proposal of sEMG signal analysis that is applicable in fuzzy control models, for real-time performance which should give the patient the sensing of control over the bionic hand. In the model presented, it is assumed the functionality of the control of hand grasping, with respect to object resistance, i.e. the aim is to reduce the speed of the device used for hand grasping functionality with the increase of potential object resistance in real-time: during the sEMG signal acquisition.

The paper is organized as follows: Sect. 2 introduces background information about the fuzzy control model applied. Section 3 explains the basic steps of a sEMG processing. Section 4 presents the basic research idea proposed. Sections 5 and 6 provide results and discussion and finally, Sect. 7 draws conclusion.

2 Basic Notions

In this section, the preliminaries of fuzzy sets [17] and fuzzy control systems of Mamdani type [10] are presented.

2.1 Type-1 Fuzzy Set

A type-1 fuzzy set A consists of a domain X of real numbers together with a function $\mu_A : X \rightarrow [0, 1]$, [17] i.e.:

$$A =_{df} \int_X \mu_A(x)/x , \quad x \in X \tag{1}$$

here the integral denotes the collection of all points $x \in X$ with associated membership grade $\mu_A(x) \in [0, 1]$. The function μ_A is also known as the membership function of the fuzzy set A, as its values represents the grade of membership of the elements of X to the fuzzy set A. The idea is to use membership functions as characteristic functions (any crisp set is defined by its characteristic function) to describe imprecise or vague information. This possibility along with the corresponding defined mathematical apparatus, initiated a number of applications. Assuming discrete domain, the basic set operations: union and intersection of two fussy sets A and B, are defined as follows: $\forall_{x \in X} \mu_{A \cup B}(x) =_{df} \boldsymbol{max}\{\mu_A(x), \mu_B(x)\}, \forall_{x \in X} \mu_{A \cap B}(x) =_{df} \boldsymbol{min}\{\mu_A(x), \mu_B(x)\}.$

2.2 Type-1 Fuzzy Logic Controller

Figure 1 shows the schematic diagram of a type-1 fuzzy controller. The main idea is that all input information are fuzzified and then processed with respect to the assumed knowledge base, inference method and the corresponding defuzzification method.

Let consider the rule base of a fuzzy logic controller consisting of N rules which take the following form:

$$R^n :_{df} \text{IF } (x_1 \text{ is } X_1^n) \text{ o } \ldots \text{ o } (x_i \text{ is } X_i^n) \text{ THEN } y \text{ is } Y_n \tag{2}$$

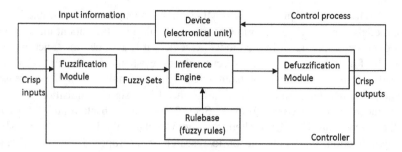

Fig. 1. Information flow within a typical type-1 fuzzy controller

where X_i^n ($i = 1,..., I$; $n = 1,..., N$) are fuzzy sets defined over corresponding domains and Y_n is an output information, which in the Mamdani model [10] is assumed as fuzzy set as well, defined over some domain Y. The binary operator 'o' is the t- or s-norm (o $\in \{\otimes, \oplus\}$; $\otimes, \oplus$: $[0, 1]^2 \rightarrow [0, 1]$) which have the commutative, associative and the monotonic properties, and have the constants 1 and 0 as unit elements, respectively. In fuzzy logic, the t-norm operator provides the characterization of the AND operator, while the s-norm provides the characterization of the OR operator [1].

Assuming an input vector $\bar{x} = \{x_1', x_2', x_3', ..., x_i'\}$, typical computations of a fuzzy system consist of the following steps:

(1) Compute the membership grades of x_i' on each X_i^n, $\mu_{X_i^n}(x_i')$, $i = 1, ..., I; n = 1,..., N$
(2) Compute the firing value of the nth rule, $F^n(\bar{x})$:

$$F^n(\bar{x}) =_{df} \mu_{X_1^n}(x_1')o...o\mu_{X_i^n}(x_i') \in [0, 1] \tag{3}$$

(3) Compute defuzzification output. The most common method is the centre of gravity (COG) method with assumed relation between the premise and the conclusion of the fuzzy rules as the *min* operator:

$$Y_{COG}(\bar{x}) =_{df} \frac{\sum_{y \in Y} \mu_{\bigcup_n Y_n'}(y) \cdot y}{\sum_{y \in Y} \mu_{\bigcup_n Y_n'}(y)} \tag{4}$$

where

$$\forall_{y \in Y} \mu_{Y_n'}(y) =_{df} min\{f^n, \mu_{Y_n}(y)\}, \quad n = 1,..., N \tag{5}$$

The output value is directly related to the control process.

3   Electromyography Signal Processing

Surface electromyography delivers a non-invasive method for the objective evaluation of the electrical activity of the skeletal muscles. It provides information about the functionality of the peripheral nerves and muscles [7]. In clinical research, this information has significant influence to the preparation of relevant procedures for patient rehabilitation. Myoelectric signals refer to the system of voluntary muscle contraction, which motor units are activated at different frequencies and their contributions to the signal are added asynchronously [6]. This signal presents harmonics with frequencies ranging from 15 Hz to about 500 Hz, and amplitude from approximately 50 µV to 5 mV [4].

In our research the sEMG signal is used as basic information source for control; but, in order to make the raw signal interpretable, the corresponding sEMG signal processing is required. Delsys® Bagnoli™ EMG System for signal acquisition, along with dedicated tool for signal processing and analyses EMGWorks® Software were used. The Bagnoli acquisition device and EMG signal processing software completely cover all the steps necessary for the processing of any raw sEMG signal, as described below:

(a) Acquisition of the signal,
(b) Elimination of the energy power grid utility frequency – this is done by Band Stop filter, depending on the country (in Poland: 50 Hz),
(c) Proper identification of the sEMG signal bandwidth – this is done by Band Pass filter with parameters: 15 Hz–200 Hz (the signal is most intense in this bandwidth),
(d) RMS (Root Mean Square) filtering with time window, applying default options: window length 0.125 and window overlap 0.0625.

On Fig. 2, a sample raw sEMG signal (taken during repetitive rehabilitation exercises) and its processed form are shown, after applying steps (a)–(d).

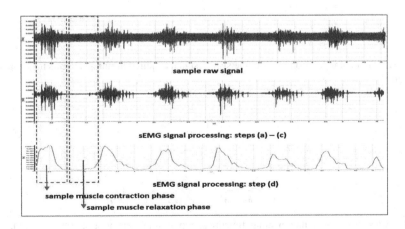

Fig. 2. Processing of a raw sEMG signal

It should be noted to process the raw signal in real-time, a corresponding time window which is slid among the time axis should be applied.

4 Basic Research Concept

The basic concept of the work presented, is to use the sEMG signal given by a human muscle to control a bionic hand. It is assumed the corresponding values from the signal, to be fuzzified and applied in a fuzzy Mamdani control model. In order to achieve this goal, we propose the following analysis of the sEMG signal:

(1) Apply polynomial regression of 3th degree, in order to generate smooth curve (for more details see 4.1) which represents the real-time sEMG signal, required for control (see Fig. 3). The polynomial is represented below by the function f.
(2) Calculate the discrete derivative in each time point t. We have applied the forward method of discrete derivative calculation:

$$\frac{df}{dt} =_{df} \frac{f(t + \Delta t) - f(t)}{\Delta t}, \tag{6}$$

where $\Delta t = 0.0625$ s. (referring to the sampling rate of the sEMG device used). The derivative is providing the information, about the rate of change of the function values that is related directly to the degree of muscle contraction in any time point.

(3) Fuzzify the derivative values, in order to provide information about the degree of function values change (see Fig. 4).

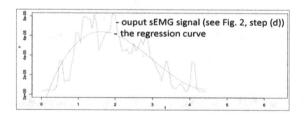

Fig. 3. Polynomial regression of a sample sEMG signal acquired during pectoralis major muscle contraction.

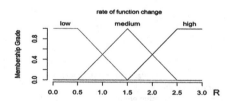

Fig. 4. 'Classical' (three membership functions, representing fuzzy sets: low, medium, high derivative values) fuzzification of the derivatives of f.

4.1   Real Time Signal Analysis

Obviously, in order to apply any regression, the whole set of measurements points given should be available. Therefore, the muscle contraction, assumed as an input for the proposed fuzzy control model, could be completed, which makes the real-time control process impossible under these assumptions.

That is why, in order to enable real-time control with regression, it is proposed the following solution of the above mentioned problem:

(1) For any patient with amputee, under a series of trainings (which provide sEMG signals of pectoralis major muscle controlled contractions), it will build a database of sEMG signals. For all the signals, a polynomial regression is applied which provides an averaged, personalised regression (see Fig. 5). This stage might be considered as a learning process, required for every patient.

(2) Next, in real-time, current regression of a sEMG signal might be constructed using data acquired in real-time in combination with the average regression curve (the remaining points of the curve, see Fig. 6 (a)–(b)). This gives the possibility to build regression curve in real-time.

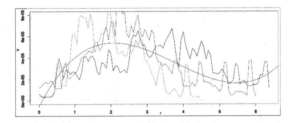

Fig. 5. The average, 'personalised' regression curve

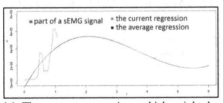

(a) The current regression, which might be used for control, is calculated by combining sEMG signal measured in real-time with the average regression curve, personalised for each patient.

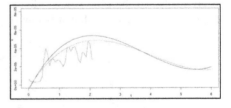

(b) Visualisation nr. 2 (sEMG signal nr. 2)

Fig. 6. (a) – (b). The process of building 'current' regression for real-time control.

Therefore, the idea of the real-time control is to use the regression curve which can be built during the contraction of the pectoralis major muscle. This muscle was chosen as it can be easily controlled by the patient.

4.2 Force Sensing Resistors

Additionally, in order to improve the control process, we propose to use a force sensor resistor. The information given by the sensor is assumed as second information source for the fuzzy controller. The natural processes idea could be stated as follows:

- the relation between degree of muscle contractions and force sensors values should be reversed, i.e. when the patient is increasing the degree of contraction, the force sensor should not give any values (the patient does not grab any object)
- and with the increasing of the applied force on the sensor, the degree of muscle contraction should decrease (the patient is releasing the muscle contraction as object is being grabbed).

In this work, as it is not aimed to present real hardware implemented bionic hand, we have simulated the values of the force sensing resistor, according to the corresponding technical documentation (Interlink Electronics®, Force Sensing Resistor® (FSR). Below, a simple force-to-voltage conversion is presented, for different values of the measuring resistor RM (Fig. 7).

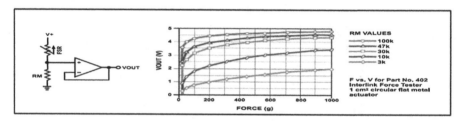

Fig. 7. FSR voltage divider; image source: http://interlinkelectronics.com/integ-guides.php

For the current experiments, a function is assumed for whose values are most closely related to the FSR voltage divider with RM = 10 kΩ. The proposed fuzzification of the FSR data is shown below (Fig. 8).

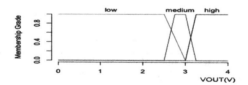

Fig. 8. Fuzzification of the output voltage.

4.3 Fuzzy Controller Output

The idea of any control system is based on present of an electronical device, which can be controlled over system assumptions. The velocity of the device might be controlled with the fuzzy controller proposed. The degree of velocity of the device should be related to the degree of bionic hand reaction (degree of hand clump). There might be

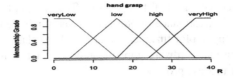

Fig. 9. Fuzzification over values of degree of control unit reaction (changes in the speed of a hypothetical motor unit).

many such devices that could be designed for this purpose, for example a typical one, often applied in bionic limbs research is the servomechanism.

The proposed model of the device performance and the corresponding fuzzification are shown in see Fig. 9, that can easily be related to the degree of performance of the control unit.

5 Experiments and Results

It should be noticed that the data of the force sensing resistor as well as the output device assumed to be controlled are simulated. The result presents possible solution for further hardware implementation; however, on the other hand, the critical issue here is

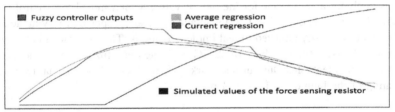

(a) Performance of system output with respect to the system inputs (assuming uniform domain and range of the functions used, for presentation purpose).

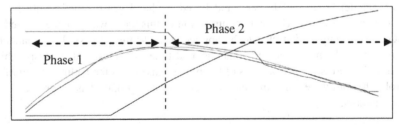

(b) Explanation of the presented data: Phase 1: Muscle contraction (increasing muscle tension) / Low values of the force sensing resistor or lack of resistance. Phase 2: Muscle contraction (muscle relaxation) / Rising values of the force sensing resistor.

Fig. 10. Values of fuzzy controller output: with the increase of the simulated force sensing resistor values along with the decreasing of the muscle tension, the outputs are decreasing which is related to reduction of controlled device reaction (for example, slowing down the controlled unit).

sEMG signal analysis concept for real-time control. The design of corresponding motor unit based on degree of performance (motor speed) that might be controlled, is trivial. In addition the values applied for fuzzification are easily to change, in order to achieve optimal performance in real hardware implementation, which will not affect the main concept of the control process proposed.

It is meant through these experiments to observe the expected performance on real sEMG signal data. Figure 10 (a)–(b) presents the relations between the sEMG signal, acquired from a patient with amputee (a female), of the assumed force sensing curve and the output device performance.

6 Discussion

This work provides a proof of concept that sEMG can be analysed in real-time and gathered data can be applied in a fuzzy control model. The applied regression over sEMG data enables to get control values on ongoing basis. Together with other sensor data, such as force sensing resistors, a control mechanism for bionic hand can be introduced. We have acquired and analysed sEMG signals from two patients with amputees (male, aged 24 and female, aged 13) and we have achieved similar results which confirm the assumed concept. It should be noticed, a simulated model was proposed to proof the design concept and to establish the necessary tools for hardware implementation in terms the simulating information (the force sensing resistor data) as well as applying assumptions for the output values which model the control process. Nevertheless, this is not a critical disadvantage as we are proposing fuzzy model, which are known to be easily adjusted to real implementations. The major idea in this work is the proposal of real-time analysis of sEMG integrated in a fuzzy control model, which is easily implementable. Our further work will be focused on the hardware implementation of the bionic hand concept proposed.

7 Conclusion

A fuzzy model control of bionic hand based on hand grasp functionality has been proposed and presented. The main contribution of this work was provided a method of sEMG signal analysis that able to be applied in real-time process. From patient perspective, it could be concluded, that the real-time sEMG signal analysis which is related to the corresponding actions of the bionic hand gives the feeling of natural hand control. The hardware implementation of the present proposal is in progress by the current authors.

Acknowledgments. This work was supported by the statutory funds of the Department of Computational Intelligence, Faculty of Computer Science and Managament, Wroclaw University of Science and Technology and partially by EC under FP7, Coordination and Support Action, Grant Agreement Number 316097, ENGINE European Research Centre of Network Intelligence for Innovation Enhancement (http://engine.pwr.edu.pl/). All computer experiments were carried out using computer equipment sponsored by ENGINE project. Additionally, the authors would like to thank the Wroclaw City Council, for the opportunity to work with the Neuro-Rehabilitation Center for the Treatment of Spinal Cord Injuries 'Akson', under the 'Mozart' city programme 2014/2015.

References

1. Bronstein, I.N., Semendjajew., K.A. Musiol, G., Mühlig, H.: Taschenbuch der Mathematik, Verlag Harri Deutsch, p. 1258 (2001)
2. Chatterjee, A., Chatterjee, R.: Matsuno., F., Endo, T.: Augmented stable fuzzy control for flexible robotic arm using LMI approach and neurofuzzy state space modeling. IEEE Trans. Ind. Electron. **55**(3), 1256–1270 (2008)
3. Crawford, B., Miller, K.J., Shenoy, P., Rao, R.P.N.: Real-time classification of electromyographic signals for robotic control. In: National Conference on Artificial Intelligence - AAAI, pp. 523–528 (2005)
4. Cram, J.R., Kasman, G.: Introduction to Surface Electromyography. Aspen Publishing, Gaterburg (1998)
5. Gauthaam, M., Kumar, S.S.: EMG controlled bionic arm. In: Proceedings of the National Conference on Innovations in Emerging Technology-2011 Kongu Engineering College, Perundurai, Erode, Tamilnadu, India, 17 & 18 February, pp. 111–114 (2011)
6. Ielpo, N., Calabrese, B., Cannataro, M., Palumbo, A., Ciliberti, S., Grillo, C., Iocco, M.: EMG-miner: automatic acquisition and processing of electromyographic signals: first experimentation in a clinical context for gait disorders evaluation. In: Proceedings of the IEEE 27th International Symposium on Computer-Based Medical Systems, pp. 441– 446 (2014)
7. Konrad, P.: The ABC of EMG: a practical introduction to kinesiological electromyography, Version 1.0 April. Noraxon INC., US (2005)
8. Lea, R.N., Iani, Y., Hoblit, J.: Fuzzy logc based robotic arm control. In: Proceedings of the Second IEEE ICFS, SF, CA, vol. 1, pp. 128–133 (1993)
9. Li, M., Jiang, Z., Wang, P., Sun, L., Ge, S.S.: Control of a quadruped robot with bionic springy legs in trotting gait. J. Bionic Eng. **11**(2), 188–198 (2014)
10. Mamdani, E.H., Assilian, S.: An experiment in linguistic synthesis with a fuzzy logic controller. Int J Man Mac Stud. **7**(1), 1–13 (1975)
11. Massa, B., Roccella, S., Carrozza, M.C., Dario, P.: Design and development of an underactuated prosthetic hand. In: Proceedings of the IEEE International Conference on Robotics and Automation (ICRA), Washington DC, May 11–15, pp. 3374–3379 (2002)
12. Pedreira, C., Martinez, J., Quiroga, R.Q.: Neural prostheses: linking brain signals to prosthetic devices. In: Proceedings on the ICROS-SICE International joint conference, Fukuoka, Japan, August (2009)
13. Shenoy, P., Miller, K.J., Crawford, B., Rao, R.P.N.: Online electromyographic control of a robotic prosthesis. IEEE Trans. Biomed. Eng. **55**(3), 1128–1135 (2008)
14. Shekhar, H., Guha, R., Juliet, A.V., Sam, J., Kumar, J.: Mathematical modeling of neuro-controlled bionic. In: Proceedings of the International Conference on Advances in Recent Technologies in Communication and Computing, pp. 576 – 578 (2009)
15. Spanias, J.A., Simon, A.M.K., Ingraham, A., Hargrove, L.J.: Effect of additional mechanical sensor data on an EMG-based pattern recognition system for a powered leg prosthesis. In: Proceedings of the 7th International IEEE/EMBS Conference on Neural Engineering (NER), Montpellier, pp. 639 – 642 (2015)

16. Tomas, S., Michal, K., Alena, K.: Fuzzy control of robotic arm implemented in PLC. In: Proceedings of the IEEE 9th International Conference on Computational Cybernetics (ICCC), pp. 45 – 49, Tihany (2013)
17. Zadeh, L.: Fuzzy sets. Inf. Control **8**(3), 338–353 (1965)
18. Zhuojun, X., Yantao, T., Yang, L.: sEMG pattern recognition of muscle force of upper arm for intelligent bionic limb control. J. Bionic Eng. **12**, 316–323 (2015)

Controllability of Positive Discrete-Time Switched Fractional Order Systems for Fixed Switching Sequence

Artur Babiarz[(✉)], Adrian Łęgowski, and Michał Niezabitowski

Institute of Automatic Control, Silesian University of Technology,
16 Akademicka St., 44-100 Gliwice, Poland
{artur.babiarz,adrian.legowski,michal.niezabitowski}@polsl.pl
http://www.ia.polsl.pl

Abstract. In the article unconstrained controllability problem of positive discrete-time switched fractional order systems is addressed. A solution of discrete-time switched fractional order systems is presented. Additionally, a transition matrix of considered dynamical systems is given. A sufficient condition for unconstrained controllability in a given number of steps is formulated and proved using the general formula of solution of difference state equation. Finally, the illustrative examples are also presented.

Keywords: Controllability · Switched system · Fractional order system

1 Backgrounds

One of the most important notion in modern mathematical control theory is controllability [5–7,15] because, for example using the assumption that the considered system is controllable, we are able to solve many problems of control theory such as optimal control, pole assignment or stabilizability. Generally, this concept means, that it is possible to steer a dynamical control system from an arbitrary initial state to an arbitrary final state using the set of admissible controls. Systematic studies of controllability were started at the beginning of sixties in XX century. Nowadays, there exist many different criteria and definitions of controllability. Moreover, the set of tools used to investigate into controllability problems becomes more and more sophisticated, for example one can consider fixed point theorems. Various criteria and definitions of controllability are presented in many publications and monographs. They depend both on constraints on the state equation and the control signal. One of the types of controllability are exact and approximate controllabilities. In finite-dimensional case these notions coincide. Whereas, in the case of infinite-dimensional systems they are not equivalent. Exact controllability means that the system can be steered to an arbitrary final state. Approximate controllability enables us to steer the system to an arbitrary small neighbourhood of the final state. Because approximate

© Springer International Publishing Switzerland 2016
N.T. Nguyen et al. (Eds.): ICCCI 2016, Part I, LNAI 9875, pp. 303–312, 2016.
DOI: 10.1007/978-3-319-45243-2_28

controllability is essentially a weaker notion than exact controllability, so exact controllability always implies approximate controllability, but the converse statement is not true, in general.

Different definitions and criteria of controllability for various systems were investigated in many publications and monographs. For example in [20] controllability problem of linear stochastic systems was studied. The authors of the paper [8,9] focused on controllability problem of jump systems. The controllability notion for switched linear system was considered for example in [1,12,13]. Controllability problems for nonlinear integer order discrete-time dynamical systems have been considered in [16]. Moreover, in the field of fractional dynamical systems [4,19,22], a lot of research concerning controllability problem were discussed [11,17,18,21].

Our previous journal paper [1] concerns controllability problem of switched standard discrete-time dynamical systems. We focus on situation when the switching sequence is constrained. In article [3] the output controllability of standard dynamical system is discussed. We study the controllability problem of fractional discrete-time dynamical system without switching in [2,14]. In mentioned articles we consider controllability problem for dynamical systems with delays in control signal and focus on the time-varying fractional positive and standard dynamical systems.

The paper is structured as follows: Sect. 2 introduces some mathematical preliminaries for modelling of positive discrete-time fractional systems. Section 3 contains the general formula of solution of positive discrete-time switched fractional order systems. In Sect. 4 we present the main result for controllability problem of positive fractional discrete time-varying delayed systems. Section 5 contains illustrative examples. Finally we summarize our results.

2 Preliminaries

Let us start with some basic notations. By $\mathcal{Z}_+$ we will denote the set of positive integers. $\Re_+^n$ represents the set of all nonnegative real vectors with dimension n, $\Re_+^{n \times m}$ is the set of all $n \times m$ matrices with nonnegative entries. A monomial matrix means the product of a permutation matrix and a non-singular diagonal matrix. A nonzero multiple of the unit vector e_i we will call a monomial vector. The non-integer derivatives can be formulated in various ways, see e.g. [19]. The definition of the fractional difference of the order α that will be used in this paper is as follows:

Definition 1 [10]. *The fractional difference of the order α is expressed by the following formula*

$$\Delta^\alpha x(k) = \sum_{j=0}^{j=k} (-1)^j \binom{\alpha}{j} x(k-j) \tag{1}$$

for $n - 1 < \alpha < n \in \mathbb{N}$, $k \in \mathcal{Z}_+$, $x(k) \in \mathbb{R}^n$ and

$$\binom{\alpha}{j} = \begin{cases} 1 & \textit{if } j{=}0 \\ \frac{\alpha(\alpha-1)...(\alpha-j+1)}{j!} & \textit{if } j{=}1,2, \ ... \end{cases} \tag{2}$$

where $\binom{\alpha}{j}$ is the so-called generalized Newton symbol.

3   A Fractional Systems

3.1   A Standard Fractional System

Consider the standard fractional discrete linear system described by the finite-dimensional state-space equations

$$\Delta^{\alpha} x_{k+1} = A x_k + B u_k \tag{3}$$

where: $x_k \in \mathbb{R}^n$ is the state vector, $u_k \in \mathbb{R}^m$ is the input (or control), A and B are $n \times n$ and $n \times m$ constant matrices, respectively. Using the definition of forward difference operator Δ we may write the Eq. (3) in the equivalent form

$$x_{k+1} + \sum_{j=1}^{k+1} (-1)^j \binom{\alpha}{j} x_{k-j+1} = A x_k + B u_k \tag{4}$$

for $k \in Z_+$.
The solution of Eq. (4) with known admissible initial condition $x_0 \in \mathbb{R}^n$ has the following form:

$$x_k = \Phi_k x(0) + \sum_{j=0}^{j=k-1} \left(\Phi_{k-j-1} B_j u_j \right). \tag{5}$$

3.2   A Positive Switched Fractional System

In this subsection we will consider a class of linear switched systems described by difference state equation:

$$\Delta^{\alpha} x_{k+1} = A_{\sigma(k)} x_k + B_{\sigma(k)} u_k \tag{6}$$

for $k \geq 0$, where: $x(k) \in \Re_+^n$ is the state vector, $u(k) \in \Re_+^m$ is the input (or control), $\sigma(k) \in \{1, 2, ..., s\} =: S$ is the switching signal. Moreover, for $\sigma(k) = i$, $A_i := A(i)$ and $B_i := B(i)$ are constant matrices of appropriate dimensions.

Definition 2. *The positive switched discrete-time fractional system (6) is called the positive if and only if $x(k) \in \Re_+^n$, for any initial conditions $x(k_0) \in \Re_+^n$ and all inputs $u(k) \in \Re_+^m$, where $\Re_+^n = \Re_+^{n \times 1}$ is the set of vectors of nonnegative entries.*

In addition, we will use the following notation

$$S_{i_0}^{(N)} = \{(i_0, i_1, ..., i_{N-1}) : i_0, i_1, ..., i_{N-1} \in S\}. \tag{7}$$

In further considerations, it will be assumed that we will focus on the fixed sequence of $S_{i_0}^{(N)}$. By $\bar{\bar{s}}$ we will denote the number of elements of $S_{i_0}^{(N)}$. Now, let us introduce the transition matrix $\Phi_{\sigma(k)}$ that can be computed using recursive formula given below:

$$\Phi_{\sigma(k+1)} = (A_{\sigma(k)} + \alpha I_n)\Phi_{\sigma(k)} + \sum_{i=2}^{i=k+1} (-1)^{j+1} \binom{\alpha}{i} \Phi_{\sigma(k-i+1)} \tag{8}$$

where: $\Phi_{\sigma(0)} = I$, I - is $n \times n$ dimensional identity matrix, additional, it assumes that $\Phi_{\sigma(k)} = 0$ for $k < 0$.

The next theorem considers a solution of switched fractional system.

Theorem 1. *The solution of positive switched fractional discrete-time system for fixed sequence $S_{i_0}^{(N)}$ and known admissible initial conditions $x_0 \in \mathcal{R}^n$ and a control signal $u_i \in \mathcal{R}^m$, $k \in \mathcal{Z}_+$ is given by the formula:*

$$x_k = \Phi_{\sigma(k)}x_0 + \sum_{i=0}^{i=k-1} \left(\Phi_{\sigma(k-i-1)} B_{\sigma(i)} u_i \right) \tag{9}$$

where: $\Phi_{\sigma(k)}$ is $n \times n$ dimensional transition matrix, $k \in \mathcal{Z}_+$.

Proof. Let us consider positive switched fractional system described by (6) for $k = 0, 1, 2, 3$ and $\sigma(k) = 0, 1, 2, 3$. Then, we get:

$$dla \quad k = 0,$$

$$x_0 = \Phi_{\sigma(0)}x_0, \qquad \Phi_{\sigma(0)} = I_n,$$

for

$$k = 1,$$

$$x_1 = \Phi_{\sigma(1)}x_0 + \Phi_{\sigma(0)}B_{\sigma(1)}u_0,$$

where:

$$\Phi_{\sigma(1)} = (A_{\sigma(0)} + \alpha I_n)\Phi_{\sigma(0)}.$$

So, we obtain:

$$x_1 = A_{\sigma(0)} + \alpha I_n + B_{\sigma(0)}u_0.$$

For $k = 2$, we can compute:

$$x_2 = \Phi_{\sigma(0)}x_0 + \Phi_{\sigma(1)}B_{\sigma(0)}u_0 + \Phi_{\sigma(0)}B_{\sigma(1)}u_1$$

where:

$$\Phi_{\sigma(2)} = (A_{\sigma(1)} + \alpha I_n)\Phi_{\sigma(1)} - \binom{\alpha}{2}\Phi_{\sigma(0)}$$

$$= A_{\sigma(1)}A_{\sigma(0)} + A_{\sigma(1)}\alpha I_n + \alpha I_n A_{\sigma(0)} + \alpha^2 I_n^2 - \binom{\alpha}{2}I_n$$

and we get:

$$x_2 = \Phi_{\sigma(2)}x_0 + A_{\sigma(0)}B_{\sigma(0)}u_0 + B_{\sigma(1)}u_1$$

For $k = 3$, we obtain:

$$x_3 = \Phi_{\sigma(3)}x_0 + \Phi_{\sigma(2)}B_{\sigma(0)}u_0 + \Phi_{\sigma(1)}B_{\sigma(1)}u_1 + \Phi_{\sigma(0)}B_{\sigma(2)}u_2$$

where:

$$\Phi_{\sigma(3)} = A_{\sigma(2)}A_{\sigma(1)}A_{\sigma(0)} + A_{\sigma(2)}A_{\sigma(1)}\alpha I_n$$

$$+ A_{\sigma(2)}\alpha I_n A_{\sigma(0)} + A_{\sigma(2)}\alpha^2 I_n^2 - A_{\sigma(2)}\binom{\alpha}{2}I_n$$

$$+ \alpha I_n A_{\sigma(1)}A_{\sigma(0)} + \alpha I_n A_{\sigma(1)}\alpha I_n + \alpha^2 I_n^2 A_{\sigma(0)} + \alpha^3 I_n^3$$

$$+ \alpha I_n\binom{\alpha}{2}I_n - \binom{\alpha}{2}(A_{\sigma(0)} + \alpha I_n)\Phi_{\sigma(0)} + \binom{\alpha}{3}I_n.$$

Finally, we can express the solution of (1) by the Eqs. (9) and (8).

4   Controllability

At the beginning we introduce some necessary definitions.

Definition 3. *The positive switched fractional system (6) is controllable on the interval $[k_0, k]$ for any admissible initial state $x(0) = x_0$ and every final state $x_f \in \mathbb{R}^n$, for fixed switching sequence $\sigma(k)$, there exists control signal $u(k)$ during $[k_0, k]$ such that $x(k) = x_f$.*

Definition 4. *Let us introduce a controllability matrix and denote it by $\mathcal{C}(\bar{s}, \sigma(k))$ or $\mathcal{C}$ in short. The controllability matrix $\mathcal{C}$ is defined by the following formula:*

$$\mathcal{C} = [\Phi_{\sigma(\bar{s}-1)}B_{\sigma(0)} \vdots \Phi_{\sigma(\bar{s}-2)}B_{\sigma(1)} \vdots \cdots \vdots \Phi_{\sigma(\bar{s}-i-1)}B_{\sigma(i)} \vdots \cdots$$

$$\cdots \vdots \Phi_{\sigma(1)}B_{\sigma(\bar{s}-2)} \vdots B_{\sigma(\bar{s}-1)}]. \tag{10}$$

Now, we can formulate the main theorem of this article.

Theorem 2. *The positive fractional switched discrete-time system (6) is controllable on interval $[k_0, k]$ if and only if*

$$rank\mathcal{C} = n \tag{11}$$

and matrix $\mathcal{C}$ contains monomial submatrix.
It means that the rank of matrix $\mathcal{C}$ is equal to rank of system (6).

Proof. From the Theorem 1 and definition of the transition matrix $\Phi_{\sigma(k)}$, for zero initial conditions and fixed k and $\sigma(k)$, we obtain:

$$x_k = \Phi_{\sigma(k)} x_0 + \sum_{i=0}^{\overline{s}-1} \left(\Phi_{\sigma(\overline{s}-i-1)} B_{\sigma(i)} u_i \right) \tag{12}$$

We can rewrite above-mentioned equation by the form:

$$x_k = \sum_{i=0}^{i=\overline{s}-1} \left(\Phi_{\sigma(\overline{s}-i-1)} B_{\sigma(i)} u_i \right) = \Phi_{\sigma(\overline{s}-1)} B_{\sigma(0)} u_0 + \dots$$

$$+ \Phi_{\sigma(\overline{s}-i-1)} B_{\sigma(i)} u_i + \dots + \Phi_{\sigma(\overline{s}-(\overline{s}-1)-1)} B_{\sigma(\overline{s}-1)} u_{\overline{s}-1} =$$

$$= \underbrace{\begin{bmatrix} \Phi_{\sigma(\overline{s}-1)} B_{\sigma(0)} \vdots \Phi_{\sigma(\overline{s}-2)} B_{\sigma(1)} \vdots \cdots \\ \vdots \Phi_{\sigma(\overline{s}-i-1)} B_{\sigma(i)} \vdots \cdots \vdots \\ \Phi_{\sigma(1)} B_{\sigma(\overline{s}-2)} \vdots B_{\sigma(\overline{s}-1)} \end{bmatrix}}_{\mathcal{C}} \begin{bmatrix} u_0 \\ u_1 \\ \vdots \\ u_i \\ \vdots \\ u_{\overline{s}-2} \\ u_{\overline{s}-1} \end{bmatrix} \tag{13}$$

Then, we have directly from the Kronecker-Capelli's Theorem that the system (12) has a solution for any x_k if and only if rank of matrix $\mathcal{C}$ is equal to n.

5 Examples

Example 1. Let us focus on the fractional system described by the following formula:

$$\Delta^\alpha x_{k+1} = A_{\sigma(k)} x_k + B_{\sigma(k)} u_k \tag{14}$$

where: $k = 0, 1, 2$ and

$$A_1 = \begin{bmatrix} 0 & \frac{1}{2} \\ \frac{1}{2} & 0 \end{bmatrix}, \quad A_2 = \begin{bmatrix} 1 & 0 \\ 0 & \frac{1}{2} \end{bmatrix}, \quad B_1 = \begin{bmatrix} 1 \\ 0 \end{bmatrix}, \quad B_2 = \begin{bmatrix} 0 \\ \frac{1}{2} \end{bmatrix}.$$

The switching signal is presented on Fig. 1.
Using (8) for $k = 1$ we compute

$$\Phi_{\sigma(1)} = (A_2 + I\alpha)\Phi_{\sigma(0)} = \begin{bmatrix} 1 & 0 \\ 0 & \frac{1}{2} \end{bmatrix} + \begin{bmatrix} \alpha & 0 \\ 0 & \alpha \end{bmatrix} = \begin{bmatrix} 1+\alpha & 0 \\ 0 & \frac{1}{2}+\alpha \end{bmatrix}.$$

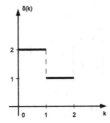

Fig. 1. The switching signal

For $k = 2$ we get

$$\Phi_{\sigma(2)} = (A_1 + I\alpha)\,\Phi_{\sigma(1)} = \left(\begin{bmatrix} 0 & \frac{1}{2} \\ \frac{1}{2} & 0 \end{bmatrix} + \begin{bmatrix} \alpha & 0 \\ 0 & \alpha \end{bmatrix}\right) \begin{bmatrix} 1+\alpha & 0 \\ 0 & \frac{1}{2}+\alpha \end{bmatrix} - \begin{pmatrix} \alpha \\ 2 \end{pmatrix} =$$

$$\begin{bmatrix} \alpha(\alpha+1) - \frac{\alpha(\alpha-1)}{2} & \frac{\alpha}{2} - \frac{\alpha(\alpha-1)}{2} + \frac{1}{4} \\ \frac{\alpha}{2} - \frac{\alpha(\alpha-1)}{2} + \frac{1}{2} & \alpha(\alpha+\frac{1}{2}) - \frac{\alpha(\alpha-1)}{2} \end{bmatrix}.$$

Now, we built the controllability matrix $\mathcal{C}$ expressed by the Definition 4:

$$\mathcal{C} = \left[\Phi_{\sigma(1)}B_{\sigma(0)} \vdots B_{\sigma(1)}\right]$$

and

$$\Phi_{\sigma(1)}B_{\sigma(0)} = \begin{bmatrix} 1+\alpha & 0 \\ 0 & \frac{1}{2}+\alpha \end{bmatrix} \begin{bmatrix} 0 \\ \frac{1}{2} \end{bmatrix} = \begin{bmatrix} 0 \\ \frac{\frac{1}{2}+\alpha}{2} \end{bmatrix}.$$

Finally, we obtain:

$$\mathcal{C} = \left[\Phi_{\sigma(1)}B_{\sigma(0)} \vdots B_{\sigma(1)}\right] = \begin{bmatrix} 1 & 0 \\ 0 & \frac{\frac{1}{2}+\alpha}{2} \end{bmatrix}.$$

The $rank\mathcal{C} = 2$ and it contains monomial submatrix then the positive switched fractional discrete-time system is controllable for fixed switching sequence.

Example 2. Next example shows the same systems but the switching sequence is given on Fig. 2. The positive switched fractional system is described by the following formula:

$$\Delta^{\alpha}x_{k+1} = A_{\sigma(k)}x_k + B_{\sigma(k)}u_k \tag{15}$$

where: $k = 0, 1, 2$ and

$$A_1 = \begin{bmatrix} 0 & \frac{1}{2} \\ \frac{1}{2} & 0 \end{bmatrix}, \qquad A_2 = \begin{bmatrix} 1 & 0 \\ 0 & \frac{1}{2} \end{bmatrix}, \qquad B_1 = \begin{bmatrix} 1 \\ 0 \end{bmatrix}, \qquad B_2 = \begin{bmatrix} 0 \\ \frac{1}{2} \end{bmatrix}.$$

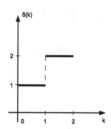

Fig. 2. The switching signal

Using (8) for $k = 1$ we compute

$$\Phi_{\sigma(1)} = (A_1 + I\alpha)\, \Phi_{\sigma(0)} = \begin{bmatrix} 0 & \frac{1}{2} \\ \frac{1}{2} & 0 \end{bmatrix} + \begin{bmatrix} \alpha & 0 \\ 0 & \alpha \end{bmatrix} = \begin{bmatrix} \alpha & \frac{1}{2} \\ \frac{1}{2} & \alpha \end{bmatrix}.$$

For $k = 2$ we get

$$\Phi_{\sigma(2)} = (A_2 + I\alpha)\, \Phi_{\sigma(1)} = \left(\begin{bmatrix} 1 & 0 \\ 0 & \frac{1}{2} \end{bmatrix} + \begin{bmatrix} \alpha & 0 \\ 0 & \alpha \end{bmatrix} \right) \begin{bmatrix} \alpha & \frac{1}{2} \\ \frac{1}{2} & \alpha \end{bmatrix} - \binom{\alpha}{2} =$$

$$\begin{bmatrix} \alpha(\alpha + 1) - \frac{\alpha(\alpha-1)}{2} & \frac{\alpha}{2} - \frac{\alpha(\alpha-1)}{2} + \frac{1}{2} \\ \frac{\alpha}{2} - \frac{\alpha(\alpha-1)}{2} + \frac{1}{4} & \alpha(\alpha + \frac{1}{2}) - \frac{\alpha(\alpha-1)}{2} \end{bmatrix}.$$

Now, we built the controllability matrix $\mathcal{C}$ expressed by the Definition 4:

$$\mathcal{C} = \left[\Phi_{\sigma(1)} B_{\sigma(0)} \vdots B_{\sigma(1)} \right]$$

and

$$\Phi_{\sigma(1)} B_{\sigma(0)} = \begin{bmatrix} \alpha & \frac{1}{2} \\ \frac{1}{2} & \alpha \end{bmatrix} \begin{bmatrix} 1 \\ 0 \end{bmatrix} = \begin{bmatrix} \alpha \\ \frac{1}{2} \end{bmatrix}.$$

Finally, we get:

$$\mathcal{C} = \left[\Phi_{\sigma(1)} B_{\sigma(0)} \vdots B_{\sigma(1)} \right] = \begin{bmatrix} \alpha & 0 \\ \frac{1}{2} & \frac{1}{2} \end{bmatrix}.$$

The $rank\mathcal{C} = 2$ but it doesn't contains all monomial submatrix, then the positive switched fractional discrete-time system is not controllable for fixed switching sequence.

6   Conclusions

The presented results are related to the controllability of linear switched fractional discrete-time system. Through the use of solutions for linear switched fractional discrete-time system, sufficient conditions for local controllability for fixed switching sequence were obtained. Furthermore, the Grünwald-Letnikov definition of fractional derivative is used. The classical Kalman's controllability conditions have been extended to the class of linear switched fractional discrete-time system.

Moreover we have presented examples which have shown that the controllability of the same positive switched system depends on fixed switching sequence. There are many possibilities to use and extend the obtained results. It should be also noted that the presented results concerning the controllability problem can be extended to switched positive discrete time-varying fractional order systems and linear switched fractional discrete time-varying systems.

Acknowledgment. The research presented here was done by first and third author as part of the project funded by the National Science Centre in Poland granted according to decision DEC-2014/13/B/ST7/00755. Moreover, the work of the second author was supported by Polish Ministry for Science and Higher Education under internal grant BKM/506/RAU1/2016 t.1 for Institute of Automatic Control, Silesian University of Technology, Gliwice, Poland. Finally, the calculations were performed with the use of IT infrastructure of GeCONiI Upper Silesian Centre for Computational Science and Engineering (NCBiR grant no POIG.02.03.01-24-099/13).

References

1. Babiarz, A., Czornik, A., Klamka, J., Niezabitowski, M.: The selected problems of controllability of discrete-time switched linear systems with constrained switching rule. Bull. Polish Acad. Sci. Tech. Sci. **63**(3), 657–666 (2015)
2. Babiarz, A., Klamka, J.: Controllability of discrete linear time-varying fractional system with constant delay. In: 13th International Conference of Numerical Analysis and Applied Mathematics (ICNAAM). AIP Conference Proceedings, Rhodes, Greece, 23–29 September, pp. 480058(1)–480058(4) (2015)
3. Babiarz, A., Czornik, A., Niezabitowski, M.: Output controllability of the discrete-time linear switched systems. Nonlinear Anal. Hybrid Syst. **21**, 1–10 (2016)
4. Badri, V., Tavazoei, M.S.: On tuning fractional order [proportional-derivative] controllers for a class of fractional order systems. Automatica **49**(7), 2297–2301 (2013)
5. Bashirov, A.E., Kerimov, K.R.: On controllability conception for stochastic systems. SIAM J. Control Optim. **35**(2), 384–398 (1997)
6. Bashirov, A.E., Mahmudov, N.I.: On concepts of controllability for deterministic and stochastic systems. SIAM J. Control Optim. **37**(6), 1808–1821 (1999)
7. Benchohra, M., Ouahab, A.: Controllability results for functional semilinear differential inclusions in Fréchet spaces. Nonlinear Anal. Theor. Methods Appl. **61**(3), 405–423 (2005)
8. Czornik, A., Świerniak, A.: On controllability with respect to the expectation of discrete time jump linear systems. J. Franklin Inst. **338**(4), 443–453 (2001)

9. Czornik, A., Świerniak, A.: On direct controllability of discrete time jump linear system. J. Franklin Inst. **341**(6), 491–503 (2004)
10. Kaczorek, T.: Selected Problems of Fractional Systems Theory, vol. 411. Springer Science & Business Media, Heidelberg (2011)
11. Klamka, J., Babiarz, A.: Local controllability of semilinear fractional order systems with variable coefficients. In: 20th International Conference on Methods and Models in Automation and Robotics (MMAR 2015), pp. 733–737, August 2015
12. Klamka, J., Czornik, A., Niezabitowski, M.: Stability and controllability of switched systems. Bull. Polish Acad. Sci. Tech. Sci. **61**(3), 547–555 (2013)
13. Klamka, J., Niezabitowski, M.: Controllability of switched linear dynamical systems. In: 18th International Conference on Methods and Models in Automation and Robotics (MMAR 2013), pp. 464–467, August 2013
14. Klamka, J., Niezabitowski, M.: Controllability of the fractional discrete linear time-varying infinite-dimensional systems. In: 13th International Conference of Numerical Analysis and Applied Mathematics (ICNAAM). AIP Conference Proceedings, Rhodes, Greece, 23–29 September, pp. 130004(1)–130004(4) (2015)
15. Klamka, J.: Controllability of Dynamical Systems. Kluwer Academic Publishers, Dordrecht (1991)
16. Klamka, J.: Controllability of nonlinear discrete systems. Appl. Math. Comput. Sci. **12**(2), 173–180 (2002)
17. Klamka, J.: Local controllability of fractional discrete-time semilinear systems. acta mechanica et automatica **5**, 55–58 (2011)
18. Mozyrska, D., Pawłuszewicz, E.: Controllability of h-difference linear control systems with two fractional orders. Int. J. Syst. Sci. **46**(4), 662–669 (2015)
19. Podlubny, I.: Fractional-order systems and $pi^\lambda d^\mu$-controllers. IEEE Trans. Autom. Control **44**(1), 208–214 (1999)
20. Sikora, B., Klamka, J.: On constrained stochastic controllability of dynamical systems with multiple delays in control. Bull. Polish Acad. Sci. Tech. Sci. **60**(2), 301–305 (2012)
21. Sikora, B.: Controllability of time-delay fractional systems with and without constraints. IET Control Theor. Appl. **10**(3), 320–327 (2016)
22. Tejado, I., Valério, D., Pires, P., Martins, J.: Fractional order human arm dynamics with variability analyses. Mechatronics **23**(7), 805–812 (2013)

A Trace Clustering Solution
Based on Using the Distance Graph Model

Quang-Thuy Ha[1(✉)], Hong-Nhung Bui[1,2], and Tri-Thanh Nguyen[1]

[1] Vietnam National University (VNU), VNU-University of Engineering
and Technology (UET), No. 144, Xuan Thuy, Cau Giay, Hanoi, Vietnam
{ntthanh, thuyhq}@vnu.edu.vn, nhungbh79@gmail.com
[2] Banking Academy of Vietnam, No.12, Chua Boc, Dong Da, Hanoi, Vietnam

Abstract. Process discovery is the most important task in the process mining. Because of the complexity of event logs (i.e. activities of several different processes are written into the same log), the discovered process models may be diffuse and unintelligible. That is why the input event logs should be clustered into simpler event sub-logs. This work provides a trace clustering solution based on the idea of using the distance graph model for trace representation. Experimental results proved the effect of the proposed solution on two measures of Fitness and Precision, especially the effect on the Precision measure.

Keywords: Event log · Process mining · Fitness measure · Precision measure · Process discovering · Trace clustering · Distance graph model

1 Introduction

Process discovery is the most important task in process mining. There exists some algorithms for discovering process models form event logs, such as α (Wil M. P. van der Aalst and Boudewijn F. van Dongen [1]), $\alpha+$ (A.K.A de Medeiros et al. [9]), $\alpha++$ (Lijie Wen et al. [17]), and other algorithms [2]. Due to the complexity of event logs, the discovered models may be diffuse and unintelligible. That is why the two-phase approach is proposed for process model discovering. In the first phase, the input event log is refined, in which clustering algorithms are popularly used. In the second phase, process discovering algorithms are run on the refined event log to find out the model. There exists some works following this approach [4, 5, 8, 10, 13, 15, 16].

The distance graph model for text processing has been proposed by Charu C. Aggarwal and Peixiang Zhao in 2013 [3]. Distance graphs of order k ($k = 0, 1, 2, \ldots$) for a document (a string of words) D based on the corpus C is a useful representation of D for text mining tasks [3, 7].

Because of the similar between the graph structure of process model and the Distance graph model, this work focuses on a trace clustering solution based on the idea of using the distance graph model for trace representation. This study is oriented to contribute a new solution to trace clustering.

The rest of this article is organized as follows: In the next section, a trace clustering solution on using the distance graph model is showed. This framework includes three phases: "Trace representation and Clustering", "Process discovery", and "Model

© Springer International Publishing Switzerland 2016
N.T. Nguyen et al. (Eds.): ICCCI 2016, Part I, LNAI 9875, pp. 313–322, 2016.
DOI: 10.1007/978-3-319-45243-2_29

Evaluation". Experiments and remarks are described in the third section. In the fourth section, related work is introduced. And conclusions are shown in the last section.

2   A Trace Clustering Solution Based on the Distance Graph Model

2.1   The Problem

The paper proposes a solution to trace clustering in event logs based on the distance graph model [3]. The problem is described as follows.

Let **A** be the activity-name universe in an organization and $A \subseteq \mathbf{A}$ be the set of all activity-names for a business process in the organization. A trace σ is a sequence of activities, i.e., $\sigma \in A^+$ (where A^+ is a set of non empty sequences of activities in A). Let L be a simple event log of a business process containing a set of traces constructed from A. Process discovery algorithms transform event logs into process models represented in a process modeling language, e.g. Petri nets (WorkFlow nets: WF-nets), BPMN (Business Process Modeling Notation), or YAWL (Yet Another Workflow Language), etc. There exists some algorithms for discovering process models form event logs, such as α [1], $\alpha+$ [9], $\alpha++$ [17], and others [2].

For example, let L = [abdeh, adceg, acdefbdeg, adbeh, acdefdcefcdeh, acdeg] (where a = "register request", b = "examine thoroughly", c = "examine casually", d = "check ticket", e = "decide", f = "reinitiate request", g = "pay compensation", h = "reject request") be an event log for the requests for compensation business process within an airline. Figure 1 describes the WorkFlow net discovered the event log L by applying the α algorithm [2].

Due to the complexity of event logs, the discovered process models may be diffuse and unintelligible. That is why the two-phase approach is proposed for process model discovering. In the first phase, the input event log is refined, in which clustering algorithms are popularly used. In the second phase, process discovering algorithms are run on the refined event log to find out the process model [6].

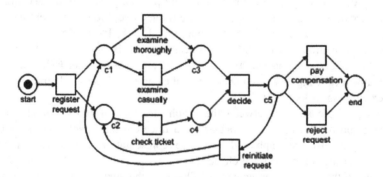

Fig. 1. WorkFlow net discovered by the α-algorithm based on L [2]

2.2   The Distance Graph Model

As mentioned in the introduction section, the distance graph model ("A distance graph of order k for a document D drawn from a corpus C") for text processing was proposed by Charu C. Aggarwal and Peixiang Zhao in 2013. Figure 2 illustrates the distance graphs of orders 0, 1, and 2 for the well-known nursery rhyme "Mary had a little lamb" [3]. As stated in [3], the most common method of representing a document D is a vector of distinct terms generated from the corpus C, where each component of the vector is the frequency of a certain term appearing in D. Charu C. et al. proposed to *convert a distance graph into a vector-space representation*, i.e. each directed edge in the distance graph is used to create a new "token" or "pseudo-word". For example, the edge from MARRY to LITTLE (in the distance graph order 2) is used to create a new pseudo-word MARRY-LITTLE; the pseudo-word created from the edge from LAMB to itself (in the distance graph order 2) is LAMB-LAMB. The frequency of the edge is used to denote the frequency of the pseudo-word. These new pseudo-words preserve the order of words in the document, thus, when combined with distinct terms in the corpus C, they enhance the semantic of the document representation in the form of a vector.

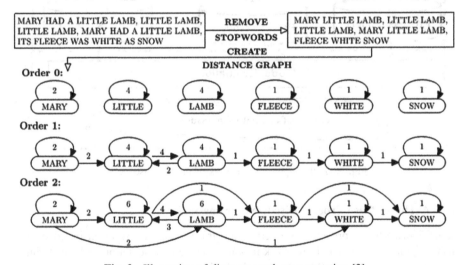

Fig. 2. Illustration of distance graph representation [3]

Charu C. Aggarwal and Peixiang Zhao showed some interesting features of distance graph model, as well as the effectiveness of the model applied for text classification. Since the order of activities within a trace plays an important role, one characteristic of distance graph which is considered to be suitable for trace representation is its ability to preserve the order of words in a document in the form of directed edges.

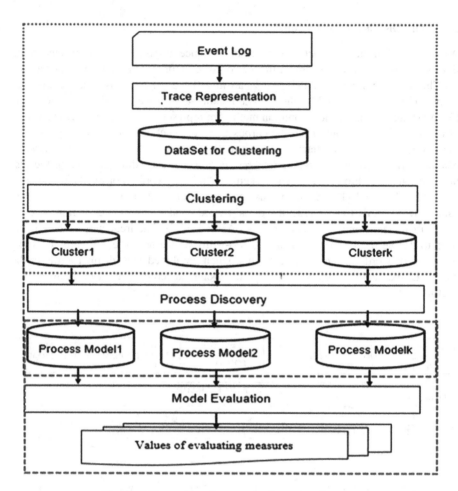

Fig. 3. A three-phase framework of process discovery

2.3   A Three-Phase Process Discovery Framework

Figure 3 describes a process discovery using trace clustering solution based on the distance graph model. The framework includes "Trace representation and Clustering", "Process discovery", and "Model evaluation" Phases.

Trace representation and Clustering Phase includes two steps. In the Trace Representation step, a dataset for clustering is created, in which a data point is a vector of distance graphs (with different orders) of a trace in the event log.

The set A of activities in the event log is considered as the set of "distinct words" in the corpus C, and a trace in the event log is considered as a document D, thus distance graphs for a trace can be constructed. For the given trace $<a\ c\ d\ e\ f\ d\ b\ e\ h>$,

- Order 0 distance graph is: $a(1)$, $c(1)$, $d(2)$, $e(2)$, $f(1)$, $b(1)$, $h(1)$, where the number denotes the *frequency* of directed edges from the node to itself. This graph contains 7 unconnected components.

- Order 1 distance graph is constructed from order 0 graph $a(1)$, $c(1)$, $d(2)$, $e(2)$, $f(1)$, $b(1)$, $h(1)$ by adding more edges: $ac(1)$, $cd(1)$, $de(1)$, $ef(1)$, $fd(1)$, $db(1)$, $be(1)$, $eh(1)$, where the number denotes the frequency.
- Order 2 distance graph is constructed from order 1 graph $a(1)$, $c(1)$, $d(2)$, $e(2)$, $f(1)$, $b(1)$, $h(1)$, $ac(1)$, $cd(1)$, $de(1)$, $ef(1)$, $fd(1)$, $db(1)$, $be(1)$, $eh(1)$ by adding more edges: $ad(1)$, $ce(1)$, $df(1)$, $ed(1)$, $fb(1)$, $de(1)$, $bh(1)$, where the number denotes the frequency.
- etc.

We followed the method of [3] to decompose a distance graph into a set of features for vector representation with a small modification. A feature is either the vertex or the *directed* edge of the graph. Our modification is to *ignore the edge from a vertex v to itself* (i.e. edge *vv*) in distance graph order 0, since every vertex in the graph order 0 always has an edge from itself to itself (*self-loop*). In addition, an edge from vertex to itself, in a trace, should indicate an activity is repeated. For the above order 1 distance graph of the trace <*a c d e f d b e h*>, the set of features is {*a, c, d, e, f, b, h, ac, cd, de, ef, fd, db, be, eh*}. The frequency of the feature in each trace is preserved in vector representation. Since a higher order distance graph of a trace includes all lower distance graphs using this representation, only the highest order distance graph is enough to represent the trace with *consideration to distinguish the self-loop of distance graph order 0 with the self-loop of higher order*. With this representation, if two graphs share common sub-graphs, it will be preserved in the representation. Obviously, for another trace <*a c d e f b h*>, its set of features {*a, c, d, e, f, b, h, ac, cd, de, ef, fb, bh*} is a subset of the above trace. Consequently, the two vectors will be close to each other in the vector space. Because event logs reflect the executions of business processes then all distance graphs of traces in an event log include some relation patterns in the discovered process model. That is why the number of features generated from all the traces in an event log L is significantly less than $(|A| + |A|*(|A|-1)/2)$ where $|A|$ denoted the cardinality of set A of activities.

In the Clustering step, one clustering algorithm is applied on the dataset (e.g. K-Modes and K-means algorithms). The output of the Trace Representation and Clustering Phase is a set of clusters (sub-logs) of traces (cases) of the event log.

In the Process Discovery Phase, a process discovery algorithm (i.e. α-algorithm) is applied on the clusters (event sub-logs) to get process models.

The Model Evaluation shows the effect of result process models. Though there are four common measures for evaluation, i.e. Fitness, Precision, Generalization, and Simplicity [2, 11, 12], this work considers two measures: i.e. Fitness and Precision, which had been described by A. Rozinat and Wil M.P. van der Aalst [11]. The Fitness measure indicates that the discovered model should accept the behaviors seen in the event log, and the Precision measure means that the discovered model should not accept behaviors completely unrelated to what was seen in the event log. Since these measures are calculated on each cluster, an aggregated value for whole event log should be calculated. This work selects a weighted average value as follow:

$$w_{avg} = \sum_{1}^{k} \frac{n_i}{n} w_i \qquad (1)$$

where w_{agv} is the aggregated value of the fitness or precision measure, k is the number of clusters, n is the number of traces in the event log, n_i is the number of traces in the ith cluster and w_i is the value of the measure of the ith cluster.

3 Experiments and Results

This work used the **prBm6** event log in the "Conformance Checking in the Large"[1] for experiments. The event log includes 1200 cases with 37961 events. In the Clustering step, two clustering algorithms: K-Modes and K-means were used. In Process discovery and Model evaluation phrases, ProM [19] was used. From several tests, we selected the maximum distance graph order of 2 for all the experiments.

3.1   The Experiment with K-Modes Algorithm

Since a trace is a sequence of activities, from an event log, we have a set of activities, a common trace representation was proposed: binary vector activities, i.e. a vector component is 1 if the trace contains a certain activity, otherwise 0 [2, 8]. To evaluate the model, binary trace vector based on activity representation was implemented as a baseline. The experiment results are described in the Table 1. We consider the values of measures of Average-Fitness and Average-Precision (1) in the cases of the vector-based and the Distance graph order 2-based trace representation in columns titled "Avg" in the table. After several runs, we found out the suitable number of clusters for the data set is 3.

Experiments on the Distance graph order 1-based also are implemented. All experimental results on the vector-based, the Distance graph order 1-based, and the Distance graph order 2-based trace representations are also showed in the Fig. 4.

3.2   The Experiment with K-Means Algorithm

In this experiment, the K-means clustering algorithm was used to run on the vector-based and distance graph-based trace representation. The experiment results are described in the Table 2. We also calculated the values of measures of Average-Fitness and Average-Precision (1) for activity-based (Vector) and the Distance graph-based (Distance graph) trace representation in columns titled "Avg" in the table.

Experiments on the Distance graph order 1-based also are implemented. All experimental results on the vector-based, the Distance graph order 1-based, and the Distance graph order 2-based trace representations are also showed in the Fig. 5.

---

[1] http://data.3tu.nl/repository/uuid:44c32783-15d0-4dbd-af8a-78b97be3de49.

Table 1. Using the K-modes clustering algorithm: the fitness and precision for all event sublogs (clusters) in the activity-based (Vector) and the distance graph order 2-based (Distance Graph) trace representation

Method Measure	Vector				Distance Graph order 2			
	Clus1	Clus2	Clus3	Avg	Clus1	Clus2	Clus3	Avg
#Traces	326	621	253	1200	326	559	315	1200
Fitness	0.9636	0.9450	0.9629	**0.9539**	0.9637	0.9876	0.9500	**0.9713**
Precision	1	0.6926	0.5868	**0.7538**	1.0	0.9914	0.6974	**0.9165**

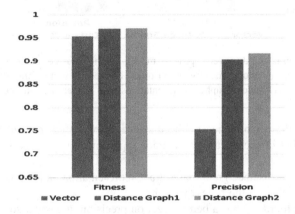

Fig. 4. Comparison of the discovered process models on the measures of Fitness and Precision between Activity-based (Vector), Distance graph order 1-based (Distance Graph1), and Distance graph order 2-based (Distance Graph2) Representations with K-Modes clustering algorithm.

Table 2. Using the K-means Clustering Algorithm: The Fitness and Precision for all event sublogs (clusters) in the Activity-based (Vector) and the Distance gpaph-based (Distance Gpaph) trace representation

Method Measure	Vector				Distance Graph order 2			
	Clus1	Clus2	Clus3	Avg	Clus1	Clus2	Clus3	Avg
#Traces	326	621	253	1200	326	475	399	1200
Fitness	0.9637	0.9787	0.9450	**0.9675**	0.9637	0.9680	0.9787	**0.9704**
Precision	1	0.6408	0.9763	**0.8091**	1	0.7106	0.9908	**0.8824**

3.2.1 Discussions

There are some findings from the results showed in Tables 1 (Fig. 4) and Table 2 (Fig. 5) as follows:

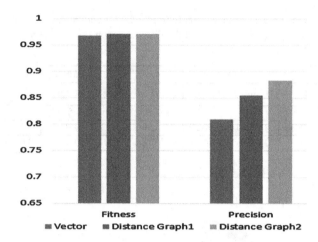

Fig. 5. Comparison of the discovered process models on the measures of fitness and precision among activity-based (Vector), distance graph order 1-based (Distance Graph1), and distance graph order 2-based (Distance Graph2) representations with K-means clustering algorithm.

- In all cases, the performance of the distance graph based trace representation is better than that of the vector based trace representation on fitness and precision measures.
- The effect of the distance graph based trace representation on the precision measure is higher than that on the fitness measure.
- Distance graph order 2 has a better effect on precision in comparison with distance graph order 1.

4 Related Work

G. Greco et al. [8] proposed a clustering solution on traces in event log. They used a vector representation for traces and the K-means algorithm. This work is the first study on trace clustering within the process mining domain.

R. P. Jagadeesh Chandra Bose [6], R. P. Jagadeesh Chandra Bose et al. [4, 5] proposed trace clustering solutions based on using some control-flow context information. i.e. "context-aware". The Levenshtein distance technique was used.

De Weerdt et al. [15] proposed a two phase solution to combine of trace clustering and text mining for process discovering. In the first phase, a MRA-based semi-supervised clustering technique (the SemSup-MRA algorithm) was applied. After that, there are two kinds of clusters, clusters of standard behaviors, and clusters of atypical behaviors. In the second phase, process mining and text-data mining techniques were applied. After [15], De Weerdt et al. [16] proposed the ActiTraC algorithm, a three-phase algorithm for clustering an event log into a collection of event logs (clusters). The ActiTraC algorithm includes three phases: Selection, Look ahead, and Residual trace resolution. They also developed the ActiTraCMRA algorithm, a further version of the ActiTraC algorithm.

T. Thaler et al. [14] provided a survey of trace clustering techniques. They also analyzed and compared the investigated trace clustering techniques.

This work is the first study on using the distance graph model [3] for trace clustering.

5 Conclusions

This work provided a trace representation solution based on the distance graph model [3] for clustering of traces in the event logs. Experiments showed that the distance graph based is more effective than activity based trace representation.

In this work, experiments are limited. There are several tasks needed to do in the future. Firstly, other distance measures between graphs, e.g. distance in graph theory [18] should be studied to directly cluster traces in the form of graphs. Secondly, more clustering algorithms, especially graph-based clustering algorithms, should be considered. Thirdly, more event log datasets should be experimented to confirm the reliability of the method.

Acknowledgments. This work was supported in part by VNU Grant QG-15- 22.

References

1. van der Aalst, W.M., van Dongen, B.F.: Discovering workflow performance models from timed logs. In: Han, Y., Tai, S., Wikarski, D. (eds.) EDCIS 2002. LNCS, vol. 2480, pp. 45–63. Springer, Heidelberg (2002)
2. Van der Aalst, W.M.P.: Process Mining: Discovery, Conformance and Enhancement of Business Processes. Springer, Heidelberg (2011)
3. Aggarwal, C.C., Zhao, P.: Towards graphical models for text processing. Knowl. Inf. Syst. **36**(1), 1–21 (2013)
4. Bose, R.C., van der Aalst, W.M.: Trace clustering based on conserved patterns: towards achieving better process models. In: Rinderle-Ma, S., Sadiq, S., Leymann, F. (eds.) BPM 2009. LNBIP, vol. 43, pp. 170–181. Springer, Heidelberg (2010)
5. Bose, R.P.J.C., van der Aalst, W.M.P.: Context aware trace clustering: towards improving process mining results. In: SDM 2009, pp. 401–412 (2009)
6. Bose, R.P.J.C.: Process Mining in the Large: Preprocessing, Discovery, and Diagnostics. Ph. D. thesis. Eindhoven University of Technology (2012)
7. Dai, Xin-Yu., Cheng, C., Huang, S., Chen, J.: Sentiment classification with graph sparsity regularization. In: Gelbukh, A. (ed.). LNCS, vol. 9042, pp. 140–151. Springer, Heidelberg (2015)
8. Greco, G., Guzzo, A., Pontieri, L., Saccà, D.: Discovering expressive process models by clustering log traces. IEEE Trans. Knowl. Data Eng. **18**(8), 1010–1027 (2006)
9. de Medeiros, A.K.A., van Dongen, B.F., van der Aalst, W.M.P., Weijters, A.J.M.M.: Process mining: extending the alpha-algorithm to mine short loops. BETA Working Paper Series (2004)
10. de Medeiros, A.K.A., Guzzo, A., Greco, G., van der Aalst, W.M., Weijters, A., van Dongen, B.F., Saccà, D.: Process mining based on clustering: a quest for precision. In: Hofstede, A. H., Benatallah, B., Paik, H.-Y. (eds.) BPM Workshops 2007. LNCS, vol. 4928, pp. 17–29. Springer, Heidelberg (2008)

11. Rozinat, A., van der Wil, M.P.: Aalst. Conformance checking of processes based on monitoring real behavior. Inf. Syst. **33**(1), 64–95 (2008)

12. Buijs, J.C., van Dongen, B.F., van der Aalst, W.M.: On the role of fitness, precision, generalization and simplicity in process discovery. In: Meersman, R., Panetto, H., Dillon, T., Rinderle-Ma, S., Dadam, P., Zhou, X., Pearson, S., Ferscha, A., Bergamaschi, S., Cruz, I.F. (eds.) OTM 2012, Part I. LNCS, vol. 7565, pp. 305–322. Springer, Heidelberg (2012)

13. Song, M., Günther, C.W., van der Aalst, W.M.: Trace clustering in process mining. In: Ardagna, D., Mecella, M., Yang, J. (eds.) Business Process Management Workshops. LNBIP, vol. 17, pp. 109–120. Springer, Heidelberg (2009)

14. Thaler, T., Ternis, S.F., Fettke, P., Loos, P.: A comparative analysis of process instance cluster techniques. In: Wirtschaftsinformatik 2015, pp. 423–437 (2015)

15. De Weerdt, J., van den Broucke, S.K.L.M., Vanthienen, J., Baesens, B.: Leveraging process discovery with trace clustering and text mining for intelligent analysis of incident management processes. In: IEEE Congress on Evolutionary Computation, pp. 1–8 (2012)

16. De Weerdt, J., vanden Broucke, S.K.L.M., Vanthienen, J., Baesens, B.: Active trace clustering for improved process discovery. IEEE Trans. Knowl. Data Eng. **25**(12), 2708–2720 (2013)

17. Wen, L., van der Aalst, W.M.P., Wang, J., Sun, J.: Mining process models with non-free-choice constructs. Data Min. Knowl. Discov. **15**(2), 145–180 (2007)

18. Deza, M.M., Deza, E.: Distances in Graph Theory. Springer, Heidelberg (2014)

19. http://www.processmining.org/prom/start

On Expressiveness of TCTL$_h^\Delta$ for Model Checking Distributed Systems

Naima Jbeli$^{(\boxtimes)}$, Zohra Sbaï, and Rahma Ben Ayed

École Nationale d'Ingénieurs de Tunis, Université de Tunis El Manar,
BP. 37 Le Belvédère, 1002 Tunis, Tunisia
{naima.jbeli,zohra.sbai,rahma.benayed}@enit.rnu.tn

Abstract. Systems analysis is becoming more and more important in different fields such as network applications, communication protocols and client server applications. This importance is seen from the fact that these systems are faced to specific errors like deadlocks and livelocks which may in the major cases cause disasters. In this context, model checking is a promising formal method which permits systems analysis at early stage, thus ensuring prevention from errors occurring. In this paper, we propose an extension of timed temporal logic TCTL with more powerful modalities called $TCTL_h^\Delta$. This logic permits to combine in the same property clocks quantifiers as well as features for transient states. We formally define the syntax and the semantics of the proposed quantitative logic and we show via examples its expressive power.

Keywords: Distributed systems · Model checking · $TCTL_h^\Delta$ · Quantitative verification

1 Introduction

Concurrent and distributed systems design and development are generally complex. In fact, there is always a high possibility that subtle errors cause erroneous behavior. These errors lead to catastrophic consequences especially in terms of money and time as well as in some cases for human life. The behavior analysis consists of a powerful technique which helps to discover behavioral problems at an earlier stage which is the design time. In addition, from a theoretical side, there is a clear need for a sound mathematical basis for the systems design.

Moreover, the presence of time delays or durations adds another dimension to the complexity of concurrent and distributed systems. In fact, the consideration of temporal properties in the specification of the behavior of competing systems provides a higher degree of expressiveness and flexibility.

Here is crucial the need for a verification of quantitative properties of concurrent systems which is a complex task requiring powerful models and efficient analysis techniques [4]. Model checking is one of the most popular verification techniques to check concurrent systems [1,3]. Indeed, the behavior of a system is represented by a finite transition system, and the properties to be verified are

© Springer International Publishing Switzerland 2016
N.T. Nguyen et al. (Eds.): ICCCI 2016, Part I, LNAI 9875, pp. 323–332, 2016.
DOI: 10.1007/978-3-319-45243-2_30

expressed in a temporal logic (LTL, CTL, MITL, TCTL, etc.). Model checking is an efficient technique compared to other methods including theorem proving as it provides automatic system verification with allowing the user to view counter-example in case the property is violated.

In this work, our main contribution is to propose an extension of timed temporal logic TCTL which is expressive and decidable. For this, we study the different timed temporal logics proposed in the literature and we define a new TCTL fragment named $TCTL_h^\Delta$ which is a combination of $TCTL^\Delta$ and $TCTL_h$. Then, we analyze the expressiveness of the proposed timed temporal logic via some examples.

The following sections are organized as follows. Section 2 shows the preliminaries needed in the rest of the paper. Section 3 introduces our new timed temporal logic $TCTL_h^\Delta$ detailing its syntax and semantics as well as a motivating example. In Sect. 4, we study some expressiveness results of $TCTL_h^\Delta$. We conclude the paper and present perspectives in Sect. 5.

2   Preliminaries

Aiming to ensure quantitative verification, we investigate in studying the timed temporal logics. Time Computation Tree Logic (TCTL) is an arborescent timed logic that adds quantitative information on delays between actions. This logic is constructed from logical connectors, atomic propositions and timed combiners that use the modality U (until). TCTL is an extension of Computation Tree Logic (CTL) proposed in [1], in order to set out time modalities which add quantitative information over time. The TCTL formulas are constructed from constraints on explicit clocks.

Definition 1 (Syntax of TCTL). *The syntax of the TCTL logic is defined as follows:* $\varphi, \psi :: = P_1 \mid P_2 \mid ... \mid \neg\varphi \mid \varphi \wedge \psi \mid E\varphi U_{\sim c}\psi \mid A\varphi U_{\sim c}\psi$
Where: $\sim \in \{<, \leq, =, \geq, >\}$, $c \in \mathbb{N}$ *and* $P_i \in AP$ *(Atomic Propositions).*

We can use other compositional temporal operators:

- $EF_{\sim k}\varphi = E(\text{true } U_{\sim k}\varphi)$ (Possibility),
- $EG_{\sim k}\varphi = \neg AF_{\sim k}\neg\varphi$ (All locations along an execution),
- $AF_{\sim k}\varphi = A(\text{true } U_{\sim k}\varphi)$ (Locations along all executions),
- $AG_{\sim k}\varphi = \neg EF_{\sim k}\neg\varphi$ (All locations along all executions).

Now, we give the formal semantics of TCTL. In fact, the formulas specified in this logic are evaluated on a configuration s of a Timed Transition System. $TTS = (Q, Q_0, \sum, \rightarrow)$ [10] where:

- Q is a set of states,
- $Q_0 \subseteq Q$ is the set of initial states,
- $\sum$ is a finite set of actions disjoint from $\mathbb{R}_{\geq 0}$,
- $\rightarrow \subseteq Q \times (\sum \cup \mathbb{R}_{\geq 0}) \times Q$ is a set of edges.

A timed structure is defined by $S_t=(TTS, \prod^t)$, where TTS is a timed transition system [11] and $\prod^t$ is a set of timed paths. A timed path of $\prod^t$ is a finite or infinite sequence of elements with the form $(q_0,t_0)... (q_n,t_n)...$ where:

- For each index i, there is a label l where $(q_i, t_i, l, q_{i+1}) \in Tr$ and $t_i \leq t_{i+1}$,
- The finite sequence finishes with (q_n,t_n) if q_n has no successor.

Let $q \in Q$ and a transition $t \in T$(the set of transition). Satisfying relation $(q,t) \models \varphi$ is inductively defined by:

- $(q,t) \models T$,
- $(q,t) \models p$ iff $p \in \rho(q)$,
- $(q,t) \models \neg\varphi$ iff $(q,t) \not\models \varphi$,
- $(q,t) \models \varphi_1 \wedge \varphi_2$ iff $(q,t) \models \varphi_1$ and $(q,t) \models \varphi_2$,
- $(q,t) \models A(\varphi_1 \ U_{\sim k} \ \varphi_2)$ iff for all paths $((q_0,t_0)...(q_n,t_n)...) \in \prod^t$ such that $q_0 = q$ and $t_0= t$, there exists t' such that $(t'\sim t + k \wedge t' \geq t, (q_n,t')\models \varphi_2)$, and for all $0 \leq i \leq n$ and $t'' \in [t, t'], (q_i, t'') \models \varphi_1$,
- $(q,t) \models E(\varphi_1 \ U_{\sim k} \ \varphi_2)$ iff there exists a path $((q_0,t_0)...(q_n,t_n)...) \in \prod^t$ such that $q_0 = q$ and $t_0= t$, there is t' such that $(t' \sim t + k \wedge t' \geq t, (q_n,t')\models \varphi_2)$, and for all $0 \leq i \leq n$ and $t'' \in [t,t'], (q_i,t'') \models \varphi_1$.

The timed temporal logic TCTL allows writing temporal properties by quantifying time. We choose this logic because the TCTL model checking is decidable and PSPACE-complete for time Petri nets and timed automata [2].

Timed automata [2] are formed of a finite control structure and manipulate a finite set of real variables which are called clocks to specify time constraints. The time is added to the classical model of automata as clocks and predicates on these clocks. These predicates are of two types: invariants and guards.

Definition 2 (Timed Automata) [2,5]. *A timed automata is a tuple A= (Q, q_{init}, X, $\sum$, $\rightarrow_A$, Inv, l) where:*

- *Q is a finite set of locations,*
- *$q_{init} \in Q$ is the initial state,*
- *X is a finite set of clocks,*
- *$\sum$ is a finite alphabet which denotes a finite set of actions (or events),*
- *$\rightarrow_A \subseteq Q \times C(X) \times \sum \times 2^X \times Q$ is a finite set of action transitions to (q, g, a, r, q') $\in \rightarrow_A$ we write $q \xrightarrow{g,a,r}_A q'$. The constraint of clocks associated: g is called guard, a is a letter of the alphabet $\sum$ and r is the set of clocks which are reset.*
- *Inv: $Q \rightarrow C(X)$ Invariant is an application which allows to associate for each control state a conjunction of constraints clocks with the form $x \sim c$ with $\sim\in \{<,\leq\}$.*
- *l: $Q \rightarrow 2^{AP}$ is an application which allows to label each control configuration by a set l(q) of atomic propositions.*

A state (or configuration) of a timed automata is a pair (q, v), where q is a state of control and v is a clock valuation. For a timed automata, the initial configuration is (q_{init}, v_0) with $v_0(x) = 0 \; \forall \; x$ clock. We note $v \models g$ which means that the valuation v is used to check the guard where $v \in \mathbb{R}_+^X$ and $g \in C(X)$ (which is the set of clocks constraints) as well as means that this check must also yield "true".

3   Timed Temporal Logic $TCTL_h^\Delta$

In this section, we begin with exposing our study on TCTL extensions leading to our proposal, then we study a motivating example before presenting the proposed timed temporal logic.

3.1   Study of TCTL Extensions

Regarding the development of temporal logics in model checking, timed temporal logics have been proposed to extend classical temporal logics with timing constraints. There are several ways of expressing such constraints, a standard one consists in constraining temporal modalities. Hence, various TCTL extensions are proposed in the literature. Among these extensions, we studied ATCTL [12], $TCTL^\Delta$, $TCTL^{ext}$, $TCTL^a$, $TCTL^0$ [17,18], $TCTL_h$ [13,14], RTCTL [8], TCTL with internal clock ($TCTL_c^{int}$) and TCTL without imbrication ($TCTL_p$) [16], $TCTL_c$ and $TCTL_s$ [15], PTCTL [7] and PRTCTL [9].

After studying these TCTL extensions, we can resume that the quantitative logic TCTL is extended with parameters [7,9] or with external clock variables [14] or also with modalities to verify properties with abstraction transients [17,18].

Among these extensions, we propose to investigate in $TCTL_h$ where we can use variables to express timing constraints and $TCTL^\Delta$ where we note abstraction transient states. $TCTL_h$ [13] is a method to integrate quantitative aspects in temporal logic which consists in adding clocks formulas (these clocks increase synchronously with time), an operator to reset (in) and a simple constraint $x \sim c$ or $x - y \sim c$ with $x, y \in H$. Thus, the reset followed by a constraint $x \sim c$ measures the delay between two states of the system.

Now, when specifying temporal properties, the authors in [17] define the timed temporal logic $TCTL^\Delta$ which extends TCTL with new modalities U^k where $k \in \mathbb{N}$.

After this study of the different extensions of TCTL, we set a goal to find fragments of the logic $TCTL^\Delta$ which are larger but which remain decidable.

3.2   Motivating Example

The timed automata of the Fig. 1 describes a barrier which opens within 10 s at the approach of a car and remains open during 120 s before closing. The automaton leaves from the closed initial state. When a car approaches the barrier (signal app?), it passes in the state opening to get to the opened state by handing

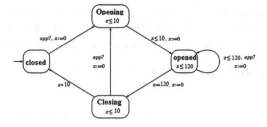

Fig. 1. Modeling of the barrier example

the x clock to zero after a delay lower or equal to 10 s. In this state, the time elapses and the approach of other cars (app?) hands each time the x clock to zero. when the x clock is 120, the barrier begins to close (the closing state). It fall back on after 10 s (the closed state), unless another car is approaching.

After modeling this example, let us focus on specifying the following formula: "Always for any execution along every interval time which duration is 120 s, the barrier can close only during at least 10 s of the opening state at the approach of a car". To deeply develop this example, we divide the work in two parts.

If we take $TCTL_h$, we can write this formula:

$$AG\ (Opened \rightarrow x\ in\ (AF(Closing\ and\ x \leq 120)))$$

Unfortunately, this formula does not properly correspond to the property related to our example since the *in* operator resets the x clock to zero when we encountered a state verifying *opened*, so it is sufficient to check that $x \leq 120$ when encountering a state verifying *closing* to ensure that the duration is less than 120 s. So, with this temporal logic we can not adequately express this property because, specifically, the barrier can close only during for at least 10 s.

Now let us consider $TCTL^\Delta$, we can write this formula:
$A(Opened)\ P^{10}(Closing)$. This formula means that for any execution if "Closing" lasts long enough then the state "Opened" also lasted long enough in the past. Unfortunately, this formula does not properly correspond to our property because for any execution along every interval time its duration is 5 units of time.

Our contribution at this step is to find a larger subset of $TCTL^\Delta$ logic which still decidable. In this paper, we propose to consider the timed temporal logic $TCTL^\Delta$ in its local semantics as well as giving the possibility to explicitly manipulate clocks in formulas (which is ensured by the timed temporal logic $TCTL_h$). We expanded the quantitative TCTL logic with two new modalities: durations and activation delays.

3.3 $TCTL_h^\Delta$ Syntax and Semantics

Our aim is to ensure a quantitative verification of properties clarifying the durations and the activation delays. These new modalities allow to ignore all events which do not occur continuously during k time units ($k \in \mathbb{N}$). We show that this is ensured via a composition of both TCTL extensions: $TCTL^\Delta$ and $TCTL_h$. Formally, we define the timed temporal logic $TCTL_h^\Delta$ as follows.

Definition 3 (Syntax of $TCTL_h^\Delta$). $TCTL_h^\Delta$ *formulas are outlined with the following grammar:*

$$\varphi, \psi ::= P_1 \mid P_2 \mid ... \mid \neg\varphi \mid \varphi \wedge \psi \mid E\varphi U\psi \mid A\varphi U\psi \mid x \text{ in } \varphi \mid$$
$$x \sim c \mid x - y \sim c \mid E\varphi\ U^k\psi \mid A\varphi\ U^k\psi$$

where: $\sim \in \{<, \leq, =, \geq, >\}$, $c, k \in \mathbb{N}$, $x, y \in H$ *and* $P_i \in AP$.

We define now the semantics of $TCTL_h^\Delta$ formulas on a state of a TTS and valuation v ($v: H \to \mathbb{T}$) for clocks of the set H.

Definition 4 (Semantics of $TCTL_h^\Delta$). *The following clauses define the truth value of a formula $TCTL_h^\Delta$ on a state s of a TTS $T = (\sum, Q, q_0, Tr, \rho)$ and a valuation $v : H \to \mathbb{T}$, denoted $s, v \models \varphi$:*

- $s, v \models x \sim c$ iff $v(x) \sim c$
- $s, v \models x - y \sim c$ iff $v(x) - v(y) \sim c$
- $s, v \models x$ in φ iff $s, v[x \leftarrow 0] \models \varphi$
- $s, v \models E_\varphi U_\psi$ iff $\exists \rho \in Exec(s)$ with $\rho = \sigma \cdot \rho'$ and $s \overset{\sigma}{\mapsto} s'$ s.t. $s', v + Time(\sigma)$ $\models \psi$ and $\forall s'' <_\rho s'$, s.t. $\rho = \sigma' \cdot \rho''$ and $s \overset{\sigma'}{\mapsto} s''$, we have $s'', v + Time(\sigma') \models \varphi$
- $s, v \models A_\varphi U_\psi$ iff $\forall \rho \in Exec(s)$ $\exists \sigma \in Pref(\rho)$ s.t. $s \overset{\sigma}{\mapsto} s'$, $s', v + Time(\sigma) \models \psi$ and $\forall s'' <_\rho s'$, s.t. $s \overset{\sigma'}{\mapsto} s''$, $s'', v + Time(\sigma') \models \varphi$
- $s, v \models E_\varphi\ U_\psi^k$ iff $\exists \rho \in Exec(s)$ with $\rho = \sigma \cdot \rho'$ and $s \overset{\sigma}{\mapsto} s'$; $\exists \lambda \in SC(\rho)$ s.t. $\mu(\lambda) > k$ and $s', v + \lambda \models \psi$ and $\lambda' \in SC(\rho)$ s.t. $\lambda' <_\rho \lambda$ and $\forall s'' <_\rho s'$, s.t. $\rho = \sigma' \cdot \rho''$ and $s \overset{\sigma'}{\mapsto} s''$, we have $s'', v + \lambda' \models \varphi \Rightarrow \hat{\mu}(\lambda') \leq k$
- $s, v \models A_\varphi\ U_\psi^k$ iff $\forall \rho \in Exec(s)$ $\exists \sigma \in Pref(\rho)$ s.t. $s \overset{\sigma}{\mapsto} s'$; $\exists \lambda \in SC(\rho)$ s.t. $\mu(\lambda) > k$ and $s', v + \lambda \models \psi$ and $\lambda' \in SC(\rho)$ s.t. $\lambda' <_\rho \lambda$ and $\forall s'' <_\rho s'$ s.t. $s \overset{\sigma'}{\mapsto} s''$, we have $s'', v + \lambda' \models \varphi \Rightarrow \hat{\mu}(\lambda') \leq k$

We note that a transition (s, α, s') which is denoted by $s \overset{\alpha}{\mapsto} s'$. We associate a duration defined by α if $\alpha \in \mathbb{R}_+$ and 0 otherwise. The transition is called action transition when $\alpha \in \sum$ (with $\sum$ the set of alphabet of transition labels), otherwise it is called duration transition.

A path of a timed transition system (TTS) is an infinite sequence of consecutive transitions: $\rho = s_0 \overset{\alpha_0}{\mapsto} s_1 \overset{\alpha_1}{\mapsto} s_2...$

A run (or an execution) of a timed automata A is an infinite path [18] $s_0 \to_{T_A} s_1 \to_{T_A} s_2 \ldots$ in T_A such that (1) time diverges and (2) there are infinitely many action transitions. We note that a run can be described as an alternating infinite sequence $s_0 \overset{d_1}{\to} {}_a s_1 \overset{d_2}{\to} {}_a \ldots$ with $d_i \in \mathbb{R}$. Given an execution ρ goes through any state s' which is reachable from s_i by a delay transition of duration $d \in [0, d_i]$.

A state (or configuration) of a A is a pair (q,v), such as q is the current location and v is the current clock valuation. (q_{init}, v_0) is the initial state of A with $v_0(x) = 0$ for any clock x in X.

Let Exec(s) be the set of all executions from state s, with an execution ρ: $(q_0, v_0)\xrightarrow{d_1}\rightarrow_a (q_1, v_1)\xrightarrow{d_2}\rightarrow_a...$ of A, then we associate a sequence of absolute dates such that $t_0 = 0$ and $t_i = \sum_{j\leq i} d_j$, for $i \geq 1$.

A state (q,v) can occur several times along an execution ρ, the notion of position makes possible to distinguish them: every occurrence of a state is associated with a unique position. Let p a position, the corresponding state is noted by s_p. The standard notions of suffix, prefix and sub-run apply to paths in TTS: for a position $p \in \rho$, $\rho^{\leq p}$ is the prefix leading according to p as well as $\rho^{\geq p}$ is the suffix issued from p. So, a sub-run σ from p to p' is denoted by $p \overset{\sigma}{\mapsto} p'$.

We note that the set of positions is totally ordered by $<_\rho$ along the run ρ. Let two positions p and p', we can say that p precedes strictly p' (written as p $<_\rho$ p') along ρ iff there exists a finite sub-run σ of an execution ρ s.t. $p \overset{\sigma}{\mapsto} p'$ and σ contains at least one action transition or one non zero delay transition. Then, we can write $\sigma <_\rho p$ such that for any position p' in the sub-run σ, we have p' $<_\rho$ p.

Let p a position in ρ, the prefix $\rho^{\leq p}$ has a duration time($\rho^{\leq p}$), which defined as the sum of all delays along $\rho^{\leq p}$. Since time diverges along a run, we have: there exists p $\in \rho$ for any t $\in \mathbb{R}$ with Time($\rho^{\leq p}$) > t.

Given a subset P $\subseteq \rho$ of positions in ρ, the authors in [5] defined a natural measure $\hat{\mu}(P) = \mu\{\text{Time}(\rho^{\leq p}) \mid p \in P\}$, where μ is Lebesgue measure on the set of real numbers.

Let us return to the example presented in Sect. 3.2, we show that the desired property can be now specified in $TCTL_h^\Delta$ by the following formula:

$$AG(x \text{ in } (A(\neg Closing) \ U^{10} \ (Opened \ and \ x \leq 120)))$$

This formula means that: along any interval with a duration up to 120 s, the barrier can not close during at least 10 s after the opening state at the approach of another car.

4   Expressiveness of $TCTL_h^\Delta$

The logic $TCTL^{ext}$ defined in [5] is the restriction of $TCTL^\Delta$ where the parameter k is always 0. As the modality EU^0 cannot be expressed in TCTL [5], $TCTL^\Delta$ is clearly more expressive then TCTL. $TCTL^a$ is a TCTL logic with an almost everywhere until modality U^a. We obtain for instance formulae like $AG^a\phi$, meaning that property ϕ is true almost everywhere. [5] prove that the modality U^a cannot be expressed with TCTL operators and conversely that U^a cannot express TCTL modalities. We can cite the following expressiveness results:

- TCTL $< TCTL^{ext}$ and $TCTL^a < TCTL^{ext}$ [5],
- $TCTL^a < TCTL^0$ and $TCTL < TCTL^0$ [18],
- $TCTL < TCTL_h$ [6].

4.1 $TCTL^\Delta < TCTL_h^\Delta$

We consider the example of leaking gas burner [18]. This system is represented by an automaton with two states as shown in Fig. 2. The gas burner can be in two states: (1) The state q_0 (initial state): is labeled with the atomic proposition L (as leaking) which represents the situation when the gas burner is leaking. (2) The state q_1: is labeled with the atomic proposition $\neg L$ (as not leaking) which represents the situation when the gas burner isn't leaking. The leaks are detected and stopped in less than 1 s, once the leak is stopped. The burner cannot restart to leak within 30 s. The clock x measures the delays: the constraint $x \leq 1$ in state q_0 is an invariant, the constraints $x \leq 1$ and $x \leq 30$ associated to transitions are guards as well as the notation. $x = 0$ corresponds to reset the clock x.

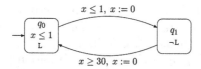

Fig. 2. Modeling of the gas burner

We want to check the following property: after an elapsed time of 60 s, would the system leak for up to 20 % of the total elapsed time?

This property can be expressed with the $TCTL^\Delta$ by the following formula:
$AG\big(A\,(\neg L)\,U_{\geq 60}^{12}\,T\big)$

$TCTL_h^\Delta$ is "more temporal" in the sense that it uses real clocks in order to assert temporal constraints. A $TCTL_h^\Delta$ formula can "reset" a formula clock at some point, and later compare the value of that clock to some integer. The property above would then be written as follows:

$AG\left(x\ in\ \big(A\,(\neg L)\,U^{12}\,(T \wedge x \geq 60)\big)\right)$ This formula means that along any

interval with a duration of at least 60 s, the burner leaked for a period less than 12 s. It is clear that any $TCTL^\Delta$ formula can be translated into an equivalent $TCTL_h^\Delta$ one.

Now, what about the opposite?
There is one $TCTL_h^\Delta$ formula which cannot be expressed with $TCTL^\Delta$:
$x\ in\ EF^k(P_1 \wedge x < 1 \wedge EG(x < 1 \Rightarrow \neg P_2))$ This formula expresses that it is possible to reach a verifying state P_1 along an interval of duration k t.u. and within a delay of 1 t.u., from which there is an execution where there is no checking state P_2 before x is equal to 1.

So we conclude that $TCTL_h^\Delta$ is more expressive then $TCTL^\Delta$. Now, let us focus on $TCTL_h^\Delta$ and $TCTL^{ext}$. $TCTL^{ext}$ is obtained by adding the two modalities $EU_{\bowtie c}^a$ and $AU_{\bowtie c}^a$ to TCTL. For example, $AG_{\geq c}^a\phi$. This formula means that along every run, the set of positions at which ϕ does not hold has a measure equal to c t.u. ($c \in \mathbb{N}$), i.e. ϕ holds almost everywhere along all paths. We can transform this formula by adding implicitly a clock x to measure the separating deadlines between actions adding thus expressiveness. For this, the specification with $TCTL_h^\Delta$ logic is as follows: $x\ in\ (AG^a\phi \wedge x \geq c)$ So, we can conclude that $TCTL_h^\Delta$ can express any property of $TCTL^{ext}$.

4.2 $TCTL_h < TCTL_h^\Delta$

Let us study the property: Any occurrence of a signal is followed by a response which lasts a sufficiently long time (>2) around a position ≤ 5. The specification of this property with $TCTL_h^\Delta$ is:
$AG(signal \Rightarrow x\ in\ AF^2(x \leq 5 \wedge reponse))$ This formula means that: along any interval with a duration up to 5 units of time, the response cannot occur during at least two units of time after the signal, i.e. that the clock value x is less than or equal to 5 when an answer occurs. The $TCTL_h$ temporal logic is the fragment of $TCTL_h^\Delta$ without EU^k and AU^k. So we cannot use it to express this formula. So, we conclude that $TCTL_h^\Delta$ is more expressive than $TCTL_h$.

Through the results presented in this section, we conclude that $TCTL_h^\Delta$ is characterized by its expressiveness over the quantitative temporal logics proposed in the literature.

5   Conclusion

Concurrent and distributed systems have become, over the last years, ubiquitous in the industry. To ensure quantitative verification of these systems, we proposed in this paper a timed temporal logic named $TCTL_h^\Delta$ and we detailed its syntax and semantics. We showed by means of examples the expressive power of this new timed temporal logic over the existing TCTL extensions.

Future work includes the application of the proposed formal verification method in a particular set of real time systems for which the quantitative analysis is of much importance. In addition, we will investigate more in the application of $TCTL_h^\Delta$ to model check time Petri nets.

References

1. Alur, R., Courcoubetis, C., Dill, D.: Model-checking in dense real-time, 2–34 (1993). Berlin
2. Alur, R., Dill, D.: A theory of timed automata. Theoretical Comput. Sci. (TCS) **126**, 183–235 (1994)
3. Alur, R., Henzinger, T.A.: A really temporal logic. J. ACM **41**(1), 181–203 (1994)
4. Alur, R., Henzinger, T.A.: Automatic symbolic verification of embedded systems. IEEE Trans. Softw. Eng. **22**, 181–201 (1996)
5. Mokadem, H.B., Bérard, B., Bouyer, P., Laroussinie, F.: A new modality for almost everywhere properties in timed automata. In: Abadi, M., de Alfaro, L. (eds.) CONCUR 2005. LNCS, vol. 3653, pp. 110–124. Springer, Heidelberg (2005)
6. Bérard, B.: Model checking temporisé. In: Roux, O.H., Jard, C. (eds.) Approches formelles des systemes embarqués communicants. Hermes/Lavoisier (2008)

7. Bruyère, V., Raskin, J.F.: Real-time model-checking: parameters everywhere. Log. Methods Comput. Sci. **3**(1:7) (2007)
8. Emerson, E.A., Mok, A.K., Sistla, A.P., Srinivasan, J.: Quantitative temporal reasoning. In: Automatic Verifiation Methods for Finite State Systems, Grenoble, France (1992)
9. Emerson, E.A., Trefler, J.R.: Parametric Quantitative Temporal Reasoning. University of Texas, USA (1999)
10. Hadjidj, R., Boucheneb, H.: On-the-fly TCTL model-checking for time Petri nets. Theor. Comput. Sci. **410**(42), 4241–4261 (2009)
11. Henzinger, T., Manna, Z., Pnueli, A.: Timed transition systems (1992)
12. Jansen, D.N., Wieringa, R.: Reducing the extensions of CTL with actions, real time. Technical report, Universiteit Twente: CTIT, Enschede, December 2000
13. Laroussinie, F.: Model checking temporisé: Algorithmes efficaces et complexité. Master's thesis, ENSCachan, Décembre (2005)
14. Laroussinie, F., Turuani, M., Schnoebelen, P.: On the expressivity and complexity of quantitative branching-time temporal logics. Theor. Comput. Sci. **297**, 297–315 (2003)
15. LSV. Dossier scientifique. Ecole Normale Supérieure de Cachan (2004)
16. Mathieu, S.: Méthodes qualitatives et quantitatives pour la détection d'information cachée. Ph.D. thesis, Université Pierre et Marie Curie (2011)
17. Bel Mokadem, H.: Verification des proprietes temporisees des automates programmables industriels. Ph.D. thesis, ECOLE NORMALE SUPERIEURE DE CACHAN (2007)
18. Mokadem, H.B., Bérard, B., Bouyer, P., Laroussinie, F.: Timed temporal logics for abstracting transient states. In: Graf, S., Zhang, W. (eds.) ATVA 2006. LNCS, vol. 4218, pp. 337–351. Springer, Heidelberg (2006)

Innovations in Intelligent Systems

Cost Optimizing Methods for Deterministic Queuing Systems

Martin Gavalec and Zuzana Němcová$^{(\boxtimes)}$

Faculty of Informatics and Management, University of Hradec Králové,
Rokitanského 62, 50003 Hradec Králové, Czech Republic
{martin.gavalec,zuzana.nemcova}@uhk.cz
http://www.uhk.cz

Abstract. The paper discusses two proposed methods for the cost optimization of the deterministic queuing system based on the control of the queue lengths. The first method uses the evaluation of actual states at the particular service places according to their development. The decision is then based on the comparison of the criteria of productivity and the expended costs. The suggested change in the system setting with the highest priority is then accomplished. The second method is based on the simulation of the future states and on this basis the appropriate time and type of the modification of the system setup is suggested.

Keywords: Optimization · Deterministic queuing system · Tandem network

1 Introduction

Queuing systems can be found in many real systems and many of us come into contact with them in everyday life. Moreover, in today's information age society, in context of the interconnected communication and data sharing, the importance of the understanding and prediction of the system behavior increases. We can find the applications, for example, in the packet network design and optimization, the design of the traffic lights at crossroads and in the manufacturing process of the production systems. The first application of the queuing theory can be traced back to the beginning of the 20th century, when the telephone networks has been designed and analyzed.

Queuing network modelling have been investigated by many researchers. The general overview of modelling the production and transfer lines using queuing networks can be found in [8]. Stochastic approach in queuing theory have been introduced in [3,5,9]; in [2] is provided the deep insight into the flow problems in networking using deterministic approach.

In general, there are two types of units in such systems, the requests and the service places. The basic idea of this concept is that the requests come into the system in order to be served at particular service places. The intensity of the requests arrival to the system can naturally vary in time. The service times

© Springer International Publishing Switzerland 2016
N.T. Nguyen et al. (Eds.): ICCCI 2016, Part I, LNAI 9875, pp. 335–344, 2016.
DOI: 10.1007/978-3-319-45243-2_31

of the servers at the service places can also differ. In dependence on the ratio of these quantities, the queues can appear and on the other hand, some of the servers in the system can become idle. Sometimes, the occurrence of the queues and/or the idle servers is undesirable because it can pose additional indirect costs. This situation can result in searching for better possible system setting.

The arrangement of the service places is important as well; the service places can be concatenated in both, the series and/or the parallel manner. Queuing systems can have various structures, from those with the simplest layout (for example the cash desk at the gasoline station) to those with the complex organization (for example assembly lines).

Particular characteristics describing the behavior of the system can be computed during the system analysis in dependence on the user's intentions. The characteristics can be divided into several categories [1]: time characteristics of the requests, characteristics related to the number of the requests, probabilistic characteristics and cost characteristics. Assessing of these quantities is important especially by the system design or modernization. It helps to project the corresponding extent of the system so that its performance is optimal.

The study focuses on the queuing systems working in discrete time, the so-called discrete-event systems (DES). The basic characteristics can be found in [4,10]. These systems are usually man-made and can be characterized by a complex hierarchical structure. The state transitions of the system are initiated by the events (DES can be also called as event-driven systems although this depends on whether the state transitions are synchronized by a clock or if they occur asynchronously) and these events occur in a discrete time moments. In other words, with each transition of the system some event can be associated. The event can be represented by a start/end of some activity, for example, a completion of the product, customer arrival, or machine breakdown. Between these events the time lags of different length can be observed (the lags can be of deterministic or stochastic character).

In further sections the deterministic concept of the system is considered. It can be understood as a special case of stochastic approach. Using this approach it is possible to compute the parameters of the system, and thus it is also possible to predict the system future development.

2 Deterministic Open Linear Queuing System

Consider an open linear queuing system (in literature also known as the tandem queuing network, see [7]) with $n + 1$ service places, see Fig. 1. The incoming requests have to pass through the series of all servers and then leave the system. There can be arbitrarily long queues in front of each server before the system starts. The system works in the so-called stages; the stage is the time period during which the system setting stays unchanged. The time variable during stage takes values $t = 1, 2, ..., T$, where T is the length of the stage. Manager of the system has following information about the system setting:

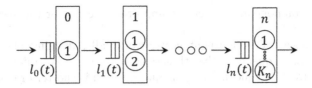

Fig. 1. Open linear system

- z_i, the basic service time of the server(s) at service place i,
- K_i, the number of identical servers at the service place i,
- l_i, the length of the queue at the service place i.

The service intensity at particular service place is then computed as $\sigma_i = K_i/z_i$. The intensity of arrivals is represented by the service intensity of the entrance server (indexed by $i = 0$). This server can be considered as a gatekeeper, which controls the access of the requests. It is assumed, that the queue in front of the entrance server is long enough; the gatekeeper admits the requests to the system with a constant speed.

Service times of the servers can be changed. The cause of the change can be either internal or external. The internal reason means that the manager of the system decide to change some of the service times, quantities, that can be directly influenced (the change in the value z_i, where $i > 0$). The external reason represents the change in the arrivals intensity of the requests coming into the system (the change in z_0, respectively K_0, in the model). It is important to note that the possibility of unexpected change in the intensity of arrivals as well as the need to cope with the different types of queues can foil the effort to set the service intensities to the intensity of arrivals at the start of the system.

Within the stage the queues can behave differently depending of the sequence of service intensities at particular service places. The change in queue length, whether the queue is increasing, decreasing or constant, is dependent not only on the service intensity of previous service place but also on the situation at all preceding service places. The time variable plays its role as well. For example, the queue in front of the ith service place can temporarily increase in transient time if the service place $i - 1$ is faster but some of the preceding service places is even slower than the ith service place, i.e. $\sigma_{i-k} < \sigma_i < \sigma_{i-1}$. In this case, the queue can increase for some time period (depending on the actual lengths of the queues), because the preceding service place sends the requests in higher intensity. In the long term the queue will decrease, because the intensity of arrivals to the ith service place is then considered as the service intensity of the slowest preceding service place.

3   Production Costs

The optimization is made in each stage separately - with respect to actual system setting and the function of total production costs of the stage. The function comprises the components that create important factors for the system performance.

Let us denote the constituent costs: work, idle, queuing and change by W_i, I_i, Q_i, C_i. Each queue at the ith service place can have different maximal limit denoted by M_i; it means that the queue which is shorter or equal to M_i is still tolerable.

Work costs are proportionally dependent on the actual number of servers at the ith service place. Idle costs are spent in case when the service capacity at the service place i is not fully used. It is proportionally dependent on the degree of such idleness. Queuing costs result from the excessively long queues, i.e. that are over the maximal tolerable limit M_i. These costs are also proportionally dependent on the degree of exceeding this limit. The last, change costs, are one-shot costs (added at the end of the stage) that are spend for the factual change in the system setting (at the service place j).

The total production costs of the stage are then computed according to the following formula:

$$
P(T, j) = \sum_{t=1}^{T} \left(\sum_{i=1}^{n} W_i K_i + \sum_{\substack{i=1 \\ l_i(t) < \frac{K_i}{z_i}}}^{n} I_i \left(\frac{K_i}{z_i} - l_i(t) \right) + \sum_{\substack{i=1 \\ l_i(t) > M_i}}^{n} Q_i \left(l_i(t) - M_i \right) \right) + C_j
$$

$$(1)$$

4   System Types

We can distinguish four types of systems according to the character of service times and requests flow. Both can be either continuous or discrete. The description of individual cases follows:

continuous service times: in this case there is exactly one server at each service place ($K_i = 1$); the change in system setting is made by acceleration or deceleration of servers' speed so that $z_i > 0$,

discrete service times: this approach allows to add or take away another identical server (with equivalent value of service time) at some service place in order to change the system setting; the condition $K_i \geq 1$ must be satisfied,

continuous requests flow: this means that there is no need to wait until the whole request is served - the served part of the request can immediately leave the service place and fall into the following queue after being served,

discrete requests flow: means that request can fall into the next queue soon after its completion as a whole.

Method for queue lengths computation can vary for different types of systems. The computation of basic characteristics as well as more detailed description can be found in [6].

For simplicity we assume that any change of the service times is performed exactly at one server by one time unit, and all more complex changes are performed as a series of such simple changes. With respect to computational complexity we consider the change costs as constant. In the following text two optimization methods are suggested. For the examples and calculations the system with continuous service time and continuous request flow is considered. The model of the system was designed using the language VBA and the tool MS Excel.

5   Suggested Methods

5.1   Method Based on the Evaluation of the System States

This method applies the Markovian property, i.e. the future development of the system states depends on its current state and finite number of previous states. The past history has been completely summarized in the current state and the system has no memory to the intent that it is not known how the current state was reached.

The method uses the evaluation of the system states. The evaluation criteria take into account just the present state and the previous one (but not the sequence of preceding events), thus we can say, that considered system is memoryless.

The suggestions for the changes in system settings are given every time unit (the so-called turn), so there is no need to penalize the change of the system setting via the addition of the change costs (in the calculation the function of total production costs), thus the parameter $C_j = 0$. Also due to continuous changes each stage is of length $T = 1$.

Remark 1. Note, that the length of the stage is adjusted to 1, therefore the turn and the stage are equivalent concepts for this method.

The following sequence of steps is performed at the end of each stage for each service place:

1. compute the characteristics, i.e. length of the queue $l_i(t)$, queue tendency $d_i(t)$, particular costs for turn, total costs for turn,
2. weight the urgency of the change, weights are denoted by $G_i(t)$,
3. suggest strategies,
4. choose the appropriate action,
5. apply the changes.

The length of the ith queue is computed according the type of the system, it depends on the above mentioned character of the requests. In general, the queue length in time (t) is equal to the queue length in time $(t-1)$ subtracted by the quantity of requests that left the queue and added by the quantity of requests that have fallen into this queue.

Queue tendency, $d_i(t)$, reflects how will be the actual queue changed with respect to the current setting; it is dependent on the $l_i(t)$, $l_{i-1}(t)$ and also on the $\sigma_i(t)$ and $\sigma_{i-1}(t)$. The value expresses the increment or decrement of actual queue and also gives the information about the intensity of this variation. If $d_i(t) > 0$, the queue is increasing; for $d_i(t) < 0$ it holds that the queue is decreasing and $d_i(t) = 0$ means that the length of the queue will not change in next turn. Computation of this parameter is again dependent on the character of the requests.

The value of particular costs for turn (considering the idle and the queuing costs) will be positive if either the $l_i(t) < K_i(t)/z_i$ in case of idle costs, or if $l_i(t) - M_i > 0$ in case of queuing costs. The total costs for turn are then computed as a sum of particular costs.

After the computation of the above mentioned basic characteristics is made, the particular situations at the service places are weighted. The weights are expressed by the multiple of Q_i, I_i according to the importance of the situation. If the queue tends to grow and the limit M_i will be exceeded in u turns (the value depends on the need to provide prompt reactions upon changes in the system), then the weight $G_i(t) = Q_i \cdot x'$, where $x' \in \langle 0, 1 \rangle$ is a coefficient expressing the urgency of the reaction. If the queue tends to grow and the limit M_i is already exceeded, then the weight is intensified by the addend. $G_i(t) = Q_i \cdot (l_i(t) - M_i) + Q_i$. Formulas for weights of the queue tending to fall are constructed similarly. The overview of formulas for computation $G_i(t)$ and suggested strategies (do nothing, accelerate z_i, accelerate preceding (z_{i-1}), decelerate) for different combinations of the values $l_i(t)$ and $d_i(t)$ are shown in the Table 1.

Table 1. Weights and strategies

	$l_i(t) = 0$	$0 < l_i(t) < M_i$	$l_i(t) > M_i$
$d_i(t) > 0$	$Q_i \cdot x'$	$Q_i \cdot x''$	$Q_i \cdot (l_i(t) - M_i) + Q_i$
	do nothing	accelerate	accelerate
$d_i(t) < 0$	$I_i \cdot y'$	$I_i \cdot y''$	$Q_i \cdot (l_i(t) - M_i)$
	acc. preceding	decelerate	accelerate
$d_i(t) = 0$	$I_i \cdot z'$	0	$Q_i \cdot (l_i(t) - M_i)$
	acc. preceding	do nothing	accelerate

Remark 2. Values of coefficients in the Table 1 should fulfill conditions $x' < x''$ and $y' > y''$ (according to the urgency of the situations).

With the knowledge of particular weights $G_i(t)$ it is possible to decide which of the situations is the most urgent - it is the situation with the highest value of weight, and according to this comparison, the suggested strategy can be implemented.

Example 1. The work of the evaluation method is illustrated in the following figures. The system with the continuous service times and requests flow is considered. During the computation of particular and total costs we can omit the work costs, because these costs are proportionally dependent on the number of servers at the service place and for this type of the system there is exactly one server at each service place. Initial parameters are set to $z_i(0) = \{3, 5, 4, 3, 6, 6, 7, 6, 4, 4\}$, $l_i(0) = \{2, 5, 6, 7, 6, 6, 8, 7, 6, 6\}$, $M_i = 3$, $Q_i = I_i = 10$.

To show the method in action, Fig. 2 depicts the 38th step of the optimization and values of computed parameters. One of the queues tends to grow (the queue in front of the fifth server), some of them tend to fall and some remain stable (the queues in front of the servers $6, 8, 10$). We can observe, that the greatest weight is $G_5(38) = 45, 5$. It means that the service time at the fifth service place will be accelerated in the next turn because the maximal tolerable limit of the queue length ($M_5 = 3$) have been already exceeded ($l_5(38) = 6, 55$) and queue tends to grow.

38	$z_i(38)$	$l_i(38)$	queue tendency	Idle costs	Queuing costs	Total costs for turn	$G_i(38)$	Strategy	Total	
1	3	2,							Total cumulative costs	10 893,5
2	2	5,65	-0,1667	0,	26,5	26,5	26,5	accelerate		
3	1	4,6	-0,5	0,	16,	16,	16	accelerate	Total for turn	127,6667
4	0,5	5,25	-1,	0,	22,5	22,5	22,5	accelerate		
5	2	6,55	1,5	0,	35,5	35,5	45,5	accelerate		
6	2	3,	0,	0,	0,	0,	0	do nothing		
7	1	3,6333	-0,5	0,	6,3333	6,3333	6,3333	accelerate		
8	1	2,6667	0,	0,	0,	0,	0	do nothing		
9	0,5	2,9	-1,	0,	0,	0,	7,5	deccelerate		
10	0,5	5,0833	0,	0,	20,8333	20,8333	20,8333	accelerate		

Fig. 2. 38. Stage of the optimization

Figure 3 depicts the evolution of costs for turn and the total costs (total costs are connected to the secondary axis in the graph) in comparison to the evolution of the system costs of unoptimized system. At the start, the costs are nearly the same, and then the optimization takes effect. It can be seen, that after 67 turns the system reaches the state where the queues are of acceptable length (not too long and/or not empty) and thus do not create any undesirable costs.

5.2   Method Based on the Simulation of Future States

Second method is based on the simulation of the system future development for several possible settings and on the computation of the function of average production costs for each of these simulations. Unlike the previous method, the suggestion to change the system setting is not given every time unit (for some types of systems it is undesirable to change the system setting continuously), thus the length of the stage is $T \geq 1$. The intention is to let the system perform

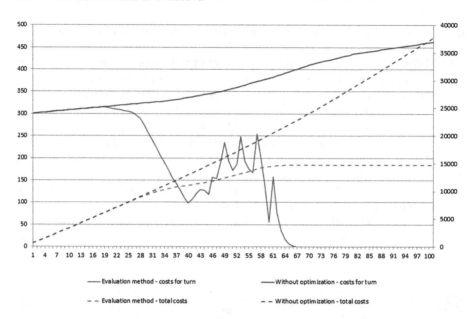

Fig. 3. The evolution of costs

for some time period and influence it only in time when necessary - not too soon but also not too late - therefore also $C_j \geq 0$. For computational simplicity we assume the change at exactly one service place at the same time.

Assuming that the stage will end in $T = t$, the average production costs during the stage are computed according to the following formula:

$$E(T, j, \delta) = \frac{P(T, j)}{T}. \tag{2}$$

Parameter j indicates the service place where the change in the system setting is made, δ expresses the intensity of this modification (for example, $j = 5$ and $\delta = 3$ means, that the change is considered at fifth service place and in dependence on the type of service times described above, either the service time of the server is decelerated by $+3$ time units or the number of servers is increased by $+3$ units).

At the beginning of each stage the rth set of functions $E(T, j, \delta)^{(r)} = \frac{P(T, j)}{T}$ is computed (in other words, index r indicates the number of stage). This set represents the evolution of the system average production costs for all possible intended settings. Optimal setting is the one that corresponds to the function of the set which contains the global minimum of the rth set. This minimum is very important, because the time when this function reaches its minimum is the time convenient for next change of the system setting (because from this moment the average costs are increasing) - and the question What will be the change like? will be answered by choosing the right function from next, $(r+1)$th, set of functions.

Example 2. The method is again illustrated on the example of the system with continuous both, the service times and request flow. Initial parameters are set to $\sigma_i(0) = \{3, 5, 4, 3, 6, 6, 7, 6, 4, 4\}$, $l_i(0) = \{2, 5, 6, 7, 6, 6, 8, 7, 6, 6\}$, $M_i = 3$, $I_i = Q_i = 10$, $C_i = 30$. For simplicity, the changes in service times are considered to be unit ($\delta = \pm 1$).

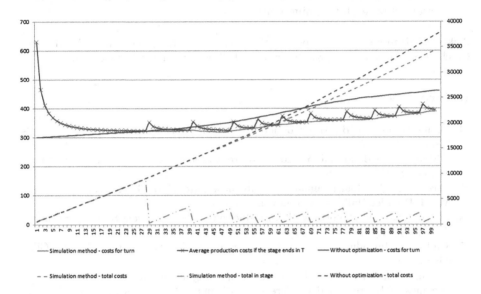

Fig. 4. The evolution of costs

The time evolution of costs for 100 turns are depicted in Fig. 4. The graph shows costs for turn and total costs in comparison to the evolution of costs of the system without optimization (total costs are connected to secondary axis in the graph).

The graph also shows the function of average production costs if the stage ends in T (purple one with marks). This function consists of parts of particular curves related to the chosen settings during the optimization process. During the first stage the function reaches its minimum after 28 turns. For the time $t = 29$, the next set of functions was computed and one of the settings was chosen. The new stage began. The part of the curve related to chosen setting corresponds to the part of the purple curve for $29 < t < 39$. Again, for time $t = 40$ the next set of functions was computed.

6   Conclusion

Two methods for cost optimization of the tandem queuing network are suggested. Each method is applicable for different system type. These two methods should not be mutually compared due to different frequency in changing the system setting and penalization in the form of change costs, that are completely omitted in case of the first method. It can be seen, that in case of the Method based on the evaluation, the reactions on the issues that can arise is really very quick and pointed. The optimization with use of the Method based on the simulation is suitable especially for systems with long run, where frequent changes in setting are unwanted (the system setting is influenced only in time, when the average production costs reach the minimum).

Acknowledgments. The support of the Specific research project at FIM UHK and Czech Science Foundation project Nr. 14-02424S is gratefully acknowledged.

References

1. Baum, D., Lothar, B.: An Introduction to Queueing Theory. Springer, Netherlands (2005)
2. Thiran, P., Le Boudec, J.-Y. (eds.): Network Calculus. LNCS, vol. 2050. Springer, Heidelberg (2001)
3. Buzacott, J.A., Shanthikumar, J.G.: Stochastic Models of Manufacturing Systems, vol. 4. Prentice Hall, Englewood Cliffs (1993)
4. Cassandras, C.G., Lafortune, S.: Introduction to Discrete Event Systems. Springer, New York (2008)
5. Cinlar, E.: Introduction to Stochastic Processes. Prentice Hall, Englewood Cliffs (1975)
6. Gavalec, M., Němcová, Z.: Deterministic queuing models in a service line. In: Proceedings of International Conference Mathematical Methods in Economics, pp. 227–232 (2014)
7. Lawrence, J.A., Pasternack, B.A.: Applied Management Science. John Wiley & Sons, New York (1997)
8. Papadopoulos, H.T., Heavey, C.: Queueing theory in manufacturing systems analysis and design: a classification of models for production and transfer lines. Eur. J. Oper. Res. **92**(1), 1–27 (1996)
9. Tempelmeier, H.: Practical considerations in the optimization of flow production systems. Int. J. Prod. Res. **41**(1), 149–170 (2003)
10. Zeigler, B.P., Praehofer, H., Kim, T.G.: Theory of Modeling and Simulation: Integrating Discrete Event and Continuous Complex Dynamic Systems. Academic Press, San Diego (2000)

User Authentication Through Keystroke Dynamics as the Protection Against Keylogger Attacks

Adrianna Kozierkiewicz-Hetmańska[(✉)], Aleksander Marciniak, and Marcin Pietranik

Department of Information Systems, Wroclaw University of Science and Technology, Wrocław, Poland
{adrianna.kozierkiewicz,marcin.pietranik}@pwr.edu.pl, 185871@student.pwr.edu.pl

Abstract. This paper addresses an authentication's scheme based on behavioural biometric method which is users' keystroke style. During an initial interaction with some system a proposed method identifies user's typing pattern and eventually creates a search template that will be further used during the authentication. It is based on an encryption of random characters into a typed password by injecting a set of emulated keystrokes when the actual typing occurs. In a decoding phase, the algorithm searches for characters for which the user's typing time is the most suitable within the assigned typing template. The article contains an overview of the developed method along with an analysis of its usability and an experimental evaluation based on assumed criteria of the false acceptance rate (FAR), the false rejection rate (FRR) and the equal error rate (EER). The obtained results have been compared with the existing method.

1 Introduction

In the modern world the rapid growth of the everyday technology led to the greater interest in information systems and their continuous development. Many of these systems allows a user to have his own account, e.g. bank accounts, email, social networking profiles etc. This variety of possibilities, where the user needs to be distinguished among other users, entails the problem of an authentication, which consists of the assignment of unique credentials to every person. The most commonly used credentials are a pair of a unique username and a password.

In case of the theft of this data, an adversary is able to gain access to the system resources of its prey. Depending on the type of system, the thief can view and send e-mails, order bank transactions, modify personal information and so on. The login details can be obtained by means of, for example, *social engineering*, capturing typed data (*key-loggers*), attacks called "*a man in the middle*", an observation by the arm (*shoulder surfing*) in a coffee shop etc.

A remedy for such threat is applying in the authentication process not only raw, textual credentials but also some kind of physical verification, for example, the RFID card. The downside of this approach is a necessity of carrying this physical device

© Springer International Publishing Switzerland 2016
N.T. Nguyen et al. (Eds.): ICCCI 2016, Part I, LNAI 9875, pp. 345–355, 2016.
DOI: 10.1007/978-3-319-45243-2_32

every time the user wants to log into the system. An alternative solution is incorporating biometrical data. These methods give "*something that a user is*", therefore the system that requires users' authentication is provided with data describing personal properties of a certain person. These properties may involve for example a pattern of user's blood vessels, his fingerprint pattern, some geometrical characteristics of his hand etc.

Biometric techniques can be divided into two categories: *physical* (e.g. facial recognition, analysis of blood vessels, fingerprints, retinal scanners) and *behavioural* (analysis method of typing on a keyboard, voice analysis etc.). The latter are mainly used to authenticate users, and the former are used for both verification and identification.

Despite obvious advantages these type of authentication methods imply two problems, first of which is a requirement of using a biometrical sensor connected to the computer which is indented to be used during the actual authentication. The second problem concerns the fact that aforementioned sensors may introduce potential distortions during the process. The same person can have a different eye colour depending on the quality of a scanning device, its calibration or the lighting in a room. Therefore, the authentication system needs to accepts some level of acceptable errors. In other words – the authentication or the identification of users needs to depend on the degree of similarity between the data stored in the system and those obtained from the sensor, e.g. data must be similar in at least 80 % level of resemblance.

It is obvious that the described difficulties are necessary to overcome. One of the most frequently used input devices are keyboards and it is obviously simple to notice that every person has a unique typing style, so why not to base the authentication method on this remark? The big advantage of behavioural techniques based on typing style is the possibility of its implementation based solely on a software layer. Therefore, it can be massively used on any kind of device (standalone computer, mobile phones, tablets etc.).

The downside of this technique is the fact that, the manner and the speed of typing on a keyboard changes with time, so after a certain period, the user can start having problems accessing the system. Therefore, the type-analysis based authentication systems should be able to modify and evolve stored patterns that are used to authenticate their users.

The main contribution of the following article is a new method for authenticating users based on an analysis of the dynamics with which they type on their keyboards. This approach makes it difficult to reveal the user's password by generating additional random characters during time gaps between single keystrokes and injecting them during the typing. Therefore, the actual password is masked and an adversary is left with meaningless set of characters. A new method has been verified and tested using standard low rate of errors sustainable (EER) method and the obtained results show that it can increase the safety of users' credentials. The main goal of our work is to propose an easy method that ensures the high level of safety and protect against keylogger attacks.

The paper is organised as follows. Section 2 contains a review of relevant publication about the topic of interest. Section 3 is a broad and detailed description of our method. Section 4 contains an overview of a preliminary verification procedure along with obtained results and their analysis. Section 5 concludes the paper and sheds some light on an upcoming research directions.

2 Related Works

Nowadays, protecting users' data only with simple credentials such as a login and a password is in sufficient due to rapidly increasing sensitivity of information stored online [10]. Biometric based methods of users' authentication and further authorization is a frequently research area of knowledge. Due to the general availability of devices (e.g. mobile phones) that can provide required data that characterize certain users it can be broadly adapted and transparently incorporated in a variety of application. The most common input device is constantly an ordinary keyboard (despite its diversity in terms of application in mobile devices, dedicated terminals, laptops etc.). We have noticed that every user types with slightly different style (e.g. using all five fingers, only two, constantly staring at the keyboard etc.). Following a remark that the user is not what he types but how he types [9], we have asked ourselves a question why not to create an authentication framework that would incorporate this kind of characteristics?

Our straight inspiration was a patent [14] in which authors present a method for training a computing system based on a personal style of using a keyboard. The method is built on top of storing the key press times and the flight times between single characters to unequivocally characterize a user and his typing styles. A survey of related methods can be found in [2] where authors have focused on free-text keystroke systems and their ability to provide a non-intrusive continual verification of users' identities during their interaction with some protected system. The main advantage of the proposed solution is its ability to constantly monitor and verify authenticated uses, therefore providing a valuable balance between security and usability. In [8] recent advances in a similar solution have been presented. Authors idea is based on so-called "differences of feature difference vectors". The classification algorithm is more user-cantered than previously developed and has been experimentally verified, proving its usefulness by halving the previously reported best equal error rate from literature.

On the other hand, in [4] authors have prepared an extended survey on a background of the psychological basis behind the use of keystroke dynamics. Authors claim that typing behaviour is difficult to impersonate and incorporating a typing cadence of some user can cheaply increase a security of the systems. Therefore, in [3] an overview of keystroke dynamics characteristics can be found. The defined attributes contain such parameters as pressing time, releasing time, latency, dwell time, flight time etc. In the proposed approach users are identified with four unique characters and every sample is additionally timestamped. Eventually, in a database a system stores not only credentials, but also calculated values which are further used during a matching process where a strict comparison of collected reference and actual typing patterns are confronted.

Authors of [15] covered issues concerning calculation complexity and performance overhead of such methods. Furthermore, they have proposed a new biometric authentication system which is based on k-nearest-neighbour classification method that was additionally verified using Receiver Operating Characteristic (ROC) curves. Eventually the presented methodology acquired the equal error rate value of about 1 %.

The issue concerning recognition of individual typing pattern has been also touched in [5] where an application specific individual key stroke pattern profile is determined for certain users. Such profile contains trained averages calculated for a series of typing features that have been identified and broadly described. Despite the fact that this framework has been used for stress detection the general idea is adaptable to a variety of other applications.

De Magalhães, Revett and Santos in [6] have described and developed "a light-weight algorithm", which they co-created. It is characterized by simple implementation through the use of basic foundations: arithmetic mean, median and standard deviations of the times introduced between successive characters typed during entering the password by the user.

In [12, 13] a presentation of a keystroke based authentication framework aided by a handwriting-based authentication focused strictly on mobile devices. In [1] a method of an authentication framework powered by neural-networks is presented. Author has developed an approach that collects key codes and inter key times for passwords during a registration phase and consecutive login steps. Eventually, this data is used to generate RGB histograms and train neural networks in order to provide a two-stage authentication based on both credential evaluation and a typing style with which the aforementioned credentials were typed.

Despite high level of interest in the recent literature, we claim that none of the found solution cover a situation in which a user's computer is compromised by some kind of a keylogger. In such case, even the most sophisticated approach (e.g. presented in [1] or [2]) will not prevent stealing sensitive data. It is commonly known that ordinary users frequently reuse the same password several times in different systems, therefore a successful attack entails an adversary gaining access to several systems at once (due to the fact that keyloggers may be used to only to register users' password but their whole online activity). In our framework, beside the authentication based on keystroke dynamics, users' passwords are additionally blurred by injecting random characters in between the gaps of single keystrokes. Therefore, our system can be treated as a double protection of sensitive data from even complex attack such as snoop-forge-replay attack presented in [11]. Moreover, it does not enforce only biometrical authentication, which increase the transparency in deployment of the proposed framework in the protected system.

3   The Authentication Method Based on Behavioural Biometrics

Many methods known from the literature are limited to user's authentication or identification based only on the style of typing on the keyboard. However, such solutions do not protect against adversary attacks that uses key-loggers.

In this paper we present the method which is free from the mentioned disadvantage [7]. The whole procedure contains following stages: user's registration, login phase which consist of: identifying user's typing pattern, determination of time gaps between keystrokes, filling the gaps by injecting an additional random character and the eventual decryption phase. The general idea of user's authentication through keystroke dynamics is presented in Fig. 1.

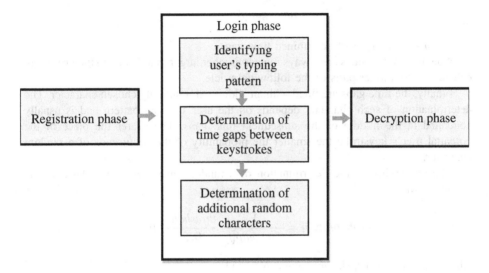

Fig. 1. The general idea of user authentication through keystroke dynamics

3.1 Registration Phase

During the initial interaction with some system a proposed method identifies and collects biometric data. A user is asked to provide a login and to type his password k times in a row. A time between the single keystroke is stored as a two dimensional matrix:

$$registration\_data[i,j] \quad \begin{matrix} i \in [0, k-1] \\ j \in [0, n-1] \end{matrix} \tag{1}$$

where: n - length of the password, k - the number of registration data.

3.2 Login Phase

Initially, based on provided registration data, the system identifies the user's typing pattern as follows:

$$template[i] = \frac{\sum_{i=0}^{k-1} registration\_data[i,j]}{k} \qquad (2)$$

where: $i \in [0, n-1]$.

The next step depends on the determination of time gaps between subsequent keystrokes, which are created based on the user's typing pattern. The largest delay should occur nearer the place of the potential user's keystroke:

$$time\_gap[j]_{j\in[0,m-1]} \qquad (3)$$

where: m - the number of determined time gaps.

Note that these values are always integer number larger than 1, due to the procedure enlisted in the further parts of the following article.

Finally, the time gaps are filled with an additional, randomly chosen character. The determination of such character depends on the user's typing pattern and is usually generated in the middle of the subsequent time gaps. The nearer the place of the potential user's keystroke the smaller the probability of the occurrence of a random character.

The probability of the determination of a random character is calculated in the following way:

$$probability[j] = \frac{1}{time\_gap[j]} \cdot \frac{threshold}{0,5} \cdot 100\,\% \qquad (4)$$

where: $threshold \in (0,1]$, $time\_gap \in N^+\backslash\{1\}$.

The accepted threshold influences the algorithm's sensitivity. If the threshold is close to 0 then no additional character is determined. The higher threshold affects the higher number of random characters, which can be desired in case of short time gaps. After this stage the password is encrypted as a sequence of characters given by users and determined by the method in random way.

3.3   Decryption Phase

Decryption of the password is done based on a stored template. It is needed to distinguish the characters provided by the user from characters determined by the algorithm. The first character of the encrypted password is the first character of decrypted password. The subsequent characters are discovered during the following procedure.

The method seeks for the characters which are near the border of time gaps because in this place the characters from user's password should occur. In other words, the algorithm searches for characters for which the user's typing time is the most suitable within the assigned typing template.

The procedure of user's authentication through keystroke's dynamics consists of the following steps:

The user authentication through dynamic keystroke scheme

1. Registration phase

  a. Collecting biometric data:
     registration_data[i][j]$_{j\in[0,k-1],\ j\in[0,n-1]}$

2. Login phase

  a. For each i∈[0,n-1] create user's typing pattern:
     template[i]=($\sum_{j\in[1,k]}$registration_data[i][j])/k

  b. Determine time gaps between keystrokes time_gap[j]$_{j\in[0,m]}$:
     temporary_gap=[]
     for each i∈[0,n-1]{
       interval=template[i], limit=interval/2, break=0
       do{
         break=int(interval/limit)*2
         add break to temporary_gap
         limit=limit-break
       }while(limit-break>0)
       if break>0
        add break to temporary_gap
       break_list=[j]$_{j\in MAX(1,\ ((temporary\_gap.length-1)*2)+1)}$

       index=int(break_list.length/2)
       for each delta∈[0, temporary_gap.length-1]{
        break_list[index+delta]=temporary_gap[delta]
        break_list[index-delta]=temporary_gap[delta]
       }
       add break_list to time_gap
     }

  c. User types the sequence of characters as the desired
    password

  d. For each time_gap[j], j∈[0,m] determine an additional
    character according the probability:
    probability[j]=(2*threshold/time_gap[j])*100%

  e. Create the vector times[j]$_{j\in[0,m]}$ to store times of putting
    passwords' characters and randomly chosen characters

  f. Create the vector encrypt_pass[j]$_{j\in[0,h1]}$ to store an
    encrypted password as the sequence of characters provided
    by the user and by the algorithm

3. Decryption phase

  a. encrypt_pass[0] is the first character of the decrypted
    password

  b. Decryption of the subsequent characters of the password:
    decrypted_password=''
    for each i∈[0,n-1]{
      sum=0
      do{
        sum=+times[j]
        j++
      }while(sum<template[i])
      if (|template[i]-sum|<template[i]-sum-times[j-1]|{
        decrypted_password.append(encrypt_pass[j-1])
      } else {
        decrypted_password.append(encrypt_pass[j-2])
      }
    }

  c. Calculate cryptographic hash function for the decrypted
    password

  d. User's authentication

4   Results of Experiments

The method described in the previous section has been implemented in the web-based simulator. For this task Java Enterprise Edition was used along with Spring and Hibernate frameworks. The proposed method has been evaluated based on the assumed criteria in comparison with the existing procedure called a lightweight algorithm [6].

In our experiment 20 users created 43 different users' accounts. In the first step, for each user's account, the biometric data were collected. Users were asked to type twelve times their passwords. Next, those data were sent to the server, initially analyzed and processed. In the final step all login data were published (only a login and a password) on the public website and volunteers tried to authenticate in the system using this publicly known login data. Users were not forced to use the same class of passwords in terms of their security. As a result, we have compared typing of very complex and safe key-phrases with simple ones such as "qwerty". For the full description of verification procedure, that we have conducted, please refer to [7].

An experiment designed in a described way simulated the scenario during which an adversary has stolen user's login data and tried to use them for authentication. After each trial, the obtained results were processed by the method described in Sect. 3 and by the lightweight algorithm.

During the experiment, 273 trials of authentication for own accounts and 455 trials of authentication for other people accounts have been noted. Those data were used to calculate the false acceptance rate (FAR), the false rejection rate (FRR) and finally the equal error rate (EER). Figure 2 presents preliminary results obtained during the experiment.

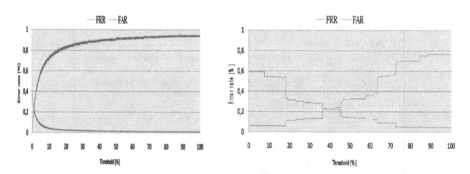

Fig. 2. FRR and FAR for the proposed method (on the left side) and for the lightweight algorithm (on the right side)

The equal error rate has been calculated for both methods and for the different number of registration data. In the initial phase of our experiment it was assumed, that the biometric data are determined based on typing a password repeatedly twelve times. Table 1 contains the results of both methods in case of the decreasing number of password typing repetition.

Table 1. EER for both methods and for the different number of registration data

The number of registration data k	EER for proposed method	EER for lightweight algorithm
12	22,71	22,34
11	21,69	16,18
10	22,73	17,45
9	23,04	16,79
8	22,53	19,05
7	22,5	18,15
6	20,17	20,82
5	22,38	19,77
4	19,82	19,92
3	21,31	21,51
2	23,03	9,84
1	25	x

The number of registration data has not significantly influenced the effectiveness of the proposed method. However, for the lightweight algorithm, it is not possible to determine EER in case of the single registration data.

For the default number of the password typing repetition equal 12, EER for the both methods are similar: 22,71 % for the lightweight algorithm and 22,34 % for the proposed method. From this results we can draw a conclusion that both methods ensure the safety of users' credentials on a similar level, but the lightweight algorithm cannot prevent the user from potential attacks done by keyloggers.

5   Conclusions and Future Work

The authentication's scheme based on a behavioural biometric method which is users' keystroke styles were proposed. It consists of the registration phase where biometric data are collected, identifying user's typing pattern, determination of time gaps between keystrokes and password encryption by adding a set of random characters. The authentication itself is done during the decoding phase where the algorithm searches for characters for which the user's typing time is the most suitable within the assigned typing template.

This method has been compared with a lightweight algorithm that is well known in the literature. The preliminary experimental results demonstrated that their effectiveness is similar for the equal error rate criterion which implicates that both of them protect users' accounts from an unauthorized access to the similar degree.

However, the proposed schema has an advantage over the lightweight algorithm. In our method some additional characters are added to the entered set of characters which entails the protection against an adversary attack using the keyloggers. Additionally, the described solution is easy to apply as the protection for the web application users.

It can also be implemented as the computer software completely transparent for other systems and applications.

In the nearest future we would like to focus on the deeper analysis of the proposed method. Due to the limitation of this paper the statistical analysis of the obtained results has not been presented and we claim that the analysis of the complexity of provided passwords should be done as well using, for example, data provided and described in [16]. Additionally, we would like to examine methods of tracking of biometrical data evolution.

References

1. Alpar, O.: Keystroke recognition in user authentication using ANN based RGB histogram technique. Eng. Appl. Artif. Intell. **32**, 213–217 (2014)
2. Alsultan, A., Warwick, K.: Keystroke dynamics authentication: a survey of free-text methods. Int. J. Comput. Sci. Issues **10**(4), 1–10 (2013)
3. Bajaj, S., Kaur, S.: Typing speed analysis of human for password protection (based on keystrokes dynamics). Int. J. Innovative Technol. Exploring Eng. (IJITEE) **3**(2), 88–91 (2013)
4. Banerjee, S.P., Woodard, D.L.: Biometric authentication and identification using keystroke dynamics: a survey. J. Pattern Recogn. Res. **7**(1), 116–139 (2012)
5. Gunawardhane, S.D.W., De Silva, P.M., Kulathunga, D.S.B., Arunatileka, S.M.K.D.: Non-invasive human stress detection using key stroke dynamics and pattern variations. In: 2013 International Conference on Advances in ICT for Emerging Regions (ICTer), Colombo, pp. 240–247 (2013). doi:10.1109/ICTer.2013.6761185
6. De Magalhães, P., dos Santos, H.: An improved statistical keystroke dynamics algorithm. In: Proceedings of the IADIS MCCSIS 2005 (2005)
7. Marciniak, A.: Uwierzytelnianie użytkowników oparte o analizę dynamiki pisania na klawiaturze. Master thesis (2016, in Polish)
8. Monaco, J.V., Bakelman, N., Cha, S.H., Tappert, C.C.: Recent advances in the development of a long-text-input keystroke biometric authentication system for arbitrary text input. In: Intelligence and Security Informatics Conference (EISIC) 2013, pp. 60–66 (2013). doi:10. 1109/EISIC.2013.16
9. Monrose, F., Rubin, A.D.: Keystroke dynamics as a biometric for authentication. Future Gener. Comput. Syst. **16**(4), 351–359 (2000)
10. Nag, A.K., Roy, A., Dasgupta, D.: An adaptive approach towards the selection of multi-factor authentication. In: IEEE Symposium Series on Computational Intelligence, pp. 463–472 (2015). doi:10.1109/SSCI.2015.75
11. Rahman, K.A., Balagani, K.S., Phoha, V.V.: Snoop-forge-replay attacks on continuous verification with keystrokes. IEEE Trans. Inf. Forensics Secur. **8**(3), 528–541 (2013). doi:10. 1109/TIFS.2013.2244091
12. Trojahn, M., Ortmeier, F.: Toward mobile authentication with keystroke dynamics on mobile phones and tablets. In: 2013 27th International Conference on Advanced Information Networking and Applications Workshops (WAINA), pp. 697–702 (2013). doi:10.1109/ WAINA.2013.36
13. Trojahn, M., Arndt, F., Ortmeier, F.: Authentication with time features for keystroke dynamics on touchscreens. In: De Decker, B., Dittmann, J., Kraetzer, C., Vielhauer, C. (eds.) CMS 2013. LNCS, vol. 8099, pp. 197–199. Springer, Heidelberg (2013). doi:10.1007/978-3-642-40779-6_17

14. Wu, T., Guo, J., Rice, L.: Method and system for biometric keyboard. U.S. Patent No. 8,134,449 (2012)
15. Zack, R.S., Tappert, C.C., Cha, S.-H.: Performance of a long-text-input keystroke biometric authentication system using an improved k-nearest-neighbor classification method. In: Proceedings of 2010 4th IEEE International Conference Theory Applications and Systems (BTAS), pp. 1–6 (2010). doi:10.1109/BTAS.2010.5634492
16. Vural, E., Huang, J., Hou, D., Schuckers, S.: Shared research dataset to support development of keystroke authentication (2014). doi:10.1109/BTAS.2014.6996259

Data Warehouses Federation as a Single Data Warehouse

Rafał Kern[(✉)]

Institute of Informatics, Wroclaw University of Technology,
Wybrzeze Wyspianskiego 27, 50-370 Wroclaw, Poland
rafal.kern@pwr.edu.pl

Abstract. In this paper author presents an experiment, which shows that it is possible to form a federation of data warehouses that may simulate effectively one, "super" data warehouse. There is no need to create complete ETL tool to load data from source data warehouses into one, dedicated data warehouse. Good relations between global schema and local schemas extracted during schema integration are indispensable to create an effective federation.

Keywords: Data warehouses · Federation · Data integration

1 Introduction

Having a single, trusted source of truth is a crucial requirement of many analytical tools. Traditional, relational databases turned out to be insufficient tool for analytical purposes. Therefore, data warehouses were introduced. The "data warehouse" term was described in the literature in many ways. In [1] was described as *"sophisticated, highly specialized database systems optimized for decision support rather than transaction support"*. In most cases the purpose of data wareouse implementation is to gather and process large amount of data in order to fulfill some needs, which may be divided into the following categories: On-Line Analytical Processing, Decision Support System, Executive Information System and Business Intelligence.

Data warehouse should provide comprehensive data set from the domain, understood as an area of activity of an organization.

What if the organization grows? What happens, when one bank buys another one. Each of them have existing clients, business processes and brand. The board is interested in having single source of the truth as soon as it is possible. But integration of such kind of systems is very difficult and expensive. Hardly ever data from one system may be easily migrated to another one. The users got used to one brand. They would be suspicious when suddenly a new log-in window occurred. It takes a lot of time to implement one generic system which provides data from two old ones. And what if before the migration ends another system/data source should be migrated? Those factors show clearly that dedicated solution might not be a good one. As an alternative some distributed approach

© Springer International Publishing Switzerland 2016
N.T. Nguyen et al. (Eds.): ICCCI 2016, Part I, LNAI 9875, pp. 356–366, 2016.
DOI: 10.1007/978-3-319-45243-2_33

should be considered. Organizations implement federation of data warehouses in order to increase analytical abilities by broadening the spectrum of already gathered data. The architecture allows to use the existing infrastructure. The effectiveness of federation should be almost as good as a dedicated data warehouse, but much less expensive. This paper shows that it is possible to create a federation with an effectiveness close to effectiveness of a single, dedicated data warehouse.

The paper is organized as follows: Sect. 2 contains analysis of related works, Sect. 3 contains necessary definitions and notions. In Sect. 4 an overview and some results details are described. Summary and some propositions for future works are presented in Sect. 5.

2   Related Works

This paper is a continuation of a topic described in papers [5–7]. The procedures of schema integration, query decomposition and responses integration were implemented and used for the experiment mentioned in introduction. Comparing to previous works experiments' results, additional metrics propositions and distance functions were added. The integration procedure needs access to the context of data in each component. This information are collected during schema integration and ETL processes [4].

A good model of *DIS (Data Integration System)* was given in [8]. It seems quite useful for data integration in federations and may be enriched by some semantic data (like in [1]). As it was mentioned in [2], the heterogeneity may occur on schema or instance levels. Operating on incomplete, inconsistent, corrupted or outdated data may make the problem even deeper. The last example luckily does not exists in field of data warehouses. No data is out of date. It only need a properly defined time dimension. Moreover, in most cases the data warehouse operates on numerical data which significantly narrows the scope of possible conflicts situations. Conflicted data may be moved into common canonical model [3]. A good review od inconsistency solving methods was presented in [10].

It is extremely hard to propose one, complex data integration procedure because different data types must be handled in different way. It is much easier to create a set of integration methods. The quality of each data set can be treated as its representativeness. High data quality is necessary to provide high level of services provided by federation. In most cases the reliability is based on different distance metrics, should take into account the user-defined wages [9] and may deal with incomplete data [11].

Most solutions from literature basis on results of schema integration results. During this process most inconsistencies are solved. The global schema is defined as the largest common schema. Therefore, each element of this common schema has it representative in each component schema. A different assumption was made in this paper. The global federation schema consists of elements that have their representatives only in some of the components. This creates an opportunity to create much more complex queries in the future. Moreover, it allows to make only some adjustment to the federation schema when the set of components changes. In other cases the federation schema must be rebuild again.

Another interesting fact emerged: when component data warehouses store large number of data it is very expensive to send detailed results as responses to queries. In order to optimize its performance some aggregation methods should be applied on the data warehouse level. When data warehouses of various sizes sends their responses to federation layer, there is a risk that the integration result will be interfered by responses of many small data warehouses. Therefore, some factors should be applied to each data warehouse, in order to balance the final federation response.

3   Basic Notions

In this section some basic definition are given:

Definition 1. Data warehouse federation *Data warehouse federation is defined as a tuple:*

$$\hat{F} = (F, H, U, q, l) \tag{1}$$

where:

F - *federation schema*
H - *set of data warehouses*
U - *user/external interface*
q - *query decomposition procedure*
l - *responses integration procedure*

The end-user interface U contains defined query language L_Q.

Definition 2. Data warehouse federation schema *Data warehouse federation schema is defined as a tuple:*

$$F = \{D_0, D_1, ..., D_n\} \tag{2}$$

where:

D_0 - *facts table schema*
D_j - *dimensions tables schemas for $j \in [1, n]$. Between dimension table D_j and facts table there occurs a relationship of type $1 - \infty$ which means, that one row from dimension table may be associated with more than one row from facts table, but one facts table is associated with exactly one row from given dimension table.*

Each attribute from federation facts table or dimension table must have its representative in at least one source data warehouse.

As it was mentioned above, the final federation response may be balanced by some factors. It is believed that the data warehouses metrics might be used. Some examples are listed below:

- **Power** - ratio between number of facts in given data warehouse and the whole federation.

$$p = \frac{Card(H_i)}{Card(F)} \tag{3}$$

where:

$p \in\, <0,1>\, Card(H_i)$ - facts number in data warehouse H_i

$Card(F)$ - facts number in federation
- **Coverage** - the coverage of federation schema by schema of a given data warehouse.

$$c = \frac{AC(H_i) + MC(H_i)}{AC(F) + MC(F)} \tag{4}$$

where:

$AC(H_i)$ - number of attributes in dimensions of data warehouse H_i

$MC(H_i)$ - number of measures in H_i

$AC(F)$ - dimensions attributes number in federation

$MC(F)$ - measures number in federation
- **Dimension significance** - ratio between unique values from dimension and its relative from federation.

$$s_D = \frac{Card(D_j^i)}{Card(D_j)} \tag{5}$$

where:

$Card(D_j^i)$ - number of rows in data warehouse's dimension D_j^i

$Card(D_j)$ - number of rows in federation's dimension related to D_j^i
- **Diversity** - number of different combination of rows from all data warehouse's dimensions.

$$d = Card(D_1^i) * Card(D_2^i) * ... * Card(D_{\alpha i}^i) \tag{6}$$

where:

$Card(D_1^i)$ - number of rows in dimension D_1^i

$Card(D_{\alpha i}^i)$ - number of rows in dimension $D_{\alpha i}{}^i$

αi -number of dimensions of data warehouse D^i
- **Complexity** - ratio between *Power* and *Diversity*.

During experiment described below the *Power* metric was used and it gave promising results. Usefulness of the other metrics need to be verified with real data warehouses.

4   Experiment

4.1   Overview

This experiments shows, that it is possible to create a federation of data warehouses that simulates one "super" data warehouse which contains all data stored in mentioned data warehouses. Moreover, effectiveness of this federation is slightly worse (max. 5 %) than effectiveness of this "super" data warehouse. The effectiveness is understood as relative difference between the ideal value obtained from pattern data warehouse and the federation (see Eq. 7). The time needed to deliver federation solution is much shorter than mentioned, one data warehouse [4, 12].

There was a lot of difficulties with obtaining very large set of real-world data which is quite similar (from semantic point of view). This condition is very important because when the structures of source data warehouses are not similar enough, the schemas integration step will be very simple. That is why the next steps would be trivial. These factors decided, that in this experiment randomly generated test data was used.

Small standard deviation of dimensions and attributes distribution is caused by necessity of proper and valuable query decomposition. If it was large, the dimensions mappings list would be very short. So only few source data warehouses would be queried by sub-queries generated during decomposition. Therefore, the responses integration also would be a trivial one. On the other hand, small differences between dimensions numbers is caused by latest trends in data warehouses. New data warehouses have less dimensions but the facts tables are more complex. Large number of dimensions caused multiple joining operations.

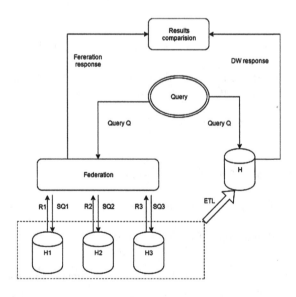

Fig. 1. Experiments overview

As a result, the query processing is longer, but a lot of disk space could be spared. Todays, when the disk space is getting cheaper, the data warehouses designers are more focused on query processing time, than the disk usage.

Figure 1 presents the general idea of the experiment.

The experiment consists of following steps:

1. Generate source data warehouses according to following parameters:
   (a) Number of source data warehouses: 10
   (b) Number of dimensions: N(6,3) - normal distribution, mean = 6, standard deviation = 3.
   (c) Number of dimensions attributes: N(6,2)
   (d) Number of measures: N(5,2)
2. Generate data warehouses instances and fill them with data according to parameters:
   (a) Number of rows in dimension tables: N(7500,3000)
   (b) Number of rows in facts tables: N(75000,30000)
   (c) Attributes values: N(400,100)
   (d) Measures values: N(X, 100), where X is randomly generated value from (3000, 600000)
3. Integrate source data warehouses schemas and generate global schema
4. Generate pattern data warehouse on the federation schema basis
5. Load data from source data warehouses to pattern data warehouse
6. Send query to federation
7. Send query to pattern data warehouse
8. Compare results from federation and pattern data warehouse.

$$dist(R_p, R) = \frac{\sum_{i=1}^{n}(\frac{\sum_{j=1}^{s} diff(R_p^{ij} - R^{ij})}{s})}{n} \qquad (7)$$

where:

R_p - pattern warehouse response

R - federation response

n - rows number in pattern data warehouse response

R_p^{ij} - value of j element in i row pattern data warehouse response

R^{ij} - value of j element in i row federation response

s - number of columns in single row of response.

Each type of element has its own calculation method. In a single row each element (which may be associated with columns in schema) may be numeric value or simple characters chain. That is why in this experimental environment two different calculation method were implemented.

For numeric values:

$$diff(x,y) = \frac{x-y}{x} \qquad (8)$$

where:

x - ideal value

y - tested value

For text values:

$$diff(x,y) = \frac{c}{l} \qquad (9)$$

where:

x - ideal value

y - tested value

c - number of positions where words x nad y are different

l - number of characters of the longer word from x and y

The introduction of such differentiated distance functions was dictated by the need to make the best representation of real data distribution in data warehouses. If all records were treated as vectors of numeric coordinates it would be possible to introduce the cosine distance. Unfortunately, it would be too far-reaching simplification. For example: if we handle the '1172' and '1275' values only as numbers, then the distance between them would be relatively small (0.08). However, if it is an identity code or postal code, then treating it as a number has not much in common with reality. More intuitive would be verifying characters compliance on each position.

The dimensions, attributes and measures names are generated randomly, thus they are like: *"dimension0"*, *"dimension1"*, *"attribute1"* and so on. The threshold values used in queries are generated properly to values used for filling-up source data warehouses with data.

4.2   Results

In [7] several responses types were described. During this experiment only one of them - the 'scalar' was checked. The query below in a real world example may be interpreted as average value of annual sale for good in given price range and some other conditions specified in *WHERE* section.

```
SELECT AVG( facts . measure0 )
FROM fdw . facts
JOIN dimension0 ON dimension0 . id = facts . dimension0−id
JOIN dimension1 ON dimension1 . id = facts . dimension1−id
JOIN dimension2 ON dimension2 . id = facts . dimension2−id
JOIN dimension4 ON dimension4 . id = facts . dimension4−id
WHERE
dimension0 . attrib14 > 200
AND dimension1 . attrib8 > 320
AND dimension2 . attrib15 > 441
AND dimension4 . attrib11 > 600
```

The Fig. 2 shows, that the difference exceeded 0.1 only twice. In other 98 cases the difference fluctuates much lower than this threshold. The vertical axis does not represent the sequence of experiment results. They have been sorted ascending by values. The responses were balanced by the *Power* factor.

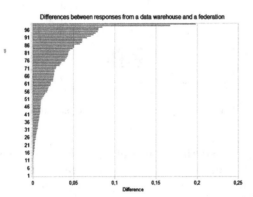

Fig. 2. Differences between federation and single data warehouse responses for response type 'Scalar' and aggregation function AVG

Collected results were verified with statistical tools. Following two hypothesis have been formulated:

- H_0 - the relative difference between the value obtained from the federation and the value from the single data warehouse is equal to 5 %.
- H_1 - the relative difference between the value obtained from the federation and the value from the single data warehouse is less than 5 %.

Initially, the data distribution was checked with Lilieforse test. It proved, that the results distribution is not a normal one. Figure 3 shows that clearly.

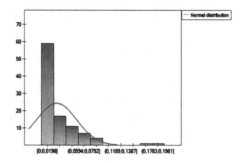

Fig. 3. Liliefors test for query type: scalar

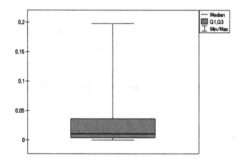

Fig. 4. Boxplot for query type: scalar

The vertical axis presents the number of results in given range. The horizontal axis represents the ranges. In this case the t-student test can not be used. Therefore Wilcoxon test was chosen to verify the hypothesis.

$$ST_{Wilcoxon} = \frac{\sum_{i=1}^{n} R_i}{\sqrt{\sum_{i=1}^{n} R_i^2}} \tag{10}$$

$$R_i = sgn(D_i) * rank|D_i| \tag{11}$$

In the definition above sgn the standard $signum$ function and $rank$ adds ranks to observed differences. Ranks are successive natural numbers assigned to ascending sorted differences. Thus, they may be treated like a measure how observed differences impact on the final results.

Wilcoxon test is a non-parametric test and the distribution type does not need to be known. On the other hand the experiment results (X_i and Y_i) should

Table 1. Wilcoxon tests summary

Significance level	0.05
Sample size	100
Group median	0.0105
Hypothetical median	0.05
Positive rank sum	506
Negative rank sum	4544
T-Statistic	506
p-value (exact)	0.0001

be comparable. On the Fig. 4 it is easy to notice that the media is placed very close to 0 and the third quartile (Q3) lies beneath hypothetical median. Because the probability that hypothetical median is equal to result is lower than the significance level was assumed, that null hypothesis H_0 may be rejected and the alternative hypothesis H_1 may be confirmed. Moreover, the sum of negative ranks is significantly greater than the sum of positive ranks. This confirms that the relative difference between responses of a data warehouse and a federation is smaller than 5 %. In the Table 1 more information about the test are given.

5   Future Works and Summary

This paper continues development of model described previously in [5–7]. It was implemented and checked in dedicated experimental environment. These experiments show, that it is possible to build a federation with effectiveness close to an effectiveness of single, dedicated, "super" data warehouse. The effectiveness difference is smaller than 5 %. Data warehouses are used mostly for high-level analysis, where the general tendency or rough estimation do not need exact values. Therefore, it is believed that this level of precision is good enough. Nevertheless, combination of different metrics and response types should increase the effectiveness as high as possible and that is the goal for further research. This model should be also checked using real business data instead of randomly generated.

References

1. Berger, S., Schrefl, M.: From federated databases to a federated data warehouse system. In: Proceedings of the 41st Annual Hawaii International Conference on System Sciences, HICSS 2008, pp. 394–404 (2008)
2. Dong, X.L., Berti-Equille, L., Srivastava, D.: Data fusion: resolving conflicts from multiple sources. In: Wang, J., Xiong, H., Ishikawa, Y., Xu, J., Zhou, J. (eds.) WAIM 2013. LNCS, vol. 7923, pp. 64–76. Springer, Heidelberg (2013)
3. Fan, W., Lu, H., Madnick, S.E., Cheung, D.: Discovering and reconciling value conflicts for numerical data integration. Inf. Syst. **26**(8), 635–656 (2001)

4. Jindal, R., Acharya, A.: Federated data warehouse architecture. White paper, Wipro Technologies (2003)
5. Kern, R., Ryk, K., Nguyen, N.T.: A framework for building logical schema and query decomposition in data warehouse federations. In: Jedrzejowicz, P., Nguyen, N.T., Hoang, K. (eds.) ICCCI 2011, Part I. LNCS, vol. 6922, pp. 612–622. Springer, Heidelberg (2011)
6. Kern, R., Stolarczyk, T., Nguyen, N.T.: A formal framework for query decomposition and knowledge integration in data warehouse federations. Expert Syst. Appl. **40**(7), 2592–2606 (2013)
7. Kern, R., Dobrowolski, G., Nguyen, N.T.: A method for response integration in federated data warehouses. In: Camacho, D., Kim, S.-W., Trawiński, B. (eds.) New Trends in Computational Collective Intelligence. Studies in Computational Intelligence, vol. 572, pp. 63–73. Springer, Switzerland (2015)
8. Lenzerini, M.: Data integration: a theoretical perspective. In: Proceedings of the Twentyfirst ACM SIGMOD-SIGACT-SIGART Symposium on Principles of Database Systems, PODS 2002, pp. 233–246 (2002)
9. Motro, A., Anokhin, P.: Fusionplex: resolution of data inconsistencies in the integration of heterogeneous information sources. Inf. Fusion **7**(2), 176–196 (2006)
10. Nguyen, N.T.: Advanced Methods for Inconsistent Knowledge Management. Advanced Information and Knowledge Processing. Springer, New York (2007)
11. Chau, V.T.N., Nguyen Hua Phung, V., Tran, T.N.: Making kernel-based vector quantization robust and effective for incomplete educational data clustering. Vietnam J. Comput. Sci. **3**(2), 93–102 (2016)
12. Waddington, R.: An Architected Approach to Integrated Information. White paper, Kalido (2004)

A Solution for Automatically Malicious Web Shell and Web Application Vulnerability Detection

Van-Giap Le, Huu-Tung Nguyen, Dang-Nhac Lu, and Ngoc-Hoa Nguyen[✉]

VNU University of Engineering and Technology, Hanoi, Vietnam
{giaplv_57,tungnh_57,nhacld.di11,hoa.nguyen}@vnu.edu.vn

Abstract. According to Internet Live Stats, it is evident that organizations and developers are underestimating security issues on their system. In this paper, we propose a protective and extensible solution for automatically detecting both the Web application vulnerabilities and malicious Web shells. Based on the original THAPS, we proposed E-THAPS that has a new detecting mechanism, improved SQLi, XSS and vulnerable functions detecting capabilities. For malicious Web shell detection, taint analysis and pattern matching methods are selected as the main approach. The broad experiment that we performed showed our outstanding results in comparison with other solutions for detecting the Web application vulnerabilities and malicious Web shells.

Keywords: Web application vulnerability · Malicious Web shell · Taint analysis · Pattern matching · SQLi detection · XSS detection

1 Introduction

In April 2016, according to Internet Live Stats, there is an enormous amount of attacked Websites every day, causing both direct and significant impact on nearly 3.36 billion Internet users [7]. Even with security specialists, in some cases, still having troubles when coming up with unfamiliar systems because of the complexity of testing processes and the vast amount of testing cases. As a result, the number of hacked Websites per day is linearly increased: from 25.000 Hacked Website per day on April 2015 to 54.700 Hacked Website per day on April 2016 [7].

These current issues in Web application security raised a need for a solution that allows Web developers and security researchers detect security-related problems in the easiest way. In this paper, we propose an extensible solution for automatically detecting Web application vulnerabilities and malicious Web Shells, called **GuruWS**, which uses white-box testing techniques. We focus on the Web applications built by the PHP language because the proportion of PHP in server-side programming languages is remaining very high through many years as W3Techs: about 82.3 % of all the websites [8].

© Springer International Publishing Switzerland 2016
N.T. Nguyen et al. (Eds.): ICCCI 2016, Part I, LNAI 9875, pp. 367–378, 2016.
DOI: 10.1007/978-3-319-45243-2_34

The rest of this paper are organized as following: In Sect. 2 we refer to some basic principles and related works. Section 3 details our extensible solution for automatically detecting the Web vulnerabilities and malicious Web Shells. In Sect. 4 we summary our experiment to verify and benchmark our approach. The last section is dedicated to some conclusions and future works.

2 Background and Related Work

2.1 Vulnerability Scanning in PHP Web Applications

There are two main approaches in finding PHP application vulnerabilities by testing: black-box testing and white-box testing. The former one is particularly prefered to find flaws in Web applications. This method operates by launching attacks against an application using the fuzzing technique. They are both time and resource consuming in practical because of fuzzing limitations. For the white-box testing, it is not commonly used for finding security flaws in Web applications. The main reasons can be listed as the limited detection capability of white-box analysis tools, the heterogeneous programming environments, and the complexity of applications [1]. However, with many efforts to fade these limitations away, many white-box vulnerability scanners are released and popularly using by millions of customers at these days such that:

- **RIPS** [9] is a PHP vulnerability scanner using static analysis. As our practical experiments, RIPS scan very fast, yet, the False/Positive rate is still quite high, and it also lacks in object-oriented supporting [3] which is an advantage feature of **GuruWS**.
- **THAPS**[1] is a very efficient scanner which applies symbolic execution as its static analysis approach and performs a taint analysis as the post process to detect flaws. Symbolic execution is a term in computer science, which denotes the process of analyzing what inputs cause each part of a program to execute. To identify vulnerabilities, the taint analysis identifies user-controllable variables and how they propagated through the application. With every user-controllable variable, every time it reaches a potentially dangerous function (a vulnerable sink) without being properly sanitized first, a vulnerability is reported [2].

2.2 Malicious Web Shell Detection

A Web Shell is defined as a script that can be run on a Web server to enable remote administration of the infected server. For detecting malicious Web Shells, we can use different approaches such (i) pattern matching, (ii) combining lexical analysis and taint analysis, and (iii) using statistical methods. Here are some typical ideas for the Web Shell detection:

---

[1] https://bitbucket.org/heinep/thaps/.

– **Web Shell Detector** [10] is a Python tool that helps on detecting Web Shells. This product is a quite good solution as it is easy in using, developing and customizing. However, the Web Shell pattern set in Web Shell Detector database is old and also very limited. Moreover, it is not able to detect tiny Web Shells as well as self-written Web Shells, due to the taint analysis mechanism lacking.
– **NeoPI**[2] is a Python script, uses statistical techniques to detect obfuscated and encrypted content within source code. Its approach is based on the recursive scanning and ranking of all files in the base directory [4]. This solution requires experiments in Web security major to validate if it is Web Shell or not.

3 Solution

The **GuruWS** system architecture can be illustrated in the following Fig. 1.

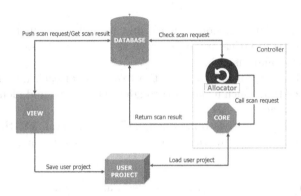

Fig. 1. *GuruWS* system architecture

– **Core** consists of **grVulnScanner** and **grMalwrScanner** modules. Each module runs simultaneously as each dependent and extensible service. In short:
  • **grVulnScanner** is a white-box Web application vulnerabilities scanner. The foundation of it is E-THAPS (Enhanced THAPS) which is the improved version of THAPS.
  • **grMalwrScanner** is objective to detect malicious Web application files based on the pattern matching and taint analysis methods.
– **UserProject** is the place where stores extracted users projects which will be the inputs for **grVulnScanner** and **grMalwrScanner** modules.
– **View** aims to support users access GuruWS's features, help them upload their compressed Web source codes and acquire results in a convenient way.
– **Database** is used to store scan requests, scan process status and scan results.
– **Allocator** takes the role of performing an efficient and flexible interaction between **Core** and **Database**. It has to get scan requests from **Database** and then to call **Core**'s components to handle these requests.

---

[2] https://github.com/Neohapsis/NeoPI.

3.1   grMalwrScanner

The primary objective of **grMalwrScanner** is supporting the developers, Web-masters and security specialists to detect malicious files in their project. In their perspective, they will proactively give it the source codes for getting the answer to two questions: (i) *Does their application contain malicious files?* and (ii) *If malicious files exist, where these files were located on their application?*

To satisfy all of their pretensions, our first step is to detect simple Web Shells. Because of their flexibility, we decided to use taint analysis method. For the second one, there are many general Web Shells, which were protected by encoding themselves challenge the method based on taint analysis. Recognizing that there is a limited number of popular Web Shells belonged to this type, we propose another method. The later depends on patterns from their identities.

One key idea in our works is to use all of the available approaches for the corresponding type of Web Shells.

Taint Analysis: This method is performed as following: firstly, the code is split into tokens (the lexical analysis process) to make it easier to manipulate and perform post analysis.

Then, **grMalwrScanner** analyses the token list of each file only once (to improve the speed) in which it passes through the token list and identifies important tokens by name.

Thus, potential dangerous functions (PDFs) are determined, then all significant arguments of these functions will be traced back to their 'source', that includes:

- **Other inputs:** get_headers(), get_browser() and so on.
- **User inputs:** $_GET, $_POST, $_COOKIE and $_FILES as well as other $_SERVER and $_ENV variables.
- **Server parameters:** HTTP_ACCEPT, HTTP_KEEP_ALIVE and so on.
- **File input:** fgets(), dlob(), readdir() and so on.
- **Database input:** mysql_fetch_array(), mysql_fetch_object() and so on.

There are some basic principles in **grMalwrScanner** taint analysis process:

- The source is always marked as tainted.
- The string created from tainted variables is also marked as tainted.
- With a function (not belonged to secure functions or PDFs), if it has any tainted input arguments, its return value will be marked as tainted.
- With every function in PDF list, there will be a set of corresponding securing functions. Hence, when significant arguments of a PDF is traced back, any argument passed through a securing function will make an untainted return value even though this is a tainted variable.

Regarding taint analysis approach, **grMalwrScanner** system just supports in detecting PHP Web Shells at the current time.

Pattern Matching: After investing in the wide-range Web Shells collecting (in various type and programming language namely ASP, PHP, Perl, Python and so on) from reliable sources: https://sourceforge.net/p/laudanum/code/25/tree/ and some repositories on https://github.com/: /tennc/webshell, /shiqiaomu/ webshell-collector, /tdifg/WebShell, /BlackArch/webshells, /JohnTroony/ other-webshells, /lhlsec/webshell, /fuzzdb-project/fuzzdb, /JohnTroony/php-webshells. We intended selecting and then distributing this data set into two parts with the ratio of 7:3. The bigger part then will be used for building pattern set purpose, and the rest part is for testing the efficiency of **grMalwrScanner**.

The pattern set dutifully follows YARA rules because of the flexibility, simpleness, and powerfulness of YARA. Consisting both sub-parts of Web Shells and particular tricky patterns, our pattern set is very efficient in detecting on-the-wild Web Shells; this will be verified in the experiment section of this paper.

Besides, we analyzed and wrote another Web Shell statistical analysis module for **GuruWS** in PHP. Our module reads source codes and returns a ranking table which is based on the probability of being a Web Shell of files. We built this module as an optional feature for security experts.

3.2 grVulnScanner

The foundation of **grVulnScanner** is the static analysis module improved from THAPS approach [2]. THAPS can automatically detect Cross-Site Scripting (XSS) and SQL Injection (SQLi) vulnerabilities. However, THAPS merely supports in identifying these two flaws; that is its primary limitation. Realizing a need to increase THAPS performance for a better capacity, we created E-THAPS (Enhanced THAPS) which is the improved version of THAPS. E-THAPS is entirely superior to the original THAPS, in which it is:

- Able to detect further Web application flaws: Command Injection, Object Injection, File Inclusion, XPATH injection, Arbitrary Eval Code Injection.
- Being equipped improved SQLi and XSS vulnerability detecting mechanism.
- Having other modifications for better performance.

The implementation of new detecting mechanism can be briefly demonstrated in Fig. 2:

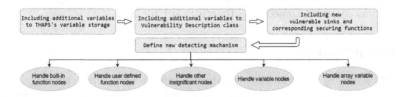

Fig. 2. Process of implementing new detecting mechanism

- Handle built-in function nodes: define how a vulnerability should be detected with new vulnerable sinks and corresponding securing functions.
- Handle user defined function nodes: include new variables in variable storage to all user defined function types.
- Handle variable/array variable nodes: consider whether variable/array variable comes from user input, if yes, mark it as tainted for all new variables in variable storage.
- Handle other insignificant nodes: further modifications to other nodes (Conditional nodes, Looping nodes, Logical nodes, Return nodes, etc.)

All new detecting mechanisms work properly like the old ones and also support in detecting flaws in object-oriented programming.

Improved SQLi Detecting Mechanism: THAPS have some limitations in analyzing projects having new SQL built-in functions, therefore, MySQL Improved potentially vulnerable function was added to SQLi detecting mechanism.

There are some further functions were included in securing functions set for SQLi: *mysqli_real_escape_string()*, *mysqli_escape_string()*.

Additionally, *mysqli_query()* function was implemented to SQL sink because with tainted parameters, it can lead to SQLi vulnerability.

These simple, yet efficient implementations significantly improved the ability to detect SQLI flaws of THAPS, that will be proved in the Experiment section.

Improved XSS Detecting Mechanism: THAPS's XSS sinks were defined via PHP-Parser (a library generating abstract syntax trees from PHP codes) nodes:

- PHPParser_Node_Stmt_Echo node: represents for *echo()* function.
- PHPParser_Node_Expr_Print node: represents for *print()* function.
- PHPParser_Node_Expr_Exit node: represents for *exit()* function).

However, there are other PHP built-in functions which can be defined as XSS sinks like: *print_r()*, *die()*, *printf()*, *vprintf()*.

By implementing these functions to THAPS's XSS sinks, the XSS detecting capacity of THAPS significantly increased. Furthermore, *highlight_string()* function was included to securing functions set for XSS.

Other Modifications

- *Update securing functions in conditional statements:* This securing function set contains a list of functions used in conditional statements (if/ elseif/ else) to ensure that passing parameter is safe for all sinks. There were some further functions included to this set: *is_bool()*, *is_null()*, *is_finite()*.

- *Update securing functions for every vulnerability:* This securing function set contains a list of functions which return untainted values with all passing tainted/untainted parameters. These untainted values are safe with all vulnerable sinks. There were 39 further functions included to this set in total including: *strftime(), md5_file(), sha1_file()* and so on.
- *Update functions that insecure the string again:* This function set contains a list of functions that return tainted value with all passing tainted/untainted parameters. These tainted values are presumably unsafe with all vulnerable sinks. There were 23 further functions included to this set in total including: *gzdecode, hex2bin(), recode(), gzinflate()* and so on.

These updates allow the detecting capabilities of E-THAPS being better than THAPS. We will validate this affirmation in the next section.

4   Experiment

Based on the proposed solution, *GuruWS* system is built and implemented on our department's server running CentOS 7.1 with Intel Xeon CPU E5-2630L and 2 GB of RAM. *GuruWS* system is now available on our official site http:// guruws.tech/ for public use.

To measure our solution's competency, we separately evaluated two main modules of **GuruWS**:

4.1   grMalwrScanner's Evaluation

To evaluate the ability to detect Web Shell of **grMalwrScanner**, we calculate the True Positive (TP) [5] and False Positive (FP) rate [6].

After investing in the wide-range Web Shells collecting from reliable sources (as the **Solution** section denoted) and then intended selecting, we distribute this data set into two parts with the ratio of 7:3. The second part was used to measure the efficient of both **grMalwrScanner** and other products.

VirusTotal[3] is an online service that supports analyze suspicious files, included viruses, worms, and Web application ones through the detection of other anti-virus products. To compare our solution with other prevalent anti-virus products, we built an Evaluation System, illustrated as Fig. 3.

Our solution interacts with VirusTotal's antivirus products via its Public API. In **Uploader**, there is a script which gathers Web Shells from Web Shell test set, then sends every Web Shell to the VirusTotal system via HTTP POST request. Each Web Shell will have a *scan_id* belonging to VirusTotal's response. **Uploader** keeps these *scan_id* and wait for the scanning progresses of VirusTotal. After the scanning progress is finished, **Analyser** uses *scan_id* to retrieve the corresponding result (contains a list of AntiViruses and their answer when scanning these Web Shells in JSON format).

---

[3] https://virustotal.com/.

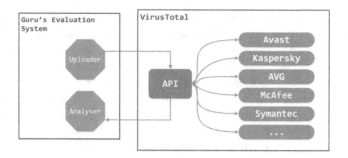

Fig. 3. GuruWS's evaluation system model

In addition, to evaluate the possibility of mistakenly detect Web Shells of **grMalwrScanner**, we also calculate the False/Positive rate when scanning 150 official Wordpress plugins[4] (contains 14527 files).

- Number of Web Shells in building pattern set: 1733 (70 %)
- Number of Web Shells in TP test set: 693 (30 %)
- Number of clean files in FP test set: 14527

Table 1 shows the final results we obtained from **grMalwrScanner** and 56 others scanners:

It is quite clear to see the differences in both TP and FP rate of **grMalwrScanner** compared with other products although they were tested on the same test set. **grMalwrScanner** achieved the outstanding TP rate (81,24 %), much higher than the rest. Additionally, there is no mistake in **grMalwrScanner** when detecting untainted files, the number of FP case is 0.

4.2 grVulnScanner's Evaluation

The test set being taken here is the series of Web application challenges which belongs to the WhiteHat Capture The Flag (CTF) contests [11]. These Web challenges trustworthily simulate the real vulnerable Web applications. In every test, a comparison is made between the original version of THAPS, our E-THAPS, and the RIPS.

To evaluate, we observe the number of TP and FP case in every test case scanning process. A TP case is defined as the case when the PVF leads to a vulnerability and the scanner report a vulnerability. Meanwhile, when the PVF does not lead to a vulnerability and the scanner still report a vulnerability, we have a FP case. If TP figure is high and FP is small, then the scanner performed well.

---

[4] https://wordpress.org/plugins/browse/popular/.

Table 1. Number of detected Web Shell and TP-FP rates

Solutions	Number of found Web Shells	TP rate	Number of detected clean files	FP rate
grMalwrScanner	563	81.24 %	0	0.0 %
Avast	493	71 %	0	0.0 %
Qihoo-360	491	71 %	2	0.014 %
GData	462	67 %	0	0.0 %
Ikarus	458	66 %	0	0.0 %
AhnLab-V3	425	61 %	0	0.0 %
AegisLab	423	61 %	0	0.0 %
ALYac	411	59 %	0	0.0 %
ESET-NOD32	404	58 %	0	0.0 %
F-Secure	396	57 %	0	0.0 %
BitDefender	393	57 %	0	0.0 %
Ad-Aware	393	57 %	0	0.0 %
Avira	391	56 %	0	0.0 %
Arcabit	382	55 %	0	0.0 %
nProtect	381	55 %	0	0.0 %
Emsisoft	381	55 %	0	0.0 %
MicroWorld-eScan	380	55 %	0	0.0 %
Tencent	356	51 %	0	0.0 %
NANO-Antivirus	331	48 %	0	0.0 %
Comodo	324	47 %	0	0.0 %
AVG	322	46 %	0	0.0 %
TrendMicro	308	44 %	0	0.0 %
TrendMicro-HouseCall	308	44 %	0	0.0 %
ClamAV	275	40 %	0	0.0 %
Baidu	265	38 %	0	0.0 %
Bkav	262	38 %	0	0.0 %
Sophos	249	36 %	0	0.0 %
VBA32	240	35 %	0	0.0 %
McAfee-GW-Edition	228	33 %	0	0.0 %
McAfee	218	31 %	0	0.0 %
Cyren	212	31 %	0	0.0 %
Fortinet	201	29 %	0	0.0 %
DrWeb	199	29 %	0	0.0 %
CMC	169	24 %	0	0.0 %

(Continued)

Table 1. *(Continued)*

Solutions	Number of found Web Shells	TP rate	Number of detected clean files	FP rate
Symantec	160	23 %	0	0.0 %
Kaspersky	158	23 %	0	0.0 %
F-Prot	141	20 %	0	0.0 %
AVware	139	20 %	1	0.007 %
ViRobot	108	16 %	0	0.0 %
Microsoft	102	15 %	0	0.0 %
VIPRE	83	12 %	0	0.0 %
TotalDefense	75	11 %	0	0.0 %
CAT-QuickHeal	68	10 %	0	0.0 %
Jiangmin	61	9 %	3	0.021 %
Antiy-AVL	58	8 %	0	0.0 %
Agnitum	57	8 %	0	0.0 %
Rising	53	8 %	3	0.021 %
K7GW	25	4 %	0	0.0 %
K7AntiVirus	25	4 %	0	0.0 %
Panda	24	3 %	0	0.0 %
TheHacker	10	1 %	0	0.0 %
Zillya	7	1 %	0	0.0 %
Web Shell Detector	5	1 %	0	0.0 %
ByteHero	0	0 %	0	0.0 %
Zoner	0	0 %	0	0.0 %
Malwarebytes	0	0 %	0	0.0 %
Alibaba	0	0 %	0	0.0 %

Web Challenge in WhiteHat Contest 7: THAPS, in this test, showed the worst performance when it was unable to detect any vulnerability. RIPS showed the best performance with 3 detected flaws while E-THAPS detected Objection Injection flaw, File Inclusion flaw and missed the Possible Flow Control flaw (Table 2).

Web Challenge in WhiteHat Contest 8: In this test, it is evident that E-THAPS and RIPS were equal in performing, however, better than THAPS in detecting SQLi and File Inclusion flaws. There also appeared some False Positive cases in these vulnerabilities as shown in Table 3.

Table 2. WhiteHat Contest 7 scanning results

Scanner	Objection injection	File inclusion	Possible flow control
RIPS	1	1	1
THAPS	0	0	0
E-THAPS	1	1	0

Table 3. WhiteHat Contest 8 scanning results

Scanner	Cross site scripting		SQL injection		File inclusion	
	TP	FP	TP	FP	TP	FP
RIPS	1	2	2	1	1	0
THAPS	1	2	0	0	0	0
E-THAPS	1	2	2	1	1	0

Table 4. WhiteHat Contest 10 scanning results

Scanner	Cross site scripting		SQL injection	
	TP	FP	TP	FP
RIPS	2	0	1	1
THAPS	2	0	0	0
E-THAPS	2	0	1	0

Table 5. WhiteHat Grand Prix 2014 scanning result

Scanner	Cross site scripting		SQL injection	
	TP	FP	TP	FP
RIPS	1	1	11	0
THAPS	2	0	0	0
E-THAPS	2	0	11	0

Web Challenge in WhiteHat Contest 10: This time, E-THAPS showed the best performance with 3 identified vulnerabilities: 2 XSS and 1 SQLi. Additionally, in SQLi flaw, E-THAPS, not like RIPS, didn't have any FP case (Table 4).

Web Challenge in WhiteHat Grand Prix 2014: E-THAPS, again, showed the best performance which has the highest TP and lowest FP figured in both XSS and SQLi vulnerabilities. It is evident that E-THAPS can detect SQLi flaws much more efficient than the original one (Table 5).

5 Conclusion and Future Works

This paper presents an extensible solution, **GuruWS**, which allows users to automatically detect the vulnerabilities and malicious Web shells for Web applications. Based on THAPS, we proposed E-THAPS for the module **grVulnScanner**. It has a new detecting mechanism: improved SQLi, XSS detecting mechanism; updated securing functions in conditional statements, functions for every vulnerability, and functions that insecure the string again. E-THAPS showed the best performing which has the highest Found and lowest Found/Confirm figure in both XSS and SQLi vulnerabilities. For the detection of malicious files, it is based on the taint analysis and pattern matching methods. Our in-depth experiment allows to confirm the **grMalwrScanner** achieved the outstanding True/Positive rate, much higher than the rest.

In the next stage, we aim to make **grVulnScanner** become a gray-box scanner, improve Web Shell pattern sets, optimize these old ones and add more flexible, yet strong enough rules for detecting stealthy Web Shells as well.

References

1. Kals, S., Kirda, E., Kruegel, C., Jovanovich, N.: SecuBat: a web vulnerability scanner. In: 15th International Conference on World Wide Web, pp. 247–256 (2006)
2. Jensen, T., Pedersen, H., Olesen, M.C., Hansen, R.R.: THAPS: automated vulnerability scanning of PHP applications. In: Jøsang, A., Carlsson, B. (eds.) NordSec 2012. LNCS, vol. 7617, pp. 31–46. Springer, Heidelberg (2012)
3. Dahse, J.: RIPS - a static source code analyser for vulnerabilities in PHP scripts. In: Seminar Work at Chair for Network and Data Security (2010)
4. Sasi, R.: Web backdoors - attack, evasion and detection. In: C0C0N Sec Conference (2011)
5. Nguyen, N.-H.: Iris recognition for biometric passport authentication. VNU J. Sci. Nat. Sci. Technol. **26**(1), 14–20 (2010)
6. Le, H.H., Nguyen, N.H., Nguyen, T.T.: Exploiting GPU for large scale fingerprint identification. In: Nguyen, N.T., Trawiński, B., Fujita, H., Hong, T.-P. (eds.) Intelligent Information and Database Systems. LNCS, vol. 9621, pp. 688–697. Springer, Heidelberg (2016)
7. http://www.internetlivestats.com/. Accessed 26 April 2016
8. Web technology surveys. http://w3techs.com/technologies/overview/programming_language/all/. Accessed 15 April 2016
9. Dahse, J., Holz, T.: Static detection of second-order vulnerabilities in web applications. In: 23rd USENIX Security Symposium (USENIX Security 14), pp. 989–1003 (2014)
10. Starov, O., Dahse, J., Ahmad, S., Holz, T., Nikiforakis, N.: Thieves, no honor among: a large-scale analysis of malicious web shells. In: 25th International Conference on World Wide Web, pp. 1021–1032 (2016)
11. Global websecurity whitehat contest. https://ctftime.org/ctf/112

Data Evolution Method in the Procedure of User Authentication Using Keystroke Dynamics

Adrianna Kozierkiewicz-Hetmanska[(⊠)], Aleksander Marciniak,
and Marcin Pietranik

Department of Information Systems,
Wroclaw University of Science and Technology, Wrocław, Poland
{adrianna.kozierkiewicz,marcin.pietranik}@pwr.edu.pl,
185871@student.pwr.edu.pl

Abstract. Due to the rapid development of Internet and web-based application the number of system which an ordinary user needs to interact grows almost proportionally. People are expected to make bank transfers, send emails using multiple mailboxes, send tax declarations, send birthday wishes solely online. What is more, sometimes only this way being available. The sensitivity of information created using online tools is unquestionable and the highest possible level of data security is therefore expected not only on a corporate level, but also it should be guaranteed to ordinary users. That is the reason why a convenient solution, that do not require any additional expensive equipment (e.g. RFID cards, fingerprint readers, retinal scanners), can assure such security is highly wanted. Therefore, a number of publications have been devoted to methods of user authentication based on their biometrical characteristics (that are obviously individual and can be easily used to encrypt users' credentials) and one potentially most accessible group of methods is build on top of analysis of users' personal typing styles. This paper is a presentation of a data evolution method used in our novel biometrical authentication procedure and contains a statistical analysis of the conducted experimental verification.

1 Introduction

Nowadays, the rapid development of commonly accessible technologies resulted in an increased interest in distributes, computer systems which offer a functionality tightly bound to their users. Many of these systems gives their users an opportunity to have their own account, e.g. bank accounts, social networking, emails and so on. To access them, a user must authenticate (pass the comparison process of his identifier (e.g. a username) and an assigned password that has been created during a registration process. Then, given credentials must pass the process of authorization, which involve granting proper access rights and available actions that can be done in the system. Processes of authentication and authorization are sometimes treated as one and referred to as logging. This situation entails that the users are exposed to the threat of losing their credentials and therefore, their accounts and related data may become corrupted and misused. Ways of obtaining logging data needed to pass the authentication process are at least a few.

© Springer International Publishing Switzerland 2016
N.T. Nguyen et al. (Eds.): ICCCI 2016, Part I, LNAI 9875, pp. 379–387, 2016.
DOI: 10.1007/978-3-319-45243-2_35

In the following paper we will focus on preventing data loss due to tracking keystrokes that users typed on their keyboards. This kind of attack involves using a keylogger software which monitors and records keystrokes and sometimes can be met as a hardware counterpart. What is important to emphasize is the fact that keyloggers are not only used by potential intruders, but also by legitimate software which for example finds frequent keystroke sequences and suggests convenient keyboard short-cuts. Obviously, this class of software works on the same principle as those prepared by the potential adversaries. It not only gathers information about the pressed keys, but is able to take screenshots when the text is entered. Eventually all of the collected data are sent to the adversary.

From the intruder's point of view, the main advantage of the keyloggers is the ability to reveal the real password, because they record the plaintext, which is entered by the user directly. All of the possible data encryption occurs afterwards so even the most secured transfer protocol will not prevent from losing vulnerable data. What is more, adversaries may also come into the possession of not only the user's passwords, but also his correspondence, a history of visited websites etc.

The remedy for the described problem may involve using biometrically enhanced method of sending credential data. Such approach increases the safety of the data because it tightly connects what has been typed with the person that has been typing.

There are two main ways of using biometrical data. The first one is biometrical identification is based on a comparison of biometric data describing a user with patterns available in the system. In the case of a compliance with one of such patterns, the user is identified and eventually authenticated. The second approach is based on a direct comparison of biometric pattern with a pattern assigned to the user that have already been identified beforehand using different identification method. This approach is called biometric verification.

In this paper we focus on improving our authentication scheme which is based on the behavioural biometric technology involving personal typing style. The motivation for this work was the lack of a method of reducing the negative effects of authentication on a computer with an unwanted keylogger installed. In our accompanying paper [9] we have developed a novel method of biometric authentication based on the analysis of typing on a keyboard, which masks the entered text in a real time by emulating additional random keystrokes performed during the actual typing. This idea is based on a remark that every user can be characterised with unique gaps that he does between pressing single keys. The following article focuses on developing an evolution algo-rithm that processes collected biometric data and modifies the pattern assigned to the user based on the analysis of how the user's typing style evolves (e.g. the user may increase the typing speed when he becomes more familiar with the keyboard, fre-quently typed phrases etc.). Additionally, we present more extensive results of verifi-cations of our authentication scheme.

The article is organised as follows. Section 2 is a brief overview of related works that has been done in the field and described in the available literature. Section 3 contains short explanation of our methods. Section 4 is a review of experiments that we have conducted in order to validate our ideas. Section 5 contains conclusions and a brief overview of our upcoming research.

2   Related Works

As stated in [2] keystrokes dynamics has a strong psychological background and is highly individual to every user. All of the methods described in literature are based on obtaining a common pattern according to which a user interacts with a keyboard and then compare it to the templates stored in a database of choice [4]. According to [15] these templates contain expressive attributes that can be used to describe typing styles – among other they include latencies, dwell times, key hold times, total duration, total typing time etc. A selected subset of these attributes form aforementioned typing pattern that are further analyzed.

The simplest approach involves statistical methods such as hypothesis testing, t-tests absolute distance measures (weighted and unweighted) [8]. In [6] the analysis of vector space has been proposed and gathered experimental outcomes has been promising.

More advanced methods apply artificial neural networks to classify inter-character times [17]. Other solutions are based on a variety of pattern recognition methods [16] for example by using a three step approach to improve the performance of keystroke identification [18].

Another group of methods are built on top of search heuristics and combination of algorithms [1] such as particle swarm optimization. In [14] an interesting application of genetic algorithms can be found. Other works involve, for example, a combination of neuro-fuzzy algorithms (e.g. Fuzzy-ARTMAP [3]) to classify users based on their keystroke dynamics. Montalvao et al. [13] investigated the histogram equalization of time intervals on the performance of the overall methodology. Other approaches include using the Markov chain algorithms in which the prior probability vectors are replaced by appropriate histograms [5].

The first experiment of user's authentication based on keystroke dynamics that has been conducted in an uncontrolled environment. Its results can be found in [12]. The outcomes have been analyzed using error rates such as False Acceptance Rate (FAR), False Rejection Rate (FRR) and Equal Error Rate (EER) [7]. Thereafter, a majority of considered methods have been tested using these factors, which are also frequently used to define requirements for systems that need to guarantee a certain level of data security.

3   The Data Evolution Method for Authentication Scheme Based on Keystroke Dynamics

In our accompanying paper [9] we have described our novel user authentication method using keystroke dynamics. The proposed scheme protects users against adversary attacks with keyloggers. The whole procedure [11] contains the following stages: user's registration, login phase which consist of identifying user's typing pattern, determination of time gaps between keystrokes and filling these gaps with additional random characters and an eventual decryption phase.

Firstly, the biometric data are collected. Based on them a user's typing pattern is identified as the mean value of gathered biometric data. Next, between each keystroke the appropriate time gaps are determined. The length of these time gaps depend on a place where a break between single keystrokes is made. The determined time gap is longer near a place of a potential user's keystroke and extends closer to the middle of such break.

Finally, the time gaps are filled with additional characters chosen in a random way. The probability of occurrence of such random character is influenced by the length of the time gap and an assumed threshold. The longer time gap and the smaller threshold - the smaller number of a random characters. After the login phase the password is encrypted as the sequence of characters given by a user and determined in a random way by the method. A password is decrypted based on a stored user's typing pattern. The general idea of the user's authentication through keystroke dynamics is presented on Fig. 1 and the broader description of mentioned scheme can be found in [9].

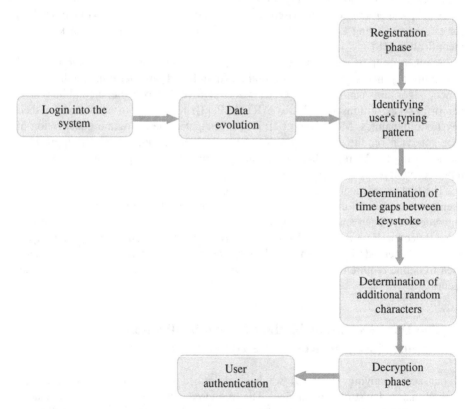

Fig. 1. The general idea of user authentication through keystroke dynamics

In this paper the authentication scheme is enriched with the data evolution method. The user typing pattern changes as the time passes. After each trial of user's login, a new biometric data is collected and based on it the new typing pattern is designated.

However, one should determine whether or not the old biometric data should be replaced by a new one?

Let us assume that *registration_data* is the two-dimensional matrix containing the times between a single keystroke for each trial of password's typing. Moreover, let us assume that the typing pattern is calculated based on *k-th* trial. In [5] authors have proposed that the data from the *registration_data* matrix is replaced according to the FIFO (First In-First Out) rule. It means that the oldest biometric data is removed.

In the following work we propose a new evolution method which is based on the removal of the worst biometric data. Firstly, the Manhattan distance between the updated template and vector of times between keystrokes obtained during the last authentication is calculated. The ranking is done by finding an element of the registration dataset for which the Manhattan distance has the lowest value. This element is replaced with new data obtained in the last trial of authentication. An overview of the developed procedure can be found below.

The data evolution method

```
Input: registration_data,
       new_data- vector of times beetwen keystrokes obtained
       during the last authentication
Output: uptaded registration_data

BEGIN

   FOR i=0 to k   DO
   {
        test=registration_data;
        test[i]=new_data;
        Create user's typing pattern for each a∈[0,n-1]:
        template[a]=(∑ⱼ∈[1,k]registration_data[a][j])/k
        difference[i]=0;
        FOR j=0 to template.lenght DO
           {
              difference[i]=difference[i]+|template[j]-
     new_data[j]|
           }

   }
   Find the index i where difference[i] is minimal
   Update the registration_data[i]=new_data;

END
```

4   Results of Experiments

The authentication schema based on keystroke dynamics along with the data evolution model proposed in the previous section have been implemented in the web-based simulator using Java Enterprise Edition [11]. Our experiment has involved *20* users for whom we have created different users' accounts. We have assumed that the typing pattern is obtained after $k = 12$ trials of user's actually typing his password. During the experiment volunteers tried to authenticate in the system using publicly known credentials (a login and a password). In this way, *273* trials of authentication for own accounts and *455* trials of authentication for other people accounts have been collected. The proposed authentication method has been compared with a lightweight algorithm described in [10]. Firstly, we present deeper analysis of the novel authentication method based on keystroke dynamics and further, the results of verification of the data evolution method proposed in this paper.

4.1   Statistical Analysis

In our accompanying paper [9] only preliminary experimental results have been presented. In this paper we would like to draw some reliable conclusions from the statistical analysis. Data collected during experiments were divided into two separate tables: Tables 1 and 2.

Table 1. The number of true and false acceptance and true and false rejection for lightweight algorithm

	Acceptance of authentication	Rejection of authentication
Owner of account	207	66
Foreign person	102	353

Table 2. The number of true and false acceptance and true and false rejection for the proposed algorithm

	Acceptance of authentication	Rejection of authentication
Owner of account	212	61
Foreign person	101	354

Based on collected date in Tables 1 and 2 the ROC Curves have been prepared and presented in Fig. 2.

We would like to check whether or not our authentication scheme and the lightweight algorithm have similar effectiveness. For this purpose, we have tested areas under ROC Curve for both of them (Table 3).

The statistical analysis has been made for the assumed significance level equal *0.05*. We have obtained the statistical test value equal *0.439* and the *p-value* equal *0.6615* therefore, we cannot reject the null hypothesis that the areas under ROC curve for both methods are equal. Therefore, we have statistically confirmed our previous hypothesis [9], that the new authentication scheme based on keystroke dynamics ensures the safety of users' credentials on a similar level like the lightweight algorithm.

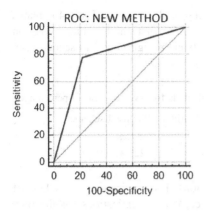
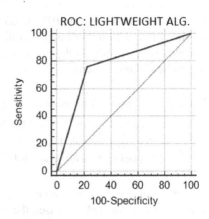

Fig. 2. ROC Curves for the new authentication scheme (on the left side) and the lightweight algorithm (on the right side).

Table 3. The AUC for both methods

	ROC curve for the new authentication method	ROC curve for the lightweight
AUC	0,777	0,767
Standard deviation	0,016	0,0163

4.2 Password Analysis

The analysis of the data collected during the experiment allows to conclude that the strongest passwords used in our proposed scheme contain long and varied time gaps. In the traditional authentication methods, the longer and more complicated passwords are required to increase safety. In our analysis the safest password was "*pawel*" because it contained *1336 ms* break between keystroke of two last characters. For this password only *0.83 %* of trials of all authentication attempts on this account as another person finished successfully.

The worst user's password created in the experiment was "*qazws*". This five-letter password is written using a simple scheme, which is known for a large number of people. This password is very popular and simple, therefore, between subsequent keystrokes only very small breaks have occurred. Shorter breaks between keystrokes decrease the probability of the determination of a random character, therefore the encryption of the password is poorer and easier to break. In our experiment almost *20 %* of trials of unauthorized authentication were successful.

4.3 Effectiveness of Data Evolution Methods

In this part of our paper we present a verification of the effectiveness of the data evolution method. Test users were requested to authenticate in the system for several times. After each successful attempt, their typing patterns were modified according to the procedure described in [10] and our method presented in Sect. 3. Table 4 contains results obtained from both methods in two use cases. At first, the two data evolution methods were applied to the outcomes of the lightweight algorithm and secondly, in our authentication scheme.

Table 4. The growth of the number of the correct credentials for own account (in percentage).

Data evolution method	The lightweight algorithm	Proposed scheme
Replacing the oldest data	11,55 %	5,89 %
Replacing the worst data	9,41 %	12,3 %

From Table 4 we can draw two major conclusions. The application of any kind of data evolution method increases the number of correct authentications for the own account. In other words, the data evolution methods caused reduction of the False Rejection Rate. The best results were obtained for our authentication scheme with data evolution method described in Sect. 3. As it was mentioned, the user's typing pattern has changed with the passage of time. Many repetitions of typing the password have shortened the time between keystrokes. Therefore, the effective methods of data evolution are desired.

5 Conclusions and Future Works

This paper described the improved authentication's scheme based on a behavioural biometric method which is users' keystroke styles. The main idea of the proposed method is based on the encryption of user's password by adding additional random characters. Decryption of such encoded password relies on searching characters for which the user's typing time is the most suitable within the assigned typing template. However, the user's typing pattern can change with time. Therefore, the data evolution method has been proposed and verified.

The experimental results have confirmed the effectiveness of the proposed method. The statistical analysis pointed out that the new authentication scheme protects users' accounts from an unauthorized access on the similar level like the lightweight algorithm. However, the proposed schema has an advantage over the lightweight algorithm. Some additional characters added in a random way to the decrypted password protects it against an adversary attack using keyloggers.

Moreover, password collected and tested during experiment have demonstrated that for the sake of the security the length of time between keystroke is more important than the length of password itself.

According to our guesses the user typing pattern can change with time. Therefore, the application of any data evolution method which is used to calculate the typing template improves the overall effectiveness of authentication methods. In other words, data evolution method increases the number of correct authentications for own account. The method proposed in this paper have reduce the False Rejection Rate for about 12,3 %.

In our future work we would like to examine our methods with the bigger number of volunteers. Moreover, the analysis of the effectiveness of described methods for different values of the threshold should be performed. Due to the fact, that the method can also be implemented as a completely transparent layer to other, 3rd party systems or any adversaries, we would like to prepare a software as the protection against attacks with keyloggers.

References

1. Azevedo, G.L., Cavalcanti, G.D., Filho, E.C.B.: Hybrid solution for the feature selection in personal identification problems through keystroke dynamics. In: International Joint Conference on Neural Networks, IJCNN 2007, pp. 1947–1952. IEEE (2007)
2. Banerjee, S.P., Woodard, D.L.: Biometric authentication and identification using keystroke dynamics: a survey. J. Pattern Recogn. Res. **7**, 116–139 (2012)
3. Carpenter, G.A., Grossberg, S.: Adaptive Resonance Theory. Springer, Heidelberg (2011)
4. Gaines, R., Press, S., Lisowski, W., Shapiro, N.: Authentication by keystroke timing. Rand Report (1980)
5. Gunetti, D., Picardi, C.: Keystroke analysis of free text. ACM Trans. Inf. Syst. Secur. (TISSEC) **8**(3), 312–347 (2005)
6. Guven, A., Sogukpinar, I.: Understanding users' keystroke patterns for computer access security. Comput. Secur. **22**(8), 695–706 (2003)
7. Jain, A.K., Ross, A., Prabhakar, S.: An introduction to biometric recognition. IEEE Trans. Circ. Syst. Video Technol. **14**(1), 4–20 (2004)
8. Joyce, R., Gupta, G.: Identity authentication based on keystroke latencies. Commun. ACM **33**(2), 168–176 (1990)
9. Kozierkiewicz-Hetmańska A., Marciniak A., Pietranik M.: User authentication method based on keystroke dynamics, Paper submitted on ICCCI 2016 conference (2016)
10. De Magalhães, S.T., Revett, K., Santos, H.: Password secured sites-stepping forward with keystroke dynamics. In: International Conference on Next Generation Web Services Practices, NWeSP 2005. IEEE (2005)
11. Marciniak A.: Uwierzytelnianie użytkowników oparte o analizę dynamiki pisania na klawiaturze. Master thesis (2016, in Polish)
12. Monrose F., Rubin A.: Authentication via keystroke dynamics. In: Proceedings of the 4th ACM Conference on Computer and Communications Security, pp. 48–56 (1997)
13. Montalvao, J., Almeida, C.A.S., Freire, E.O.: Equalization of keystroke timing histograms for improved identification performance. In: 2006 International Telecommunications Symposium, pp. 560–565. IEEE (2006)
14. Revett, K.: A bioinformatics based approach to behavioural biometrics. In: Frontiers in the Convergence of Bioscience and Information Technologies, FBIT 2007, pp. 665–670. IEEE (2007)
15. Robinson, J.A., Liang, V.M., Chambers, J., MacKenzie, C.L.: Computer user verification using login string keystroke dynamics. IEEE Trans. Syst. Man Cybern. Part A Syst. Hum. **28**(2), 236–241 (1998)
16. Theodoridis, S., Koutroumbas, K.: Pattern Recognition. Elsevier (2009)
17. Yong, S., Lai, W.-K., Goghill, G.: Weightless neural networks for typing biometrics authentication. In: Negoita, M.G., Howlett, R.J., Jain, L.C. (eds.) KES 2004. LNCS(LNAI), vol. 3214, pp. 284–293. Springer, Heidelberg (2004)
18. Yu, E., Cho, S.: Keystroke dynamics identity verification—its problems and practical solutions. Comput. Secur. **23**(5), 428–440 (2004)

Cooperative Strategies for Decision Making and Optimization

Multiagent Cooperation for Decision-Making in the Car-Following Behavior

Anouer Bennajeh[1]([✉]), Fahem Kebair[1], Lamjed Ben Said[1],
and Samir Aknine[2]

[1] SOIE, Institut Supérieur de Gestion de Tunis-ISGT, Université de Tunis, 41,
Avenue de la Liberté, Cité Bouchoucha, 2000 Bardo, Tunis, Tunisia
anouer.bennajeh@gmail.com, kebairf@gmail.com,
bensaid_lamjed@yahoo.fr
[2] LIRIS, Université Claude Bernard Lyon 1-UCBL,
43, Bd du 11 novembre 1918, 69622 Villeurbanne Cedex, France
samir.aknine@univ-lyonl.fr

Abstract. This paper presents a decision-making model for determining the velocity and safety distance values basing-on anticipation of the simulation parameters. Thus, this paper is composed of two parts. In the first one, we used a bi-level bi-objective modeling to address the problem of decision-making with two objectives, which are, maximize the safety distance and maximize the velocity, in order to define a link between the increase of velocity and the road safety in the car-following behavior. In the second part, we resolve our modeling basing-on a multi-agent cooperation approach by applying of the Tabu search algorithm. The simulation results showing the advantages of our approach, such as, the use of the multi-agent cooperation approach reflects the high number of tested solutions in a very short search time, which guarantees the high quality of selected solution for each simulation step.

Keywords: Car-following behavior · Safe distance model · Bi-objectives modeling · Decision-making · Multiagent system · Tabu search algorithm

1 Introduction

The safe distance model is developed for the first time by Kometani and Sasaki in 1959 [9]. In fact, this model is based on the calculation of the security distance by using the equations of physical movement. The principal objective of this model is to avoid the collision between the following and the leading vehicle. One of the widely used safe distance model is defined in [5], which combines a free-flow driving model with a stopping-distance. But, despite the important advantages of this model, where it has been implemented widely in micro-simulation software packages such as SISTM [13], AIMSUN [3] and DRACULA [10], it has a disadvantage resulting in its very strict restriction on the car-following behavior, since the following vehicle can move only when it has exactly a calculated safety distance with the leading vehicle. However, it is impossible to have this harmony of velocity between vehicles. Hence, it is obvious that this restriction is not consistent with the real traffic condition.

© Springer International Publishing Switzerland 2016
N.T. Nguyen et al. (Eds.): ICCCI 2016, Part I, LNAI 9875, pp. 391–401, 2016.
DOI: 10.1007/978-3-319-45243-2_36

In recent years, there is more research on the safety distance of car-following behavior in order to improve the traditional work [5]; as the research work [12] that proposes a new safety distance model based on various speed relationships between vehicles in order to improve the calculation of the safety distance for each simulation step. Furthermore, the work [15] proposed a safe distance model by formulating the safety distance as a function of the velocity difference and safety distance. Indeed, most of the research works [5, 12, 15] have the same problems. Firstly, all these research works did not treat the driver behaviors, which influences considerably on the reality of the simulation. Secondly, according to [4], the calculation of the safety distance is based on many parameters, which are: the reaction time, the decision time, the action time, the weight of the vehicle, the speed of gravity, the air density, the projection area, the air resistance factor, the efficiency of braking, the friction coefficient, the decay factor and the slope of the road. However, excepting the reaction time, all of the above mentioned parameters are not used during the calculation of the safety distance in the research works [5, 12, 15]. Finally, these research works used a fixed reaction time, which is far from the reality, where the duration of the reaction time is influenced by the driver behaviors. For example, the duration of reaction time for a common driver is 0.6–0.9 s [4].

The overall objective of our research is to integrate the driver behaviors in a computing model, which led to the construction of a realistic road traffic simulation. In particular, we are interested in this paper to model drivers with normative behaviors. Indeed, this paper intends to construct a new decision-making model to manage the decision-making of the driver, by adopting an approach based on the anticipation of the simulation parameters for the next simulation step.

Modeling and implementing the car-following microscopic driving behavior requires a technology that guarantees the autonomy, the reactivity, the adaptability and the interaction. The software agent technology [14] meets perfectly these criteria and is positioned as an appropriate solution to simulate the road traffic with a microscopic approach. Thus, we adopted this technology to model the drivers and to resolve the research of the best realizable solution using a multi-agents cooperation approach.

Basing-on the theoretical background of our research, we present in the next section our modeling with more details. In the third section, we present the resolution of our modeling based on the Tabu search algorithm and a multi-agent cooperation. In the fourth section, we present the simulation results of our model. Finally, we conclude by discussing the first obtained results of our approach and future works.

2   Bi-Level Bi-Objective Modeling

The multi-objective optimization problem consists in finding the values for a set of decision variables that satisfy a set of constraints and that optimize a vector function whose elements represent the objectives. These functions are a mathematical representation of performance criteria that are usually in conflict [11].

The driver behaviors are different according to their objectives. Indeed, each driver translates its objectives through actions that help him to achieve them. In this modeling, the actions of drivers are categorized under two objectives. The first one is the reduction of the travel time and the second one is the road safety. To achieve the first objective, the driver increases the velocity of its vehicle, but with the car-following behavior this action influences on the second objective by reducing to the safety distance between the following and the leading vehicle. However, according to traffic rule, if there is an increase of the velocity, there is an increase of the safety distance, which is not the case with the car-following behavior. Thus, the two objectives of the following agent driver are totally contradictory to the car-following behavior, which we used a multi-objective modeling to define our problematic by focusing on a normative driver behavior.

The following driver agent tries to reduce the travel time by increasing the velocity of its vehicle, which presents in our modeling the objective function "maximize the velocity"; and simultaneously, it tries to take into account the road safety by keeping a safety distance with the leading vehicle, which presents the objective function "maximize the safety distance". Thus, the objective functions of each following driver agent reflect its choice for the velocity and the safety distance that meet their needs for each simulation step. We expressed the first objective function "maximize the safety distance" by Eq. (1) for the simulation step $T+1$.

$$\text{Max } D_{sec\_x}(T+1) = (T_R + T_D + T_A) \times V_x(T+1) + (W/(2 \times G \times \rho \times A_f \times Cd))$$
$$\times \ln\left(1 + \left((\rho \times A_f \times Cd)/2 \times (V_x(T+1))^2\right)/((\eta \times \mu \times W) + (f_r \times W \times \cos\theta) + (W \times \sin\theta))\right) \quad (1)$$

Where, T_R is the reaction time, T_D is the decision time, T_A is the action time, W is the weight of the vehicle, G is the speed of gravity, ρ is the density of air, A_f is the projection area, C_d is the air resistance factor, η is the efficiency of braking, μ is the friction coefficient, f_r is the decay factor and θ is the slope of the road. In order to ensure a realistic simulation, we used the simulation parameters of [4] for these parameters. $V_x(T+1)$ is the velocity that will be adopted during the next simulation step $T+1$ and it is the only decision variable for this objective function.

Indeed, the calculation of the safety distance is based on a combination between the secure stopping distance $(W/(2 \times G \times \rho \times A_f \times C_d)) \times \ln(1 + ((\rho \times A_f \times C_d)/2 \times (V_x(T+1))^2)/((\eta \times \mu \times W) + (f_r \times W \times \cos\theta) + (W \times \sin\theta)))$ and the reaction distance $(T_R + T_D + T_A) \times V_x(T+1)$. This objective function based on the safety distance defined by [4].

Concerning the second objective function "maximize the velocity", we expressed this function by Eq. (2).

$$\text{Max } V_x(T+1) = V_x(T) + a\, t + m \quad (2)$$

Where, "a" is a decision variable that presents the acceleration during the increasing speed and the deceleration in the opposite direction, "t" is the duration of the simulation step and "m" is the precision margin that influence on the precision of velocity of the following vehicle since it is impossible to adopt exactly the calculated velocity by a real driver. In fact, the first objective function contains a decision variable "$V_x(T+1)$" which represents the velocity that will be adopted during the next simulation step $T+1$. Furthermore, the decision variable "$V_x(T+1)$" is designed as an objective function in Eq. (2). Therefore, the modeling of our problematic will be a bi-level bi-objective modeling.

Beginning by defining the constraints of the lower objective function of our modeling (maximize the velocity). In fact, the increasing of the velocity relative with the drivers' behaviors, where according to [1] the drivers with normative behaviors interest firstly to the traffic rules in order to ensure the road security. Thus, we defined eight constraints that ensure the security with this objective function. Indeed, these constraints appear depending two simulation parameters, which are, "D_{sec_x}" the calculated safety distance and "D_{xy}" the gap between the following vehicle X and the leading vehicle Y. Consequently, there are three possible states that the following agent driver must take into account.

Starting by the first state "$D_{sec\_x}(T) > D_{xy}(T)$", where the following vehicle X is in an unsecure situation because the safety distance calculated during the simulation step T is reduced until it enters in the dangerous zone. In this state, the driver agent of following vehicle X must avoid colliding with the leading vehicle Y, by ensuring that it does not go reduce more its safety distance during the next simulation step $T + 1$. Thus, it reduces the velocity until that will be lower than the velocity of the leading vehicle Y. The modeling of this constraint translates according to the speed relationships between vehicles. The inequality (3) presents this modeling.

$$a < \left(V_y(T) - V_x(T)\right)/t \qquad (3)$$

According to the inequality (3), the following driver agent X decelerates to avoid collision with the leading vehicle Y, but in order to present a realistic simulation, the acceleration and deceleration values must be realistic. Therefore, there exist many works in the literature as [5, 8, 10] those present the acceleration and deceleration field. The selection of the deceleration value "a" should be between d_{min} and d_{max}.

$$d_{min} < \; = a < \; = d_{max} \qquad (4)$$

Moving to the second state "$D_{sec\_x}(T) < D_{xy}(T)$", the following vehicle X is far to the leading vehicle Y during the simulation step T. In this state, the following driver agent X can increase its velocity in order to ensure its objective to reduce the travel time. The modeling of this constraint is expressed by the inequality (5).

$$a > \left(V_y(T) - V_x(T)\right)/t \qquad (5)$$

According to the inequality (5), the following driver agent X may exceed the velocity of the leading vehicle Y, but at the same time, we must ensure that this increase will take place under two constraints: the road safety and the reality of the simulation. Thus, the translation of the road safety is expressed by the velocity relationships between vehicles and the gap $D_{xy}(T)$ of the simulation step T. This constraint is expressed by the inequality (6).

$$a < \left(D_{xy}(T)/t^2\right) + \left(V_y(T)/t\right) - \left(V_x(T)/t\right) \qquad (6)$$

The reality of simulation during the increase of velocity translates by the selection of the acceleration value "a" that should be between a_{min} and a_{max}.

$$a_{min} < \, = a < \, = a_{max} \qquad (7)$$

Our modeling is based on drivers with normative behaviors. According to [1], this type of driver respects firstly the traffic rules. Hence, even with our objective "maximize the velocity", the translation of this objective should be achieved by actions that respect the traffic rules. In this context, our modeling should ensure the respect of the maximum speed rule of the traffic zone. This constraint is expressed by the inequality (8).

$$a < \, = (V_{max} - V_x(T))/t \qquad (8)$$

Finally, the third state "$D_{sec\_x}(T) = D_{xy}(T)$", the following driver agent X tries to keep a perfect gap of the simulation step T for the next simulation step. In this state, the following driver agent X may act by two ways. In the first one, it can maintain the velocity of the simulation step T in order to ensure the objective maximize the velocity. In the second one, it can reduce its velocity in order to guarantee the road safety objective, since the leading driver agent Y can suddenly reduce its speed during the next simulation step. The modeling of this constraint translates by the velocity relationships between vehicles. This constraint expressed by the inequality (9).

$$a < \, = \big(V_y(T) - V_x(T)\big)/t \qquad (9)$$

The choice of the acceleration value is based on two scenarios. In the first one, the acceleration value should be equal to zero in order to maintain the same velocity of the leading driver agent. In the second one, the acceleration value varies between d_{min} and d_{max}. To choose between the two scenarios, we used a probability "p" for the first scenario and a probability "q" for the second.

$$\begin{cases} a = 0 & \text{with a probability p} \\ & \text{Or} \\ d_{min} < \, = a < \, = d_{max} & \text{with a probability q} = 1 - p \end{cases} \qquad (10)$$

Turning to present the constraints of the upper objective function Eq. (1), where we have two constraints. The first constraint presents the velocity that will be adopted in the simulation step $T+1$, where this constraint is in the same time an objective function, the Eq. (2) expresses this constraint. The second constraint ensures the road safety by avoiding the longitudinal collision, where the safety distance should be strictly greater than zero. Inequality (11) expresses this second constraint.

$$D_{sec\_x} > 0 \qquad (11)$$

3 Resolution of Modeling

3.1 Tabu Search Algorithm

Tabu search algorithm was proposed by Fred Glover in 1986 [6]. Since then, the method has become very popular, thanks to its successes to solve many problems.

In fact, the research Tabu algorithm is characterized by the rapidity during search [2]. Thus, this characteristic may help us; especially that we need to find the best solution with a research time equal to the decision time of driver agent.

We used the Tabu search in our decision strategy to select the best velocity and safety distance that present the realizable solution. In our search strategy, we used a Tabu list called Tabu list OUT; it contains the acceleration or deceleration values already selected and the acceleration or deceleration values that do not respect the constraints of our modeling. In this context, the values of the Tabu list OUT should not be selected during the research. At each simulation step, the list will be reset to zero. Moreover, the length of the Tabu list OUT is dynamic which adjusts itself during the search. In addition, the stopping criterion of the Tabu search algorithm is the calculation time. Indeed, while we used the Tabu search algorithm as a strategy for decision-making by searching the best solution that answers to the needs of the following driver agent, then we chose a random search time between 0.15 and 0.25 s, which is the stopping criterion to select the best realizable solution. This interval represents the decision time for an ordinary behavior driver [4].

3.2   A Research Based on a Multi-agent Cooperation

To select the best realizable solution that will be adopted during simulation step $T + 1$, the driver agent tries each acceleration and deceleration value that presents the decision variable of the second objective function "the maximization of the traffic speed". Thus, it chooses the best realization solution. Indeed, in this scenario, we applied the intensification technique, where according to [7]; this technique memorizes a list of high quality solutions and return to one of these solutions. But, since the stopping criterion of the Tabu search algorithm is the decision time, which is a very short duration, the driver agent can't try many acceleration and deceleration values, especially with very precise values that increase the range of values, which does not allow the research towards unexplored areas. Therefore, we divided the space of acceleration and deceleration values to a set of subspaces. Thus, by exploiting the parallel execution of the multi-agents system (MAS) technology, we affected for each subspace an agent called Tabu agent that applies the Tabu search algorithm in its space. Furthermore, the Tabu agents communicate between them during the research, in order to improve the realizable solution. Finally, when the decision time is achieved, the Tabu agent, that has the best realizable solution, sends its solution to the driver agent to execute it.

The values of various simulation parameters (a: the acceleration or deceleration value, $V_x(T)$: the velocity, $D_{sec\_x}(T)$: the calculated safety distance, $D_{xy}(T)$: the gap between the following driver and the leading driver) of the starting solution of the Tabu search algorithm are already calculated at the simulation step T and they will be sent by driver agent for each Tabu agent in order to make a decision concerning the safety distance and the velocity for the next simulation step $T + 1$. However, we assign a high value to the parameter D that presents the evaluation criteria for each realizable solution. Moreover, the simulation parameters (a', V'_x, D'_{sec_x}, D'_{xy}, and D') present the realizable solution before deciding to save it in the list IN or not. Furthermore, the simulation parameters (a'', V''_x, D''_{sec_x}, D''_{xy} and D'') present the realizable solution sent by another Tabu agent. Whereas, the simulation parameters (a, $V_x(T+1)$, $D_{sec\_x}(T+1)$, $D_{xy}(T+1)$ and D) present the realizable solution that will be saved in the list IN.

Algorithm. Tabu search algorithm applied by each Tabu agent

```
While  CurrentTime < DecisionTime  do
   a'◄─randomly select a value of the deceleration and
       acceleration domain.
   if a' ∉ TabuListOUT then
      TabuListOUT.add(a');
      V'ₓ ◄─calculate of the velocity by Eq. (2);
      D'sec_x ◄─calculate of the safety distance by Eq. (1);
      D'xy ◄─calculate of the gap by Eq. (13);
      D' ◄─calculate of the evaluate criterion by Eq. (12);
      if D > D' then
         Vₓ(T+1)◄─V'ₓ; Dsec_x(T+1)◄─D'sec_x; Dxy(T+1)◄─D'xy;
         D◄─D'; a ◄─a';
         foreach (TabuAgent : {AgentList}) do
            SEND (Vₓ(T+1), Dsec_x(T+1), Dxy(T+1), D, TabuAgent);
            // Send this best realizable solution to
               the others Tabu agents
   if (BOX != null) && (BOX.D'' < D) then
      //BOX is the dialog box for each Tabu agent
      Vₓ(T+1)◄─V''ₓ; Dsec_x(T+1)◄─D''sec_x; Dxy(T+1)◄─D''xy;
      D◄─D''; a◄─a'':
SEND (Vₓ(T+1), Dsec_x(T+1), Dxy(T+1), Driver_agent) ;
// Send the simulation parameters that presents the best
   realizable solution to the driver agent
```

The evaluation criteria D allows the selection of the best realizable solution that contains the best velocity and safety distance, and that corresponds to the objectives of the driver agent for the simulation step $T + 1$. Indeed, the realizable solution with the smallest value of evaluation criteria D will be selected for the next simulation step $T + 1$. The calculation of the evaluation criteria D is expressed by the Eq. (12).

$$D = \mid D_{xy}(T+1) - D_{sec\_x}(T+1) \mid \tag{12}$$

Where, the calculation of the gap $D_{xy}(T+1)$ is expressed by the Eq. (13).

$$D_{xy}(T+1) = D_{xy}(T) + (V_y(T) \times t) - (V_x(T) \times t) \tag{13}$$

4 Simulation

The objective of the proposed modeling is the anticipation of the simulation parameters for the next simulation step, in order to make a decision that ensures a real link between the reduction of the travel time and the guarantee of the road safety. In this context, we simulated our model with two cars in an urban zone with bottling conditions, where the leading vehicle circulates with a very slow speed. Therefore, our following agent driver should react with these conditions by basing during the decision-making on the concept of anticipation of the simulation parameters. For the simulation parameters of Eq. 1, we used the simulation parameters of Mitsubishi Free car 2.0 [4], which are, W: 1735 kg, G = 9.81(m/s²), ρ = 1.25, Af = 2.562m², Cd = 0.4, η = 0.6, μ = 0.8, fr = 0.015 and μ = 0.8, TA = 0.05–0.15 s, TD = 0.15–0.25 s. The duration for each simulation step, represented by the parameter t is 1 s. Furthermore, to ensure a realistic simulation of the acceleration or deceleration behavior, we used the acceleration and deceleration interval as defined in [8], where the acceleration values are between 0.9 m/s² and 3.6 m/s², and the deceleration values vary between 0.9 m/s² and 2.4 m/s².

According to the results that we have obtained by the simulation, as it is presented in Fig. 1, we note the homogeneity between the calculated safety distance $D_{sec\_x}(T+1)$ and the gap $D_{xy}(T+1)$ for each simulation step, where we have an acceptable distance margin between 0.0011 m and 0.9641 m. In fact, by basing on multi-agents cooperation with Tabu search algorithm, the distance margin interval is improved, which expresses the real link with our objectives. The Fig. 2 models the distance margin interval that presents the evaluation criterion "D" for the realizable solutions.

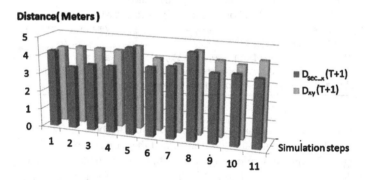

Fig. 1. Comparison between the calculated safety distance and the gap

In reality, it is impossible to keep a safety distance strictly equal to the calculated safety distance during circulation [13]. In this context, we have three curves on Fig. 3, the blue curve presents the leading vehicle velocity fixed to 3 m/s, the green curve presents the following vehicle based on MAS, the red curve presents the following vehicle without MAS. Indeed, with car-following driving behavior, the following driver tries to keep a gap around the calculated distance by controlling the velocity of its vehicle against the leading vehicle velocity. Thus, with a stable leading vehicle velocity, the following vehicle must have a velocity almost stable and close to the leading vehicle velocity. Thus, according to Fig. 3, the research results with a multi-agents cooperation are best to the research results without MAS.

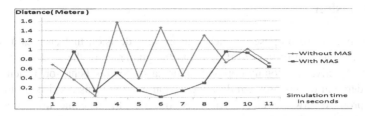

Fig. 2. Comparison of the evaluation criteria values D

According to results on Fig. 3, we consider the variance of the following vehicle velocity. Thus, this simulation reflects as much as possible the reality of our following driver agent reaction compared with research work [5] that presents results not real [13].

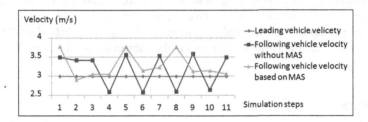

Fig. 3. Comparison of the velocity for each simulation step (Color figure online)

Indeed, the best result of the research based on a multi-agent cooperation is reflected by the number of tested acceleration and deceleration values, where by exploiting to the parallel execution of different Tabu agents in the same time, the number of realizable solution is increased, which improve the best realizable solution that will be adopted during the circulation. Figure 4 models the number of tested acceleration and deceleration values during 11 simulation steps for each research approach.

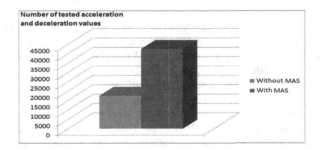

Fig. 4. Comparison of the number of tested acceleration and deceleration values

5 Conclusion

The objective of our proposed model is to make a decision that ensures a real link between the reduction of the travel time and the guarantee of the road safety. In fact, these two objectives are totally opposed to the car-following behavior, which requires the consideration of this problem during decision-making. This consists in finding the best compromise between velocity and safety distance. In this context, we proposed an approach based on the anticipation of the simulation parameters. Thus, to model the anticipation concept, we used a bi-level bi-objective modeling resolved by a Tabu search algorithm. Furthermore, the software agent technology has been used to model and to simulate drivers, in order to make them able to sense and to react according to their environment, changes thanks to their autonomy and reactivity features. Moreover, we used this technology to make a decision by the selection of the best realizable solution basing-on a multi-agent cooperation approach. The model implementation and experimentation provides promising results, since we obtained acceptable distance margins between the calculated safety distance and the gap for each simulation step.

The current model concerns only the driver with normative behaviors. Thus, the integration of the non-normative behaviors is the subject of our future work.

References

1. Arnaud, D., René, M., Sylvain, P., Stéphane, E.: A behavioral multi-agent model for road traffic simulation. Eng. Appl. Artif. Intell. **21**, 1443–1454 (2008)
2. Bajeh, A.O., Abolarinwa, K.O.: Optimization: a comparative study of genetic and Tabu search algorithms. Int. J. Comput. Appl. (0975–8887) **31**(5) (2011)
3. Barcelo, J., Ferrer, J., Grau, R., Florian, M., Chabini, E.: A route based version of the AIMSUN2 micro-simulation model. In: 2nd World Congree on ITS, Yokohama (1995)
4. Chen, Y.-L., Wang, C.-A.: Vehicle safety distance warning system: a novel algorithm for vehicle safety distance calculating between moving cars, 1550-2252/$25.00 ©16 IEEE (2007)
5. Gipps, P.G.: A behavioural car following model for computer simulation. Transp. Res. B **15** (2), 105–111 (1981)
6. Glover, F.: Tabu search-part I. ORSA J. Comput. **1**(3), 0899-1 499/89/0103-0190 $01.25 (1989)
7. Hao, J.K., Galinier, P., Habib, M.: Méthaheuristiques pour l'optimisation combinatoire et l'affectation sous contraintes. Revue d'Intelligence Artificielle, Vol: No. 1999 (1999)
8. ITE: Transportation and Traffic Engineering Handbook, 2nd Edn. Institute of Transportation Engineers, Prentice-Hall, Inc. New Jersey (1982)
9. Kometani, E., Sasaki, T.: Dynamic behaviour of traffic with a nonlinear spacing-speed relationship. In: Proceedings of the Symposium on Theory of Traffic Flow, Research Laboratories, General Motors, New York, pp. 105–119 (1959)
10. Liu, R., Van, V.D., Wating, D.P.: DRACULA: dynamic route assignment combining user learning and microsimulation. In: Proceedings of PTRC Summer Annual Conference, Seminar E, pp. 143–152 (1995)
11. Pacheco, J., Marti, R.: Tabu search for a multi-objective routing problem. J. Oper. Res. Soc. **57**(1), 29–37 (2006)

12. Qiang, L., Lunhui, X., Zhihui, C., Yanguo, H.: Simulation analysis and study on car-following safety distance model based on braking process of leading vehicle. In: Proceedings of the 8th World Congress on Intelligent Control and Automation, 978-1-61284-700-9/11/$26.00 ©2011 IEEE (2011)
13. Wilson, R.E.: An analysis of Gipps's car-following model of highway traffic. IMA J. Appl. Math. **66**, 509–537 (2001)
14. Wooldridge, M., Jennings, N.R.: Intelligent agents: theory and practice. Knowl. Eng. Rev. **10**(2), 115–152 (1995)
15. Yang, D., Zhu, L.L., Yu, D., Yang, F., Pu, Y.: An enhanced safe distance car-following model. J. Shanghai Jiaotong Univ. (Sci.) **19**(1), 115–122 (2014)

GRASP Applied to Multi–Skill Resource–Constrained Project Scheduling Problem

Paweł B. Myszkowski[⊠] and Jędrzej J. Siemieński

Department of Computational Intelligence,
Wrocław University of Technology, Wrocław, Poland
pawel.myszkowski@pwr.edu.pl

Abstract. The paper describes an application of Greedy Randomized Adaptive Search Procedure (GRASP) in solving Multi–Skill Resource-Constrained Project Scheduling Problem (MS-RCPSP). Proposed work proposes a specific greedy–based local search and schedule constructor specialised to MS-RCPSP. The GRASP is presented as the better option to classical heuristic but also as a faster and successful alternative to another metaheuristic. To compare results of GRASP to others approaches, various methods are proposed: methods of constructing scheduling based on the greedy algorithm, randomized greedy approach, and HAntCO. The research was performed using all instances of benchmark iMOPSE dataset and the results compared to best–known methods.

1 Introduction

Resource–Constrained Project Scheduling Problem (RCPSP) is one of the most investigated problems in computer science [5]. Moreover, it is overconstrained NP-hard problem, one of the most complicated problems of operational research. It is because of its practical nature, where results can be applied in many kinds of enterprises: logistic, healthcare, chemistry, production and much more. The goal of the RCPSP is to assign scarce resources to defined tasks to satisfy defined objectives. In most cases, the main objective is to minimize the schedule duration, often called as a makespan. However, there could also be other objectives, regarding the specific domain or requirements, e.g. cost, tardiness, resource usage, and many others.

One of the extensions of RCPSP is to include the skills domain to define the Multi–Skill RCPSP (MS-)RCPSP. In this problem, resources own some skills pool while every task requires some skill to be performed. Therefore, not every resource can be assigned to every task. It makes the solution space more constrained, but it also make the problem more realistic, especially in the field of project management in software development or production industry. Because of the combinatorial and NP–hard nature of the problem, it is very hard to find the optimal solution in reasonable time. As the solution space is often huge, scientists look for methods that would be capable of finding the reasonable solution in

© Springer International Publishing Switzerland 2016
N.T. Nguyen et al. (Eds.): ICCCI 2016, Part I, LNAI 9875, pp. 402–411, 2016.
DOI: 10.1007/978-3-319-45243-2_37

short, acceptable time. It encouraged scientists to look on heuristics and meta–heuristic methods, e.g. Evolutionary Algorithms, Swarm Intelligence methods, Tabu Search or Simulated Annealing. Especially the second group of methods provide very good results.

However, the non–deterministic character of metaheuristics and potential long computing time make that researchers often are focused on simpler heuristics. These methods usually cannot provide as good solutions as ones obtained from metaheuristic methods but their execution time is more acceptable. Therefore, greedy algorithms or other simple heuristics (like priority rules) [10] are investigated. The GRASP metaheuristic, presented in this paper, bases straight on the greedy heuristic, also using some elements of randomized sampling and adaptation methods, combining the simplicity of heuristics with the robustness of metaheuristics. This approach has been applied to MS–RCPSP and tested on the iMOPSE benchmark dataset [11].

In this work the GRASP application MS–RCPSP is investigated. The GRASP can be analyzed as the next step among random solution method, greedy approach and random greedy method. It is more efficient than those methods as it works iterative and generates one solution. GRASP can be considered as an alternative for more complex global optimization methods such as TS, SA or EA. Moreover, it can be linked to them improving its effectiveness: it can be used as initalisation method, local search in hybrids or boosting solution's neighborhood searching.

The main goal of this paper is to present the GRASP–based approach in solving MS–RCPSP, focusing on its robustness and effectiveness in comparison to other investigated methods. The rest of the paper is organized as follows. Section 2 presents short survey of existing works (state of the art). In Sect. 3 the problem statement with constraints and requirements are described. The proposed approach is given in Sect. 4. Finally, Sect. 5 shows performed experiments and results. Section 6 concludes the article and presents potential directions of future work.

2   Related Work

Heuristic approaches for solving MS–RCPSP were widely researched in past years, focused mainly on heuristics based on the priority rules [3, 10]. Priority rules are used to determine the order in which tasks are taken to the schedule builder – an algorithm that assigns provided tasks to given resource into the given timeslot, according to the algorithm rules. There could be different priority rules but the most often investigated in the literature base on the task's duration [10], its predecessors [3], start/finish time [3] or resource utilization [10].

Each of that heuristic can provide different order of tasks to be scheduled by schedule builder. However, those methods are mostly deterministic. Therefore, every launch without changing configuration should provide the same result. What is more, this kind of methods are very fast, because scheduling mechanism is often very fast, to the most time– and memory–consuming methods are those related to sorting the tasks according to given criteria.

Metaheuristics are very often used to solve RCPSP because of its NP–hard nature. Evolutionary Algorithms [5,8], Tabu Search [12,16], Simulated Annealing [4] and hybrids (e.g. [17]) are well explored and widely applied to solve MS–RCPSP. There are also many swarm intelligence approaches, solving (MS–) RCPSP. Ant Colony Optimization (ACO) or hybrid ACO [9], while PSO approaches can be found in [18], and other metaheuristics.

There is still lack of papers regarding multi–objective Multi–Skill extension of RCPSP. Some approaches solving MS–RCPSP in project duration domain [1,15] or project cost domain [6] could be found. On the other hand, there are methods solving classical RCPSP extended by cost domain but without skills considerations. Such research has been presented in [7,17]. Hence, we have decided to combine those two elements: multi–objective optimization and multi–skill domain for project scheduling problem.

Although classical RCPSP is deeply investigated and numerous approaches could be easily compared using PSPLIB instances, it is very hard to find multiobjective MS–RCPSP methods working on datasets that could be regarded as a benchmark. Some papers describe instances artificially generated, e.g. [15], while some others propose methods of PSPLIB dataset adaptation, e.g. [1,6]. However, both of those approaches for handling MS–RCPSP benchmark data are not supplied by any published dataset instances. Hence, we used the proposed IMOPSE benchmark dataset [11].

2.1   GRASP Applied to RCPSP

A GRASP is a multi–start and iterative process consists of two main steps performing in every iteration. The first step is a construction one when a feasible solution is created. Then in the second step, called local–search step, the local search is launched for the solution found in the first step and the final solution from those two steps is stored as a solution of given iteration. In the construction step, a solution is generated iteratively element–by–element from element candidate list, according to the greedy function. This function determines how good is to select given element. In every iteration, benefits related with given element are updated. That is why the method is called as *adaptive*. Not the best candidate has to be selected in every construction iteration, but close to the best one (restricted candidate list).

In the approach presented in [13] GRASP has been used in scheduling project networks, where resources are renewable. For the purpose of finding the minimal resource tardiness penalty cost the GRASP approach has been hybridized with the path relinking algorithm, developed in reactive and non–reactive versions. Additionally, bias probability function has been used to improve obtained results, what is confirmed in presented experiments results.

A slightly different approach has been presented in [14], where RCPSP approach has been combined with Scatter Search and justification. The justification is a heuristic that shifts the start date of given task without enlarging the whole project schedule makespan. Scatter search is a heuristic of building a new schedule from the subset of good schedules found in previous iterations, while the quality is not measured by the makespan or another objective, rather than by the diversity contributed by solution added to the set. This approach has been tested on PSPLIB dataset instances, providing relatively good and promising results.

Another application of GRASP in scheduling has been presented in [2] where this method has been applied in solving RCPSP with stochastic activity durations. Classical GRASP has been extended by double justification – scheduling tasks as late as possible without enlarging the makespan of the schedule and then scheduling tasks as soon as possible without shortening the makespan. The eligible task is selected according to the latest finish time or randomly to increase the randomness of the approach. What is more, the diversity has been enhanced in the procedure by taking the first task from the inverted list of elite set, if the probability condition is satisfied. This method has been tested on PSPLIB dataset instances, while obtained results show the robustness of this approach.

3   Problem Statement

The goal of the MS–RCPSP is to find feasible and optimal schedule. Optimal schedule is the one with the minimal objective function value f, while the feasible means – satisfying all constraints, related with tasks, resources, skills and precedence relations. Formally, the problem can be stated as follows:

$$f : \Omega \to \mathbb{R}, min(f) \tag{1}$$

where Ω is the feasible solution space (feasible schedules), while the f is the given objective function. The most common objective functions used in MS–RCPSP are f_C – schedule's performance cost optimization and f_τ – the project schedule duration's (makespan) optimization. In some approaches those two objective functions are merged and linear combined [9]. In this paper the schedule's duration optimization is considered, hence the f_τ would be used.

The feasible schedule (PH) consists of set of tasks $J = 1...n$, set of resources $K = 1..m$ and the set U defining, which task is assigned to given resource in specified time unit. Formally PH could be described as follows:

$$PH := (J, K, U) \tag{2}$$

To define the sets of tasks J and resources K, the set of skills Q ($|Q| = k$) has to be defined before. For given task in MS–RCPSP, the set of skills Q is defined as follows:

$$Q := \{q_i : \forall_{i \in \mathbb{N}, i < k} \ q_i := (h_{q_i}, l_{q_i})\} \tag{3}$$

while:

$$h_{q_i}, \ l_{q_i} \in \mathbb{N} \tag{4}$$

where: h_q is a q skill type and l_q is a q skill level. Having the skills set defined, the set of tasks J can be defined as follows:

$$J := \{j_i : \forall_{i \in \mathbb{N}, i < n} \ j_i := (d_{j_i}, S_{j_i}, F_{j_i}, q_{j_i}, P_{j_i})\} \tag{5}$$

While:
$$d_{j_i}, S_{j_i}, F_{j_i} \in \mathbb{N}, F_{j_i} = S_{j_i} + d_{j_i}, q_{j_i} \in Q, P_{j_i} \subset J, \tag{6}$$

where: d_j is a duration of task j, S_j is a start time of task j, F_j is a finish time of task j, q_j is a skill required by task j, P_j is a set of j–task predecessors. For a given task, its set of predecessors can be defined as follows:

$$\forall_{j \in J} \; \forall_{i \in P_j} \; F_i \leq S_j \tag{7}$$

Equation 7 defines constraint that a given task j cannot be started before its all predecessors P_j would be finished. Now, the set of resources K can be defined as follows:

$$K := \{k_i : \forall_{i \in \mathbb{N}, i < m} \; k_i := (S_{k_i}, Q_{k_i})\} \tag{8}$$

while:

$$S_{k_i} \in \mathbb{N} \setminus \{0\}, Q_{k_i} \subseteq Q, \tag{9}$$

where: S_k is a salary of resource k and Q_k is a set of skills owned by resource k. Every resource must have at least one skill. Formally it is defined as follows:

$$\forall_{k \in K} \; Q_k \neq \emptyset \tag{10}$$

For the k resource the subset of J_k tasks is defined as a subset of tasks that requires one of the skill owned by resource k. In other words, J_k is a subset of tasks that can be performed by resource k. Formally it can be stated as follows:

$$\forall_{k \in K}, \; \exists_{J_k \subseteq J}, \; \forall_{j \in J_k}, \; \exists_{q \in Q_k} \; h_q = h_{q_j} \wedge l_q \geq l_{q_j} \tag{11}$$

Time in the project schedule is defined as a discrete unit $t \in \mathbb{N}$. Therefore the project schedule duration τ (makespan) can be defined as the minimal element of set of timeslots that are equal or smaller of the every task's finish time. Formally it can be defined as follows:

$$\tau := min\{t \in \mathbb{N} : \forall_{j \in J} \; F_j \leq t\} f_\tau(PH) := \tau \tag{12}$$

For given j task, k resource and timestamp t, the variable $U_{j,k}^t \in \{0,1\}$ informing, whether the j task is assigned to k resource in the timestamp t. If this statement is true, then the value of $U_{j,k}^t$ is equal to 1, value 0 otherwise. Therefore we introduced the constraint that any resource can be assigned to no more than one task in given timestamp and that any task can be assigned to only one resource.

$$\forall_{j \in J}, \; \exists!_{t \leq \tau}, \; \exists!_{k \in K} \; U_{j,k}^t = 1 \tag{13}$$

The above constraint allows to define the set U as follows:

$$U := \{U_{j,k} \in \{0,1\} : j \in J, k \in K\} \tag{14}$$

According to given definitions, the execution time of performing the schedule can be defined as follows:

$$f_t(PH) := \Sigma_{k \in K} \Sigma_{j \in J} \; d_j s_k U_{j,k} \tag{15}$$

Such definition reduces MS–RCPSP to makespan minimization (optimization of project schedule duration) and in this work it is taken into consideration.

4   Proposed GRASP–Based Approach

GRASP is a multi–start metaheuristic in which each iteration is divided into two phases: construction and local search [13]. The main pseudocode of the GRASP is presented in Pseudocode 1. The GRASP procedure works iterative (*Iter*) to find near the best schedule in the first phase (construction), then the solution neighborhood is explored to find the best (locally) solution.

Pseudocode 1. Pseudocode of GRASP applied to MS–RCPSP

```
procedure GRASP ( Iter, LSIter, NSize, TasksSeq, alpha )
  BestSchedule:={}
  for i=1 to Iter
    Tlist:=createFeasibleRandomList(TasksSeq);
    schedule := RGB(TList, alpha);
    schedule' := LS(schedule, Tlist, LSIter, NSize);
    if f(BestSchedule) < f(schedule') BestSchedule:=schedule';
return BestSchedule.
```

The construction phase is based on a randomized greedy algorithm (see Pseudocode 2.). In each step of construction, the locally optimal choice is made. The procedure bases on a given sequence of the tasks (tasks sequence, *TasksSeq*) and for each task (*T*) a set of capable resources is created (*R*). Then all resources from *R* are ranked by the influence they have on the solution, creating a candidates list (*CL*), which is reduced by *alpha* $\in [0, 1]$ parameter. Next, the best-ranked resource is assigned to the task.

Pseudocode 2. Pseudocode of randomized greedy–based schedule builder (RGB)

```
procedure RGB ( TasksSeq, alpha )
  schedule :={};
  for all T in TasksSeq
    R  := getCapableResources(T);
    CL := rankResources(R);
    RCLsize := alpha*sizeof(CL);
    RCL := getBestResources(CL, RCLSize);
    schedule.assign(t,R);
  return schedule.
```

The construction method bases on task sequences, as different sequences of tasks, produces different final schedule. Thus, it allows to *stochastically sample* the solution landscape. The GRASP improves stochastic sampling by adding a local search phase (see Pseudocode 3.) after each construction phase. The goal of the local search is to explore a local optimum in the neighborhood (*nbd*) of the base solution *schedule* based on initial sequence of tasks *TaskSeq*. To define the neighborhood in the solution space, a neighborhood in the legal tasks sequence (*moves*) space is defined. A task sequence is legal when the order of the task does not break the precedence relation. A neighbor of task sequence *nbd* is created by swapping two tasks (*move*). A *move* is legal, when swapping the tasks does not break the precedence relation. The neighborhood's size is restricted to a given number (*NSize*) of moves which are picked randomly. Next, each *TaskSeq''* is used to build schedule and is evaluated by f. The whole LS procedure is repeated *LSIter* times and returns the best-found schedule.

Pseudocode 3. Pseudocode of Local Search (LS) phase

```
procedure LS ( schedule, TasksSeq, LSIter, NSize )
  BestSchedule:=schedule;
  BestTaskSeq:=TasksSeq;
  for i=1 to LSIter do
    ////// neighborhood (nbd) creation ////
    nbd:={};
    TasksSeq':=BestTaskSeq;
    moves:=getAllLegalMoves(TasksSeq');
    moves:=getRandomSubSequence(moves,NSize);
      for all m in moves do
        TasksSeq'.makeMove(m);
        nbd.add(TasksSeq');
    ////// neighborhood (nbd) exploration ////
    for all TaskSeq'' in nbd do
      schedule'':=RGB(TaskSeq'',alpha);
      if f(schedule'') > f(schedule)
              BNschedule:=schedule'';
              BestTaskSeq:=TaskSeq'';
      if f(BestSchedule) > f(BNschedule) BestSchedule:=BNschedule;
  return BestSchedule.
```

Proposed GRASP–based approach is nodeterministic metaheuristic which uses four parameters. The results of GRASP tuning and its influence to gained results are presented in next section.

5   Experiments and Results

The main goal of presented experiments is to investigate the effectiveness of proposed GRASP application to MS–RCPSP in comparison to alternative methods. To compare experiments results benchmark *IMOPSE* dataset is used.

5.1   IMOPSE Dataset and Setup

The benchmark *iMOPSE* dataset contains 36 project instances and is created [11] on the basis of real–world instances. The iMOPSE dataset is used in basic form [10,16], however extended iMOPSE is used in [9]. In this work the extended iMOPSE dataset is used without calendar restrictions in duration optimization (*DO*) mode.

The main motivation of experiments is to gain the best GRASP results and compare to other methods. We decided to use the comparison to classical Greedy Algorithm and Randomized Greedy Algorithm. To compare results to reference method we used HAntCO [9], where results are computed using the same assumptions: only DO optimization mode and no calendar restrictions. We used standard Java language programming to develop all experiments.

The GRASP($Iter$, $LSIter$, $NSize$) uses three[1] main parameters: number of GRASP iterations ($Iter$), number of local search iterations ($LSIter$) and size of neighborhood ($NSize$) used in local search procedure. We examined several GRASP configurations but only 3 configurations we selected as interesting: GRASP(40,10,40), GRASP(12,40,40) and GRASP(40,40,40). Comparison of gained results is presented in next section. To investigate non–deterministic GRASP effectiveness experiments procedure requires each repeated 50 times to compare averaged results and prove the statistical significance in Wilcoxon signed-rank test for all 36 instances of the iMOPSE dataset.

5.2   Results Discussion

Due to the limited space of paper, in Table 1. We presented only selected results of experiments. Two classical methods based on greedy algorithm give the only benchmark for GRASP because results cannot successfully compete to GRASP. Both provide comparable solutions. The GRASP(40,10,40) results show that lower number of $LSIter$ makes solution worse. The results of second GRASP(12,40,40) configuration give a clue if reduced number of GRASP $Iter$ has the influence to method's results. However, GRASP gains much better results than reference methods in almost each case in each examined configuration. Averaged results of two GRASP configurations GRASP(40,10,40) and GRASP(12,40,40) show that effectiveness is similar because GRASP(40,10,40) uses average 348.98 ± 8.70 and second one 349.72 ± 4.4. The best configuration GRASP(12,40,40) gives in 29 cases, but statistical significance is in 24 cases. For three instances results are worse (D4, D5 and D6) – these instances are very complex by nature as the method needs more iterations to leave local optima. Thus, in this case, it showed that reduced GRASP parameter $Iter = 12$ has the negative influence.

The best-examined configuration we found GRASP(40,40,40) where the average iMOPSE schedule lasts 341.37 ± 2.58. Results only in 5 cases are comparable to GRASP(12,40,40), but for 31 instances are better, what proves statistical significance Wilcoxon signed–rank test results ($W_{0.05} = 646 > W_c = 208$).

To compare GRASP results[2] to reference method we selected HAntCO [9]. Results presented in Table 1. Show that GRASP can compete to other metaheuristics because solutions gained by GRASP (average project duration equals to 341 ± 2.58) for each iMOPSE instance is shorter that HAntCO (405 ± 5.11).

---

[1] parameter $alpha = 0.3$ used in RGB procedure was established experimentally.

[2] All best found GRASP solutions we published on iMOPSE project homepage: http://imopse.ii.pwr.edu.pl.

Table 1. Summary and comparision of GRASP results using iMOPSE dataset

instance	Greedy			GreedyRandomized			GRASP(40,10,40)			GRASP(12,40,40)			GRASP(40,40,40)			HAntCO[9]		
	Best	Avg	Std	Best	Avg	Std	Best	Avg	Std	Best	Avg	Std	Best	Avg	Std	Best	Avg	Std
100_10_26_15	293	300.6	5.71	280	293.4	6.41	260	267	4.31	250	255.3	3.29	251	252.8	3.19	266	270	2.2
100_10_27_9_D2	244	249.9	5.89	252	261.7	5.8	223	227.2	2.82	222	225	2.19	221	222.7	1.62	294	306	6.7
100_10_47_9	276	281.4	2.94	285	289.9	4.53	264	267	1.73	263	265.5	1.75	263	264.1	0.7	297	302	4
100_10_48_15	280	288.9	6.39	276	288.9	7.23	257	262.4	3.41	255	259.5	1.86	256	257.2	1.17	278	286	4.4
100_10_64_9	275	287.1	5.65	289	298.7	6.77	261	262.9	1.51	254	258.9	2.55	255	257.5	1.91	287	296	5.1
100_10_65_15	286	298.4	9.08	287	304.5	7.92	260	265.6	4.65	262	265.2	2.18	256	260.8	1.94	281	287	3.5
100_20_22_15	156	165.1	5.17	154	162.3	5.2	139	142.1	2.3	136	140	2.57	134	137.1	1.58	161	170	3.6
100_20_23_9_D1	172	175.4	5.59	172	174.9	4.72	172	172	0	172	172	0	172	172	0	219	223	3
100_20_46_15	199	221.9	8.79	209	217.9	4.83	180	187.2	6.03	172	179.7	3.87	170	174.1	3.08	194	199	2.6
100_20_47_9	165	174.7	5.71	165	172.1	3.86	142	146.8	2.48	139	142.6	2.01	133	139.9	2.21	180	187	3.7
100_20_65_15	222	240.2	8.69	220	246	10.15	213	213.5	1.5	213	213	0	213	213	0	218	220	2.7
100_20_65_9	151	160.5	5.04	150	163.8	6.06	134	139.3	2.28	135	137.7	1.9	135	135	1.41	180	185	3.1
100_5_20_9_D3	413	423.2	6.45	412	425.7	6.48	401	403.8	1.33	401	402	0.63	401	402	0.63	437	442	5
100_5_22_15	508	514.7	3	512	520.6	4	504	504.5	0.5	503	504	0.63	503	503.6	0.49	504	505	0.8
100_5_46_15	615	625.3	10.95	610	627.3	8.71	565	570.3	3.35	557	561.1	2.91	552	555.9	2.59	604	604	0
100_5_48_9	541	553.3	7.48	526	543.7	8.91	511	513.1	1.3	509	510.5	0.81	510	510.2	0.75	521	523	1.6
100_5_64_15	528	522.2	11.41	538	550.6	8.51	505	511	3.1	502	504.6	2.11	501	502.7	0.78	516	523	2.9
100_5_64_9	502	516.5	7.63	518	528.1	5.72	494	496.5	1.2	494	495.9	1.3	494	494.3	0.64	507	515	3.9
200_10_128_15	524	551.3	13.04	540	559.4	10	504	508.1	2.91	493	502.4	6.26	491	496.4	4.22	522	528	3.1
200_10_135_9_D6	689	747.9	27.32	691	732.1	27.38	619	656.4	19.11	672	699.8	13.37	584	617.4	20.31	1115	1133	11.4
200_10_50_15	584	597.9	8.94	570	583.8	8.67	531	545.5	6.84	522	536.4	7.64	524	528.1	3.78	529	543	8
200_10_50_9	528	546.2	10.58	531	548.7	8.14	508	512.2	2.27	506	511.3	3.95	506	508.3	1.42	546	556	4.8
200_10_84_9	535	537.5	1.75	537	543.8	3.52	528	529.2	1.08	527	528.4	0.8	526	527.3	0.78	571	581	6.9
200_10_85_15	513	522.7	4.86	517	529.6	8.21	498	502.1	2.77	498	499.8	1.47	496	498.3	1.1	526	538	6.5
200_20_145_15	295	328.8	13.58	293	310.1	7.34	277	286.1	4.93	273	281	5.12	262	271.4	4.18	309	314	4.4
200_20_150_9_D5	971	1019.8	26.54	927	1005.1	37.4	916	948.9	25.6	912	982.8	40.28	900	912.5	13.58	1177	1234	27.6
200_20_54_15	343	352.9	6.86	341	352.1	6.52	306	315.9	5.05	308	313.6	3.5	304	308	3.66	336	349	6.6
200_20_55_9	266	270	2.72	271	277.8	2.64	259	260.4	1.02	258	258.6	0.87	257	258.2	0.6	313	317	3.1
200_20_97_15	351	366.1	8.98	366	372	4	347	347	0	347	347	0	347	347	0	356	368	4.8
200_20_97_9	266	274.4	4.98	278	284.7	4.52	257	259.7	1.73	254	257.1	2.12	253	256	1.55	326	332	4.1
200_40_130_9_D4	516	555.8	16.96	513	513.5	1.2	513	514.1	2.12	530	573.9	25.09	513	513	0	642	675	19.2
200_40_133_15	190	212.7	11.23	194	201	4.77	171	176.6	4.22	165	174.5	4.98	163	169.9	4.06	214	221	3.4
200_40_45_15	164	171	5.67	168	177.3	3.8	164	164	0	164	164	0	164	164	0	206	212	2.6
200_40_45_9	164	172.9	5.28	170	176.4	5.02	148	152.4	2.69	144	148.3	2.69	144	146.5	1.63	209	215	3.8
200_40_90_9	172	192.7	9.83	176	185	4.1	152	160.4	4.08	145	153.5	4.86	148	153.2	2.79	211	214	1.8
200_40_91_15	182	208.3	12.37	186	195.1	6.25	161	172	6.93	158	164.9	2.84	153	159	4.54	207	212	2.9
average	363	379	8.70	365	378	7.48	343	349	3.81	342	350	4.40	338	341	2.58	396	405	5.11

The difference is significant statistically what proves the result of Wilcoxon signed–rank test results ($W_{0.05} = 666 > W_c = 208$). Results of experiments showed that GRASP using all iMOPSE instances is 15.8 % efficient in average than HAntCO.

6   Conclusions and Future Work

In this paper GRASP has been proposed as a successful application to MS–RCPSP. The experiments have been conduced using benchmark IMOPSE dataset optimizing of project duration. There have been investigated three basic configurations of GRASP and gained results have been compared to two greedy–based heuristic and reference HAntCO. All results have been repeated and averaged, compared and investigated by statistical test to prove the statistical significance.

Results of presented experiments show that GRASP is quite a good alternative to heuristic and also can be an inspiration for metaheuristic construction. Particularly, very promising can be a multistart strategy that can be used to non–population metaheuristic (like TS, SA etc.) to improve effectiveness. At least but not last, GRASP can be used as part of the metaheuristic as effective local search method or inital population constructor. Another application of GRASP is connected to hybridize metaheuristic such as EA to build boost evolution. Moreover, MS-RCPSP in its nature is multiobjective and in this direction we want to extend GRASP application.

References

1. Al-Anzi, F., Al-Zamel, K., Allahverdi, A.: Weighted multi-skill resources project scheduling. J. Sof. Eng. App. **3**, 1125–1130 (2010)
2. Ballestin, F., Leus, R.: Resource-constrained project scheduling for timely project completion with stochastic activity durations. Prod. Oper. Manage. **18**(4), 459–474 (2009)
3. Boctor, F.: Heuristics for scheduling projects with resource restrictions and several resource-duration modes. Inter. J. Prod. Res. **31**(1), 2547–2558 (1993)
4. Bouleimen, K., Lecoc, H.: A new efficient simulated annealing algorithm for the resource-constrained project scheduling problem and its multiple mode version. Eur. J. Oper. Res. **149**, 268–281 (2003)
5. Hartmann, S., Briskorn, D.: A survey of variants, extensions of the resource-constrained project scheduling problem. Eur. J. Oper. Res. **207**, 1–14 (2010)
6. Li, H., Womer, K.: Scheduling projects with multi-skilled personnel by a hybrid milp/cp benders decomposition algorithm. J. Sched. **12**, 281–298 (2009)
7. Luna, F., Gonzalez-Alvarez, D., Chicano, F., Vega-Rodriguez, M.: A scalability analysis of multi-objective metaheuristics. Appl. Soft Comput. **15**, 136–148 (2014)
8. Mendes, J., Goncalves, J., Resende, M.: A radom key based genetic algorithm for the resource constrained project scheduling problem. Comp. Oper. Res. **36**, 92–109 (2009)
9. Myszkowski, P., Skowroński, M., Olech, Ł., Oślizło, K.: Hybrid ant colony optimization in solving multi-skill resource-constrained project scheduling problem. Soft Comput. **19**(12), 3599–3619 (2015). doi:10.1007/s00500-014-1455-x
10. Myszkowski, P., Skowroński, M., and Podlodowski, Ł.: Novel heuristic solutions for multi-skill resource-constrained project scheduling problem. In: Federated Conference on Computer Science and Information Systems, pp. 159–166 (2013)
11. Myszkowski, P., Skowroński, M., Sikora, M.: A new benchmark dataset for multi-skill resource-constrained project scheduling problem. In: Federated Conference on Computer Science and Information Systems, ACSIS, vol. 5, pp. 129–138 (2015)
12. Poppenborg, J., Knust, S.: A flow-based tabu search algorithm for the rcpsp with transfer times. OR Spectrum (2015). doi:10.1007/s00291-015-0402-2
13. Ranjbar, M.: A hybrid grasp algorithm for minimizing total weighted resource tardiness penalty costs in scheduling of project networks. Inter. J. Indust. Eng. Prod. Res. **23**(3), 231–234 (2012)
14. Rivera, J., Moreno, L., Diaz, F., Pena, G.: A hybrid heuristic algorithm for solving the resource constrained project scheduling problem (rcpsp). Revista EIA **10**(20), 87–100 (2013)
15. Santos, M., Tereso, A.: On the multi-mode, multi-skill resource constrained project scheduling problem-computational results. In: ICOPEV 2011, pp. 93–99 (2011)
16. Skowroński, M., Myszkowski, P., Adamski, M., Kwiatek, P.: Tabu search approach for multi-skill resource-constrained project scheduling problem. In: Federated Conference on Computer Science and Information Systems, pp. 153–158 (2013)
17. Yannibelli, V., Amandi, A.: Hybridizing a multi-objective simulated annealing algorithm with a multi-objective evolutionary algorithm to solve a multi-objective project scheduling problem. Expert Syst. App. **40**, 2421–2434 (2013)
18. Zhang, K., Zhao, G., Jiang, J.: Particle swarm optimization method for resource-constrained project scheduling problem. In: Conference on ICEMI 2009, pp. 792–796 (2009)

Biological Regulation and Psychological Mechanisms Models of Adaptive Decision-Making Behaviors: Drives, Emotions, and Personality

Amine Chohra$^{(\boxtimes)}$ and Kurosh Madani

Images, Signals, and Intelligent Systems Laboratory (LISSI/EA 3956),
Paris-East University (UPEC), Senart Institute of Technology,
Avenue Pierre Point, 77127 Lieusaint, France
{chohra,madani}@u-pec.fr

Abstract. The aim of this paper is to suggest a framework for adaptive agent decision-making modeling of biological regulation and psychological mechanisms. For this purpose, first, a perception-action cycle scheme for the agent-environment interactions and deduced framework for adaptive agent decision-making modeling are developed. Second, motivation systems: drives (homeostatic regulation), personality traits (five-factor model), and emotions (basic emotions) are developed. Third, a neural architecture implementation of the framework is suggested. Then, first tests related to a stimulation-drive (from a moving object), for two different agent personalities, and the activation level of emotions are presented and analyzed. The obtained results demonstrate how the personality and emotion of the agent can be used to regulate the intensity of the interaction; predicting a promising result in future: to demonstrate how the nature of the interaction (stimulation-drive, social-drive, ...) influences the agent behavior which could be very interesting for cooperative agents.

Keywords: Complex systems · Decision-making · Agent-environment interactions · Perception-action cycle · Adaptive goal-directed behavior

1 Introduction

The cognitive science implies the cognition (computation, "information processing psychology", manipulation of data structures stored in memory, formal operations carried out on symbol structures), perception, and action [1–3]. Its basic aim is identifying the functional architecture of cognition, in terms of rules and representations as well as a form that is more analog and more biologically plausible, that mediate thought.

Researchers from *artificial intelligence*, *computer science*, *brain and cognitive science*, and *psychology* have been oriented, by the end of 1980s, towards a new field to build intelligent machines called *embodied cognitive science* or *new artificial intelligence* or *behavior-based artificial intelligence* [4].

© Springer International Publishing Switzerland 2016
N.T. Nguyen et al. (Eds.): ICCCI 2016, Part I, LNAI 9875, pp. 412–422, 2016.
DOI: 10.1007/978-3-319-45243-2_38

The brain does not run 'programs': it does something entirely different, i.e., it does not do mathematical proofs, but *controls behavior*, to ensure our survival [4]. The researchers from these various disciplines agreed that *intelligence* always manifests itself in *behavior* and consequently that we must understand the behavior [4]. In fact, a particular attention must be given on thinking and high-level cognition focusing on the interaction with the real world. This interaction is always mediated by a body, i.e., the intelligence needs to be '*embodied*'. This, has rapidly changed the research disciplines of artificial intelligence and cognitive science towards a new research field which is exerting more and more its influence on psychology, neurobiology, and ethology, as well as engineering science.

By another way, throughout recorded human intellectual history, there has been active debate about the nature of the role of emotions or "passions" in human behavior [5], with the dominant view being that passions are a negative force in human behavior [6]. By contrast, some of the latest research has been characterized by a new appreciation of the positive functions served by emotions [7]. An appreciation for the positive functions is not entirely new in behavioral science. Darwin, in 1872, was one of the first to hypothesize the adaptive mechanisms through which emotion might guide human behavior [8].

The great interest and large investigations in decision-making research associated with the emergence of behavioral decision theory, then, largely ignored the role played by the irrationality part (related to affect, in general) in decision-making [5]. However, with the research developments particularly psychology-related fields from 1990s, a great interest have been oriented towards the role of the irrationality part (related to affect) in decision-making.

The aim of this paper is to suggest a framework for adaptive agent decision-making modeling of biological regulation and psychological mechanisms. It integrates drives, personality traits, and emotions in order to:

- use this framework as a test bed to test, analyze, and compare different pertinent models of drives developed from biological regulation and survival of social organisms, personality traits developed from the field of personality psychology, and emotions, central aspects of biological regulation,
- analyze the impacts (effects), to asses the variation consequences of different personality and emotion aspects on the decision agents make,
- emphasize the adaptive behaviors emerging from agent-environment interactions.

For this purpose, first, a perception-action cycle scheme for the agent-environment interactions and deduced framework for adaptive agent decision-making modeling are developed in Sect. 2. Second, motivation systems: drives, personality traits, and emotions are developed in Sect. 3. Third, a neural architecture implementation of the framework is suggested in Sect. 4. Then, in Sect. 5 first tests related to a stimulation-drive, for two different agent personalities, and the activation level of emotions are presented and analyzed.

2 Agent-Environment Interactions (Perception-Action Cycle)

The perception-action cycle scheme for the agent-environment interactions, suggested in Fig. 1(a), is inspired from [9].

The agent and environment are coupled in two ways:

– an informational function (which maps properties of the agent-environment system into informational variables, in accordance with laws of ecological perception-action approach to the control of behavior):

$$\mathbf{i} = \lambda(\mathbf{e}), \tag{1}$$

where $\mathbf{i}$ is a vector of informational variables, $\mathbf{e}$ is a vector of environmental variables;

– an effector function (which transforms the vector of action variables into muscle activation patterns that produce forces in the environment, action is thus characterized as a relation defined over the agent, causal forces, and the environment):

$$\mathbf{f} = \beta(\mathbf{a}), \tag{2}$$

where $\mathbf{f}$ is a vector of external forces, and $\mathbf{a}$ is a vector of agent state variables (which describes the current state of the action system).

Thus, the adaptive, goal-directed behavior emerges from these local interactions between an agent governed by the control laws Ψ and an environment governed by the physical laws Φ such as:

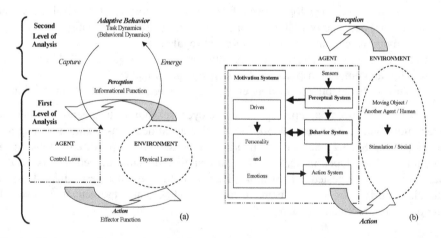

Fig. 1. (a) Agent-environment interactions (perception-action cycle). (b) A framework for agent decision-making modeling in behavior-based systems.

$$\dot{\mathbf{a}} = \Psi(\mathbf{a}, \mathbf{i}),$$
$$\dot{\mathbf{e}} = \Phi(\mathbf{e}, \mathbf{f}). \tag{3}$$

Indeed, adaptive behavior emerges from task dynamics or behavioral dynamics (information-based approach to perception, dynamical systems approach to action).

From this, the deduced framework for agent decision-making modeling in behavior-based systems is then suggested in Fig. 1(b).

This framework integrates particularly the motivation systems: drives (homeostatic regulation), personality traits (five-factors model), and emotions (basic emotions).

3 Motivation Systems

In this Section, the motivation systems consists of:
- an homeostatic regulation system implementing the drives of the agent,
- an emotion system implementing the emotions of the agent,
- a personality system implementing the personality traits of the agent.

3.1 Drives (Homeostatic Regulation)

The design of homeostatic regulation system is inspired by ethological views of the analogous process in animals [10, 11]. However, it is a simplified and idealized model of those discovered in living systems. The drive features are: its temporally cyclic behavior with three regimes (under-stimulated, homeostatic, and overwhelming), i.e., if no stimulation, a drive will tend to increase in intensity unless it is satiated. This is analogous to an animal's degree of hunger or level of fatigue, both following a cyclical pattern [7, 12, 13].

3.2 Personality (Five-Factor Model)

Social psychologists believe that human behavior is determined by both a person's characteristics and the social situation. They also believe that the social situation is frequently a stronger influence on behavior than are a person's characteristics [14]. From this purpose, it is very important to integrate the personality traits in the suggested framework.

The model in personality which appears to represent a major conceptual and empirical advance in the field of personality psychology is the five-factor model in personality (Extraversion, Agreeableness, Conscientiousness, Neuroticism, and Openness to experience) developed in [15].

3.3 Emotions (Basic Emotions)

Emotions are not a luxury, they play a role in communicating meanings to others, and they may also play the cognitive guidance role [7].

In fact, emotions are another important motivation system for complex organisms. They seem to be centrally involved in determining the behavioral reaction to environmental (often social) and internal events of major significance for the needs and goals of a creature [16, 17].

4 Neural Architecture Implementation

In the design and achievement of the architecture of the suggested framework for agent decision-making modeling, first, carry out the aspects of computation and cognition (issues in the foundations of cognitive science) related to cognitive functional architecture, respecting the biological and psychological nature under the cognitive impenetrability condition (non influence by purely cognitive factors as goals, beliefs, inferences, tacit knowledge, ...) is of great importance [1, 2]. This allows the fixed capacities of mind (called its functional architecture) avoiding the particular representations and algorithms used on specific occasions. This in turn requires that the fixed architectural function and the algorithms be independently validated in order to examine the fundamental distinction between a behavior governed by rules and representations, and a behavior that is merely the result of the causal structure of the underlying biological system.

Second importance is related to the neural aspect of functional architecture which is highly distributed network of interacting neurons [3, 13, 18].

The basic computational process, implemented as a value based system (influences graded in intensity, instead of simply being "on" or "off"), is modeled by its activation level A_i which is computed by Eq. (4).

$$A_i = (\sum_{j=1}^{n} \omega_{ji}\, i_j) + b, \tag{4}$$

where the inputs i_j are integer values, the weights ω_{ji}, and the bias b, over the number of inputs n. The process is active when the activation level A_i exceeds a threshold T.

The weights can be either positive or negative; a positive weight corresponds to an excitatory connection and a negative weight corresponds to an inhibitory connection.

Each perceptual unit, drive, emotion, personality, behavior, and motor process is modeled as a different type that is specifically tailored for its role in the overall architecture.

4.1 Perceptual Units

The antecedent conditions come through the perceptual system where they are assessed with respect to the agent's "well being" and active goals. Thus, the perceptual system, which is inspired from [11], is built of perceptual units:

– related to stimulus (from moving object, another agent, human) as shown in Fig. 2
  (a), i.e., there is a set of perceptual units defined for each stimulus, that indicate its

presence (time: short, medium, long), absence (time: short, medium, long), nature (moving object, another agent, or human), quality (intensity of stimulus: too low, just right, too high), desirability (desired or not desired),

- related to drives (stimulation drive, social drive, ...) as shown in Fig. 2(b), i.e., there is a perceptual unit defined for each regime (under-stimulated, homeostatic, overwhelming) of each drive, to represent how well each drive is being satiated,
- related to behaviors as shown in Fig. 3(a), i.e., there is a set of perceptual units defined for each behavior, that indicate whether its goal has been achieved or not, and if not, then for how long.

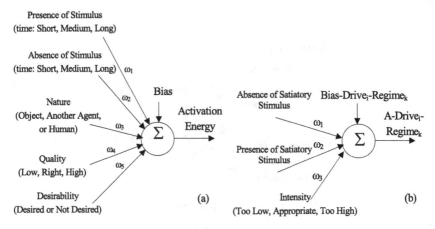

Fig. 2. (a) Perceptual unit implementation related to stimulus. (b) Perceptual unit implementation related to drives.

Note that the perceptual unit implementation is designed for each regime k (under-stimulated, homeostatic, overwhelming) of each drive i (stimulation-drive, social-drive, ...).

4.2 Drives

The drive model given in Fig. 3(b) concerns the Stimulation-Drive leading to a cyclic behavior of a drive, where t is a temporal input; given no stimulation, a drive will tend to increase in intensity ($A_{Stimulation-Drive}$) unless it is satiated (homeostatic regime). Note that a similar drive model will be implemented in future concerning Social-Drive.

4.3 Personality

The five-factor model in personality, suggested in Fig. 4, is inspired from [15, 19, 20] implying five broad dimensions which are used to describe human personality: Openness to experience (Op), Agreeableness (Ag), Conscientiousness (Co), Extraversion (Ex), and Neuroticism (Ne).

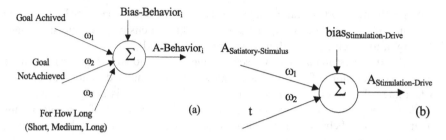

Fig. 3. (a) Perceptual unit implementation related to behaviors. (b) Drive implementation ($A_{Stimulation-Drive}$).

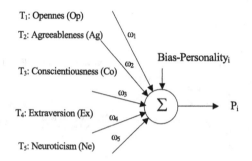

Fig. 4. Personality model.

Thus, combined personality value P_i affecting a behavior is defined in Eq. (5) as:

$$P_i = \left(\sum_{j=1}^{5} \omega_{ji} T_j\right) + b, \tag{5}$$

where T_j denotes the intensity of each personality parameter, and ω_{ji} the influence (inverse influence -1, no influence 0, direct influence $+1$) on a particular behavior.

4.4   Emotions

The relations between emotions and behavioural responses, i.e., under what conditions certain "emotions" and behavioural responses arise, are given in Table 1. This table is derived from the evolutionary, cross-species, and social functions hypothesized by [8, 16, 17]. Then, the perceptual behavioral or motivational information is tagged (arousal, valence, stance) with affective information [7]. Note that the stance is related to how approachable the percept is to the agent. Moreover, each regime of a drive biases arousal and valence differently, contributing to the activation of different emotions.

Table 1. Relations between emotions and behaviors under antecedent conditions.

Antecedent conditions	Emotion	Behavior
Difficulty in achieving goal	Anger	Complain
Presence of an undesired stimulus	Disgust	Withdraw
Presence of a threatening (overwhelming)	Fear	Escape
Prolonged presence of a desired stimulus	Calm	Engage
Success in achieving goal of active behavior	Joy	Display pleasure
Prolonged absence of a desired stimulus	Sadness	Display sorrow
A sudden, close stimulus	Surprise	Startle response
Appearance of a desired stimulus	Interest	Orient
Need of an absent and desired stimulus	Boredom	Seek

Table 2. Influence and intensity of personality parameters, Openness to experience (Op), Agreeableness (Ag), Conscientiousness (Co), Extraversion (Ex), and Neuroticism (Ne).

Agent		Op	Ag	Co	Ex	Ne
Personality1:	Influence	0	−1	0	0	+1
	Intensity		2			4
Personality2:	Influence	0	−1	0	0	+1
	Intensity		3			7

5 Tests and Results

In this Section, first tests related to a Stimulation-Drive, from a Stimulus of a moving object, presented in Fig. 5, for two different agent personalities, and the activation level of emotions are presented and analyzed. Influence and intensity parameters presented in Table 2 concern two different personalities (since personality trait of Neuroticism has been found to influence avoidance behavior [21], Escape in Table 1 in our concern, and Agreeableness to inversely influence it):

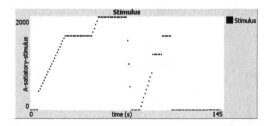

Fig. 5. The used Stimulus (A-Satiatory-Stimulus).

In Fig. 7, the activation of emotions: Sadness (corresponding to prolonged absence of a desired stimulus, as shown in Table 1), followed by Interest (appearance of a desired stimulus), then Fear (presence of a threatening, corresponding to overwhelming stimulus, as shown in Fig. 6), followed by Interest (appearance of a desired stimulus), and finally Sadness (prolonged absence of a desired stimulus).

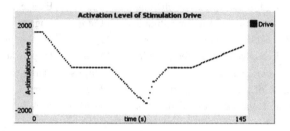

Fig. 6. Stimulation-Drive results (from the Stimulus, Fig. 5).

Note that in Fig. 7, Fear appears and crosses Interest the first time, around t = 65 s, for Agent-Personality1.

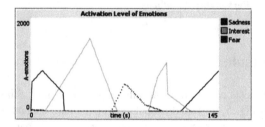

Fig. 7. Activation of emotions in case of agent-personality1 (from Stimulation-Drive Fig. 6).

In Fig. 8, Fear appears and crosses Interest the first time, around t = 63 s, for Agent-Personality2. Note that in this case Fear appears and crosses Interest earlier than in the case of the Agent-Personality1; moreover, Fear in the case of the Agent-Personality2 reaches more intensity than in the case of the Agent-Personality1.

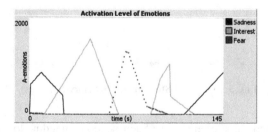

Fig. 8. Activation of emotions in case of agent-personality2 (from the same Stimulus, Fig. 5).

This can be explained by the following:

- in Personality 1, from Eq. (5):

$$P_i = (-1) * (2) + (+1) * 4 + b = 2 + b;$$

- in Personality 2, from Eq. (5):

$$P_i = (-1) * (3) + (+1) * 7 + b = 4 + b.$$

This means that Personality 2 will demonstrate the avoidance behavior, Escape, (earlier and with great intensity) than in the case of Personality 1.

Note that, as personality traits of one agent, remain invariant throughout execution, the corresponding behavioral P_i is computed only once at the beginning of execution.

Thus, the personality effect is indirect, i.e., it influences emotion generation rather than the behaviors themselves.

6 Discussion and Conclusion

In this paper, a framework for adaptive agent decision-making modeling, deduced from a perception-action cycle scheme, has been suggested. Then, motivation systems: drives (homeostatic regulation), personality traits (five-factor model), and emotions (basic emotions) have been developed. Afterwards, a neural architecture implementation of the framework has been suggested.

The first tests related to a stimulation-drive (from a moving object), for two different agent personalities, and the activation level of emotions are presented and analyzed. The obtained results demonstrate how the personality and emotion of the agent can be used to regulate the intensity of the interaction; predicting a promising result in future: to demonstrate how the nature of the interaction (stimulation-drive, social-drive, ...) influences the agent behavior which could be very interesting for cooperative agents.

After investigating the behavior system, the action system, and the social-drive to test the interactions agent-agent and human-agent, it is interesting to investigate different cooperative strategies with different emotion regulation strategies [22], and the learning from interaction [4, 23].

References

1. Pylyshyn, Z.W.: Computation and cognition: issues in the foundations of cognitive science. Behav. Brain Sci. **3**, 111–169 (1980)
2. Pylyshyn, Z.W.: Computation and Cognition: Towards a Foundation for Cognitive Science. The MIT Press, Cambridge (1984)

3. Dawson, M.R.W.: From embodied cognitive science to synthetic psychology. In: Proceedings of the First IEEE International Conference on Cognitive Informatics, pp. 13–22 (2002)
4. Pfeifer, R., Scheier, C.: Understanding Intelligence. MIT Press, Cambridge (1999)
5. Loewenstein, G., Lerner, J.S.: The role of affect in decision-making. In: Davidson, R.J., Scherer, K.R., Goldsmith, H.H. (eds.) Handbook of Affective Sciences, pp. 619–642. Oxford University Press, Oxford (2003)
6. Elster, J.: Alchemies of the Mind: Rationality and the Emotions. Cambridge University Press, Cambridge (1999)
7. Damasio, A.R.: Descartes' Error: Emotion, Reason, and Human Brain. Putnam, New York (1994)
8. Darwin, C.: The Expression of the Emotions in Man and Animals, 3rd edn. Oxford University Press, New York (1998). Original Work Published in 1872
9. Warren, W.H.: The dynamics of perception and action. Psychol. Rev. **113**(2), 358–389 (2006)
10. McFarland, D., Bosser, T.: Intelligent Behavior in Animals and Robots. MIT Press, Cambridge (1993)
11. Breazeal, C.: Emotion and sociable humanoid robots. Int. J. Hum Comput Stud. **59**, 119–155 (2003)
12. Walter, W.G.: The Living Brain. W. W. Norton & Co., New York (1963)
13. Ashby, W.R.: Design for a Brain: The Origin of Adaptive Behaviour, 2nd edn. Wiley, Chapman & Hall, New York, London (1960)
14. McDougall, W.: An Introduction to Social Psychology, 14th edn. Batoche Books, Kitchener (2001)
15. McAdams, D.P.: The five-factor model in personality: a critical appraisal. J. Pers. **60**(2), 328–361 (1992). Duke University Press
16. Izard, C.: Human Emotions. Plenum, New York (1977)
17. Plutchik, R.: The Emotions. University Press of America, Lanham (1991)
18. Pickering, A.: The Cybernetic Brain: Sketches of Another Future. The University of Chicago Press, Chicago (2010)
19. Moshkina, L., Arkin, R.C.: On TAMEing robots. In: International IEEE Conference on Systems, Man, and Cybernetics, vol. 4, pp. 3949–3956 (2003)
20. Egges, A., Kshirsagar, S., Thalmann, N.M.: A model for personality and emotion simulation. Knowl. Based Intell. Inf. Eng. Syst., 1–8 (2003)
21. Elliot, A.J., Thrash, T.M.: Approach avoidance motivation in personality: approach and avoidance temperaments and goals. J. Pers. Soc. Psychol. **82**(5), 804–818 (2000). American Psychological Association
22. Abro, A.H., Manzoor, A., Tabatabei, S.A., Treur, Y.: A computational cognitive model integrating different emotion regulation strategies. In: Annual International Conference on Biologically Inspired Cognitive Architectures, vol. 71, pp. 157–168. Elsevier (2015)
23. Chohra, A.: Embodied cognitive science, intelligent behavior control, machine learning, soft computing, and FPGA integration: towards fast, cooperative and adversarial robot team (RoboCup). Technical Fraunhofer (GMD) report. Germany, no. 136 (2001)

A Constraint-Based Approach to Modeling and Solving Resource-Constrained Scheduling Problems

Paweł Sitek[(✉)] and Jarosław Wikarek

Department of Information Systems, Kielce University of Technology,
Kielce, Poland
{sitek, j.wikarek}@tu.kielce.pl

Abstract. Constrained scheduling problems are common in manufacturing, project management, transportation, supply chain management, software engineering, computer networks etc. Multiple binary and integer decision variables representing the allocation of resources to activities and numerous specific constraints on these variables are typical components of the constraint scheduling problem modeling. With their increased computational complexity, the models are more demanding, particularly when methods of operations research (mathematical programming, network programming, dynamic programming) are used. By contrast, most resource-constrained scheduling problems can be easily modeled as instances of the constraint satisfaction problems (CSPs) and solved using constraint programming (CP) or others methods. In the CP-based environment the problem definition is separated from the methods and algorithms used to solve the problem. Therefore, a constraint-based approach to resource-constrained scheduling problems that combines an OR-based approach for problem solving and a CP-based approach for problem modeling is proposed. To evaluate the applicability and efficiency of this approach and its implementation framework, illustrative examples of resource-constrained scheduling problems are implemented separately for different environments.

Keywords: Constraint programming · Mathematical programming · Resource-constrained scheduling problem · Knowledge-based approach

1 Introduction

Nowadays, in highly competitive economy, decision-making problems often involve the optimal allocation of different types of constrained resources to activities over the time. This allocation must satisfy a set of different, sometimes conflicting, constraints (precedence, resource, capacity, temporal, etc.). In the above case, we can talk about a general form of resource-constrained scheduling problem (RCSP), which covers different organizational forms and environments, such as job-shop, flow-shop, open-shop, project and multi-project, mixed and hybrid environments. The RCSP consists in deciding when, where and how to execute each activity (job). As these problems appear widely in many real life decisions in transportation, supply chain, distribution, logistic,

© Springer International Publishing Switzerland 2016
N.T. Nguyen et al. (Eds.): ICCCI 2016, Part I, LNAI 9875, pp. 423–433, 2016.
DOI: 10.1007/978-3-319-45243-2_39

manufacturing, computer networks, and construction engineering, it is important to find an effective and efficient method to model and solve them.

The methods based on operational research, such as mathematical programming (MP) [1], dynamic programming and network programming seem to be a natural choice, as confirmed by numerous publications [2, 3]. However, the number of decision variables including integer variables in the models and 0–1 allocation variables increases computational complexity of the problems and muddles their structure. Mathematical programming allows modeling only linear and integer constraints [1], whereas complex industrial problems may contain constraints with different structures and features, which result from legal, business-related and safety-related conditions. These constraints are usually non-linear, logical, etc.

Most resource-constrained scheduling problems can easily be represented as instances of the constraint satisfaction problem (CSP) [4, 5]: given a set of decision variables, a set of possible values (variable domain) for each decision variable, and a set of constraints between the variables, assign a value to each variable so that all the constraints are satisfied. Formally, a CSP is defined as a triple (Xi, D^i_{om}, C^i_{st}), where Xi is a set of variables (i-index of the variable), D^i_{om} is a domain of values, and C^i_{st} is a set of constraints. Constraint satisfaction problems (CSPs) are the subject of intense study in the fields of artificial intelligence, operations research and soft computing [4].

CSPs are typically solved using a different form of search. The most commonly used methods include variants of backtracking algorithms, constraint propagation, and local search. Constraint programming (CP) and constraint logic programming (CLP) use the CSP techniques.

CP/CLP-based programming environments can be defined as programming methods based on the following main assumptions [2]:

- The problem to be solved is explicitly represented in terms of decision variables and constraints over these variables.
- In a constraint-based program, the explicit problem definition is clearly separated from the method/algorithm used to solve the problem.
- The problem definition and a set of decisions are translated into constraints.
- A purely deductive process referred to as "constraint propagation" is used to propagate the effects and impacts of the constraints.
- Constraint propagation is applied each time a new decision is made, and this process is clearly separated from the decision-making (solving) method/algorithm.

CP/CLP-based programming environments are very effective in solving binary constraints (involving a pair of decision variables). More than two decision variables makes the CP/CLP efficiency decrease dramatically. Moreover, discrete optimization is not a strong point of CP/CLP.

Based on [4–8] and our previous study on the integration of CP/CLP and MP [9–12], some strengths and weaknesses of the aforementioned environments are observed. The CP/CLP and MP integration can help model and solve decision and optimization problems for resource-constrained scheduling that are difficult to solve with one of the two methods alone. This integration was complemented with problem transformation [9, 10].

The main contribution as well as the motivation of this study was to apply a constrained-based approach to the modeling and solving of resource-constrained scheduling problems. The proposed approach combines constraint programming for modeling and mathematical programming for solving scheduling problems. This approach is enriched with the transformation of the problem used as a presolving method (authors' concept). Moreover, constrained-based implementation framework and hybrid model for RCSP are proposed.

2 Resource-Constrained Scheduling Problems

Constrained scheduling processes are common to many different management and engineering areas. A scheduling process responds to "What", "How" and/or "Where" and "When" something will be done. One of the most important types of scheduling problems is a resource-constrained scheduling problem (RCSP).

This problem in its most general form [13–15] is defined as a set of activities (e.g., jobs, orders, machine operations, services) that must be done, a set of resources (e.g., machines, processors, and additional resources such as tools, employees, electricity) to execute these activities, a set of different types of constraints (e.g., capacity, allocation, precedence, temporary etc.) that must be satisfied, and a set of objectives (e.g., costs, makespan, tardiness, etc.) to evaluate the schedule's execution (Fig. 1).

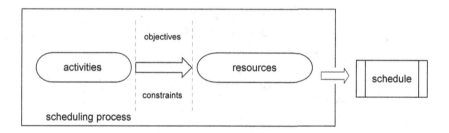

Fig. 1. The scheduling process

In practical and industrial problems, logical conditions may exist, for example, a condition, resulting from legal, business or safety-related requirements, that two additional resources or machines cannot be used simultaneously. Two main concerns have to be resolved: the choice of the best way to allocate available resources to the activities within given processing times such that all constraints are satisfied, and the choice of the best objective measures (with regard to optimization problems). Objectives and constraints are defined during the construction of the model. Constraints define the "feasibility" of the schedule while objectives define the "optimality" of the schedule.

Constraints appear in many areas and forms in scheduling problems. Precedence constraints define the order in which activities can be executed. Temporal constraints limit the times at which resources may be used and/or activities may be done. The constraints are linear, integer, binary and or logical.

3   A Constrained-Based Approach to Modeling and Solving RCSPs- The Concept and Implementation Platform

The proposed constraint-based approach to the modeling and solving of RCSPs is capable of filling the gaps and eliminate the shortcomings that occur in both MP-based and CP/CLP-based approaches when they are used separately.

This approach is based on the property that the definition of the problem is clearly separated from the methods used to solve it. A given problem is modeled with the use of CP/CLP and solved with the use of MP. The constraint-based approach is not just a simple integration of the two environments (CP/CLP and MP) [6, 7] but also contains the process model transformation [9, 12]. The transformation changes the representation of the problem using data instances (facts) and the characteristics and possibilities of CP/CLP. The result of the transformation is to reduce the number of decision variables and constraints, which is particularly important and useful in optimization because it reduces the search space.

To support and illustrate the concept of the proposed approach, an implementation platform is proposed (Fig. 2).

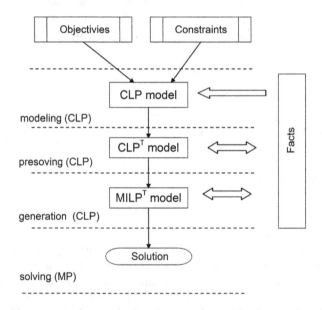

Fig. 2. The concept of constraint-based approach as an implementation platform

The implementation platform includes the following assumptions. Specification of the problem (CLP model) can be stated as integrity, binary, linear, logical and symbolic constraints with different objectives. Data instances of the problem are stored as sets of facts. Two environments, i.e. CP/CLP and MP are integrated. The implementation platform supports the transformation of the problem [12] as a presolving method.

The transformation reduces the number of variables and constraints in the CLP^T model. Following the transformation, a MILP^T (mixed integer linear programming) [1] model, based on the CLPT model, is automatically generated and solved in the MP environment.

The schema of implementation platform for the constraint-based approach is shown in Fig. 2. Using this platform, you can solve any resource-constrained scheduling problems. From a variety of tools for the implementation of the CP/CLP part in the platform, the ECL^{i}PS^e software was chosen. ECL^{i}PS^e is an open-source software system for the cost-effective development and deployment of constraint programming applications [16]. The area of mathematical programming (MP) in the platform was implemented using the SCIP [17].

4   Illustrative Example and Computational Experiments

The illustrative example shows a scheduling problem with additional resources (employees, tools, etc.) in job-shop system. Job-shop scheduling has been extensively studied in the literature reporting problems of varied degree of difficulty and complexity [13–15]. To evaluate the presented implementation platform for decision support, a number of numerical experiments were performed for the illustrative example.

The formal model for illustrative example in the form of indices, decision variables, and the main constraints (2)..(12), are shown in Table 2 (Appendix B). The objective of this model is to minimize the makespan (1). The model introduced additional constraints specifying the logical simultaneous use of machines (13) and additional resources (14). Knowledge about the problem has been presented in the form of facts. The structure of the facts and their relationships for illustrative example are shown in Fig. 3. The numerical experiments are performed for the data instance consisting of eight products (U1..U8), six machines (P1..P6), and four different sets of the additional resources (S1..S4) (for example, employees with different skills). The set of facts for illustrative example is shown in Appendix A.

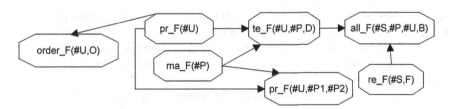

Fig. 3. The relationship between the sets of facts for illustrative example (#-ID of the fact, F-limit of additional resources, B- the number of allocated additional resources, D-exec. time).

Table 1. Obtained results for computational experiments at different values of set fs (N1..N3) and logic constraints (N4..N7)

Pn	Parameters	MILP (MP)				MILPT (Implementation platform)			
		V	C	T	Cmax	V	C	T	Cmax
N1	f1 = f2 = f3 = f4 = 10	7346	4034	345	13	1991	1377	29	13
N2	f1 = f2 = f3 = f4 = 4	7346	4034	456	14	1991	1377	38	14
N3	f1 = f2 = f3 = f4 = 2	7346	4034	462	15	1991	1377	39	15
N4	Exclusion P1 & P2	–	–	–	–	2105	1511	42	17
N5	Exclusion P1 & P3	–	–	–	–	2105	1511	42	16
N6	Exclusion S1 & S2	–	–	–	–	2181	1607	52	16
N7	Exclusion S2 & S3	–	–	–	–	2181	1607	48	17

V-the number of decision variables, C-the number of constraints, T-solution time

Table 2. Indices, parameters and constraints for mathematical model of RCSP

Indices	
p	Machine/processor $p = 1..P$
u	product/service type $u = 1..U$
s	additional resource (i.e. employees) $s = 1..S$
t	period $t = 1..T$
Parameters	
o_u	demand/order for product u
$d_{u,p}$	the time required to make a product u on the machine p
$dl_{u,p}$	if the product u is executed on the machine p than $dl_{u,p = 1}$, otherwise $dl_{u,p} = 0$
f_s	the number of additional resources s
$bu_{u,p,s}$	If additional resource s is required to make a product u on the machine p // $bu_{u,p}$, $s = 1$, otherwise $bu_{u,p,s = 0}$
$b_{u,p,s}$	how much additional resource s need to make the product u machine p
$ko_{u,p1,}$ $p2$	if the operation of the product u on the machine p1 to be performed before the operation on the machine p2 than $ko_{u,p1,p2 = 1}$, otherwise $ko_{u,p1,p2} = 0$
e_t	coefficients for conversion numbers of periods t for the variables $e_t = t$
Decision Variables	
$X_{u,p,s,t}$	if the employee s in period t makes the product u on the machine p than $X_{u,p,s,t = 1}$ otherwise $X_{u,p,s,t = 0}$
$G_{u,p}$	the number period in which the end product u realization on the machine p
$Z_{u,p,s,t}$	if the period t is the latest in which the employee s makes the product u on the machine p than $Y_{u,p,s,t = 1}$ otherwise $Y_{u,p,s,t = 0}$
C_{max}	makespan (objective function)

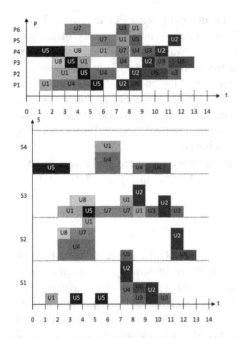

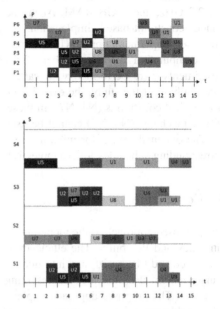

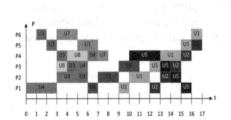

Fig. 4. Gantt charts (schedules) for machines (top) and additional resources (down) corresponding to the example N1 for f1 = f2 = f3 = f4 = 10

Fig. 5. Gantt charts (schedules) for machines (top) and additional resources (down) corresponding to the example N3 for f1 = f3 = 4 f2 = f4 = 2

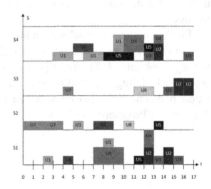

Fig. 6. Gantt chart (schedule) for machines corresponding to example N4 for f1 = f2 = f3 = f4 = 10

Fig. 7. Gantt chart (schedule) for additional resource corresponding to example N7

All numerical experiments were performed using an ordinary PC, Intel core (TM2), i-3, 2.2 GHz, and 4 GB RAM. At the beginning of the study, efficiency of the presented constraint-based implementation platform was evaluated relative to the MP-based implementations (N1..N3) for illustrative model (1)..(12) in the context of time computation and the number of decision variables and constraints. Numerous experiments were performed with varied parameter fs (additional resources). In the second phase of the study, the experiments were performed for hybrid model (1)..(14) with logical constraints (N4..N7). In these examples the two additional resources (N6, N7) or machines (N4, N5) cannot be used simultaneously.

The obtained results for minimization makespan are presented in Table 1 and the corresponding schedules for examples N1, N3 and N4, N7 in Figs. 4, 5, 6 and 7.

5 Conclusion

The proposed approach to the modeling and solving of RCSPs is characterized by high efficiency and flexibility. The efficiency is the result of integrating two environments (CP-based and MP-based) and introducing transformation as a presolving method. The transformation uses the facts representing the knowledge about the problem and its instances. The proposed approach implemented for the illustrative example resulted in a four-fold search time reduction compared to the MP-based approach (Table 1). Also, the number of variables and constraints was reduced four times and three times the initial number. The size of the search space was reduced to consequently enable addressing larger size problems within acceptable time limits. Moreover, the proposed constraint-based approach allows extremely flexible modeling of various types of problems with different types of constraints (linear, integer, logical, symbolic etc.). For instance, the introduction of logical constraints into the MP-based approach is virtually impossible.

Further studies will include the implementation of various types of other scheduling problems in the area of flexible manufacturing, transport systems, capacity vehicle routing problems [18], multi-project scheduling problems [19], resource capacity project scheduling and multi-assortment repetitive production. The hybrid implementation platform will also be supplemented with the paradigm of Fuzzy Logic with Answer Set Programming [20].

Appendix A Sets of Facts for Illustrative Example

```
%ma_F(#P). - machine
 ma_F(P1). ma_F(P2). ma_F(P3). ma_F(P4). ma_F(P5). ma_F(P6).
%pr_F(#U). - product
 pr_F(U1). pr_F(U2). pr_F(U3). pr_F(U4).
 pr_F(U5). pr_F(U6). pr_F(U7). pr_F(U8).
%te_F(#U,#P,D). - technology
 te_F(U1,P1,1). te_F(U1,P2,2). te_F(U1,P3,1). te_F(U1,P4,2).
 te_F(U1,P5,1). te_F(U1,P6,1). te_F(U2,P1,1). te_F(U2,P2,1).
 te_F(U2,P3,1). te_F(U2,P4,1). te_F(U2,P5,1). te_F(U3,P2,1).
 te_F(U3,P3,1). te_F(U3,P4,1). te_F(U3,P5,1). te_F(U3,P6,1).
 te_F(U4,P1,3). te_F(U4,P2,2). te_F(U4,P3,1). te_F(U4,P4,1).
 te_F(U5,P1,1). te_F(U5,P2,1). te_F(U5,P3,1). te_F(U5,P4,3).
 te_F(U6,P1,1). te_F(U6,P2,2). te_F(U6,P3,2).
 te_F(U7,P4,1). te_F(U7,P5,2). te_F(U7,P6,2).
 te_F(U8,P3,1). te_F(U8,P4,2).
%re_F (#S,F). - resources
 re_F(S1,20). re_F(S2,20). re_F(S3,20). re_F(S4,20).
%all_F(#S,#P,#U,B) - allocation
 all_F(S1,U1,P1,2). all_F(S2,U1,P1,2). all_F(S3,U1,P2,2).
 all_F(S4,U1,P2,2). all_F(S2,U1,P3,2). all_F(S4,U1,P4,2).
 all_F(S1,U1,P5,2). all_F(S3,U1,P5,2). all_F(S3,U1,P6,2).
 all_F(S4,U1,P6,4). all_F(S1,U2,P1,4). all_F(S2,U2,P1,4).
 all_F(S3,U2,P2,4). all_F(S4,U2,P2,4). all_F(S1,U2,P3,4).
 all_F(S3,U2,P4,4). all_F(S2,U2,P5,4). all_F(S3,U2,P5,4).
 all_F(S3,U3,P2,2). all_F(S4,U3,P2,2). all_F(S1,U3,P3,2).
 all_F(S3,U3,P4,2). all_F(S1,U3,P5,2). all_F(S2,U3,P5,2).
 all_F(S3,U3,P5,2). all_F(S2,U3,P6,2). all_F(S4,U3,P6,2).
 all_F(S1,U4,P1,4). all_F(S2,U4,P1,4). all_F(S3,U4,P2,4).
 all_F(S4,U4,P2,4). all_F(S1,U4,P3,4). all_F(S4,U4,P4,2).
 all_F(S1,U5,P1,2). all_F(S2,U5,P1,2). all_F(S3,U5,P2,2).
 all_F(S4,U5,P2,2). all_F(S1,U5,P3,2). all_F(S4,U5,P4,2).
 all_F(S1,U6,P1,2). all_F(S2,U6,P1,2). all_F(S3,U6,P2,2).
 all_F(S4,U6,P2,2). all_F(S2,U6,P3,2). all_F(S3,U7,P4,2).
 all_F(S2,U7,P5,2). all_F(S3,U7,P5,2). all_F(S2,U7,P6,2).
 all_F(S3,U7,P6,2). all_F(S2,U8,P3,2). all_F(S3,U8,P4,2).
%pr_F(#U,#P1,#P2). - precedence
 pr_F(U1,P1,P2). pr_F(U1,P2,P3). pr_F(U1,P3,P4).
 pr_F(U1,P4,P5). pr_F(U1,P5,P6). pr_F(U2,P1,P2).
 pr_F(U2,P2,P3). pr_F(U2,P3,P4). pr_F(U2,P4,P5).
 pr_F(U3,P6,P5). pr_F(U3,P5,P4). pr_F(U3,P4,P3).
 pr_F(U3,P3,P2). pr_F(U4,P1,P2). pr_F(U4,P2,P3).
 pr_F(U4,P3,P4). pr_F(U5,P4,P3). pr_F(U5,P3,P2).
 pr_F(U5,P2,P1). pr_F(U6,P1,P2). pr_F(U6,P2,P3).
 pr_F(U7,P6,P5). pr_F(U7,P5,P4). pr_F(U8,P3,P4).
order_F(#U,O).
 order_F(U1,1). order_F(U2,1). order_F(U3,1). order_F(U4,1).
 order_F(U5,1). order_F(U6,1). order_F(U7,1). order_F(U8,1).
```

Appendix B Illustrative Example-Formal Model

$$\text{Min } C_{max} \tag{1}$$

$$G_{u,p} \leq C_{max} \forall u = 1..U, p = 1..P \tag{2}$$

$$\sum_{s=1}^{S} \sum_{t=1}^{T} dl_{u,p} \cdot bu_{u,p,s} \cdot X_{u,p,s,t} = d_{u,p} \cdot o_u \forall u = 1..U, p = 1..P \tag{3}$$

$$\sum_{u=1}^{U} \sum_{s=1}^{S} X_{u,p,s,t} \leq 1 \forall p = 1..P, t = 1..T \tag{4}$$

$$\sum_{u=1}^{U} \sum_{p=1}^{P} b_{u,p,s} \cdot X_{u,p,s,t} \leq f_s \forall s = 1..S, t = 1..T \tag{5}$$

$$X_{u,p,s,t-1} - X_{u,p,s,t} \leq Z_{u,p,s,t-1} \forall u = 1..U, p = 1..P, s = 1..S, t = 2..T \tag{6}$$

$$\sum_{t=1}^{T} Z_{u,p,s,t} = 1 \forall u = 1..U, p = 1..P, s = 1..S \tag{7}$$

$$G_{u,p} \geq e_t \cdot X_{u,p,s,t} \forall u = 1..U, p = 1..P, s = 1..S, t = 1..T \tag{8}$$

$$G_{u,p2} - b_{u,p2} \geq G_{u,p1} \forall u = 1..U, p1, p2 = 1..P : ko_{u,p1,p2} = 1 \tag{9}$$

$$G_{u,p} \in C \forall u = 1..U, p = 1..P \tag{10}$$

$$X_{u,p,s,t} \in \{0,1\} \forall u = 1..U, p = 1..P, s = 1..S, t = 1..T \tag{11}$$

$$Z_{u,p,s,t} \in \{0,1\} \forall u = 1..U, p = 1..P, s = 1..S, t = 1..T \tag{12}$$

$$\text{Exclusion\_M}(p1, p2) \exists p1, p2 = 1..P : p1 \neq p2 \tag{13}$$

$$\text{Exclusion\_R}(s1, s2) \exists s1, s2 = 1..S : s1 \neq s2 \tag{14}$$

References

1. Schrijver, A.: Theory of Linear and Integer Programming. Wiley, New York (1998)
2. Joseph, Y.-T.L., Anderson, J.H.: Handbook of Scheduling: Algorithms, Models, and Performance Analysis. Chapman & Hall/CRC, Boca Raton (2004). ISBN:1584883979
3. Błażewicz, J., Ecker, K.H., Pesch, E., Schmidt, G., Węglarz, J.: Handbook on Scheduling. From Theory to Applications. Springer, Heidelberg (2007). ISBN:978-3-540-28046-0
4. Rossi, F., Van Beek, P., Walsh, T.: Handbook of Constraint Programming (Foundations of Artificial Intelligence). Elsevier Science Inc, New York (2006)
5. Apt, K., Wallace, M.: Constraint Logic Programming Using Eclipse. Cambridge University Press, Cambridge (2006)
6. Milano, M., Wallace, M.: Integrating operations research constraint programming. Ann. Oper. Res. **175**(1), 37–76 (2010)
7. Achterberg, T., Berthold, T., Koch, T., Wolter, K.: Constraint integer programming, a new approach to integrate CP and MIP. In: Perron, L., Trick, M.A. (eds.) CPAIOR 2008. LNCS, vol. 5015, pp. 6–20. Springer, Heidelberg (2008)

8. Bocewicz, G., Banaszak, Z.: Declarative approach to cyclic steady states space refinement: periodic processes scheduling. Int. J. Adv. Manuf. Technol. **67**(1–4), 137–155 (2013)
9. Sitek, P., Wikarek, J.: A hybrid approach to the optimization of multiechelon systems. Math. Probl. Eng., Article ID 925675. Hindawi Publishing Corporation (2014). doi:10.1155/2014/925675
10. Sitek, P., Nielsen, I.E., Wikarek, J.: A hybrid multi-agent approach to the solving supply chain problems. Procedia Comput. Sci. **35**, 1557–1566 (2014). Knowledge-Based and Intelligent Information & Engineering Systems 18th Annual Conference, KES-2014
11. Sitek, P., Wikarek, J.: A hybrid framework for the modelling and optimisation of decision problems in sustainable supply chain management. Int. J. Prod. Res. 1–18 (2015). doi:10.1080/00207543.2015.1005762
12. Sitek, P.: A hybrid CP/MP approach to supply chain modelling, optimization and analysis. In: Proceedings of the 2014 Federated Conference on Computer Science and Information Systems, Annals of Computer Science and Information Systems, vol. 2, pp. 1345–1352 (2014). doi:10.15439/2014F89
13. Guyon, O., Lemaire, P., Pinson, Ă., Rivreau, D.: Solving an integrated job-shop problem with human resource constraints. Ann. Oper. Res. **213**(1), 147–171 (2014)
14. Blazewicz, J., Lenstra, J.K., Rinnooy Kan, A.H.G.: Scheduling subject to resource constraints: classification and complexity. Discrete Appl. Math. **5**, 11–24 (1983)
15. Lawrence, S.R., Morton, T.E.: Resource-constrained multi-project scheduling with tardy costs: comparing myopic bottleneck, and resource pricing heuristics. Eur. J. Oper. Res. **64**(2), 168–187 (1993)
16. Eclipse - The Eclipse Foundation open source community website. www.eclipse.org. Accessed 20 Apr 2016
17. SCIP. http://scip.zib.de/. Accessed 20 Apr 2016
18. Toth, P., Vigo, D.: Models, relaxations and exact approaches for the capacitated vehicle routing problem. Discrete Appl. Math. **123**(1–3), 487–512 (2002)
19. Coelho, J., Vanhoucke, M.: Multi-mode resource-constrained project scheduling using RCPSP and SAT solvers. Eur. J. Oper. Res. **213**, 73–82 (2011)
20. Wang, J., Liu, C.: Fuzzy constraint logic programming with answer set semantics. In: Zhang, Z., Siekmann, J. (eds.) KSEM 2007. LNCS, vol. 4798, pp. 52–60. Springer, Heidelberg (2007). doi:10.1007/978-3-540-76719-0_9

Some Remarks on the Mean-Based Prioritization Methods in AHP

Konrad Kułakowski[1]($\boxtimes$) and Anna Kędzior[2]

[1] Faculty of Electrical Engineering, Automatics,
Computer Science and Biomedical Engineering,
AGH University of Science and Technology, Al. Mickiewicza 30,
30-059 Cracow, Poland
konrad.kulakowski@agh.edu.pl
[2] Faculty of Applied Mathematics, Al. Mickiewicza 30, 30-059 Cracow, Poland
kedzior@agh.edu.pl

Abstract. EVM (eigenvector method) and GMM (geometric mean method) are probably the two most popular priority deriving techniques for AHP (Analytic Hierarchy Process). Although much has already been discussed about these methods, one frequently repeated question is: what do they have in common? In this paper we show that both these methods can be constructed based on the same principle that the priority of one alternative should correspond to the weighted mean of priorities of other alternatives. We also show how the accepted principle can be used to construct priority deriving methods for the generalized (non-reciprocal) pairwise comparisons matrices.

Keywords: Decision making · MCDM · AHP · Eigenvalue method · Geometric mean method · Pairwise comparisons

1 Introduction

Analytic Hierarchy Process (AHP) is a hierarchical multiple-criteria decision-making technique proposed by *Thomas Saaty* [23]. The core of *AHP* is the pairwise comparisons *(PC)* method. Its usefulness has been confirmed by numerous examples [8,25]. It has also become a cause of vigorous discussions [1,5] and prompted researchers into further exploration. Examples of the recent works carried out in this area are the *Rough Set* approach [7], fuzzy *PC* relation handling [6,20], incomplete *PC* relation [2], non-numerical rankings [11], the nature of inconsistency [3,13], rankings with reference values [15,17,18] and others. A more complete discussion of the *PC* method in the context of *AHP* and decision-making systems can be found in [24].

2 Preliminaries

2.1 Pairwise Comparisons Method

Having many alternatives to choose from, it is very often hard to pick the best one. On the other hand, when one compares only two alternatives the choice

© Springer International Publishing Switzerland 2016
N.T. Nguyen et al. (Eds.): ICCCI 2016, Part I, LNAI 9875, pp. 434–443, 2016.
DOI: 10.1007/978-3-319-45243-2_40

of the better one is much more straightforward. Therefore, according to the PC method, all the alternatives must be compared to each other, then on the basis of these comparisons the final result is synthesized. AHP uses quantitative comparisons. This means that the person who compares two alternatives has to indicate which of them and to what extent it is better. Assessments take the form of positive real numbers. The set of paired comparisons is usually represented as a positive and real matrix $A = [a_{ij}]$ where the entry a_{ij} corresponds to the result of comparing the i-th and j-th alternatives (1).

$$A = \begin{pmatrix} 1 & a_{12} & \cdots & a_{1n} \\ a_{21} & 1 & \cdots & a_{2n} \\ \vdots & \vdots & \ddots & \vdots \\ a_{n1} & a_{n2} & \cdots & 1 \end{pmatrix} \tag{1}$$

The values a_{ij} represent the relative importance of the alternatives $c_1, \ldots, c_n$. Therefore, according to the best knowledge of evaluators, it should hold that $a_{ij} = w(c_i)/w(c_j)$, where $w(c_i), w(c_j)$ denotes the real weight (priority) of the i-th and j-th alternatives. It is easy to see that if every a_{ij} for $i, j, k = 1, \ldots, n$ equals the ratio $w(c_i)/w(c_j)$, then it holds that $a_{ij} a_{jk} a_{ki} = 1$.

Definition 1. *The matrix A is said to be* consistent *if*

$$a_{ij} a_{jk} a_{kj} = 1, \quad \text{for every } i, j, k = 1, \ldots, n \tag{2}$$

Since comparing c_i to c_j results in a_{ij}, one may expect that the reverse comparison c_j to c_i leads to $a_{ji} = 1/a_{ij}$.

Definition 2. *The matrix A is said to be* reciprocal *if $a_{ij} = 1/a_{ji}$ for every $i, j = 1, \ldots, n$.*

Although the reciprocity assumption sometimes seems to be very natural, non-reciprocal matrices are also sometimes considered [9]. These matrices will be called *generalized PC* matrices. It is worth noting that every consistent PC matrix must also be reciprocal, but not reversely. Hence, reciprocity is a weaker assumption than consistency.

2.2   Priority Estimation Methods

The set of numerical assessments given as the PC matrix A is provided by experts. Thus, in practice, A is sometimes inconsistent, non-reciprocal or even incomplete. While the latter two problems can sometimes be overcome (for instance, by asking experts to complete the matrix), the first one is an inherent part of human nature [12]. Therefore, the method used for priority derivation must take into account the possible inconsistency of A. The two most popular of these are: the *Eigenvector method (EVM)* and the *Geometric Mean Method*

(GMM) [10]. According to the first method, the priorities $w(c_1), w(c_2), \ldots, w(c_n)$ are determined as:

$$w(c_i) = \frac{1}{S} w_{\max}(c_i) \quad \text{where} \quad S = \sum_{j=1}^{n} w_{\max}(c_j) \tag{3}$$

where $w_{\max} = [w_{\max}(c_1), \ldots, w_{\max}(c_n)]^T$ is the principal eigenvector[1] of A. The priority of an alternative is meaningful only in relation to the priorities of other alternatives. Therefore, in practice, the ratio $w(c_k)/w(c_r)$ is more interesting than in the exact values of $w(c_k)$ and $w(c_r)$. Thus, in view of the convenience, a priority vector w is rescaled so that the sum of its components is 1. It is possible to prove that when A is consistent (Definition 1) then $w_{max}(c_i) = a_{ij} w_{max}(c_j)$ for all $i, j = 1, \ldots, n$ [16].

For the purpose of *EVM, Saaty* introduced the consistency index (4) that allows the level of inconsistency of the given *PC* matrix A to be estimated.

$$CI = \frac{\lambda_{max} - n}{n - 1} \tag{4}$$

The value of *CI* for the given $n \times n$ matrix A depends on its size and its principal eigenvalue λ_{max}.

In the second approach the priorities of alternatives correspond to the geometric means of rows of A.

$$w(c_i) = \frac{1}{S} \left(\prod_{j=1}^{n} a_{ij} \right)^{\frac{1}{n}} \quad \text{where} \quad S = \sum_{i=1}^{n} \left(\prod_{j=1}^{n} a_{ij} \right)^{\frac{1}{n}} \tag{5}$$

In addition, in this case the priority vector is rescaled so that all its entries sum up to 1. As before, one can easily demonstrate that consistency (2) implies $w(c_i) = a_{ij} w(c_j)$ for all $i, j = 1, \ldots, n$. Crawford [4] also introduced the geometric consistency index, given as:

$$GCI = \frac{2}{(n-1)(n-2)} \sum_{i=1}^{n-1} \sum_{j=i+1}^{n} \left(\log \frac{a_{ij} w_j}{w_i} \right)^2 \tag{6}$$

Later in this article, we will look more closely at *EVM* and *GMM*. Although at first glance they do not have a lot in common, we show that both methods can be derived from the same premises. Besides the two mentioned, there are many other less popular methods. Further information about these and other methods can be found in [10,14].

3   Common Roots of EVM and GMM

Let us recall that, according to the *PC* approach, a_{ik} (entry of A) should express the relative value of the i-th alternative c_i with respect to the k-th alternative c_k.

---

[1] The vector associated with the principal eigenvalue (spectral radius) of A.

Therefore, one would expect that $w(c_i)/w(c_k) = a_{ik}$, i.e. $w(c_i) = a_{ik}w(c_k)$, and similarly $a_{ki}w(c_i) = w(c_k)$. In particular, it is desirable that

$$a_{ki}w(c_i) = w(c_k) = a_{kj}w(c_j) \tag{7}$$

for every $i, j, k = 1, \ldots, n$. Due to the possible inconsistency, it is possible that $a_{ki}w(c_i) \neq a_{kj}w(c_j)$. Therefore the question arises as to what $w(c_k)$ should be? Since the values $a_{ki}w(c_i)$ for $i = 1, \ldots, n$ can vary from each other, the natural (and probably a straightforward) proposition is to strive for a situation in which $w(c_k)$ is the mean of elements in the form of $a_{ki}w(c_i)$.

3.1 EVM

The above consideration suggests the adoption of *mean heuristics*, according to which every $w(c_k)$ should correspond to the mean of $a_{k1}w(c_1), \ldots, a_{kn}w(c_n)$. For the purpose of *EVM*, let us assume *arithmetic mean heuristics* (AMH):

$$\frac{1}{n} \sum_{i=1}^{n} a_{ki}w(c_i) = w(c_k) \tag{8}$$

In most cases, the priority of one alternative is meaningful only in the context of priorities of other alternatives. Hence, the most essential thing is the ratio between the priorities $w(c_k)$ and $w(c_r)$:

$$\frac{w(c_k)}{w(c_r)} = \frac{\sum_{i=1}^{n} a_{ki}w(c_i)}{\sum_{i=1}^{n} a_{ri}w(c_i)} \tag{9}$$

In order to calculate w we would have to solve the following equation system:

$$\sum_{i=1}^{n} a_{1i}w(c_i) = n \cdot w(c_1)$$
$$\cdots\cdots\cdots\cdots \tag{10}$$
$$\sum_{i=1}^{n} a_{ni}w(c_i) = n \cdot w(c_n)$$

that can be written in the form of the following matrix equation:

$$Aw = nw \tag{11}$$

Of course, when the matrix A is consistent, i.e. it holds that (7), the Eq. 11 has a positive and real solution[2]. When A is not consistent, the situation becomes complicated. Let us consider (11) in the form

$$(A - n \cdot I)w = 0 \tag{12}$$

---

[2] Note that accepting $w(c_n) = 1$, due to (7), we immediately get a solution of (11) in the form $[a_{1n}, a_{2n}, \ldots, a_{n-1,n}, 1]$.

where I is the identity matrix. The above equation has a non-zero solution only if $(A - n \cdot I)$ is singular [22, p. 10]. Thus, in practice, it is possible that there is no non-trivial solution for (11). On the other hand, only the real and positive solutions of (11) can be used as the ranking vector. Thus, in the original article [23] *Saaty* proposed a more general approach, according to which the priorities of alternatives are determined by the real and positive solution of the following equation:

$$Aw = \lambda w \tag{13}$$

The value λ is an eigenvalue of A, whilst w is the appropriate eigenvector. There might be many eigenvectors and eigenvalues of A. However, when A is positive and real it has exactly one positive and real eigenvalue associated with the positive and real eigenvector [21]. This eigenvalue is also the largest eigenvalue of A. Let λ_{max} be the largest (principal) eigenvalue, and w_{max} be the associated (principal) eigenvector of A. The *EVM* as proposed by Saaty [23] adopts w_{max} as the solution of (Eq. 13). In other words, our initial *AMH* (8) needs to be modified by adding the scaling factor γ:

$$\gamma \frac{1}{n} \sum_{i=1}^{n} a_{ki} w_{max}(c_i) = w_{max}(c_k), \quad where \quad \gamma = \frac{n}{\lambda_{max}} \quad and \quad k = 1, \ldots, n \tag{14}$$

Fortunately, for the ratio between the priorities of alternatives the scaling factor does not matter. Hence, we still have:

$$\frac{w_{max}(c_k)}{w_{max}(c_r)} = \frac{\sum_{i=1}^{n} a_{ki} w_{max}(c_i)}{\sum_{i=1}^{n} a_{ri} w_{max}(c_i)} \tag{15}$$

Therefore, it becomes clear that the same idea underlines the use of both (8) and (14) as the heuristics leading to *EVM*.

3.2 GMM

In the previous section, it was shown that adopting heuristics according to which $w(c_k)$ is the *arithmetic mean* of elements: $a_{k1}w(c_1), \ldots, a_{kn}w(c_n)$ leads to the definition of *EVM*. Someone may ask, however, why the geometric mean is not used. Thus, let us consider the *geometric mean heuristics (GMH)* according to which $w_{gm}(c_k)$ is the geometric mean of all the components in the form $a_{ki}w_{gm}(c_i)$ for $i = 1, \ldots, n$:

$$w_{gm}(c_k) = \left(\prod_{i=1}^{n} a_{ki} w_{gm}(c_i) \right)^{\frac{1}{n}}, \quad for \quad k = 1, \ldots, n \tag{16}$$

The above equation can be rewritten in the form:

$$w_{gm}(c_k) = \left(\prod_{i=1}^{n} a_{ki} \right)^{\frac{1}{n}} \left(\prod_{i=1}^{n} w_{gm}(c_i) \right)^{\frac{1}{n}}, \quad for \quad k = 1, \ldots, n \tag{17}$$

Let us denote $\delta = (\prod_{i=1}^{n} w_{gm}(c_i))^{\frac{1}{n}}$. Then, when comparing the relationship between the priorities of c_k and c_r, where $k, r = 1, \dots, n$ we obtain:

$$\frac{w_{gm}(c_k)}{w_{gm}(c_r)} = \frac{\delta \left(\prod_{i=1}^{n} a_{ki}\right)^{\frac{1}{n}}}{\delta \left(\prod_{i=1}^{n} a_{ri}\right)^{\frac{1}{n}}} = \frac{\left(\prod_{i=1}^{n} a_{ki}\right)^{\frac{1}{n}}}{\left(\prod_{i=1}^{n} a_{ri}\right)^{\frac{1}{n}}}, \quad for\ k, r = 1, \dots, n \qquad (18)$$

In other words, for the purpose of mutual priority comparisons, we may adopt:

$$w_{gm}(c_k) = \left(\prod_{i=1}^{n} a_{ki}\right)^{\frac{1}{n}}, \quad for\ k = 1, \dots, n \qquad (19)$$

The above reasoning shows that the acceptance of GMH leads directly to GMM as presented in (Sect. 2.2). According to [4] GMM (19) is optimal with respect to minimization of the sum of multiplicative errors $e_{ij} = a_{ij} w_j / w_i$.

3.3   EVM and GMM for the Consistent PC Matrix

It is commonly known that when the PC matrix A is consistent, then EVM and GMM lead to the same priority vectors. Of course, it is the *"mean heritage"*. To see that, let us consider a consistent matrix A for which the equation (7) holds. Due to the consistency of A it holds that $\alpha_k = a_{ki} w_{max}(c_i) = a_{k1} w_{max}(c_1) = \dots = a_{kn} w_{max}(c_n)$, see: [16]. Hence,

$$w_{max}(c_k) = \alpha_k = a_{kr} w_{max}(c_r), \quad for\ every\ k, r = 1, \dots, n \qquad (20)$$

i.e.

$$\frac{w_{max}(c_k)}{w_{max}(c_r)} = a_{kr}, \quad for\ every\ k, r = 1, \dots, n \qquad (21)$$

Similarly, the consistency of A implies[3] $\beta_k = a_{ki} w_{gm}(c_i) = a_{k1} w_{gm}(c_1) = \dots = a_{kn} w_{gm}(c_n)$. Thus, in particular

$$w_{gm}(c_k) = \beta_k = a_{kr} w_{gm}(c_r), \quad for\ every\ k, r = 1, \dots, n \qquad (22)$$

i.e.

$$\frac{w_{gm}(c_k)}{w_{gm}(c_r)} = a_{kr}, \quad and\ k, r = 1, \dots, n \qquad (23)$$

Finally,

$$\frac{w_{max}(c_k)}{w_{max}(c_r)} = \frac{w_{gm}(c_k)}{w_{gm}(c_r)}, \quad and\ k, r = 1, \dots, n \qquad (24)$$

---

[3] It is enough to notice that:

$$\frac{w_r}{w_k} = \frac{\left(\prod_{j=1}^{n} a_{rj}\right)^{\frac{1}{n}}}{\left(\prod_{j=1}^{n} a_{kj}\right)^{\frac{1}{n}}} = \left(\prod_{j=1}^{n} \frac{a_{rj}}{a_{kj}}\right)^{\frac{1}{n}} = \left(\prod_{j=1}^{n} a_{rj} a_{jr} a_{rk}\right)^{\frac{1}{n}} = \left(\prod_{j=1}^{n} a_{rk}\right)^{\frac{1}{n}} = a_{rk}$$

for every $k, r = 1, \dots, n$.

which means that, for the consistent PC matrix A the EVM and GMM methods lead to the two priority vectors which may differ at most by a constant multiplier. Hence, both methods form two identical priority vectors.

4   EVM and GMM for Inconsistent and Non-reciprocal PC Matrices

The EVM and GMM methods are generally used to calculate the priority for inconsistent but reciprocal matrices. However, the reasons for which we adopted heuristics AMH and GMH (Sect. 3) remain in force even if A is non-reciprocal. Indeed, since $w(c_k)$ is approximated by every element in the form of $a_{ki}w(c_i)$, it makes sense to require that $w(c_k)$ be the mean of its approximations $a_{k1}w(c_1), \ldots, a_{kn}w(c_n)$. Of course, the question arises as to whether in such a case it is possible to find a vector of priorities which meets the relevant equations (14) and (19). The GMM case (Sect. 3.2) is quite simple. Since, the priorities $w(c_k)$ depend only on the entries of A which are positive and real, the priority vector exists and is positive and real. GCI (6) also seems to remain unaffected. The EVM is a bit less obvious since it requires the consideration of a principal eigenvector of A. Fortunately, according to the *Perron-Frobenus* theory [19], such a vector exists and is positive and real only if A is positive and real. Hence, it appears that EVM can be successfully used also when A is non-reciprocal. Unfortunately, the *Perron-Frobenus* theory does not provide a guarantee that $\lambda_{max} \geq n$. We can only be sure that λ_{max} is positive and real. Hence, it is possible that if A is non-reciprocal CI may be negative. Since the negative values of the consistency index are meaningless, CI cannot be used with EVM and the non-reciprocal PC matrices.

Although for non-reciprocal matrices AMH and GMH also seem to be valid, these heuristics can be improved. It is enough to note that both $a_{ki}w(c_i)$ and $\frac{1}{a_{ik}}w(c_i)$ provide good approximations of $w(c_k)$. Thus, let us adopt *improved mean heuristics* according to which $w(c_k)$ is the mean of all the elements in the form: $a_{k1}w(c_1), \frac{1}{a_{1k}}w(c_1), \ldots, a_{kn}w(c_n), \frac{1}{a_{nk}}w(c_n)$.

4.1   EVM

For the purpose of EVM, let us assume *improved arithmetic mean heuristics* *(I)AMH* according to which $w(c_k)$ is the *arithmetic mean* of all the components $a_{k1}w(c_1), \frac{1}{a_{1k}}w(c_1), \ldots, a_{kn}w(c_n), \frac{1}{a_{nk}}w(c_n)$, i.e.

$$w(c_k) = \frac{1}{2n} \sum_{i=1}^{n} \left(a_{ik} + \frac{1}{a_{ki}} \right) w(c_i) = \frac{1}{n} \sum_{i=1}^{n} \frac{a_{ik} + \frac{1}{a_{ki}}}{2} w(c_i), \quad for \ k = 1, \ldots, n$$

(25)

Similarly as before, let us introduce the scaling coefficient $\gamma = \frac{n}{\lambda_{max}}$,

$$w_{max}(c_k) = \gamma \frac{1}{n} \sum_{i=1}^{n} \frac{a_{ik} + \frac{1}{a_{ki}}}{2} w_{max}(c_i), \quad for \ k = 1, \ldots, n \qquad (26)$$

so that the initial problem (25) can be reduced to finding the principal eigenvector of some positive and real non-reciprocal PC matrix:

$$\widehat{A} = \begin{pmatrix} 1 & \frac{a_{12}+\frac{1}{a_{21}}}{2} & \cdots & \frac{a_{1n}+\frac{1}{a_{n1}}}{2} \\ \frac{a_{21}+\frac{1}{a_{12}}}{2} & 1 & \cdots & \frac{a_{2n}+\frac{1}{a_{n2}}}{2} \\ \vdots & \vdots & \ddots & \vdots \\ \frac{a_{n1}+\frac{1}{a_{1n}}}{2} & \frac{a_{n2}+\frac{1}{a_{n2}}}{2} & \cdots & 1 \end{pmatrix} \tag{27}$$

Since $\widehat{A}$ is positive and real, then, according to the *Perron-Frobenus* theory, there is a unique pair of a positive and real principal eigenvector and a positive and real eigenvalue that meets (26). It is easy to note that the proposed modification to *EVM* is a simple generalization of the original method. Hence, when A is reciprocal there is no difference between the original and improved methods, in particular $\widehat{A} = A$.

4.2   GMM

In the case of GMM, the *improved mean heuristics* takes the form:

$$w(c_k) = \left(\prod_{j=1}^{n} a_{kj} w(c_j) \prod_{j=1}^{n} \frac{1}{a_{jk}} w(c_j) \right)^{\frac{1}{2n}} = \left(\prod_{j=1}^{n} \frac{a_{kj}}{a_{jk}} w(c_j) \right)^{\frac{1}{2n}} , \ for \ k = 1, \ldots, n \tag{28}$$

Thus,

$$w(c_k) = \left(\prod_{j=1}^{n} \frac{a_{kj}}{a_{jk}} \right)^{\frac{1}{2n}} \left(\prod_{j=1}^{n} w(c_j) \right)^{\frac{1}{2n}} \tag{29}$$

Comparing the priorities of c_k and c_r we obtain:

$$\frac{w(c_k)}{w(c_r)} = \frac{\left(\prod_{j=1}^{n} \frac{a_{kj}}{a_{jk}} \right)^{\frac{1}{2n}}}{\left(\prod_{j=1}^{n} \frac{a_{kj}}{a_{jk}} \right)^{\frac{1}{2n}}} \tag{30}$$

Therefore, for the purpose of prioritization, we can safely adopt:

$$w(c_k) = \left(\prod_{j=1}^{n} \frac{a_{kj}}{a_{jk}} \right)^{\frac{1}{2n}} \tag{31}$$

The above method of priority calculation (31) can be found in [9]. It has been proven that (31) is optimal with respect to the minimization of the sum of multiplicative errors. Similarly as before, the proposed modification to *GMM* is a simple generalization of the original *GMM* method.

5   Discussion and Summary

Both the presented methods, *EVM* and *GMM,* have been repeatedly debated and analysed. Various arguments were adduced for and against their use. For us, however, how much they have in common was important. In particular, we were intrigued by the fact that the quite straightforward assumption called by us the *mean heuristics* leads to both *EVM* and *GMM*. In our perception, the *mean heuristics* as presented in (Sect. 3) is quite straightforward. We realize, however, that this is our subjective view. There are many other priority deriving methods for the *PC* matrices. For most of them, the authors provide good arguments justifying their introduction. Indeed, even in the case of *EVM* and *GMM* various arguments in favour and against have been given. Let us recall a few: the *continuity* and *consistency* criteria [4], *optimality* with respect to the multiplicative errors [4,9], *condition of order preservation* [1] or the *correctness* and *effectiveness* criteria [16]. So, the *mean heuristics* cannot be an argument for or against mean based prioritization methods. We believe, however, that drawing attention to the common roots of *EVM* and *GMM* helps us to better understand their strengths and weakness, and which, in turn, may lead to the establishment of new and better priority deriving methods in the future.

References

1. Bana e Costa, C.A., Vansnick, J.: A critical analysis of the eigenvalue method used to derive priorities in AHP. Eur. J. Oper. Res. **187**(3), 1422–1428 (2008)
2. Bozóki, S., Fülöp, J., Rónyai, L.: On optimal completion of incomplete pairwise comparison matrices. Math. Comput. Model. **52**(1–2), 318–333 (2010)
3. Brunelli, M., Fedrizzi, M.: Axiomatic properties of inconsistency indices. J. Oper. Res. Soc. (2013)
4. Crawford, G.B.: The geometric mean procedure for estimating the scale of a judgement matrix. Math. Modell. **9**, 327–334 (1987)
5. Dyer, J.S.: Remarks on the analytic hierarchy process. Manage. Sci. **36**(3), 249–258 (1990)
6. Fedrizzi, M., Brunelli, M.: On the priority vector associated with a reciprocal relation and a pairwise comparison matrix. J. Soft Comput. **14**(6), 639–645 (2010)
7. Greco, S., Matarazzo, B., Słowiński, R.: Dominance-based rough set approach to preference learning from pairwise comparisons in case of decision under uncertainty. In: Hüllermeier, E., Kruse, R., Hoffmann, F. (eds.) IPMU 2010. LNCS, vol. 6178, pp. 584–594. Springer, Heidelberg (2010)
8. Ho, W.: Integrated analytic hierarchy process and its applications - a literature review. Eur. J. Oper. Res. **186**(1), 18–18 (2008)
9. Hovanov, N.V., Kolari, J.W., Sokolov, M.V.: Deriving weights from general pairwise comparison matrices. Math. Soc. Sci. **55**(2), 205–220 (2008)
10. Ishizaka, A., Lusti, M.: How to derive priorities in AHP: a comparative study. Central Europ. J. Oper. Res. **14**, 387–400 (2006)
11. Janicki, R., Zhai, Y.: On a pairwise comparison-based consistent non-numerical ranking. Logic J. IGPL **20**(4), 667–676 (2012)
12. Kahneman, D.: Thinking, Fast and Slow. New York, Farrar, Straus and Giroux (2011)

13. Koczkodaj, W.W., Szwarc, R.: On axiomatization of inconsistency indicators for pairwise comparisons. Fundamenta Informaticae **4**(132), 485–500 (2014)
14. Kou, G., Lin, C.: A cosine maximization method for the priority vector derivation in AHP. Eur. J. Oper. Res. **235**(1), 225–232 (2014)
15. Kułakowski, K.: Heuristic rating estimation approach to the pairwise comparisons method. Fundamenta Informaticae **133**, 367–386 (2014)
16. Kułakowski, K.: On the properties of the priority deriving procedure in the pairwise comparisons method. Fundamenta Informaticae **139**(4), 403–419 (2015)
17. Kułakowski, K.: Notes on the existence of a solution in the pairwise comparisons method using the heuristic rating estimation approach. Ann. Math. Artif. Intell. **77**(1), 105–121 (2016)
18. Kułakowski, K., Grobler-Dębska, K., Wąs, J.: Heuristic rating estimation: geometric approach. J. Glob. Optim. **62**(3), 529–543 (2015)
19. Meyer, C.: Matrix Analysis and Applied Linear Algebra. Society for Industrial and Applied Mathematics, SIAM, Philadelphia (2000)
20. Mikhailov, L.: Deriving priorities from fuzzy pairwise comparison judgements. Fuzzy Sets Syst. **134**(3), 365–385 (2003)
21. Perron, O.: Zur theorie der matrices. Math. Ann. **64**(2), 248–263 (1907)
22. Quarteroni, A., Sacco, R., Saleri, F.: Numerical Mathematics. Springer, New York (2000)
23. Saaty, T.L.: A scaling method for priorities in hierarchical structures. J. Math. Psychol. **15**(3), 234–281 (1977)
24. Smith, J.E., Winterfeldt, D.: Anniversary article: decision analysis in management science. Manage. Sci. **50**(5), 561–574 (2004)
25. Vaidya, O.S., Kumar, S.: Analytic hierarchy process: an overview of applications. Eur. J. Oper. Res. **169**(1), 1–29 (2006)

Bi-criteria Data Reduction for Instance-Based Classification

Ireneusz Czarnowski[1], Joanna Jędrzejowicz[2(✉)], and Piotr Jędrzejowicz[1]

[1] Department of Information Systems, Gdynia Maritime University,
Morska 83, 81-225 Gdynia, Poland
{irek,pj}@am.gdynia.pl
[2] Institute of Informatics, Gdańsk University, Wita Stwosza 57,
80-952 Gdańsk, Poland
jj@inf.ug.edu.pl

Abstract. One of the approaches to deal with the big data problem is the training data reduction which may improve generalization quality and decrease complexity of the data mining algorithm. In this paper we see the instance reduction problem as the multiple objective one and propose criteria which allow to generate the Pareto-optimal set of 'typical' instances. Next, the reduced dataset is used to construct classification function or to induce a classifier. The approach is validated experimentally.

Keywords: Classification · Clustering with bias-correction · Pareto-optimal prototypes

1 Introduction

One of the major problems facing machine learning community is finding techniques for dealing with the so-called big data problem. Sheer size of the available datasets makes it difficult or sometimes impossible to avoid excessive storage and time complexity which pose several barriers for designing efficient and reliable data mining algorithms. One of the obvious approaches allowing to deal with big datasets and, at the same time, providing data mining solutions of the acceptable accuracy is to reduce original training set by removing some instances before learning phase or to modify the instances using their new representation [12,28]. Although a variety of instance reduction methods has been so far proposed in the literature (see Sect. 2 for a brief review), no single approach can be considered as universally superior, guaranteeing satisfactory results and a minimization of the learning error or increased efficiency of the supervised learning. Therefore, the problem of selecting the reference instances, tackled by various researchers during last 15 years, remains valid and interesting field of research. It has been recognized (see for example [6,8]) that selecting reference vectors is inherently a multiple-objective problem where selection of instances affects not only the

© Springer International Publishing Switzerland 2016
N.T. Nguyen et al. (Eds.): ICCCI 2016, Part I, LNAI 9875, pp. 444–453, 2016.
DOI: 10.1007/978-3-319-45243-2_41

generalization (classification) quality of the resulting model, but also its complexity (for example, the resulting number of rules), data compression level, computational time required etc.

In this paper we also see the instance reduction problem as a multiple-objective one. The difference, in comparison with earlier approaches, is that we propose two possibly conflicting and simple instance selection criteria which help to generate the non-dominated or Pareto-optimal set of 'typical' instances, later used to construct the classification function or to induce a classifier. Thus, within the proposed approach, the training set is, eventually, reduced through applying computationally simple procedure, before the classifier is induced. The approach belongs to the class of the instance-based learning algorithms combining the similarity function, the selection of typical examples and induction of the classification function (see [1]).

Main research question addressed in the paper can be formulated as follows: does the instance selection procedure carried-out using the two proposed selection criteria improve accuracy of the classification function used at the classification stage? To answer the above question we carry-out computational experiment involving several well-known benchmark datasets. To perform classification we produce the Pareto-optimal set of instances from clusters of instances obtained by applying bias-correction fuzzy clustering algorithm [34]. Such obtained Pareto-optimal set of instances is next used by classification function to classify instances with the unknown class labels. As the classification tool we use the Naive-Bayes algorithm. The results are compared with the classification results obtained by the Naive-Bayes classifier applied to the full training dataset.

The remainder of this paper is organized as follows. Section 2 introduces current instance reduction techniques. Section 3 describes instance selection mechanism including similarity measures, clustering algorithm and procedure for producing Pareto-optimal set of instances, and provides details on the implemented classification mechanism. Section 4 describes computational experiment and its results. Finally, Sect. 5 summarizes conclusions and future work.

2   Instance Reduction Techniques

Instance selection is one of many data reduction techniques. Usually, instance selection algorithms are based on distance calculation between instances in the training set [5]. Methods based on other approaches, known as instance-based methods, remove an instance if it has the same output class as its k nearest neighbours, assuming that all neighbourhood instances will be, later on, correctly classified [30]. Both approaches have several weaknesses. They often use distance functions that are inappropriate or inadequate for linear and nominal attributes [5,30]. Besides, there is a need to store all the available training examples in the model. To eliminate the above, several approaches have been proposed, including, for example the condensed nearest neighbour (CNN) algorithm [15], the instance-based learning algorithm 2 (IB2) [1], the instance-based learning algorithm 3 (IB3) [1], the selective nearest neighbour (SNN) algorithm [23], the edited nearest neighbour (ENN) algorithm [31], the family of decremental reduction optimization procedures (DROP1-DROP5) [31], and the instance weighting approach [35]. The other group of methods (e.g. for example: the family of four instance reduction algorithms denoted respectively IRA1-IRA4 [7], the All k-NN method [26]) try to eliminate unwanted training examples using some removal criteria that need to be fulfilled. The same principle has been mentioned in [33]. The authors of [33] conclude that if many instances of the same class are found in an area, and when the area does not include instances from the other classes, then an unknown instance can be correctly classified when only selected reference vectors from such area is used.

The above reasoning also results in approaches, where the instance situated close to the center of a cluster of similar instances should be selected as a reference vectors (see, for example [2,31]). Such approach requires using some clustering algorithms like, for example, k-means or fuzzy k-means algorithm [11]. These algorithms generate cluster centers that are later considered to be the centroids, and the reduced dataset is produced.

In [10] prototypes are selected from clusters of instances and each instance has a chance to be selected as the reference instances. Another approach is to consider the so-called candidate instances situated close to the center of clusters and then to select the prototypes using the classification accuracy as a criterion [13].

In [18], the authors propose an algorithm for instance selection, Local Set-based Centroids Selector method (LSCo), where after noise removing the local sets clustering is applied for identifying clusters in the training dataset. Finally, the reduced set contains only the centroids of the resulting clusters. Another kind of algorithm attempting to select the reference instances evaluate the instances of each classes separately and keeps only the descent instances in a given (arbitrary) neighborhood [4].

In the literature one can also find others approaches to instance selection. For example, in [32] the technique for data reduction based on the idea of sampling has been proposed. Different versions of sampling-based prototype selection, including random sampling, stratified sampling, clustering sampling, inverse sampling and others, are proposed and discussed in [19].

It was proved that the instance selection belongs to the class of NP-hard problems [14]. Thus, local search heuristics and metaheuristics, like for example tabu search, simulated annealing, genetic algorithms, evolutionary algorithms, agent-based population learning algorithm, etc., seem to be the practical approach to solving the instance selection problem (see, for example [9,17,24,25,27]).

The instance selection methods can be classified on the basis of several different criteria. Raman in [22] points out that the instance selection methods can be grouped into three classes-filter methods, wrapper methods, and embedded methods. Wilson and Martinez in [31] suggested that the search for the reduced training set can be carried out in one of the three modes, incremental, decremental and batch.

The instance selection algorithms can also be classified as deterministic or non-deterministic. In deterministic ones, the final number of prototypes is controlled and determined by the user [16].

An overview of the instance selection techniques and algorithms, as well as some alternative techniques used for data reduction can be found in [10,21].

3   An Approach to Instance Selection and Classification

3.1   Bias-Correction Fuzzy Clustering

Fuzzy C-means clustering is a method which allows one row of data to belong to two or more clusters. The method is based on minimization of the objective function

$$J_m(u, c) = \sum_{i=1}^{N} \sum_{j=1}^{noCl} u_{ij}^m dist(x_i, c_j)$$

where m is a fixed number greater than 1, N is the number of data rows, $noCl$ is the number of clusters, c_j is the center of the j-th cluster and u_{ij} is the degree of membership of the i-th data row x_i in cluster c_j. The idea of bias-correction algorithm (BFCM), see [34], is to introduce a bias-correction term for reducing the effects of poor initialization. The probability mass p_i is used to represent the proportion of the cluster c_i to the $noCl$ clusters, with $\sum_{i=1}^{noCl} p_i = 1$. The total information based on fuzzy partitions is expressed as $-\sum_{i=1}^{N} \sum_{j=1}^{noCl} u_{ij}^m \ln(p_j)$, which denotes entropy. An optimal p_j is obtained as the one which minimizes entropy. In this case the objective function is defined as follows:

$$J_m(u, c, p) = \sum_{i=1}^{N} \sum_{j=1}^{noCl} u_{ij}^m dist(x_i, c_j)^2 - w \sum_{i=1}^{N} \sum_{j=1}^{noCl} u_{ij}^m \ln(p_j)$$

with w a parameter, updated in each iteration step t according to $w^{(t)} = (0.99)^t$ and subject to

$$\sum_{j=1}^{noCl} u_{ij} = 1, \ \sum_{j=1}^{noCl} p_j = 1$$

BFCM clustering is an iterative process with the update of membership factors u_{ij}, cluster centers c_j and probability mass p_i, for $i = 1, \ldots, N$ and $j = 1, \ldots, noCl$, defined by:

$$u_{ij} = \frac{(dist(x_i, c_j)^2 - w\ln(p_j))^{\frac{-1}{m-1}}}{\sum_{k=1}^{noCl}(dist(x_i, c_j)^2 - w\ln(p_k))^{\frac{-1}{m-1}}} \tag{1}$$

$$c_j = \frac{\sum_{i=1}^{N} u_{ij}^m \cdot x_i}{\sum_{i=1}^{N} u_{ij}^m} \tag{2}$$

$$p_j = \frac{\sum_{i=1}^{N} u_{ij}^m}{\sum_{i=1}^{N} \sum_{k=1}^{noCl} u_{ik}^m} \tag{3}$$

The initial values in matrix $u = (u_{ij})$ are random numbers between 0 and 1. The algorithm of bias-correction fuzzy C-means clustering is shown as Algorithm 1. Note that the computational complexity of Algorithm 1 is $O(noIt \cdot N \cdot |CL|)$, where $|CL| = |C| \cdot noCl$ is the number of clusters, $N = |TD|$ is the number of datarows in the training set and $noIt$ is the number of iterations.

Algorithm 1. Bias-correction fuzzy C-means clustering

Input: training data $TD = \bigcup_{c \in C} TD^c$, $noCl$ - number of clusters, accuracy ϵ
Output: partition of each TD^c into clusters
1: **for all** $c \in C$ **do**
2:     choose randomly $noCl$ rows from TD^c as initial centroids $c_1^{(0)}, \ldots, c_{noCl}^{(0)}$
3:     initialize probability mass $p_i^{(0)} = \frac{1}{noCl}$
4:     initialize parameter $w = 1.0$
5:     $t = 0$
6:     **repeat**
7:         $t++$
8:         modify $w* = 0.99$
9:         compute the membership matrix U using (1)
10:        compute the probability weight p_j using (3)
11:        update the cluster centers $c_j^{(t)}$ according to (2)
12:        compare $c^{(t)}$ to $c^{(t-1)}$ using a fixed norm $\| \cdot \|$
13:    **until** $\| c^{(t)} - c^{(t-1)} \| < \epsilon$
14:    distribute TD^c into $noCl$ clusters choosing for each x_i cluster c_j such that $dist(x_i, c_j^{(t)})$ is minimal
15: **end for**
16: **return** distribution of data rows into clusters and cluster centroids

---

3.2   Pareto Optimal Prototypes

After clustering training data, prototypes from each cluster are selected to reduce the training set size. Prototypes as those rows from the cluster which are not dominated by other rows from the same cluster. To define the domination relation two measures for each data row are introduced. These measures are, at the same time, instance selection criteria. The first one measures the mean distance from data from the same cluster and the second - the mean distance from rows from other classes. For a data row x let $Cl(x)$ stand for the cluster assigned to x and $Class(x)$ - for the class of x. The mentioned measures are:

$$distCl(x) = \frac{\sum_{y \in Cl(x)} dist(x, y)}{|Cl(x)|}$$

$$distGl(x) = \frac{\sum_{Class(x) \neq Class(y)} dist(x,y)}{\sum_{c \neq Class(x)} |TD^c|}$$

Definition. We say that data row y dominates data row x if

$$(distCl(y) \leq distCl(x)) \wedge (distGl(y) > dist(Gl(x)))$$

or

$$(distCl(y) < distCl(x)) \wedge (distGl(y) \geq dist(Gl(x)))$$

The relation of domination is irreflexive, antisymmetric and transitive. We say that x is not dominated in its cluster when no y from the same cluster dominates x.

Algorithm 2. Generating Pareto optimal prototypes

Input: number of clusters $noCl$, training data $TD = \bigcup_{c \in C} \bigcup_{i=1}^{noCl} TD_i^c$
Output: Pareto optimal prototypes $PR = \bigcup_{c \in C} PR^c$
1: **for all** $c \in C$ **do**
2:    **for** $i = 1, \ldots, noCl$ **do**
3:       **for** $x \in TD_i^c$ **do**
4:          calculate $distCl(x)$ and $distGl(x)$
5:       **end for**
6:       $PR_i^c = \{x \in TD_i^c |\ x$ is not dominated in $TD_i^c\}$
7:    **end for**
8:    $PR^c = \bigcup_{i=1}^{noCl} PR_i^c$
9: **end for**
10: **return** $PR = \bigcup_{c \in C} PR^c$

3.3 Classification

The following metrics were used (and compared) in the experiments. For s, t being attribute vectors of dimension n we have:

- Canberra metrics: $d_{Canberra}(s,t) = \sum_{i=1}^{n} \frac{|s_i - t_i|}{|s_i| + |t_i|}$
- cosine metrics: $d_{cosine}(s,t) = 1 - \frac{\sum_{i=1}^{n} s_i \cdot t_i}{\sqrt{\sum_{i=1}^{n} s_i^2} \cdot \sqrt{\sum_{i=1}^{n} t_i^2}}$

Algorithm 3. Classification

Input: training data $TD = \bigcup_{c \in C} TD^c$, testing data TS, $noCl$ - number of clusters
Output: accuracy of two modes of classification
1: use Algorithm 1 to distribute TD into $|C| \cdot noCl$ clusters
2: use Algorithm 2 to generate Pareto optimal prototypes $PR = \bigcup_{c \in C} PR^c$
3: perform Naive Bayes classification using TD as training data, TS as testing data, return accuracy acc^{full}
4: perform Naive Bayes classification using PR as training data, TS as testing data, return accuracy acc^{prot}
5: **return** accuracy, ROC area

---

4   Computational Experiment Results

To enable answering basic research question posed in the Introduction we have carried-out the computational experiment. In the experiment the following datasets from the UCI Machine Learning Repository [3]: Acredit (690/16/2), Breast (263/10/2), Heart (303/14/2), Heptitis (155/20/2), Ionosphere (351/35/2), Image (2086/19/2), Sonar (208/61/2), Thyroid (7000/21/3), Banana (5300/16/2) set from [20], and finally Chess (503/9/2) and Elec2 (17423,6,2) from [29]. For the experiments we used 2 distance measures Canberra and Cosine described in Sect. 3. We have been comparing classification results produced by the Naive Bayes classifier using the full datasets with the results produced by the Naive Bayes classifier induced from the reduced training datasets consisting of the non-dominated instances as explained in Sect. 3. All computation results shown in Tables 1, 2 are averages from 10 repetitions of the 10-cross-validation scheme. In all cases we set the number of clusters required to carry-out the data reduction procedure as 3. Additional experiments have shown that this value could be increased in case of big datasets, without apparent loss to the resulting classification accuracy.

Table 1. Classification results - full training dataset

Dataset	Acc	±	ROC
ACredit	0.777	0.013	0.438
Banana	0.737	0.021	0.691
Breast	0.731	0.011	0.684
Chess	0.811	0.045	0.871
Elec2	0.790	0.007	0.879
Heart	0.761	0.028	0.757
Hepatitis	0.810	0.018	0.689
Ionosphere	0.786	0.009	0.521
Image	0.604	0.022	0.680
Sonar	0.711	0.008	0.771
Thyroid	0.950	0.017	–

In Table 1 values of classification accuracy, its standard deviation and area under receiver operating characteristic (ROC) curve obtained through inducing Naive Bayes classification algorithm from the full training dataset, are shown.

In Table 2 values of classification accuracy, its standard deviation and area under receiver operating characteristic (ROC) curve obtained through inducing Naive Bayes classification algorithm from the reduced training dataset with Canberra distance measure and cosine distance, are shown.

From data shown in Tables 1, 2 it is clear that in several cases data reduction is beneficial to improving average accuracy level, increasing area under ROC

Table 2. Classification results - reduced dataset, Canberra and cosine measure

Dataset	Canberra distance				Cosine distance			
	Acc	±	ROC	Reduced by	Acc	±	ROC	Reduced by
ACredit	0.777	0.014	0.438	0 %	**0.791**	0.019	0.4493	84.10 %
Banana	**0.745**	0.023	0.626	91.40 %	0.737	0.009	0.512	51.50 %
Breast	**0.738**	0.031	0.688	74.80 %	**0.753**	0.007	0.691	75 %
Chess	0.811	0.029	0.871	0 %	0.745	0.012	0.824	80.80 %
Elec2	0.790	0.007	0.841	97.90 %	**0.796**	0.016	0.889	95.80 %
Heart	0.761	0.028	0.757	0 %	**0.782**	0.026	0.875	81.80 %
Hepatitis	0.756	0.011	0.759	76.30 %	0.81	0.018	0.689	0 %
Ionosphere	0.786	0.009	0.521	0 %	0.786	0.009	0.521	0 %
Image	0.604	0.022	0.679	1 %	**0.706**	0.031	0.69	91.40 %
Sonar	0.701	0.009	0.772	79.80 %	**0.781**	0.013	0.798	74.10 %
Thyroid	0.950	0.017	–	0 %	0.756	0.019	–	96 %

curve and reducing number of instances required to provide good quality classification results. Out of several distance measures tested within the data reduction procedure including Manhattan, Euclidean, Chessboard, Bray Curtis, Canberra and Cosine, the last two proved most effective with Cosine being a clear winner. Accuracies shown in bold in Table 2 are better than these obtained over the full training dataset. Considering that in some cases data reduction brings about slight decrease of the classification accuracy it is still worth noting that, statistically, Cosine-based data reduction assured better quality results as compared with results obtained using full datasets, in case of the investigated 11 randomly selected benchmark datasets.

5   Conclusions

Main contribution of the paper is proposing computationally simple data reduction procedure based on bias-correction fuzzy clustering and two instance selection criteria allowing for simple construction of the Pareto-optimal set of instances. Computational experiment designed to validate the approach allows to claim that the proposed instance selection procedure carried-out using the two selection criteria can, in many cases, improve accuracy of the classification function used at the classification stage. The approach contributes also to substantial decrease of the number of instances in the training set. Additional advantage of the approach is that there is only one parameter to be determined by the user, and that its value is not critical to the classification results. Further extended experiments should be carried-out to support the above findings. Especially, different classification algorithms and functions are planned to be investigated in the future.

References

1. Aha, D.W., Kibler, D., Albert, M.K.: Instance-based learning algorithms. Mach. Learn. **66**, 37–66 (1991)
2. Andrews, N.O., Fox, E.A.: Clustering for data reduction: a divide and conquer approach. Technical Report TR-07-36, Computer Science, Virginia Tech (2007)
3. Asuncion, A., Newman, D.J.: UCI Machine Learning Repository. University of California, School of Information and Computer Science (2007). http://www.ics.uci.edu/mlearn/MLRepository.html
4. Carbonera, J.L., Abel, M.: A density-based approach for instance selection. In: Proceedings of the 2015 IEEE 27th International Conference on Tool with Artificial Intelligence, pp. 768–774 (2015). doi:10.1109/ICTAI.2015.114
5. Chin-Liang, C.: Finding prototypes for nearest neighbor classifier. IEEE Trans. Comput. **23**(11), 1179–1184 (1974)
6. Czarnowski, I., Jędrzejowicz, P.: An agent-based approach to the multiple-objective selection of reference vectors. In: Perner, P. (ed.) MLDM 2007. LNCS (LNAI), vol. 4571, pp. 117–130. Springer, Heidelberg (2007)
7. Czarnowski, I., Jędrzejowicz, P.: An approach to instance reduction in supervised learning. In: Research and Development in Intelligent Systems XX, pp. 267–282. Springer, London (2004)
8. Czarnowski, I., Jędrzejowicz, P.: An approach to data reduction and integrated machine classification. New Generation Comput. **28**, 21–40 (2010)
9. Czarnowski, I.: Distributed learning with data reduction. In: Nguyen, N.T. (ed.) TCCI IV 2011. LNCS, vol. 6660, pp. 3–121. Springer, Heidelberg (2011)
10. Czarnowski, I., Jędrzejowicz, P.: A new cluster-based instance selection algorithm. In: O'Shea, J., Nguyen, N.T., Crockett, K., Howlett, R.J., Jain, L.C. (eds.) KES-AMSTA 2011. LNCS, vol. 6682, pp. 436–445. Springer, Heidelberg (2011)
11. Eschrich, S., Ke, J., Hall, L.O., Goldgof, D.B.: Fast accurate fuzzy clustering through data reduction. IEEE Trans. Fuzzy Syst. **11**(2), 262–270 (2013)
12. Garcia, S., Luengo, J., Herrera, F.: Data Preprocessing in Data Mining. Springer (2015)
13. Grudzinski, K., Duch, W.: SBL-PM: simple algorithm for selection of reference instances in similarity based methods. In: Proceedings of the Intelligence Systems, Bystra, Poland, pp. 99–107 (2000)
14. Hamo, Y., Markovitch, S.: The COMPSET algorithm for subset selection. In: Proceedings of the 19 International Joint Conference for Artificial Intelligence, Edinburgh, Scotland, pp. 728–733 (2005)
15. Hart, P.E.: The condensed nearest neighbour rule. IEEE Trans. Inf. Theory **14**, 515–516 (1968)
16. Kim, S.W., Oommen, B.J.: A brief taxonomy and ranking of creative prototype reduction schemes. Pattern Anal. Appl. **6**, 232–244 (2003)
17. Kuncheva, L.I., Bezdek, J.C.: Nearest prototype classification: clustering, genetic algorithm or random search? IEEE Trans. Syst. Man Cybern. **28**(1), 160–164 (1998)
18. Leyva, E., Gonzalez, A., Perez, R.: Three new instances selection methods based on local sets: a comparative study with several approaches from bi-objective perspective. Pattern Recogn. **48**(4), 1523–1537 (2015)
19. Liu, H., Motoda, H.: Instance Selection and Construction for Data Mining. Kluwer, Dordrecht (2001)

20. Machine Learning Data Set Repository (2013). http://mldata.org/repository/tags/data/IDA_Benchmark_Repository/
21. Olvera-Lopez, J.A., Carrasco-Ochoa, A.J., Martnez-Trinidad, J.F., Kittler, J.: A review of instance selection methods. Artif. Intell. Rev. **34**, 133–143 (2010). doi:10.1007/s10462-010-9165-y
22. Raman, B.: Enhancing Learning Using Feature and Example Selection. Texas A&M University, College Station (2003)
23. Ritter, G.L., Woodruff, H.B., Lowry, S.R., Isenhour, T.L.: An algorithm for a selective nearest decision rule. IEEE Trans. Inf. Theory **21**, 665–669 (1975)
24. Skalak, D.B.: Prototype and feature selection by sampling and random mutation hill climbing algorithm. In: Proceedings of the International Conference on Machine Learning, pp. 293–301 (1994)
25. Song, H.H., Lee, S.W.: LVQ combined with simulated annealing for optimal design of large-set reference models. Neural Netw. **9**(2), 329–336 (1996)
26. Tomek, I.: An experiment with the edited nearest-neighbour rule. IEEE Trans. Syst. Man Cybern. **6–6**, 448–452 (1976)
27. Tsai, C.F., Eberle, W., Chu, C.Y.: Genetic algorithms in feature and instance selection. Knowl. Based Syst. **39**, 240–247 (2013). doi:10.1016/j.knosys.2012.11.005
28. Lin, W.C., Tsai, C.F., Ke, S.W., Hung, C.W., Eberle, W.: Learning to detect representative data for large scale instance selection. J. Syst. Softw. **106**, 1–8 (2015)
29. Waikato. http://moa.cms.waikato.ac.nz/datasets/2013
30. Wilson, D.R., Martinez, T.R.: An integrated instance-based learning algorithm. Comput. Intell. **16**, 1–28 (2000)
31. Wilson, D.R., Martinez, T.R.: Reduction techniques for instance-based learning algorithm. Mach. Learn. **38**, 257–286 (2000). Kluwer Academic Publishers, Boston
32. Winton, D., Pete, E.: Using instance selection to combine multiple models learned from disjoint subsets. In: Instance Selection and Construction for Data Mining. Kluwer, Dordrecht (2001)
33. Wu, Y., Ianakiev, K., Govindaraju, V.: Improvements in k-nearest neighbor classification. In: Singh, S., Murshed, N., Kropatsch, W.G. (eds.) ICAPR 2001. LNCS, vol. 2013, pp. 222–229. Springer, Heidelberg (2001)
34. Yang, M.-S., Yi-Cheng, T.: Bias-correction fuzzy clustering algorithms. Inf. Sci. **309**, 138–162 (2015)
35. Yu, K., Xiaowei, X., Ester, M., Kriegel, H.P.: Feature weighting and instance selection for collaborative filtering: an information-theoretic approach. Knowl. Inf. Syst. **5**(2), 201–224 (2004)

Dynamic Cooperative Interaction Strategy for Solving RCPSP by a Team of Agents

Piotr Jędrzejowicz[(✉)] and Ewa Ratajczak-Ropel

Department of Information Systems, Gdynia Maritime University,
Morska 83, 81-225 Gdynia, Poland
{pj,ewra}@am.gdynia.pl

Abstract. In this paper a dynamic cooperative interaction strategy for the A-Team solving the Resource-Constrained Project Scheduling Problem (RCPSP) is proposed and experimentally validated. The RCPSP belongs to the class of NP-hard optimization problems. To solve this problem a team of asynchronous agents (A-Team) has been implemented using multiagent environment. An A-Team consist of the set of objects including multiple optimization agents, manager agents and the common memory which through interactions produce solutions of hard optimization problems. In this paper the dynamic cooperative interaction strategy is proposed. The strategy supervises cooperation between agents and the common memory. To validate the proposed approach the preliminary computational experiment has been carried out.

Keywords: Resource-Constrained Project Scheduling Problem · RCPSP · Optimization · Agent · A-Team

1 Introduction

The Resource Constrained Project Scheduling Problem (RCPSP) has attracted attention of many researches and a lot of exact, heuristic and metaheuristic solution methods have been proposed in the literature in recent years [1,8,10,19]. The current methods of solving these problems produce approximate solutions or can only be applied for solving instances of the limited size. Hence, searching for more effective algorithms and methods is still an active field of research. One of the promising directions of the research is using the parallel and distributed computation solutions, which are features of the contemporary multiagent systems [24].

Modern multiagent system architectures are an important and intensively expanding area of research and development. There are many of multiple-agent approaches proposed to solve different types of optimization problems in the literature. One of them is the concept of an A-Team, originally proposed in [21]. The idea of the A-Team was used to develop the software environment for solving different computationally hard optimization problems called JABAT [2,16]. JADE based A-Team (JABAT) system supports the construction of the dedicated A-Team architectures. Agents used in JABAT assure decentralization

© Springer International Publishing Switzerland 2016
N.T. Nguyen et al. (Eds.): ICCCI 2016, Part I, LNAI 9875, pp. 454–463, 2016.
DOI: 10.1007/978-3-319-45243-2_42

of computation across multiple hardware platforms. Parallel processing lead to more effective use of the available resources and ultimately, a reduction of the time of computation.

A-Team is a system comprised of the set of objects including multiple optimization agents, manager agents and the common memory which through interactions produce solutions of hard optimization problems. Several strategies controlling the interactions between agents and memories have been recently proposed and experimentally validated. The influence of such interaction strategy on the A-Team was investigated in [3]. In [13,14] the dynamic interaction strategy based on Reinforcement Learning (RL) for A-Team solving the RCPSP and MRCPSP has been described. In [15] the dynamic interaction strategy based on Population Learning Algorithm (PLA) for A-Team solving the RCPSP has been proposed. The similar topics were also considered for different multi-agent systems, e.g. [6,18,25].

In this paper the dynamic cooperative interaction strategy for the A-Team solving the RCPSP is proposed and experimentally validated. The proposed strategy is used to control the parameters and manage the process of searching for the RCPSP solutions by a team of agents. In this approach the parameters depend on the current state of the environment and the results received from the optimization agents. The most promising features of the RL and PLA based strategies have been used.

The proposed A-Team produces solutions to the RCPSP instances using four kinds of the optimization agents. They include local search, tabu search metaheuristic, crossover search and path relinking procedures.

The paper extends earlier results of authors [13,15] and reports on novel approach within the framework which have been used since our early papers on project scheduling. In such circumstances it is obvious that problem formulation, some experiment settings and computational environment description contain similarities as compared with previous papers.

The paper is constructed as follows: Sect. 2 contains the RCPSP formulation. Section 3 provides details of the proposed dynamic cooperative interaction strategy and its implementation in JABAT environment. Section 4 contains settings of the computational experiment and a discussion of the results. Finally, Sect. 5 contains conclusions and suggestions for future research.

2   Problem Formulation

A single-mode resource-constrained project scheduling problem (RCPSP) consists of a set of n activities, where each activity has to be processed without interruption to complete the project. The dummy activities 1 and n represent the beginning and the end of the project. The duration of an activity j, $j = 1, \ldots, n$ is denoted by d_j where $d_1 = d_n = 0$. There are r renewable resource types. The availability of each resource type k in each time period is r_k units, $k = 1, \ldots, r$. Each activity j requires r_{jk} units of resource k during each period of its duration, where $r_{1k} = r_{nk} = 0$, $k = 1, ..., r$. All parameters are non-negative integers.

There are precedence relations of the finish-start type with a zero parameter value (i.e. $FS = 0$) defined between the activities. In other words activity i precedes activity j if j cannot start until i has been completed. The structure of a project can be represented by an activity-on-node network $G = (SV, SA)$, where SV is the set of activities and SA is the set of precedence relationships. SS_j (SP_j) is the set of successors (predecessors) of activity j, $j = 1, \ldots, n$. It is further assumed that $1 \in SP_j$, $j = 2, \ldots, n$, and $n \in SS_j$, $j = 1, \ldots, n-1$. The objective is to find a schedule S of activities starting times $[s_1, \ldots, s_n]$, where $s_1 = 0$ and resource constraints are satisfied, such that the schedule duration $T(S) = s_n$ is minimized.

The objective is to find a minimal schedule in respect of the makespan that meets the constraints imposed by the precedence relations and the limited resource availabilities.

The RCPSP as a generalization of the classical job shop scheduling problem belongs to the class of NP-hard optimization problems [4]. The considered problem class is denoted as $PS|prec|C_{max}$ [5] or $m, 1|cpm|C_{max}$ [7].

3   A-Team with the Dynamic Cooperative Interaction Strategy

JABAT was successfully used by the authors for solving RCPSP and MRCPSP (see [11]) as well as RCPSP/max and MRCPSP/max (see [12]), where static interaction strategies have been used.

In [13–15] the dynamic interaction strategies based on Reinforcement Learning (RL) and Population Learning Algorithm (PLA) have been proposed and successfully used. In this approach the dynamic cooperative interaction strategy is proposed, where the most effective features of RL and PLA based strategies has been used.

The behavior of the A-Team in JABAT environment is controlled by the interaction strategy defined by the user. A-Team uses a population of individuals (solutions) managed by Solution Manager agent. To improve the individuals the optimization agents (OptiAgents) are used. All OptiAgents within the A-Team work together to improve individuals from the population in accordance with the interaction strategy.

To adapt JABAT to solving RCPSP problem the sets of object classes and agents were implemented. The first set contains object classes describing the problem instance. These classes are responsible for storing, reading and pre-processing of the data and generating random instances of the problem. This set includes: RCPSPTask, RCPSPSolution, RCPSPActivity and RCPSPResource representing the RCPSP instance, solution, activity and renewable resource, respectively [13].

The second set includes object classes representing the optimization agents. They are inheriting from the OptiAgent class. Optimization agents include implementation of optimization algorithms solving the RCPSP: CA, PRA, LSAm, LSAe, TSAm and TSAe described below. The prefix Opti is assigned to each agent with its embedded algorithm:

OptiLSAm - Local Search Algorithm with moving move (LSAm),
OptiLSAe - Local Search Algorithm with exchanging move (LSAe),
OptiTSAm - Tabu Search Algorithm with moving move (TSAm),
OptiTSAe - Tabu Search Algorithm with exchanging move (TSAe),
OptiCA - Crossover Algorithm (CA),
OptiPRA - Path Relinking Algorithm (PRA).

In the PRA, LSAm and TSAm the new solutions are obtained by moving the activities to new positions in the schedule, while in the LSAe and TSAe by exchanging pairs of activities. The algorithms details can be found in [13, 15].

The basic features of the proposed dynamic cooperative interaction strategy are based on the Reinforcement Learning strategy proposed in [13] and PLA-based strategy proposed in [15]. They are as follows:

- The individuals (solutions) in the population are generated randomly using different priority rules and serial forward SGS. The population is stored in the common memory.
- The individuals for improvement are chosen from the population randomly and blocked for the particular OptiAgent. Once chosen individual (or individuals) cannot be chosen again until the OptiAgent to which they have been sent returns the solution.
- The returning individual always represents the feasible solution. In the worst case it is the same solution which has been sent. It replaces its version before the attempted improvement. All solutions blocked for the considered OptiAgent are released and returned to the common memory.
- The new feasible solution is generated randomly using different priority rules and serial forward SGS with fixed probability and replaces another one. The methods of generating and replacing solutions in the population are described below.
- The environment state is remembered. This state includes: the best individual, average diversity of the population, weights and probabilities. The state is calculated every fixed number of iterations. To reduce the computation time, average diversity of the population is evaluated by comparison with the best solution only. Diversity of two individuals is evaluated as the sum of differences between activities starting times in a project.

To describe the proposed dynamic cooperative interaction strategy the following notation will be used:

P - population of individuals;
$\text{avgdiv}(P)$ - current average diversity of the population P;
$nITns$ - number of iterations after which a new environment state is calculated;
$S2imp$ - set of solutions selected for improvement.

Additionally, three probability measures have been used:

p_{mg} - probability of selecting the method mg for selection of a new individual;

p_{mr} - probability of selecting the method mr for replacing an individual in the population;

p_{ma} - probability of selecting the optimization agent ma used to improve an individual in the population.

There are four possible methods of generating a new individual:

mgr - randomly;
$mgrc$ - using one point crossover operator for two randomly chosen individuals;
mgb - random changes of the best individual in the population;
$mgbc$ - using one point crossover operator for two randomly chosen individuals from the five best individuals from the population.

For each method the weight w_{mg} is calculated, where $mg \in Mg$, $Mg = \{mgr, mgrc, mgb, mgbc\}$. The w_{mgr} and w_{mgrc} are increased where the population average diversity decreases and they are decreased in the opposite case. The w_{mgb} and w_{mgbc} are decreased where the population average diversity increases and they are increased in the opposite case.

There are three methods of replacing an individual from the population by a new one:

mrr - new solution replaces the random one in the population;
mrw - new solution replaces the random worse one in the population;
mrt - new solution replaces the worst solution in the population.

In most cases replacing the worse and worst solution intensifies exploitation while replacing the random one is effective to intensify exploration. The weight w_{mr} for each method is calculated, where $mr \in Mr$, $Mr = \{mrr, mrw, mrt\}$. The w_{mrr} is increased where the average diversity of the population decreases and it is decreased in the opposite case. The w_{mrw} and w_{mrt} are decreased where the average diversity of the population decreases and they are increased in the opposite case.

There are six optimization agents representing six optimization algorithms described above. For each of them the weight w_{ma} is calculated, where $ma \in Ma$, $Ma = \{maCA, maPRA, maLSAm, maLSAe, maTSAm, maTSAe\}$. The w_{ma} is increased if the optimization agent received the improved solution and is decreased in the other case. Additionally, the weights for $maCA$ and $maPRA$ are increased where the average diversity of the population decreases and they are decreased in the opposite case. The weights for $maLSAm$, $maLSAe$, $maTSAm$, $maTSAe$ are increased where the average diversity of the population increases and they are decreased in the opposite case. Using the most sophisticated algorithms is beneficial to intensify exploitation.

The probabilities of selecting the method are calculated as following:

$$p_{mg} = \frac{w_{mg}}{\sum_{mg \in M} w_{mg}} , \quad p_{mr} = \frac{w_{mr}}{\sum_{mr \in M} w_{mr}} , \quad p_{ma} = \frac{w_{ma}}{\sum_{ma \in M} w_{ma}} .$$

The current environment state parameters are updated after $nITns$ iterations: The update includes:

```
DCI_Strategy{
    generate the initial population P;
    calculate environment state;
    it=0;
    while(none of the stopping criteria is met){
        it++;
        select OptiAgent with probability p_ma;
        select solutions to improvement randomly and remember them in the set S2imp;
        use slected OptiAgent to improve the solutions from S2imp;
        generate a new solution S_new with probability p_mg;
        replace the individual in P by S_new with p_mr;
        if(it mod nITns = 0) calculate environment state;
    }
}
```

Fig. 1. General schema of the DCI_Strategy

- set w_{mgr}, w_{mgrc}, w_{mgb}, w_{mgbc};
- set w_{mrr}, w_{mrw}, w_{mrt};
- set w_{maCA}, w_{maPRA}, w_{maLSAm}, w_{maLSAe}, w_{maTSAm}, w_{maTSAe};
- remember the best solution;
- calculate the avgdiv(P).

The general schema of the proposed dynamic cooperative interaction strategy (DCI_Strategy) for the A-Team is presented in Fig. 1. It is worth noticing that computations are carried out independently and possibly in parallel, within the JABAT environment.

To implement this strategy in the JABAT environment the new procedure has been implemented. It allows to initialize and destroy new optimization agents according to the needs of the DCI_Strategy.

4   Computational Experiment

4.1   Settings

To estimate the effectiveness of the proposed approach the computational experiment has been carried out using benchmark instances of RCPSP from PSPLIB[1] - test sets: sm30 (single mode, 30 activities), sm60, sm90, sm120. Each of the first three sets includes 480 problem instances while set sm120 includes 600 instances. The experiment involved computation with a varying number of optimization agents, fixed population size, and the stopping criteria indicated by the environment state.

The following global parameters have been used in the computational experiment: $|P| = 30$, $nITns = 5$. Number of optimization agents vary from 6 to 12.

---

[1] See PSPLIB at http://www.om-db.wi.tum.de/psplib/.

To enable comparisons with other methods known from the literature, the number of schedules generated during computation is calculated. In the presented experiments the number of schedules $nSGS$ is limited to 5000.

The DCI_Strategy initial settings are as follow:

Initial weights :
$w_{mgr} = 25$, $w_{mgrc} = 25$, $w_{mgb} = 25$, $w_{mgbc} = 25$,
$w_{mrr} = 34$, $w_{mrw} = 33$, $w_{mrt} = 33$,
$w_{maCA} = 30$, $w_{maPRA} = 30$,
$w_{maLSAm}=10$, $w_{maLSAe}=10$, $w_{maTSAm}=10$, $w_{maTSAe}=10$;
Iteration numbers of optimization algorithms :
OptCA - $\lfloor n/2 \rfloor$, OptPRA - $\lfloor n/2 \rfloor$,
OptLSAm - $\lfloor n/3 \rfloor$, OptLSAe - $\lfloor n/3 \rfloor$, OptTSAm - $\lfloor n/3 \rfloor$, OptTSAe - $\lfloor n/3 \rfloor$;
Stopping criteria :
avgdiv$(P)<$ 0.05 or $nSGS>$5000;

To calculate weights and other initial parameters the effective approaches proposed in [13,15] have been used. The parameters values have been chosen experimentally based on earlier experiments for the RCPSP in JABAT and the preliminary experiments performed on data set sm60.

The experiment has been performed using nodes of the cluster Holk of the Tricity Academic Computer Network built of 256 Intel Itanium 2 Dual Core 1.4 GHz with 12 MB L3 cache processors and with Mellanox InfiniBand interconnections with 10 Gb/s bandwidth. During the computation one node per four optimization agents has been used.

4.2 Results

The following computational results features have been calculated and compared: Mean Relative Error (MRE) calculated as the deviation from the optimal solution for sm30 set or from the best known results for sm60, sm90 and sm120, MRE from the Critical Path Lower Bound (CPLB), Mean Computation Time (MCT) needed to find the best solution and Mean Total Computation Time (MTCT) needed to stop all optimization agents and the whole system. The number of schedules generated by SGS heuristics is limited to 5000 for all optimization agents used during search for the solution for each problem instance.

Table 1. Results for DCI_Strategy

Set	MRE from the optimal (*) or best known solution	MRE from the CPLB	MCT [s]	MTCT [s]
sm30	0.02 %*	-	2.43	41.12
sm60	0.45 %	10.87 %	15.34	66.23
sm90	0.92 %	10.39 %	25.21	74.31
sm120	2.44 %	32.17 %	86.43	182.40

Table 2. Literature reported results [1, 10]

Method	Authors	MRE	MCT [s]	CPU
Set sm30				
Decompos. & local opt	Palpant et al. [17]	0.00	10.26	2.3 GHz
Filter and fan	Ranjbar [20]	0.00	–	–
Event list-based EA	Paraskevopoulos et al. [19]	0.00	0.19	1.33 GHz
VNS–activity list	Fleszar, Hindi [9]	0.01	5.9	1.0 GHz
this approach		*0.02*	*2.43*	*1.4 GHz*
Set sm60				
Event list-based EA	Paraskevopoulos et al. [19]	10.54	16.31	1.33 GHz
Filter and fan	Ranjbar [20]	10.56	5	–
Decompos. & local opt	Palpant et al. [17]	10.81	38.78	2.3 GHz
MAOA	Zheng and Wang [25]	10.84	–	–
this approach		*10.87*	*15.34*	*1.4 GHz*
Population–based	Valls et al. [22]	10.89	3.65	400 MHz
Set sm90				
Filter and fan	Ranjbar [20]	10.11	5	–
Population-based	Valls et al. [22]	10.19	9.49	400 MHz
Decompos. & local opt	Palpant et al. [17]	10.29	61.25	2.3 GHz
Decomposition based GA	Debels, Vanhoucke [1]	10.35	–	–
this approach		*10.39*	*25.21*	*1.4 GHz*
GA–hybrid, FBI	Valls et al. [23]	10.46	0.61	400 MHz
Set sm120				
Event list-based EA	Paraskevopoulos et al. [19]	30.78	123.45	1.33 GHz
Filter and fan	Ranjbar [20]	31.42	5	–
Population-based	Valls et al. [22]	31.58	59.43	400 MHz
this approach		*32.17*	*86.43*	*1.4 GHz*
Decompos. & local opt	Palpant et al. [17]	32.41	207.93	2.3 GHz
MAOA	Zheng and Wang [25]	32.64	–	–

Each instance of the problem has been solved five times and the average results have been calculated.

All the solutions receiving during computations are feasible. The preliminary results obtained using the proposed dynamic cooperation strategy, presented in Table 1, are generally better then the results of strategies for the A-Team tested in [13, 15]. The MRE below 1 % in case of 30, 60 and 90 activities and 2.44 % in case of 120 activities have been obtained. The maximum RE is below 5 % and 7 %, respectively.

The presented results are comparable with the results known from the literature, see Table 2. However, it is worth noticing that in agent based approaches computations are performed by many processors (nodes) working in parallel. Hence, it is difficult to compare some features of the results. Results obtained by a single agent may or may not influence those obtained by other agents. Additionally, computation times include times used by agents to prepare, send and receive messages.

The results of the preliminary experiment show that the proposed implementation is efficient and using the proposed DCI_Strategy to control the A-Team is beneficial.

5   Conclusions and Future Work

The results of the computational experiment show that the proposed DCI_Strategy is the most effective interaction strategy tested for the A-Team in JABAT by authors. The A-Team using DCI_Strategy implemented in JABAT is also efficient and competitive tool for solving RCPSP instances. The obtained results are comparable with solutions presented in the literature and in some cases outperform them. It is worth mentioning that they have been obtained in a comparable computation time and number of schedules.

The presented experiment could be extended to examine different and additional parameters of the environment state and solutions as well as iteration numbers, probabilities and weights. The kind and number of optimization agents (OptiAgents) used should be interesting to investigate. Additionally, an effective method for tuning optimization agents parameters including a number of iterations needed should be developed.

References

1. Agarwal, A., Colak, S., Erenguc, S.: A neurogenetic approach for the resource-constrained project scheduling problem. Comput. Oper. Res. **38**, 44–50 (2011)
2. Barbucha, D., Czarnowski, I., Jędrzejowicz, P., Ratajczak-Ropel, E., Wierzbowska, I.: E-JABAT - an implementation of the web-based A-Team. In: Nguyen, N.T., Jain, L.C. (eds.) Intelligent Agents in the Evolution of Web and Applications. SCI, vol. 167, pp. 57–86. Springer, Heilderberg (2009)
3. Barbucha, D., Czarnowski, I., Jędrzejowicz, P., Ratajczak-Ropel, E., Wierzbowska, I.: Influence of the working strategy on A-Team performance. In: Szczerbicki, E., Nguyen, N.T. (eds.) Smart Information and Knowledge Management. SCI, vol. 260, pp. 83–102. Springer, Heidelberg (2010)
4. Błażewicz, J., Lenstra, J., Rinnooy, A.: Scheduling subject to resource constraints: classification and complexity. Discrete Appl. Math. **5**, 11–24 (1983)
5. Brucker, P., Drexl, A., Möhring, R., Neumann, K., Pesch, E.: Resource-constrained project scheduling: notation, classification, models, and methods. Eur. J. Oper. Res. **112**, 3–41 (1999)
6. Cadenas, J.M., Garrido, M.C., Muñoz, E.: Using machine learning in a cooperative hybrid parallel strategy of metaheuristics. Inf. Sci. **179**(19), 3255–3267 (2009)
7. Demeulemeester, E., Herroelen, W.: Project Scheduling: A Research Handbook. Kluwer Academic Publishers, Boston (2002)

8. Fang, C., Wang, L.: An effective shuffled frog-leaping algorithm for resource-constrained project scheduling problem. Comput. Oper. Res. **39**, 890–901 (2012)
9. Fleszar, K., Hindi, K.: Solving the resource-constrained project scheduling problem by a variable neighbourhood search. Eur. J. Oper. Res. **155**, 402–413 (2004)
10. Hartmann, S., Kölisch, R.: Experimental investigation of heuristics for resource-constrained project scheduling: an update. Eur. J. Oper. Res. **174**, 23–37 (2006)
11. Jędrzejowicz, P., Ratajczak-Ropel, E.: New generation A-Team for solving the resource constrained project scheduling. In: Proceedings of the Eleventh International Workshop on Project Management and Scheduling, Istanbul, pp. 156–159 (2008)
12. Jędrzejowicz, P., Ratajczak-Ropel, E.: Solving the RCPSP/max problem by the team of agents. In: Håkansson, A., Nguyen, N.T., Hartung, R.L., Howlett, R.J., Jain, L.C. (eds.) KES-AMSTA 2009. LNCS, vol. 5559, pp. 734–743. Springer, Heidelberg (2009)
13. Jędrzejowicz, P., Ratajczak-Ropel, E.: Reinforcement learning strategies for A-Team solving the resource-constrained project scheduling problem. Neurocomputing **146**, 301–307 (2014)
14. Jędrzejowicz, P., Ratajczak-Ropel, E.: Reinforcement learning strategy for solving the MRCPSP by a team of agents; intelligent decision technologies. In: Neves-Silva, R., Jain, L.C., Howlett, R.J. (eds.) Intelligent Decision Technologies. Smart Innovation, Systems and Technologies, vol. 39, pp. 537–548. Springer, Heidelberg (2015)
15. Jędrzejowicz, P., Ratajczak-Ropel, E.: PLA based strategy for solving RCPSP by a team of agents. Comput. Intell. Tools Process. Collective Data, J. Univ. Sci. (to appear 2016)
16. Jędrzejowicz, P., Wierzbowska, I.: JADE-based A-Team environment. In: Alexandrov, V.N., Albada, G.D., Sloot, P.M.A., Dongarra, J. (eds.) ICCS 2006. LNCS, vol. 3993, pp. 719–726. Springer, Heidelberg (2006)
17. Palpant, M., Artigues, C., Michelon, P.: LSSPER: solving the resource-constrained project scheduling problem with large nighbourhood search. Ann. Oper. Res. **131**, 237–257 (2004)
18. Pelta, D., Cruz, C., Sancho-Royo, A., Verdegay, J.L.: Using memory and fuzzy rules in a cooperative multi-thread strategy for optimization. Inf. Sci. **176**(13), 1849–1868 (2006)
19. Paraskevopoulos, D.C., Tarantilis, C.D., Ioannou, G.: Solving project scheduling problems with resource constraints via an event list-based evolutionary algorithm. Expert Syst. Appl. **39**, 3983–3994 (2012)
20. Ranjbar, M.: Solving the resource-constrained project scheduling problem using filter-and-fun approach. Appl. Math. Comput. **201**, 313–318 (2008)
21. Talukdar, S., Baerentzen, L., Gove, A., De Souza, P.: Asynchronous Teams: Cooperation Schemes for Autonomous, Computer-Based Agents. Technical report EDRC 18–59-96, Carnegie Mellon University, Pittsburgh (1996)
22. Valls, V., Ballestín, F.: A population-based approach to the resource-constrained project scheduling problem. Ann. Oper. Res. **131**, 305–324 (2004)
23. Valls, V., Ballestín, F., Quintanilla, S.: A hybrid genetic algorithm for the resource-constrained project scheduling problem. Eur. J. Oper. Res. **185**, 495–508 (2008)
24. Wooldridge, M.: An Introduction to MultiAgent Systems, 2nd edn. Wiley, New York (2009)
25. Zheng, X., Wang, L.: A multi-agent optimization algorithm for resource constrained project scheduling problem. Expert Syst. Appl. **42**, 6039–6049 (2015)

Influence of the Waiting Strategy on the Performance of the Multi-Agent Approach to the DVRPTW

Dariusz Barbucha[(✉)]

Department of Information Systems, Gdynia Maritime University,
Morska 83, 81-225 Gdynia, Poland
d.barbucha@wpit.am.gdynia.pl

Abstract. A multi-agent approach to the Dynamic Vehicle Routing Problem with Time Windows has been proposed in the paper. The process of solving instances of the problem is performed by a set of software agents. They are responsible for managing the sets of dynamic requests, allocating them to the available vehicles, and optimizing the routes covered by the vehicles in order to satisfy several requests and vehicles constraints. The paper focuses on waiting strategies which aim at deciding whether a vehicle should wait after servicing a request, before heading toward the next customer. The influence of the proposed waiting strategy on the performance of the approach has been investigated via a computational experiment. It confirmed the positive impact of the strategy on the obtained results.

Keywords: Dynamic vehicle routing problem with time windows · Waiting strategy · Multi-agent system

1 Introduction

Dynamic Vehicle Routing Problems class (DVRPs) refers to a group of routing problems that the required information about customers, vehicles, etc. is not given a priori to the decision maker but is revealed concurrently with the process of decision-making. The dynamism can be represented for example by stochastic vehicle speed depending on road's condition or stochastic demand of customers. On the other hand, recent advances in development and possibility of application of different communication and information technologies (global positioning systems, wireless networks, etc.) have allowed transportation companies to benefit from real-time information and to plan vehicles routes in more efficient way.

The most important variants of DVRPs include: classical Dynamic Vehicle Routing Problem (DVRP), where a set of customers' requests has to be served by a set of vehicles in order to minimize the cost of transport, and satisfying several customers and vehicles constraints, Dynamic Vehicle Routing Problem with Time Windows (DVRPTW), where the customers have to be visited during

© Springer International Publishing Switzerland 2016
N.T. Nguyen et al. (Eds.): ICCCI 2016, Part I, LNAI 9875, pp. 464–473, 2016.
DOI: 10.1007/978-3-319-45243-2_43

a specific time interval, Dynamic Pickup and Delivery Problem (DPDP), where goods have to be either picked-up or delivered in specific amounts in each of customer location, and Dynamic Pickup and Delivery Problem wit Time Windows (DPDPTW), a variant of DPDP with time windows. A review of different variants of DVRPs, methods of solving them and examples of practical applications can be found for example in [12].

The paper focuses on Vehicle Routing Problem with Time Windows (VRPTW). The *static* version of it can be formulated as an undirected graph $G = (V, E)$, where $V = \{0, 1, \ldots, N\}$ is a set of nodes and $E = \{(i, j)|i, j \in V\}$ is a set of edges. Node 0 is a central depot with K identical vehicles of capacity W. Each node $i \in V \setminus \{0\}$ denotes a customer characterized by a non-negative demand d_i, and a service time s_i. Moreover, with each customer $i \in V$, a time window $[e_i, l_i]$ wherein the customer has to be supplied, is associated. Here e_i is the earliest possible departure (ready time), and l_i - the latest time the customer's request has to be started to be served. The time window at the depot ($[e_0, l_0]$) is called the scheduling horizon. Each edge $(i, j) \in E$ denotes the path between customers i to j and is described by the cost c_{ij} of travel from i to j by shortest path $(i, j \in V)$. It is assumed that $c_{ij} = c_{ji}(i, j \in V)$. It is also often assumed that c_{ij} is equal to travel time t_{ij}.

The goal is to minimize the vehicle fleet size and the total distance needed to pass by vehicles in order to supply all customers (minimization of the fleet size is often considered to be the primary objective of the VRPTW). The following constraints have to be also satisfied: each route starts and ends at the depot, each customer $i \in V \setminus \{0\}$ is serviced exactly once by a single vehicle, the total load on any vehicle associated with a given route does not exceed vehicle capacity, each customer $i \in V$ has to be supplied within the time window $[e_i, l_i]$ associated with it (the vehicle arriving before the lower limit of the time window causes additional waiting time on the route), and each route must start and end within the time window associated with the depot.

The *dynamic* version of the VRPTW considered in the paper assumes that customers' requests are not known in advance but they are revealed dynamically and unpredictably during the execution of already arrived requests. Let the planning horizon starts at time 0 and ends at time T. Let $t_i \in [0, T]$ $(i = 1, \ldots, N)$ denotes the time when the i-th customer request is submitted. Following the *degree of dynamism* measure $dod = N_d/N$ [9] (N_d - number of dynamic requests, N - number of all requests), the problem considered in the paper is fully dynamic ($dod = 1$).

The main contribution of the paper is to propose an approach based on a multi-agent paradigm to the DVRPTW with a waiting strategy implemented within the approach. The paper aims at investigation of the influence of the waiting strategy on the performance of the proposed approach. The approach presented in the paper extends the multi-agent environment for solving DVRP and DVRPTW proposed by author in [1, 2].

The rest of the paper is divided on four sections. Section 2 includes a review of different waiting strategies implemented in frameworks proposed by other authors to solve different variants of VRP. Section 3 presents the multi-agent approach with the waiting strategy to the DVRPTW. Goal, assumptions and results of the computational experiment are presented in Sect. 4. Finally, Sect. 5 includes main conclusions and directions of the future research.

2   Waiting Strategies

Different strategies have been implemented within the approaches dedicated to solve Vehicle Routing Problems in order to improve their performance. The most known ones refer to *request buffering* and *vehicle waiting*. The *request buffering* strategy (proposed by Pureza and Laporte [13] and by Mitrovic-Minic et al. [11] for dynamic Pickup and Delivery Problems with Time Windows) aims at postponing a request assignment decision by storing some requests in a buffer. It means that allocation of each new request to one of the available vehicles is not performed immediately whenever the new request arrives. After arriving, the request is stored in the buffer, and it is considered to allocate at later stages. It is expected that by buffering the requests, better routing decisions are more likely to be achieved due to the larger number of accumulated requests available [13]. The *waiting strategy* [10] aims at deciding whether a vehicle should wait after servicing a request, before driving toward the next customer. This strategy is particularly important in problems with time windows, where time lags may appear between requests. It is expected that using information about the likely location of new customers, better decisions may be taken. Besides the waiting after servicing a customer, a vehicle can be also relocated (positioned) to a strategic position where probability of occurrence of new customers is higher.

The waiting strategies have been implemented in different frameworks for the DVRP [6], DVRPTW [4,5,8], and DPDPTW [10,13], where authors looked at the potential benefit of applying these strategies. Branke et al. [6] considered a standard DVRP, where one additional customer arrives at a beforehand unknown location when the vehicles are already under way. Their objective was to maximize the probability that the additional customer can be integrated into one of the fixed tours without violating time constraints. This was achieved by allowing the vehicles wait at suitable locations during their tours in order to maximize the probability that a new customer, appearing anywhere in the service region, can be integrated into one of the tours. The authors proposed several waiting strategies and an evolutionary algorithm to optimize the waiting strategy. Empirical comparison of the strategies allowed them to conclude that a proper waiting strategy can greatly increase the probability of being able to service the additional customer, at the same time reducing the average detour to serve that customer.

Another approach to increase the probability of servicing future unknown customers has been proposed by Mitrovic-Minic and Laporte [10]. They examined whether waiting strategies can reduce the total detour or the number of required vehicles for DPDPTW. They adapted a tabu search procedure proposed by Gendreau et al. [7] and suggested four waiting strategies: drive-first, wait-first, dynamic waiting, and advanced dynamic waiting. The results of computational experiment allowed them to discover benefits of applying these strategies within their approach and to point the last strategy as the most efficient.

Bent and Hentenryck [4] considered online Stochastic Multiple Vehicle Routing Problem with Time Windows in which requests arrive dynamically and the goal is to maximize the number of serviced customers. Contrary to many other algorithms which only move vehicles to known customers, they investigated waiting and relocation strategies in which vehicles may wait at their current location or being relocated to arbitrary sites. The decisions (to wait and/or to relocate) did not exploit any problem-specific features but rather were obtained by including choices in the online algorithm that are necessarily sub-optimal in an offline setting. Experimental results showed that waiting and relocation strategies may dramatically improve customer service, especially for problems that are highly dynamic and contain many late requests.

The approach proposed by Ichoua et al. [8] focused on DVRPTW. It exploited probabilistic knowledge about future events to better manage the fleet of vehicles and provide good coverage of the territory. They extended a parallel tabu search heuristic developed by Gendreau et al. [7] by allowing a vehicle to wait in its current zone instead of driving to the next planned destination located in another zone. The waiting strategy determined the interval of time a vehicle should wait at the current position, however, waiting was only allowed if the probability of a future request reached a particular threshold. Computational tests performed by authors showed that the inclusion of the waiting strategy improves the tabu search performance.

Pureza and Laporte [13] proposed a constructive-deconstructive heuristic for the DPDP with time windows and random travel times, where two kinds of strategies (request buffering and vehicle waiting) have been implemented within it. The waiting strategy based on fastest paths (WE_FP) took advantage of the fact that it was possible to reach a location earlier by using an indirect path instead of the direct path. Since the basic approach has already used a waiting strategy that allowed to arrive at the beginning of the time window of the locations, the extra time provided by faster paths was used to anticipate arrivals (when applicable), or to extend waiting after service. Comparisons of the quality of solutions obtained by an implementation of this strategy to an approach without it confirmed the advantages of this strategy both in terms of lost requests and number of vehicles.

Branchini et al. [5] proposed an adaptive granular local search heuristic for DVRP, where the strategies of vehicle waiting, positioning a vehicle in a region where customers are likely to appear, and diverting a vehicle away from its current destination have been integrated. They implemented the wait-first strategy proposed by Mitrovic-Minic and Laporte [10]. Good performance of the proposed approach has been confirmed by computational experiment performed by them on test problems derived from real-life Brazilian transportation companies.

3   Agent-Based Approach to DVRPTW

The proposed approach uses a multi-agent platform proposed by the author for simulating and solving the dynamic VRP/VRPTW [1,2]. The architecture of the platform is based on Java Agent Development Framework (JADE) [3], where several autonomous agents are defined:

- GlobalManager - an agent which initializes and destroys all agents,
- RequestGenerator - an agent which is responsible for generating (or reading from a file) new customers' requests,
- RequestManager - an agent which manages the list of customers' requests and coordinates the activity of other agents,
- Vehicle agents - represent vehicles which are responsible for serving the customers' requests.

The steps of the approach are presented in Algorithm 1. Because of the fact that all requests arrive while the process of solving the problem is ongoing, the most important activities refer to handling the newly arrived requests. Hence, the Algorithm 1 emphasizes this part of the process.

Algorithm 1. Main steps of the algorithm MAS-DVRPTW

1: The system initializes the GlobalManager agent, which next initializes other agents: RequestsGenerator, RequestsManager, and Vehicle
2: RequestsManager is waiting for events: $event = getEvent()$
3: **while** $(event\ != endOfRequests)$ **do**
4:    RequestGenerator agent generates (or reads) a new request and sends it to the RequestManager (newRequest event)
5:    if RequestManager receives a message with a new request then it creates/updates the routing plan R_s by (re-)solving the problem P_{DVRPTW} taking into account the requests do not assigned to any Vehicle agents (s is a new solution of P_{DVRPTW})
6:    On the other hand, if Vehicle agents reach the locations of their current customers, they inform the RequestManager about their readiness for serving next requests (vehicleStopAtLocation event)
7:    According to the current routing plan, the RequestManager allocates next requests to the Vehicle agents using *Centralized Dispatching Strategy* combined with *Waiting Strategy*.
8:    RequestsManager is waiting for next events: $event = getEvent()$
9: **end while**
10: All vehicles finishes their routes and drive back to the depot

---

The process of simulating and solving the DVRPTW starts with initialization of all variables (states of requests, vehicles, etc.) and creation of the above mentioned agents. When all agents report their readiness to act, the system is waiting for next events. Although several events are generated by the agents and many messages are sent between them, the most important ones refer to requests (announcing arriving a new request, end of requests, etc.), and to vehicles (messages reporting about their states, for example reaching a customer, waiting before servicing at a given location, etc.).

During the whole process of simulating and solving the DVRPTW, the new requests arrive to the system. All received requests are collected and maintained by the RequestManager agent, and next they are allocated to the Vehicle agents, which have just announced their readiness to act. The process of allocating the requests to the available Vehicle agents is performed according to the Centralized Dispatching Strategy (CDS) defined in [2].

In particular, when the newRequest event (sent by the RequestGenerator) is received by the RequestManager agent, it creates or updates the current global routing plan $R_s = [R_s^1, R_s^2, \ldots, R_s^K]$ by resolving the problem P_{DVRPTW} including the requests which have not been assigned to any Vehicle agents (s is a new solution of P_{DVRPTW}). A procedure of *cheapest insertion* of this request to the existing routes is performed. Moreover, the RequestManager performs a few local improvements of the current solution taking into account the requests which have not been yet assigned to the vehicles [2].

On the other hand, at any iteration, let $v(i)$ be a Vehicle agent, let $R^{v(i)} = [r_1^{v(i)}, r_2^{v(i)}, \ldots, r_k^{v(i)}]$ be a current route assigned to the vehicle $v(i)$ ($i = 1, \ldots, K$) (it contains the customers already visited by vehicle), and $r_k^{v(i)}$ is a customer (location) the vehicle $v(i)$ is currently driving to. After reaching a location of the customer r_k^i and finishing its service, the Vehicle agent $v(i)$ sends the message vehicleStopAtLocation to the RequestManager informing it about readiness for serving next requests. According to the current global routing plan, the RequestManager agent sends the next request to the Vehicle agent $v(i)$. The request is inserted on position $k + 1$ of the route of $v(i)$ agent.

The algorithm stops when the set of requests has been exhausted (message endOfRequests). Then all vehicles finish their mission and return to the depot.

Fundamental version of the proposed approach assumes that each Vehicle agent, after finishing its partial route at some location, immediately leaves it and starts driving to the next planned customer, according to the global routing plan. Following the constraint of the VRPTW that each customer has to be supplied within the time window associated with it, the vehicle arriving before the lower limit of the time window has to wait before servicing a customer at this location. In this case, the waiting time w_i of the vehicle *before* servicing i-th request can be calculated as $w_i = (e_i - t_i)$, where t_i is time of arrival the vehicle to the customer i.

Following the observations provided by a few authors and included in Sect. 2, it has been also decided to extend the above approach by implementing the strategy which aim at deciding whether a vehicle should wait also *after* servicing a request, before heading toward the next customer. The proposed strategy, called *waiting strategy*, is inspired by Mitrovic-Minic and Laporte (wait-first) strategy applied by them to the DPDPTW [10]. Its general assumption is that vehicle should leave current location at the latest possible departure time.

4   Computational Experiment

A computational experiment has been carried out to evaluate the influence of the waiting strategy on the performance of the proposed multi-agent approach to the DVRPTW. The performance of the approach has been measured by the number of vehicles needed to serve all requests in the predefined time and the total distance needed to pass by these vehicles.

The approach was tested on 56 classical VRPTW instances of Solomon [14] (available at [15]) including 100 customers, each, which have been transformed into the dynamic version through revealing all requests dynamically. The whole set of instances is divided into six groups (R1, R2, C1, C2, RC1, RC2) including customers with randomly generated coordinates (R1, R2), clustered coordinates (C1, C2) or both (RC1, RC2). Additionally, instances belonging to R1, C1, and RC1 have a short scheduling horizon, whereas the instances from R2, C2 and RC2 have a long scheduling horizon.

It has been assumed that requests may arrive with various frequencies, and in the experiment arrivals of the dynamic requests have been generated using the Poisson distribution with λ parameter denoting the mean number of requests occurring in the unit of time (1 h in the experiment). For the purpose of experiment it has been assumed that: λ is equal to 5, 10, 15, and 30, all requests have to be served, and the vehicle speed is set to 60 Km/h.

In order to discover the influence of the proposed waiting strategy on the obtained results, two cases have been tested and compared in the experiment:

– NO-WAIT - it assumes that each Vehicle agent does not consider waiting at its current location after servicing a request. It means that the Vehicle agent after finishing its partially route at some destination, it immediately leaves this location and starts driving to the next customer assigned to it,
– WAIT - the case where the waiting strategy has been applied. It means that each Vehicle agent after finishing its partially route at some destination, it waits a period of time and then starts driving to the next planned customer. This strategy assumes that the vehicle leaves its current location at the latest possible departure time.

For each case, each instance was repeatedly solved five times and mean results from these runs were recorded. All simulations have been carried out on PC Intel Core i5-2540M CPU 2.60 GHz with 8 GB RAM running under MS Windows 7.

The experiment results for both cases (NO-WAIT and WAIT) are presented in Table 1. The table includes the following columns: the name of the instance set, frequencies of arrivals of new requests (λ), and the results (the average number of vehicles used, the average distance traveled by all vehicles, and difference in % between results obtained for both cases).

Analysis of the results presented in the table provides several interesting conclusions. The first observation which stems from the dynamization of the problem is that the results for dynamic instances of the VRPTW are worse that the results obtained for their static counterparts (best known solutions identified by heuristics averaged for each group of instances the reader can find

Table 1. Results obtained by the proposed multi-agent approach (cases: NO-WAIT, WAIT)

Instance	λ	NO-WAIT		WAIT		Difference	
		#Vehicles	Distance	#Vehicles	Distance	#Vehicles	Distance
R1	5	13.58	2114.09	14.12	2092.95	-4%	1%
	10	13.54	1736.71	14.08	1719.34	-4%	1%
	15	13.32	1469.36	13.72	1425.28	-3%	3%
	30	13.09	1261.63	13.35	1236.40	-2%	2%
C1	5	10.97	1532.50	11.08	1517.18	-1%	1%
	10	10.55	1272.35	10.97	1246.90	-4%	2%
	15	10.00	870.94	10.50	853.52	-5%	2%
	30	10.00	853.50	10.10	836.43	-1%	2%
RC1	5	12.42	2336.13	13.17	2312.77	-6%	1%
	10	13.40	1810.14	13.53	1773.94	-1%	2%
	15	13.00	1516.53	13.13	1486.20	-1%	2%
	30	13.00	1489.81	13.13	1445.12	-1%	3%
R2	5	3.87	1827.66	3.99	1809.38	-3%	1%
	10	3.50	1191.66	3.71	1179.74	-6%	1%
	15	3.48	1171.94	3.55	1148.50	-2%	2%
	30	3.81	982.37	3.89	962.72	-2%	2%
C2	5	3.00	1103.04	3.09	1092.01	-3%	1%
	10	3.00	899.72	3.27	881.73	-9%	2%
	15	3.00	718.83	3.09	697.27	-3%	3%
	30	3.00	627.72	3.06	621.44	-2%	1%
RC2	5	4.21	2137.96	4.63	2116.58	-10%	1%
	10	4.21	1800.48	4.55	1782.48	-8%	1%
	15	4.04	1276.20	4.12	1250.68	-2%	2%
	30	4.24	1149.08	4.37	1137.59	-3%	1%

for example in [15] or [2]). Deterioration of the results often depends on the frequency of request arrivals (λ). General observation is that low ratio of request arrivals implies that the results are worse in comparison with the case where a lot of customers arrive at early stage of computation.

By comparison of the results obtained by the approach for both cases, one can conclude that waiting strategy (WAIT case) allows the algorithm to build shorter routes when compare to the NO-WAIT case for all tested instances. By allowing the vehicles to wait at their early locations, it is expected that more requests are being known at the time they leave their current locations. Moreover, the greatest reduction of the routes length has been observed for cases when new requests arrive often ($\lambda = 15, 30$) than in cases when $\lambda = 5, 10$. The observed reduction is up to 3 %.

On the other hand, the number of required vehicles, in case of WAIT strategy, is greater than the number of vehicles required by the system without implementation of the waiting strategy (NO-WAIT case). The observed increase ranges from 1 % to 10 %. The greatest increase is observed for cases when new requests arrive rather slowly ($\lambda = 5, 10$).

5 Conclusions

The multi-agent approach to the Dynamic Vehicle Routing Problem with Time Windows has been proposed in the paper. The architecture of the system includes the set of software agents with different roles and abilities. In order to increase the efficiency of the process of solving the DVRPTW, the waiting strategy has been also implemented in the system. It aims at deciding whether a vehicle should wait (or not) after servicing a request, before driving toward the next customer. The computational experiment which has been carried out on several benchmark instances allowed one to investigate the impact of waiting strategy on the performance of the system. It confirmed outperformance of the version with waiting strategy implemented within the system over the version without waiting strategy in terms of the total distance, but not necessarily in terms of the number of vehicles.

Future research will aim at an implementation of the proposed waiting strategy to other problems (for example DPDPTW), implementation of different variants of waiting strategies, and integration of waiting strategy with other reported efficient strategies (for example request buffering [11,13]).

References

1. Barbucha, D., Jędrzejowicz, P.: Agent-based approach to the dynamic vehicle routing problem. In: Demazeau, Y., Pavón, J., Corchado, J.M., Bajo, J. (eds.) 7th International Conference on Practical Applications of Agents and Multi-Agent Systems (PAAMS 2009). AISC, vol. 55, pp. 169–178. Springer, Heidelberg (2009)
2. Barbucha, D.: A multi-agent approach to the dynamic vehicle routing problem with time windows. In: Bădică, C., Nguyen, N.T., Brezovan, M. (eds.) ICCCI 2013. LNCS, vol. 8083, pp. 467–476. Springer, Heidelberg (2013)
3. Bellifemine, F., Caire, G., Greenwood, D.: Developing Multi-Agent Systems with JADE. John Wiley & Sons, Chichester (2007)
4. Bent, R., Van Hentenryck, P.: Waiting and relocation strategies in online stochastic vehicle routing. In: Veloso, M. (ed.) Proceedings of the 20th International Joint Conference on Artificial Intelligence (IJCAI-07), pp. 1816-1821 (2007)
5. Branchini, R.M., Armentano, A.V., Lokketangen, A.: Adaptive granular local search heuristic for a dynamic vehicle routing problem. Comput. Oper. Res. **36**(11), 2955–2968 (2009)
6. Branke, J., Middendorf, M., Noeth, G., Dessouky, M.: Waiting strategies for dynamic vehicle routing. Transp. Sci. **39**(3), 298–312 (2005)

7. Gendreau, M., Guertin, F., Potvin, J.-Y., Taillard, E.: Parallel tabu search for real-time vehicle routing and dispatching. Transp. Sci. **33**(4), 381–390 (1999)
8. Ichoua, S., Gendreau, M., Potvin, J.-Y.: Exploiting knowledge about future demands for real-time vehicle dispatching. Transp. Sci. **40**(2), 211–225 (2006)
9. Larsen, A.: The dynamic vehicle routing problem. Ph.d. thesis, Institute of Mathematical Modelling, Technical University of Denmark (2001)
10. Mitrovic-Minic, S., Laporte, G.: Waiting strategies for the dynamic pickup and delivery problem with time windows. Transp. Res. Part B **38**, 635–655 (2004)
11. Mitrovic-Minic, S., Krishnamurti, R., Laporte, G.: Double-horizon based heuristics for the dynamic pickup and delivery problem with time windows. Transp. Res. Part B **38**, 669–685 (2004)
12. Pillac, V., Gendreau, M., Guret, C., Medaglia, A.L.: A review of dynamic vehicle routing problems. Eur. J. Oper. Res. **225**, 1–11 (2013)
13. Pureza, V., Laporte, G.: Waiting and buffering strategies for the dynamic pickup and delivery problem with time windows. INFOR **46**(3), 165–175 (2008)
14. Solomon, M.: Algorithms for the vehicle routing and scheduling problems with time window constraints. Oper. Res. **35**, 254–265 (1987)
15. Solomon, M.: VRPTW Benchmark problems. http://w.cba.neu.edu/msolomon/problems.htm

Local Termination Criteria for Stochastic Diffusion Search: A Comparison with the Behaviour of Ant Nest-Site Selection

J. Mark Bishop[1]([envelope]), Andrew O. Martin[1], and Elva J.H. Robinson[2]

[1] TCIDA, Goldsmiths, University of London, New Cross, London, UK
{mark.bishop,andrew.martin}@tungsten-network.com
[2] York Centre for Complex Systems Analysis and Department of Biology,
University of York, York, UK
elva.robinson@york.ac.uk
http://www.tungsten-network.com/tcida/
https://www.york.ac.uk/biology/research/ecology-evolution/elva-robinson/

Abstract. Population based decision mechanisms employed by many Swarm Intelligence methods can suffer poor convergence resulting in ill-defined halting criteria and loss of the best solution. Conversely, as a result of its resource allocation mechanism, the solutions found by Stochastic Diffusion Search enjoy excellent stability. Previous implementations of SDS have deployed complex stopping criteria derived from global properties of the agent population; this paper examines two new *local* SDS halting criteria and compares their performance with 'quorum sensing' - a natural termination criterion deployed in nature by some species of tandem-running ants. We empirically demonstrate that local termination criteria are almost as robust as the classical SDS termination criteria, whilst the average time taken to reach a decision is around three times faster.

Keywords: Collective decision making · Ant nest selection · Stochastic diffusion search · Swarm Intelligence · Global search

1 Introduction

In recent years there has been growing interest in swarm intelligence, a distributed mode of computation utilising interaction between simple agents [21]. Such systems have often been inspired by observing interactions between social insects: ants, bees, termites (cf. Ant Algorithms and Particle Swarm Optimisers) see Bonabeau [10] for a comprehensive review. Swarm Intelligence algorithms also include methods inspired by natural evolution such as Genetic Algorithms [18,20] or indeed Evolutionary Algorithms [5]. The problem solving ability of Swarm Intelligence methods emerges from positive feedback reinforcing potentially good solutions and the spatial/temporal characteristics of their agent interactions.

© Springer International Publishing Switzerland 2016
N.T. Nguyen et al. (Eds.): ICCCI 2016, Part I, LNAI 9875, pp. 474–486, 2016.
DOI: 10.1007/978-3-319-45243-2_44

Independently of these algorithms, Stochastic Diffusion Search (SDS), was first described in 1989 as a population-based, pattern-matching algorithm [7]. Unlike stigmergic communication employed in Ant Algorithms, which is based on modification of the physical properties of a simulated environment, SDS uses a form of direct communication between the agents similar to the tandem running mechanism employed by some species of ants (e.g. *Temnothorax* species, [14]).

SDS is an efficient probabilistic multi-agent global search, optimisation and decision making technique [23] that has been applied to diverse problems such as site selection for wireless networks [41], mobile robot self-localisation [6], object recognition [8] and text search [7]. Additionally, a hybrid SDS and n-tuple RAM [1] technique has been used to track facial features in video sequences [8,19]. Previous analysis of SDS has investigated its global convergence [26], linear time complexity [27] and resource allocation [25] under a variety of search conditions. For a recent review of the theoretical foundations, and applications of SDS see Al-Rifaie and Bishop [2].

In arriving at a 'decision' - *halting* - standard implementations of SDS examine the *stability* of the agent population as a whole; in this manner halting is defined as a *global* property of the agent population. However such global mechanisms are both less biologically/naturally plausible and more complex to implement on parallel computational systems, than local decision making mechanisms. This paper examines the local quorum sensing behaviour observed in some natural (ant) systems and uses this as the inspiration for two new local termination mechanisms - one mechanism, 'independent termination', seeks to implement an asynchronous protocol in SDS that in many ways is quite close to the quorum sensing method used by real ants; a second method - confirmation termination - aims to implement a synchronous local termination protocol with more similarity to the conventional SDS architecture.

2   Stochastic Diffusion Search

SDS is based on distributed computation, in which the operations of simple computational units, or agents, are inherently probabilistic. Agents collectively construct the solution by performing independent searches followed by diffusion/communication of information through the population [28]. Positive feedback promotes better solutions by allocating to them more agents for their exploration. Limited resources induce strong competition from which the largest population of agents corresponding to the best-fit solution rapidly emerges.

In many search problems the solution can be thought of as being composed of many subparts and, in contrast to most Swarm Intelligence methods, SDS explicitly utilises such decomposition to increase the search efficiency of individual agents. Thus in SDS each agent poses a hypothesis about the possible solution and evaluates it *partially* [23]. Successful agents repeatedly test their hypothesis while recruiting unsuccessful agents by direct communication. This creates a positive feedback mechanism ensuring rapid convergence of agents onto promising solutions in the space of all solutions. Regions of the solution space labelled by the presence of agent clusters *with the same hypothesis* can be interpreted as good candidate solutions. A global solution is thus constructed from the interaction of many simple, locally operating agents forming the *largest cluster of agents with the same hypothesis*. Such a *cluster* is dynamic in nature, yet stable, analogous to, "*a forest whose contours do not change but whose individual trees do*", [4, 9, 29].

Algorithm 1. Classical - inactive recruitment - SDS

```
 1: procedure step(swarm, search_space)
 2:     for each agent in swarm do                              ▷ Diffuse Phase
 3:         if not agent.active then
 4:             polled_agent = swarm.random_agent()
 5:             if polled_agent.active then
 6:                 agent.hypothesis = polled_agent.hypothesis
 7:             else
 8:                 agent.hypothesis = search_space.random_hypothesis()
 9:     for each agent in swarm do                               ▷ Test Phase
10:         test_result = perform_random_test(hypothesis)
11:         agent.active = test_result
```

Central to the power of SDS (see **Algorithm 1**) is its ability to escape local minima. This is achieved by the probabilistic outcome of the partial hypothesis evaluation in combination with reallocation of resources (agents) via stochastic recruitment mechanisms. Partial hypothesis evaluation allows an agent to quickly form its opinion on the quality of the investigated solution without exhaustive testing [23].

The termination of SDS has historically been defined as a function of the *stability* of the population size of a group of *active* agents. Such methods are termed *global* halting criteria as they are a function of the number of active agents within the total population of agents. Two well documented global methods for determining when SDS should halt are the *Weak Halting Criterion* and *Strong Halting Criterion* [25, 26]; the former is simply a function of the *total* number of active agents and the latter the *total* number of active agents maintaining the *same* hypothesis.

3   Collective Decision-Making in House Hunting Ants

A model system for collective decision-making is provided by the process of house-hunting in social insects, such as cavity-nesting ants. These ants cannot modify their nest-site and instead relocate the entire colony if the need arises. The processes by which cavity-nesting ants of the genus *Temnothorax* choose a new nest site and emigrate to it has been well-explored, both empirically and theoretically, and is used as a key model of animal collective decision-making.

The ant emigration process can be summarised thus: If the home nest cavity is damaged or degraded, scouts search for new nest sites. Scouts assess available nests across a number of metrics using a weighted additive strategy [16]. If a nest is judged as unsuitable, a scout continues searching; if a nest is assessed as suitable by a scouting ant (Scout A), this scout will return towards the home nest and recruit a second scout (Scout B) [37]. Scout A will lead Scout B to the new nest by tandem-running. Scout B will then make an independent assessment of the nest, and will either reject it and keep searching, or accept the nest and spend some time in it, before returning home and recruiting a further scout. By this positive feedback process, a good quality nest will accumulate ants [34, 36]. Different ants appear to have differing thresholds for starting recruitment to a nest; this means that even low quality nests can attract some ants, but scouts will accumulate more quickly and to a higher level at higher quality nests [35, 39]. This assessment and recruitment process is terminated when scouts sense that a nest site has reached quorum. Scouts then move into a 'post-quorum' behavioural state [32]: they stop leading other scouts by tandem-running, and are no longer willing to be recruited by tandem-running themselves. Instead, they transport brood, queen, other workers to the chosen site. Transported workers do not learn the route between the home nest and the new nest, so are unable to return home, and thus cannot challenge the decision that has been implemented [32]. Transported ants therefore contribute strongly to the quorum by staying in the new nest, so once a few scouts have entered a 'post-quorum' state and started transporting, others quickly follow suit.

4 Quorum Sensing in House-Hunting Ants

Quorum sensing is widespread throughout biological systems. When a collective decision is required, a quick and effective way of moving from an information-gathering phase to an implementation phase is to use a quorum threshold. A quorum response can be said to occur when an individual's probability of exhibiting a behaviour (e.g. choosing a given option) is a sharply nonlinear function of the number of other individuals already performing this behaviour (or having chosen that option) [40]. For house-hunting ants, quorum sensing is central to the decision-making process, as it marks the transition from assessment to implementation. Terminating information-gathering promotes cohesion, which is important for ant colonies that only have one reproductively active queen. For cavity-nesting ants, scouts sense quorum by spending 1–2 min in a nest assessing the number of workers present via encounter rate, rather using than indirect cues such as pheromone concentration [31, 34]. Quorum threshold as a proportion of colony size is remarkably constant across a range of colony sizes (c3.5 %) [13], and this is intriguing, because the relationship between colony size and cavity size is not simple positive correlation: although larger colonies do inhabit larger cavities in the wild, in laboratory tests both small and large colonies prefer larger cavities, presumably to allow for growth [11, 22, 33].

Quorum sensing is a separate process from quality assessment and recruitment. This means that the quorum sensing process in effect detects an average quality assessment across many scouts, and has the potential to smooth out differences in individual nest acceptance thresholds [17]. Once quorum is reached, scouts do not re-assess quorum on subsequent visits - they will continue to bring brood even if the nest is artificially emptied of ants [31]. The quorum threshold itself is not modulated depending on the quality of the new nest [30]. If nest quality is artificially manipulated during the assessment phase of an emigration, the ants are able to respond flexibly to the new nest qualities; if quality is manipulated after quorum is reach and implementation has begun, then colonies often become 'trapped' in an inferior nest [15]. This indicates that quality is not re-assessed after quorum has been reached in these cavity-dwelling *Temnothorax* species. In contrast, a different ant species, *Diacamma indicum* recruits only by tandem-running with no clear quorum point and no adult transport [3]. Colonies of *Diacamma indicum* are able to respond flexibly to manipulated qualities at any stage of the emigration - but overall colony cohesion is lower, supporting the idea that using a quorum threshold increases cohesion, but at a cost to flexibility.

Although quorum sensing behaviour is not modulated by the quality of the options available, it is influenced by the experience and context. Naive scouts use different quorum thresholds to those used by more experienced scouts, but the direction of this difference differs between species, indicating a learning component to quorum sensing behaviour [30]. Emigrations often occur in an emergency context, but cavity-nesting ants do also sometimes emigrate even when their home nest is undamaged, if a better nest is available in the neighbouring area. This is not due to direct comparison of the quality of the two nests, but due to quality-dependent nest acceptance [36,39]. In these non-emergency migrations, scouts appear to use a quorum threshold around twice as high as in emergency migrations [12], suggesting that colonies prioritise speed over accuracy when conditions are harsher.

5   SDS Local Halting Criteria

Drawing inspiration from the behaviour of *Temnothorax* ants in their nest selection behaviour, the halting behaviour of SDS was modified such that it would emerge from purely *local* interactions of SDS agents. By analogy with the behaviour of tandem running *Temnothorax* ants (as outlined in Sects. 3 and 4 herein), in the following we propose two new variants of the process for determining when an agent should switch from the classical SDS *explore-exploit* behaviour to a new, so called, '*terminating*' behaviour which we term the *independent* and *confirmation* halting criteria.

In these variants agents can take on an additional behaviour in which they enter a new state we define 'Terminating', wherein their hypothesis becomes fixed and they subsequently seek to actively remove agents from the dynamic swarm[1] and give them their own (now fixed) termination hypothesis (analogous to the 'post-quorum' behavioural state in ants of the genus *Temnothorax*, wherein post-quorum ants literally carry other ants they encounter to the new nest site).

As this decision making process successively removes agents from the population we name this form of SDS *Reducing SDS*; in this vein a collective 'decision' is made (and the local halting condition met) when all agents are either active and/or have been removed from the population.

5.1 'Independent' Termination Behaviour

In *independent reducing SDS* we relax the assumption that all SDS agents update synchronously in iterative 'cycles' (wherein one such cycle corresponds to all agents being updated).

Algorithm 2. Independent SDS

```
1:  procedure step(swarm, search_space)
2:      swarm = shuffle(swarm)
3:      for each agent in swarm do
4:          polled_agent = swarm.random_agent()              ▷ Diffusion behaviour
5:          if Both agents are inactive then
6:              Both agents randomise hypothesis
7:          else if One agent is inactive and other is active but not terminating then
8:              Inactive agent assumes active agent's hypothesis
9:          else if One agent is terminating then
10:             Other agent is removed from the swarm
11:         else if Agents share a hypothesis then
12:             Both agents become terminating
13:         if not agent.terminating then                     ▷ Testing behaviour
14:             test_result = perform_random_test(hypothesis)
15:             agent.active = test_result
```

In independent SDS agents update independently and probabilistically[2] - which is more analogous to the behaviour of a collection of real ants - and recruitment becomes bidirectional. Considering two such interacting agents:

---

[1] Standard SDS has previously been shown to be a global search algorithm [26] - it will eventually converge to the global best solution in a given search space; by removing agents form the swarm, relative to standard SDS the number of potential agents remaining available for explore-exploit behaviour is reduced; precisely how this reduction impacts the robustness of the algorithm [with respect to erroneous convergence to sub-optimal solutions] has yet to be fully established.

[2] To facilitate the use of homogenous performance metrics, we assume that in a population of k agents, k single asynchronous updates corresponds to one standard synchronous iteration cycle.

- if neither agent is active both reselect new random hypotheses;
- one agent is inactive and other is active but not terminating then the inactive agent assumes active agent's hypothesis; in *Temnothorax* nest selection this is analogous to a 'scout' ant being recruited to a new nest hypothesis by tandem running.
- one of the agents is in *terminating* mode then the other is assigned the solution hypothesis and 'removed' from the population (playing no further part in the search); in *Temnothorax* nest selection this is analogous to an ant in post-quorum (terminating) mode carrying an ant to the new nest site.
- if the two agents meet that both have the same hypothesis then both switch to *terminating* mode; in *Temnothorax* nest selection this recruitment behaviour would serve to reinforce an ant's initial nest-judgement[3].

The above process is algorithmically outlined in **Algorithm 2**.

5.2   'Confirmation' Termination Behaviour

Since its inception in 1989 [7] a substantial body of algorithmic analysis (describing the theoretical behaviour of SDS), empirical studies and practical applications have been published (for a recent review see [2]). To more readily facilitate the future use of these results in both local termination variants and potentially to extend the reach of this analysis to some aspects of real ant behaviour, we suggest a further simplification of Independent SDS to a second reducing behaviour that more closely aligns with standard SDS diffusion; we term this mode *confirmation* reducing termination.

In 'confirmation reduction' SDS agents are once again assumed to update *synchronously* and the diffusion of information is changed to more closely resemble that of classical dual mode (passive and active) recruitment SDS [24]. In *confirmation* SDS an active agent polls random agents in the diffusion phase. Active agents become *terminating* if their polled agent is also active and both agents share a hypothesis. The agent is then locked into being active, maintaining that hypothesis. If an inactive agent polls a terminating agent, the inactive agent is removed from the population (see **Algorithm 3** for details).

6   Experiments

A series of experiments was performed to investigate the diffusion behaviour of the two new halting criteria over a variety of search parameters to establish

---

[3] *Temnothorax* ants are indeed sometimes recruited back to nests they have already visited, so there is potential for this 'reinforcement recruitment' process to play a role for ant colonies. For example, 'reinforcement recruitment' could cause ants to enter a post-quorum state at a lowered encounter rate. This would help extra rapid acceptance of a nest if there were only one new nest site available. This idea could be tested empirically, ideally in a complex arena that would promote tandem-running behaviour, allowing communication of preference.

Algorithm 3. Confirmation SDS

```
1: procedure step(swarm, search_space)
2:     for each agent in swarm do                                    ▷ Diffuse Phase
3:         polled_agent = swarm.random_agent()
4:         if agent.active then
5:             hyp_1 = agent.hypothesis
6:             hyp_2 = polled_agent.hypothesis
7:             if polled_agent.active and hyp_1 == hyp_2 then
8:                 agent.terminating == True
9:         else
10:            if polled_agent.active then
11:                if polled_agent.terminating then
12:                    swarm.remove(agent)
13:                else
14:                    agent.hypothesis = polled_agent.hypothesis
15:            else
16:                agent.hypothesis = search_space.random_hypothesis()
17:     for each agent in swarm do                                    ▷ Test Phase
18:         test_result = perform_random_test(hypothesis)
19:         agent.active = test_result
```

(a) if the algorithms' gross behaviour remains characteristic of SDS and (b) to evaluate their robustness over a variety of search parameters (which effectively characterise the quality of the putative best solution, α $(0 \leq \alpha < 1)$, relative to β, $(0 \leq \beta < 1)$, the quality of the distractor solution[4]); in the 'ant migration' problem, α is analogous to a measure of the quality of the potential new nest site and β effectively a measure of the quality of the original nest.

In all experiments the population is initialised with one agent maintaining the hypothesis representing the potential best solution and the probability of an agent randomly selecting the hypothesis of the potential best solution is set to zero; this ensures that only the *diffusion* behaviour of the algorithm is explored[5].

In the first experiment each of the three termination functions (strong, independent and confirmation) was modelled in a population of 10000 agents, one of which was active and at the solution hypothesis at time zero, with all other agents set inactive pointing to the 'noise' hypothesis. The algorithm was then evaluated 25 times from these conditions against a range of possible values of α and β (from 0 to 0.875 with a step of 0.125). The number of times the algorithm successfully halted within 250 iterations was recorded as was the mean average number of iterations before halting in these cases.

---

[4] β defines a "uniform random noise" hypothesis; an aggregate of all the possible hypotheses an agent could have other than the putative solution hypothesis.

[5] These parameters define a problem analogous to the search space being infinitely large, wherein the only way an agent can adopt the 'best' solution is to receive it via diffusion from an active agent.

Table 1. Mean average iterations before termination for three different halting criteria (*strong, independent and confirmation*) over varying quality of solutions

β	α	strong		independent		confirmation	
		i	c	i	c	i	c
0.000	0.625	151.2	17	42.0	6	46.2	14
0.000	0.750	126.8	20	25.8	16	27.8	18
0.000	0.875	118.2	21	20.0	22	21.1	21
0.125	0.625	195.0	7	52.2	12	58.4	7
0.125	0.750	130.4	11	29.6	15	34.1	16
0.125	0.875	122.0	23	22.9	21	25.8	20
0.250	0.625	216.0	1	77.1	7	88.6	5
0.250	0.750	138.7	17	35.5	16	42.2	15
0.250	0.875	125.5	22	26.0	21	31.4	22
0.375	0.625	100.0	1	232.0	1	244.0	1
0.375	0.750	165.6	14	48.5	12	56.7	16
0.375	0.875	131.5	21	30.0	19	38.5	20
0.500	0.750	212.0	5	74.2	12	87.9	7
0.500	0.875	140.5	18	38.5	13	51.5	16
0.625	0.750	100.0	4	150.0	3	238.0	1
0.625	0.875	161.9	18	50.6	14	73.5	20
0.750	0.875	211.0	7	92.4	14	142.5	13
0.875	0.875	100.0	1	–	0	–	0

In the case of strong halting SDS, halting was considered successful if the halting criterion was satisfied; Fig. 1 shows the characteristic *S-shape* convergence curve obtained deploying SDS using the Strong Halting criterion. All algorithms would also halt if all agents were active at the solution hypothesis, as this is analogous to a successful migration of agents to an optimal state; in addition the process was also halted if the algorithm had run for more than a specified number of iterations or if all the agents held the noise hypothesis. Any experiment that halted for the latter two reasons was considered **unsuccessful**.

In a second experiment the three algorithms were run against fixed values of α and β which the first experiment had shown would be likely to successfully halt. The state of all agents was recorded at every iteration and number of agents (as a proportion of the total population) in various states was graphed over time to visualise the characteristic behaviour of the halting criteria (see Fig. 2).

Table 1 lists i the average number of iterations before halting and c the number of times that the algorithm successfully halted for a SDS experiment for all three algorithms using a population of 10,000 agents across a variety of parameter values of the noise hypothesis (β) and solution hypothesis (α).

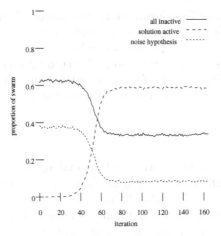

Fig. 1. Cluster size evolution over time for SDS using the *strong halting criterion*. The x-axis counts iterations, the y-axis shows cluster size as a proportion of the entire population. The positive feedback effect can be seen in the sharp S-curve of the solution cluster size.

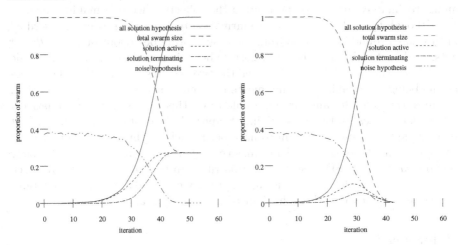

Fig. 2. Cluster size evolution over time for SDS using *confirmation SDS* (left) and *independent SDS* (right). The x-axis counts iterations, the y-axis shows cluster size as a proportion of the entire population. Both show an accelerating growth in the number of agents at the solution hypothesis followed by a similar growth of terminating agents at the solution hypothesis until the entire swarm is either active (in the case of confirmation SDS) or removed from the swarm (in the case of independent SDS).

NB. Pairs of values for α and β for which *all three* algorithms failed to converge 25 times out of 25 are not listed. Examining the results presented in **Table** 1, the following comparative observations can be made:

strong halting versus independent reduction on average the convergence time is 3.3 times faster for independent whilst its robustness is similar (strong halting is more robust in 11 cases, less robust in 6);

strong halting versus confirmation reduction on average the convergence time is around 2.8 times faster for confirmation whilst its robustness is similar (strong halting is more robust in 8 cases, less robust in 6);

7   Conclusion

This paper has looked at cooperative decision making in the Stochastic Diffusion Search algorithm and house-hunting ants. Typically, decision making in standard SDS is based on the use of a global halting function which entails *global* access to the activity of the SDS agent population as a whole. Conversely in this paper, inspired by the quorum sensing mechanism deployed by some species of ants in nest selection, we have successfully demonstrated two new *local* termination criteria for SDS which, in terms of their robustness to noise, have been demonstrated to have broadly similar behaviour to the standard SDS meta-heuristic. Furthermore, it is observed that the use of a local halting mechanism results in an approximately three-fold speed-up in the collective decision making time.

Although the independent and confirmation termination processes described in this paper found inspiration from the nest hunting behaviour of *Temnothorax* ants, we do not claim that the nest selection behaviour of these ants is isomorphic to SDS:- one critical difference between the two systems is that SDS relies on its agents being easily able to communicate their current hypothesis to each other, whereas *Temnothorax* ants are only able to do this by the slow [and relatively infrequent] process of tandem-running. Empirical observations have shown that scouting ants can judge quorum to have been reached (through encounter rate) without having followed a tandem run, so clearly *Temnothorax* ants do not solely rely on *independent*-SDS like termination rules. In this context, future research will investigate the degree to which appropriately modified SDS characterisations can be used to describe the behaviour of *Temnothorax* ants.

References

1. Aleksander, I., Stonham, T.J.: Guide to pattern recognition using random access memories. Comput. Digital Tech. **2**(1), 29–40 (1979)
2. al-Rifaie, M.M., Bishop, J.M.: Stochastic diffusion search review. J. Behav. Robot. **4**(3), 155–173 (2013)
3. Anoop, K., Sumana, A.: Response to a change in the target nest during ant relocation. J. Exp. Biol. **218**(6), 887–892 (2015)
4. Arthur, W.B.: Inductive reasoning and bounded rationality, (The El Farol Problem). Amer. Econ. Rev. **84**, 406–411 (1994)
5. Back, T.: Evolutionary Algorithms in Theory and Practice. Oxford University Press, Oxford (1996)
6. Beattie, P.D., Bishop, J.M.: Self-localisation in the 'SENARIO' autonomous wheelchair. J. Intell. Robot. Syst. **22**, 255–267 (1998)

7. Bishop, J.M.: Stochastic searching networks. In: Proceedings of 1st IEE International Conference Artificial Neural Networks, IEE Conference Publication, vol. 313, pp. 329–331. IEE, London (1989)
8. Bishop, J.M., Torr, P.H.S.: The stochastic search network. In: Linggard, R., Myers, D.J., Nightingale, C. (eds.) Neural Networks for Images, Speech and Natural Language. Chapman Hall, New York (1992)
9. Bishop, J.M., Nasuto, S.J., De Meyer, K.: Dynamic knowledge representation in connectionist systems. In: Dorronsoro, J.R. (ed.) ICANN 2002. LNCS, vol. 2415, pp. 308–313. Springer, Heidelberg (2002)
10. Bonabeau, E., Dorigo, M., Theraulaz, G.: Swarm Intelligence: from Natural to Artificial Systems. Oxford University Press, Oxford (1999)
11. Cao, T.T.: High social density increases foraging and scouting rates and induces polydomy in Temnothorax ants. Behav. Ecol. Sociobiol. **67**(11), 1799–1807 (2013)
12. Dornhaus, A., Franks, N.R., Hawkins, R.M., Shere, H.N.S.: Ants move to improve: colonies of Leptothorax albipennis emigrate whenever they find a superior nest site. Anim. Behav. **67**(5), 959–963 (2004)
13. Dornhaus, A., Holley, J.A., Pook, V.G., Worswick, G., Franks, N.R.: Why do not all workers work? colony size and workload during emigrations in the ant Temnothorax albipennis. Behav. Ecol. Sociobiol. **63**(1), 43–51 (2008)
14. Franklin, E.L.: The journey of tandem running: the twists, turns and what we have learned. Insectes Soc. **61**, 1–8 (2014)
15. Franks, N.R., Hooper, J.W., Gumn, M., Bridger, T.H., Marshall, J.A.R., Gro, R., Dornhaus, A.: Moving targets: collective decisions and flexible choices in house-hunting ants. Swarm Intell. **1**(2), 81–94 (2007)
16. Franks, N.R., Mallon, E.B., Bray, H.E., Hamilton, M.J., Mischler, T.C.: Strategies for choosing between alternatives with different attributes: exemplified by house-hunting ants. Anim. Behav. **65**, 215–223 (2003)
17. Franks, N.R., Stuttard, J.P., Doran, C., Esposito, J.C., Master, M.C., Sendova-Franks, A.B., Masuda, N., Britton, N.F.: How ants use quorum sensing to estimate the average quality of a fluctuating resource. Sci. Rep. **5**, 11890 (2015)
18. Goldberg, D.: Genetic Algorithms in Search, Optimization and Machine Learning. Addison Wesley, Reading (1989)
19. Grech-Cini, E.: Locating facial features. Ph.D. dissertation, University of Reading, Reading (1995)
20. Holland, J.H.: Adaptation in Natural and Artificial Systems. The University of Michigan Press, Ann Arbor (1975)
21. Kennedy, J., Eberhart, R.C., Shi, Y.: Swarm Intelligence. Morgan Kauffman, San Francisco (2001)
22. Kramer, B.H., Scharf, I., Foitzik, S.: The role of per-capita productivity in the evolution of small colony sizes in ants. Behav. Ecol. Sociobiol. **68**(1), 41–53 (2013)
23. De Meyer, K., Nasuto, S.J., Bishop, J.M.: Stochastic diffusion optimisation: the application of partial function evaluation and stochastic recruitment. In: Abraham, A., Grosam, C., Ramos, V. (eds.) Stigmergic Optimization. Studies in Computational Intelligence, vol. 31, pp. 185–207. Springer, Heidelberg (2006)
24. Myatt, D., Nasuto, S.J., Bishop J.M.: Alternative recruitment strategies for SDS. In: Proceedings AISB06: Symposium on Exploration vs. Exploitation in Naturally Inspired Search, pp. 181–187. Bristol, UK (2006)
25. Nasuto, S.J.: Analysis of resource allocation of stochastic diffusion search. Ph.D. dissertation, University of Reading, Reading, UK (1999)
26. Nasuto, S.J., Bishop, J.M.: Convergence of the stochastic diffusion search. Parallel Algorithms Appl. **14**, 89–107 (1999)

27. Nasuto, S.J., Bishop, J.M., Lauria, S.: Time complexity of stochastic diffusion search. In: Heiss, M. (ed.) Proceeding International ICSC/IFAC Symposium on Neural Computation, Vienna (1998)

28. Nasuto, S.J., Dautenhahn, K., Bishop, J.M.: Communication as an emergent methaphor for neuronal operation. In: Nehaniv, C. (ed.) Computation for Metaphors, Analogy, and Agents. LNAI, vol. 1562, pp. 365–379. Springer, Heidelberg (1999)

29. Nasuto, S.J., Bishop, J.M., De Meyer, K.: Communicating neurons: a connectionist spiking neuron implementation of stochastic diffusion search. Neurocomputing **72**(4–6), 704–712 (2008)

30. Pratt, S.C.: Behavioral mechanisms of collective nest-site choice by the ant Temnothorax curvispinosus. Insectes Soc. **52**, 383–392 (2005)

31. Pratt, S.C.: Quorum sensing by encounter rates in the ant Temnothorax albipennis. Behav. Ecol. **16**, 488–496 (2005)

32. Pratt, S.C., Mallon, E.B., Sumpter, D.J.T., Franks, N.R.: Quorum sensing, recruitment, and collective decision-making during colony emigration by the ant Leptothorax albipennis. Behav. Ecol. Sociobiol. **52**(2), 117–127 (2002)

33. Pratt, S.C., Pierce, N.E.: The cavity-dwelling ant Leptothorax curvispinosus uses nest geometry to discriminate among potential homes. Anim. Behav. **62**, 281–287 (2001)

34. Pratt, S.C., Sumpter, D.J.T., Mallon, E.B., Franks, N.R.: An agent-based model of collective nest site choice by the ant Temnothorax albipennis. Anim. Behav. **70**, 1023–1036 (2005)

35. Robinson, E.J.H., Feinerman, O., Franks, N.R.: How collective comparisons emerge without individual comparisons of the options. Proc. Royal Soc. B **281**, 20140737 (2014). doi:10.1098/rspb.2014.0737

36. Robinson, E.J.H., Franks, N.R., Ellis, S., Okuda, S., Marshall, J.A.R.: A simple threshold rule is sufficient to explain sophisticated collective decision-making. PLoS One **6**, e19981 (2011)

37. Robinson, E.J.H., Smith, F.D., Sullivan, K.M.E., Franks, N.R.: Do ants make direct comparisons? Proc. Royal Soc. B **276**, 2635–2641 (2009)

38. Seeley, T.D., Visscher, P.K., Schlegel, T., Hogan, P.M., Franks, N.R., Marshall, J.A.R.: Stop signals provide cross inhibition in collective decision-making by honey bee swarms. Science **335**, 108–111 (2012)

39. Stroeymeyt, N., Robinson, E.J.H., Hogan, P.M., Marshall, J.A.R., Giurfa, M., Franks, N.R.: Experience-dependent flexibility in collective decision-making by house-hunting ants. Behav. Ecol. **22**(3), 535–542 (2011)

40. Sumpter, D.J.T., Pratt, S.C.: Quorum responses and consensus decision making. Proc. Royal Soc. B **364**(1518), 743–753 (2009)

41. Whitaker, R.M., Hurley, S.: An agent based approach to site selection for wireless networks. In: Proceeding 2002 ACM Symposium Applied Computing (Madrid), pp. 574–577. ACM, New York (2002)

Meta-Heuristics Techniques
and Applications

.

Optimizing Urban Public Transportation with Ant Colony Algorithm

Elena Kochegurova[✉] and Ekaterina Gorokhova

Tomsk Polytechnic University,
Lenin Avenue 30, 634050 Tomsk, Russian Federation
kocheg@tpu.ru, gorokhovaes@mail.ru

Abstract. Transport system in most cities has some problems and should be optimized. In particular, timetable of the city public transportation needs to be changed. Metaheuristic methods for timetabling were considered the most efficient. Ant algorithm was chosen as one of these methods. It was adapted for optimization of an urban public transport timetable. A timetable for one bus route in the city of Tomsk, Russia was created on the basis of the developed software. Different combinations of parameters in ant algorithm allow obtaining new variants of the timetable that better fit passengers' needs.

Keywords: Ant algorithm · Timetable · Transport · Optimization

1 Introduction

Nowadays there are more than 500 thousand people living in Tomsk city. Efficient operation of urban public transport is extremely important in citizens' daily life. However, some problems in the organization of the transport system exist in the city, i.e. duplication of routes and overloading of streets in the downtown [1]. That is why the problem of creating a timetable for urban public transport is relevant and must be solved in the framework of overall optimization of the transport system in Tomsk.

Timetables can be created by various methods, including [2–4]:

- heuristic algorithms;
- dynamic programming;
- branch and bound algorithms;
- metaheuristic algorithms;
- graphical method.

Metaheuristic algorithms are considered the most efficient so far. Firstly it was shown by M. Dorigo in 1996 for solving the travelling salesmen problem, the quadratic assignment problem and the job-shop scheduling problem [5]. The efficiency of metaheuristic algorithms for network design and scheduling was also attested in global review [2]. This group of algorithms includes genetic algorithm, neural networks, ant colony algorithm and others. This work investigates the application of the ant algorithm for optimization of the urban public transport timetable in Tomsk city. ACO algorithms

© Springer International Publishing Switzerland 2016
N.T. Nguyen et al. (Eds.): ICCCI 2016, Part I, LNAI 9875, pp. 489–497, 2016.
DOI: 10.1007/978-3-319-45243-2_45

are widely used for transport problems, mostly for creating routes. This work attempts to apply classical ACO in slightly different field – for creating a timetable.

2 Problem Description

The aim of this work is to create a timetable for urban public transport that would satisfy the needs of both passengers and passenger carriers [6, 7]. The main requirement from the passengers is to minimize the transportation time between bus stops:

$$\sum_{i,j} \left(tw_i^j + \sum_{k=i}^{l} \Delta t_k^j \right) \rightarrow \min, \tag{1}$$

where i is a departure point, l is an arrival point, j is a route number, tw_i^j is a waiting time for a bus on some route j at the bus stop i, Δt_k^j is a transportation time between bus stops k and $(k+1)$ on the route j.

Passenger carriers mostly care about their revenue from business. Due to the fact that ticket prices in Tomsk are regulated by the municipal government, revenue may be increased only by decreasing expenses or by increasing the number of passengers. Expenses usually include the cost of fuel, operating costs and repairs. That is why passenger carriers are interested in minimizing the amount of buses on routes while maximizing the number of passengers:

$$\sum_{j} N_j + M \rightarrow \min, \tag{2}$$

where j is a route number, N_j is the number of buses on the route j, M is the number of backup buses;

$$\sum_{n,j} pass_n^j \rightarrow \max, \tag{3}$$

where n is a bus number, j is a route number, $pass_n^j$ is the number of passengers in the bus n of the route j.

Finally, it is necessary to consider that traffic depends on season, time of day and whether it is a weekday or a festive day as this influences the transportation time.

3 Basic Notations

The following variable parameters should be determined to create the timetable:

- number of buses;
  - number of buses N_j on some route j;
  - number of backup buses M. A bus is considered backup if it is used only in case of breakdown of other buses;

- time $t_{start}^{n,j}$ when a bus n starts movement on the route j;
- time $t_{end}^{n,j}$ when the bus n ends movement on the route j; and
- downtime t_{rest}.

The transport system is also determined by some parameters that cannot be changed and should also be considered while creating the timetable:

- transportation time Δt_i^j between bus stops i and $i + 1$ on the route j;
- number of passengers $pass_n^j$ carried by the bus n on the route j during one day;
- number of bus stops BS_j on the route j;
- number BS_i^j of passengers who want to leave the bus stop i on the route j per hour;
- time tw_i^j that one passenger spends awaiting a bus at the bus stop i on the route j;
- number $passBSin_n^{i,j}$ of passengers who entered the bus n at the bus stop i on the route j;
- number $passBSout_n^{i,j}$ of passengers who left the bus n at the bus stop i on the route j;
- bus capacity TC, i.e. the maximum number of passengers that may be safely carried on the bus.

Restrictions connected with the number of passengers on every bus should also be considered:

$$\sum_{i \leq l} BSin_n^{i,j} - \sum_{i \leq l} BSout_n^{i,j} \leq TC_n \qquad (4)$$

for $\forall l$, where i is the number of the stop, j is the number of the route, and n is the number of the bus.

4   Ant Colony Algorithm

It is known that ants live in a commune. One colony may consist of millions of ants and their actions in searching for food or overcoming obstacles are usually close to the theoretically optimal [8–10]. The base of the ants' behavior is self-organization and the use of local information to make decisions. It means that there is no centralized control in an ant colony. Instead, ants use randomness, frequency, positive and negative feedbacks. Information is transferred with a pheromone –a special secret that is dropped on the ground while ants move. Other ants select the trace with the largest amount of pheromone. Pheromones vapor during the time, and that allows correcting routes according to environmental changes.

Ant algorithm is widely used for optimizing routs, for example in [11–13]. But in this paper ant algorithm was adapted to the task of creating an urban public transport timetable. The following assertions were elaborated [14, 15]:

- Pheromone trail. The pheromone is secreted while the bus moves. The amount of pheromone depends on the number of passengers who got on this bus at the bus stops. The larger number of passengers means the stronger pheromone trail. Let $F_{i,j}$ be the amount of pheromone dropped at the bus stop i on some route j.

- Pheromone vapors over time.
- Tabu list. There is a list of prohibited parameter values. In this study such parameter is the time that the bus spends to move from one stop to the next one. $T_{i1,i2}$ is the tabu to move between stops $i1$ and $i2$.
- Visibility. It is local information that represents the bus's desire to spend some exact time to move from one stop to another. The optimal time is determined according to the distance between the stops and is influenced by road conditions. However, the bus may spend time that is more or less close to the optimal. The larger the deviation from the optimal time is, the smaller visibility value is.

$$\eta(i,j) = \frac{1}{\Delta t(i,j)_{real} - \Delta t(i,j)_{opt}}, \tag{5}$$

where j is a route number, i is a bus stop number, $\Delta t(i,j)_{real}$ is some chosen transportation time between stops i and $i+1$, and $\Delta t(i,j)_{opt}$ is the optimal transportation time between stops i and $i+1$.

The output of the algorithm is a new timetable. Arrival and department time are determined for every bus and bus stop. Different timetables result from regulation of the values of two parameters: visibility and pheromone trail.

5   Numerical Test and Results

The developed algorithm was used to optimize the bus route number 24 of Tomsk. This route is quite popular and makes 87 turnarounds a day. However, the buses on this route work with significant deviations from the timetable.

A city-wide inspection of the passenger stream was carried out in Tomsk. Figure 1 demonstrates the loading of buses on the route No. 24 at every bus stop. One bus can carry up to 42 passengers simultaneously.

On the graph in Fig. 1 bus loading is calculated when a bus arrives at a bus stop, therefore, the loading value of the first bus stop equals zero. Some numbers on the graph are higher than 100 %. It means that passengers overload the bus and as a result they experience severe discomfort. Moreover, in this case buses usually move more slowly and can deviate from the timetable. There are 13 bus stops in the middle of the route where buses are loaded at more than 80 %. It means that the route is popular and carries many people every day, but also that route No. 24 should be optimized.

The route No. 24 has 29 stops and covers the distance of 33 km. The model of this route was created with the simulation software AnyLogic [16]. AnyLogic has an integrated city map, so the model was based on this map. The model of the route No. 24 is represented in Fig. 2 where yellow circles mean buses while brown circles are bus stops.

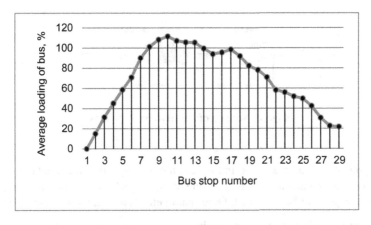

Fig. 1. Average bus loading on the route no. 24

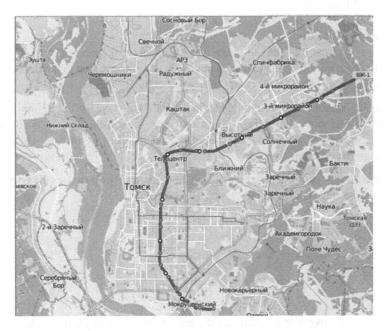

Fig. 2. Bus route no. 24 on the map of Tomsk

The model shows how buses move on the route. The process includes 3 main stages:

- A bus starts moving from a departure point;
- The bus moves to the next stop according to the route;
- The bus waits at the stop to let passengers entering and leaving the bus.

The logical scheme of the modeling process is shown in Fig. 3.

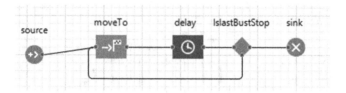

Fig. 3. Scheme of the movement process

An application was created to implement the ant algorithm for timetabling. In this application such parameters as pheromone weight, visibility weight and velocity of pheromone vapor can be regulated. Other parameter values are also defined. They are:

- The time when buses start their work;
- The time when buses finish their work;
- An initial timetable;
- The number of stops;
- The number of buses;
- The optimal time for a bus to move from one stop to the next one;
- Other variants of time intervals for a bus to move from one stop to the next one; and
- The number of people who enter and leave the bus at the bus stop.

In the application the algorithm works in the following way. At the beginning buses are at the starting stops. From these points they start movement. Time was considered to be discrete, that is why variables should be calculated every minute. Their values may change for different reasons:

- The bus arrives at the stop and picks up n passengers. This increases the pheromone trail for the bus stop by n:

$$F_{i,j} = F_{i,j} + n; \qquad (6)$$

- Pheromone vapors over time with velocity p.

$$F_{i,j}(t) = (1 - p)F_{i,j}(t - 1), p \in [0, 1]. \qquad (7)$$

This helps to take into account that the passenger stream is changing. Some hours are peak while others are off-peak.

- The bus decides how long it will take to move to the next stop of its route. By default it should move with the optimal velocity and spend $\Delta t(i, j)_{opt}$. However, the bus may decide to choose another time according to the weight of pheromone and visibility. The probability for this decision is calculated as:

$$P(\Delta t_{i,j}[k]) = \frac{F_{i,j}(t_k)^{\alpha} \cdot \eta_k(i,j)^{\beta}}{\sum\limits_{l \notin T_{i,i+1}} F_{i,j}(t_l)^{\alpha} \cdot \eta_l(i,j)^{\beta}}, \tag{8}$$

where j is a route number, i is a bus stop number, $i+1$ is the number of the next bus stop, k is the number of the time interval required to move from one stop to the other, l is the numbers of variants for the time interval required to move between the stops; t_k is the time when the bus arrives at the stop if variant k is chosen; α is a pheromone weight coefficient and β is a visibility weight coefficient.

- Values are calculated only when the bus arrives at the bus stop.

At the end of the day the number of transported passengers and working hours for the buses should be calculated. After analyzing this data a new timetable may be compared with the old one. The algorithm may work during several days in order to increase its efficiency.

Some of required data is public and easy to obtain, for example the number of stops or the initial timetable. However, the number of passengers who get on a bus on every bus stop is different in different time of the day and may be calculated only with the help of statistics. Because collection and processing of data takes much time, the application was piloted on test data. In was assumed that buses arrived at the next stop in a constant time interval that was optimal for every bus stop. This assumption makes results easy to examine.

The application generates new timetables according to the pheromone weights, visibility weights and velocity of pheromone evaporation. Table 1 shows the difference between the timetables created for different balances of weights for pheromone and visibility. In this example the velocity of pheromone evaporation was set to 0.1 and the optimal time interval between bus stops at 3 min. The table cells contain time from the beginning of algorithm operation (in minutes). It is necessary to notice that when the first bus starts movement, there is no pheromone trail at the stops. That is why the table shows the 5th bus of the route and not the first one, which starts movement at the time = 0.

Table 1. Comparing timetables for different parameter values

	Bus stop no. 15	Bus stop no. 16	Bus stop no. 17
Visibility weight = 2, Pheromone weight = 60	43	45.1	47.1
Visibility weight = 60, Pheromone weight = 2	43	46	49

In the first case the pheromone weight is much bigger than the visibility weight. That means that buses must be oriented on the pheromone trail mostly. Pheromones evaporate every minute, so the bus claims to arrive at the next stop as fast as possible.

Since 3 min were considered the optimal time interval, the fastest transportation time in the set conditions is about 2 min.

In the second case the values are set at the opposite. The visibility weight dominates over the pheromone weight, so the bus always chooses the most convenient time interval to move between the stops.

6 Conclusion

The next step of this work will be connected with using the data of proper accuracy. The application may be improved in order to calculate the total number of passengers and the loading level of transport and of city streets. Other routes should also be process in order to create a common timetable for Tomsk.

Application of ant algorithm described above may improve the existing timetable of the urban public transport in Tomsk city. The optimized timetable can help many people to spend less time waiting for a bus at a bus stop. It also provides efficient allocation of available buses for passenger carriers.

References

1. Yurchenko, M., Kochegurova, E., Fadeev, A., Piletskya, A.: Calculation of performance indicators for passenger transport based on telemetry information. In: Engineering Technology, Engineering Education and Engineering Management, pp. 847–852. Taylor & Francis Group, London (2015)
2. Guihaire, V., Hao, J.K.: Transit network design and scheduling: a global review. Transp. Res. Part A: Policy Pract. **42**(10), 1251–1273 (2008)
3. Zhao, F., Zeng, X.: Optimization of transit route network, vehicle headways and timetables for large-scale transit networks. Eur. J. Oper. Res. **186**(2), 841–845 (2008)
4. Yu, B., Yang, Z., Sun, X., Yao, B., Zeng, Q., Jeppesen, E.: Parallel genetic algorithm in bus route headway optimization. Appl. Soft Comput. **11**(8), 5081–5091 (2011)
5. Dorigo, M., Maniezzo, V., Colomi, A.: Ant system: optimization by a colony of cooperating agents. IEEE Trans. Syst. Man Cybern. Part B (Cybern.) **26**, 29–41 (1996). IEEE
6. Lazarev, A.A., Gafarov, E.R.: Teoriya raspisanii. Zadachi i algoritmi, Moskva, Moskovskii gosudarstvennyi universitet im. M.V. Lomonosova (MGU), 222 p (2011). (in Russian)
7. Eliseev, M.E., Lipenkov, A.V., Eliseev, E.M.: O modeli gorodskogo passazhirskogo transporta: modelirovanie logiki passaghira 3(90), 347–352 (2011). Transactions of Nizhny Novgorod State Technical University n.a. R.E. Alekseev
8. Dorigo, M., Gambardella, L.M.: Ant colonies for the travelling salesman problem. Biosystems **43**(2), 73–81 (1997)
9. Pedemonte, M., Nesmachnow, S., Cancela, H.: A survey on parallel ant colony optimization. Appl. Soft Comput. **11**(8), 5181–5197 (2011)
10. Martynova, Y.A., Shutova, Y.O., Martynov, Y.A., Kochegurova, E.A.: Ant colony algorithm for rational transit network design of urban passenger transport. In: International Scientific Symposium Lifelong Wellbeing in the World (WELLSO-2014), pp. 48–55, TPU Publishing House, Tomsk (2014)

11. Alba, E., Doerner, K.F., Dorronsoro, B.: Adapting the savings based ant system for non-stationary vehicle routing problems. In: Proceedings of the META 2006, Tunisia (2006)
12. Zidi, S., Maouche, S.: Ant colony optimization for the rescheduling of multimodal transport networks. In: IMACS Multiconference on Computational Engineering in Systems Applications, vol. 1, pp. 965–971. IEEE (2006)
13. D'Acierno, L., Gallo, M., Montella, B.: Ant colony optimisation approaches for the transportation assignment problem. In: 16th International Conference on Urban Transport and the Environment, pp. 37–48 (2010)
14. Gilmour, S., Dras, M.: Understanding the pheromone system within ant colony optimization. In: Zhang, S., Jarvis, R.A. (eds.) AI 2005. LNCS (LNAI), vol. 3809, pp. 786–789. Springer, Heidelberg (2005)
15. Wong, K.Y., See, P.C.: A new minimum pheromone threshold strategy (MPTS) for max–min ant system. Appl. Soft Comput. 9(3), 882–888 (2009)
16. Borshchev, A.: The Big Book of Simulation Modeling: Multimethod Modeling with AnyLogic 6. AnyLogic North America, Chicago (2013)

A Hybrid Approach Based on Particle Swarm Optimization and Random Forests for E-Mail Spam Filtering

Hossam Faris[1], Ibrahim Aljarah[1,2(✉)], and Bashar Al-Shboul[1,2]

[1] Business Information Technology Department,
King Abdullah II School for Information Technology,
The University of Jordan, Amman, Jordan
hossam.faris@ju.edu.jo
[2] The University of Jordan, Amman, Jordan
{i.aljarah,b.shboul}@ju.edu.jo

Abstract. Internet is flooded every day with a huge number of spam emails. This will lead the internet users to spend a lot of time and effort to manage their mailboxes to distinguish between legitimate and spam emails, which can considerably reduce their productivity. Therefore, in the last decade, many researchers and practitioners proposed different approaches in order to increase the effectiveness and safety of spam filtering models. In this paper, we propose a spam filtering approach consisted of two main stages; feature selection and emails classification. In the first step a Particle Swarm Optimization (PSO) based Wrapper Feature Selection is used to select the best representative set of features to reduce the large number of measured features. In the second stage, a Random Forest spam filtering model is developed using the selected features in the first stage. Experimental results on real-world spam data set show the better performance of the proposed method over other five traditional machine learning approaches from the literature. Furthermore, four cost functions are used to evaluate the proposed spam filtering method. The results reveal that the PSO based Wrapper with Random Forest can effectively be used for spam detection.

Keywords: Particle Swarm Optimization · Random Forest · Wrapper feature selection · Fitness functions · Spam

1   Introduction

Email messaging is one of the most popular ways to exchange digital messages over the internet. However with the major increase of email traffic comes a major problem which is email spam. Email spam is a term refers to unsolicited and unwanted email messages like advertisement related messages or malicious content messages. Email spam has major negative impacts on networks bandwidth, servers' storage and user time and productivity. Moreover, considerable

© Springer International Publishing Switzerland 2016
N.T. Nguyen et al. (Eds.): ICCCI 2016, Part I, LNAI 9875, pp. 498–508, 2016.
DOI: 10.1007/978-3-319-45243-2_46

percentage of spam emails are serious potential for malware infections and phishing. Therefore, companies and organizations are not just spending big amount of money to buy anti-spam systems but also affording the expenses of the support and maintenance of these systems [1].

The spam filtering problem can be considered as an example of the text categorization problem, where an email can be categorized as spam or ham [2]. However, the high-dimensionality of the problem is a main challenge when sophisticated learning algorithms are applied in text categorization [3].

In literature, researchers and practitioners investigated different approaches and technologies for developing accurate spam detection and filtering systems. Recently, most of these approaches are machine learning and data mining based techniques that rely on analysing the content of the e-mail messages then extracting knowledge to classify new coming messages [4]. In previous works, many spam filtering approaches were proposed based on applying single machine learning algorithm such as Artificial Neural Networks (ANNs) in [5-7], Naive Bayes classifiers (NB) [8], k-Nearest Neighbor (k-NN) [9] and Support Vector Machines (SVM) [10]. Other researchers investigated ensemble classifiers by combining multiple classifiers in order to obtain better predictive performance [2,11-13]. Random Forests (RF) is considered one of the best ensemble based classification methods [14]. RF generates multiple decision trees based on different random subsets drawn from the original training data with replacement. In [15], authors showed highly accurate performance of RF when applied to clustering that used to select email messages as training examples. In [16], authors showed in their experiments that RF showed superior accuracy when compared to SVM.

Another type of approaches that is getting more attention is the evolutionary algorithms like Genetic Algorithm (GA) and Particle Swarm Optimization (PSO) [1,4,17]. In the field of computer science, Evolutionary Algorithms (EA) are a branch of artificial intelligence which are inspired by the biological evolution and natural selection theories. Evolutionary algorithms have some powerful advantages that make them fit to many classical modeling and optimization problems. Such advantages include their flexibility, self-adaptation, their conceptual simplicity. EAs have ability to capture the non-linearity and dynamicity of very complex process with too many variables in interaction [18]. In general, evolutionary algorithms were investigated for spam filtering in different schemes such as in the feature selection or in the classification process [19-22].

In this paper, a machine learning technique composed of two stages is proposed for spam filtering. In the first stage feature selection is applied while in the second the classification model is constructed. For feature selection a Particle Swarm Optimization (PSO) based Wrapper is used to select the best representative set of features to reduce the dimensionality of the training data. In this work, the PSO wrapper is experimented using four different cost (i.e.; fitness) functions. In the second stage, a Random Forest spam classifier is developed using the selected features in the first stage. The proposed model is evaluated and compared to other traditional machine learning approaches from the literature including Decision Trees, Naive Bayes (NB), k-Nearest Neighbor (k-NN) and Support Vector Machine (SVM).

This paper is organized as follows. Section 2 discuss the materials and methods that used in this paper. Section 3 presents the proposed spam filtering approach. The experiments and results obtained are discussed in Sect. 4. Finally, findings of this work are concluded in Sect. 5.

2 Materials and Methods

2.1 Particle Swarm Optimization (PSO)

Swarm intelligence is one of the evolutionary computation algorithm forms, which mimics the social natural systems such as birds flocks [23], krill herds [6,24], fireflies, and glowworms [25]. The social behavior of these systems is based on the interactions between the individuals by contacting with each other to achieve some objective such as searching for optimal food sources.

PSO is a swarm intelligence algorithm that was developed by Kennedy and Eberhart in 1995 [23]. The PSO algorithm simulates the social behavior of the bird flocks or bird swarms to find the optimal food location. PSO algorithm like other evolutionary computations algorithms starts with initial population, which contains random particles such as each particle represents a candidate solution. PSO algorithm searches for best solution by updating these particles according to specific fitness function. The updating process for each particle is controlled by three fractions, the current location of the particle, the best location achieved by the particle so far, and the best location achieved by its neighbors. In PSO, particles are placed in specific search space of the problem. Each particle evaluates its position based on fitness function to determine its movement (velocity) through the available search space. The velocity calculations takes into account the best positon of the particle which is called personal best position (pBest) and best position achieved by the particles neighbors (gBest). In addition, particle's movement is influenced by its inertia, and other constants.

PSO particle positions are updated using the following equations:

$$X_i(t+1) = X_i(t) + V_i(t+1) \tag{1}$$

where X_i is the particle position i, t is the iteration number, and V_i is the velocity of particle i, which is calculated by the following equation:

$$V_i(t+1) = W \cdot V_i(t) + r_1 \cdot c_1 \cdot [pBest_i - X_i(t)] \\ + r_2 \cdot c_2 \cdot [gBest_i - X_i(t)] \tag{2}$$

where W is inertia weight, r_1 and r_2 are random numbers between 0 and 1, c_1, c_2 are constant coefficients, $pBest_i$ is the current best position of particle i, and $gBest_i$ is the current global best position of the particle's neighbors.

In this work we adopt a general form of the classical PSO called Geometric PSO (GPSO). GPSO was proposed and developed in [26]. The GPSO form differs from the standard PSO in that there is no explicit velocity and the process of updating the positions of the particle is based on a three-parent mask-based crossover (CX) and a mutation operator. Particles are represented as binary vectors in the Hamming space. The new update mechanism is given as in Eq. 3. The weights ω_1, ω_2, and ω_3 are positive and their sum equals 1.

$$X_i(t+1) = CX((X_i(t, \omega_1), (gBest, \omega_2), (pBest, \omega_3))$$ (3)

2.2 Cost Functions

Cost functions (also known as fitness functions) have great influence on the performance of the PSO algorithm. In this work, we use four different cost functions to evaluate the performance of attribute combinations. The first two cost functions are the Accuracy and F-measure which are based on the confusion matrix shown in Table 1. Confusion matrix is considered as one of the primary sources for evaluating classification models. In this matrix, TS is the number of spam e-mails which are correctly predicted as spam, FL is the number of spams which are predicted as non-spam, TL is the number of normal e-mails which are predicted as non-spam, and FS is the number of normal e-mails which are predicted as spam. The other two cost functions are the Area Under Curve (AUC) and the Root Mean Squared Error (RMSE). The four functions can be described as follows:

Table 1. Confusion matrix

		Predicted	
		Legitimate	Spam
True	Legitimate	TL	FS
	Spam	FL	TS

- **Accuracy:** measures the rate of the correctly classified instances of both classes as shown in Eq. 4.

$$Accuracy = \frac{TS + TL}{TS + FL + FS + TL}.$$ (4)

- **F-measure:** this measure combines two important ratios related to the spam class. They are the spam precision (SP) and spam recall (SR). SP is the ratio of the number of email messages correctly classified as spam to the total number of emails classified as spam, $SP = TS/(TS + FS)$. On the other hand, SR measures the ratio of the number of the email messages correctly classified as spam to the total number of the emails that are actually spam $SR = TS/(TS + FL)$. F-measure combines SP and RS as follows:

$$F - measure = \frac{2 \times SP \times SR}{SP + SR}.$$ (5)

– **Area under ROC curve (AUC):** AUC is commonly used to evaluate probabilistic classifiers. A random classifier is AUC equal to 0.5 while a perfect classifier has an AUC equals 1. AUC can be calculated by the following equation:

$$AUC = \int_0^1 \frac{TS}{S} \mathrm{d}\frac{FS}{L} = \frac{1}{S \cdot L} \int_0^1 TS \mathrm{d}FS \qquad (6)$$

– **RMSE :** is given by the following equation where y is the actual value, $\hat{y}$ is the estimated one and n is the total number of instances.

$$RMSE = \frac{1}{n} \sqrt{\sum_i (y_i - \hat{y}_i)^2} \qquad (7)$$

2.3 Random Forest Classifier

Random forest is a type of ensemble classifiers which was first introduced by Breiman in 2001 [27]. Random Forests generates a number of no pruned decision trees, then aggregates their results by majority vote. In random forests, each tree is constructed based on different bootstrap sample drawn from the data to increase the diversity of the trees. Samples that are not selected for training are collected to form other subset called "Out of bag" OOB and used to evaluate the generated decision trees.

Moreover, a random subset of features is selected for each node in the tree. Number of selected features for each node is usually much smaller than the total number of available features in the original dataset. The feature that minimizes the Gini impurity with respect to the other classes is selected for the split.

It was reported different advantages for RF in literature that include: its speed and efficiency when applied on large datasets, it doesn't overfit, no presumptions on the distribution of the data are needed and the low number of parameters which are the number of trees in the forest and the number of variables in the random subset at each node [28].

2.4 Data Set Description

The developed spam filtering model in this work is applied on a spam data set consists of 9346 records with 79 features. The data is extracted from SpamAssassin public mail corpus[1]. Each example in the data is labeled as Ham or Spam. The data includes 6951 ham emails and 2395 spam emails. The percentage of spam email forms approximately 25.6 % of the emails which makes the data imbalanced and therefor more challenging. The full description of the features can be found in [29].

---

[1] https://spamassassin.apache.org/publiccorpus/.

3   Proposed Approach

The approach proposed in this work is mainly composed of two stages. The first is the feature selection stage while the second is the classification model development. The goal of the feature selection process is to find the best representative set of features that achieves the maximum classification performance of the classifier in the second stage. To conduct this process, we apply a wrapper based feature selection method [30]. Wrappers have three main components:

- The search strategy.
- The induction algorithm.
- The evaluation functions.

In this work, PSO is used as metaheuristic algorithm to search the space of all possible subsets of features on the training part of the original dataset. Candidate feature subsets generated by PSO are evaluated using an inductive algorithm and some evaluation function. As an inductive algorithm Random Forest classifier is used while as an evaluation function, we test four different functions. In order to maximize the reliability of the evaluation process, the inductive algorithm is trained and tested using a cross validation scheme. In the second stage, the Random Forest model is trained with the best features subset in the first stage, and after that the model will be tested on another unrepresented part of the dataset (testing part) to get the final evaluation results. Figure 1 illustrates all the details of the proposed approach. PSO, selected cost functions, and RF algorithm are described in the following three subsections.

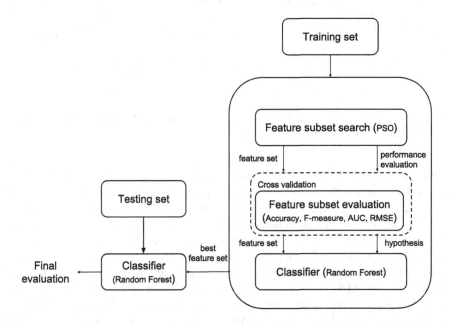

Fig. 1. Framework of the proposed approach for spam prediction

4   Experiments and Results

In order to experiment and evaluate the proposed approach in this work, WEKA 3.7 the developer version is used. Weka is a popular collection of machine learning algorithms written in Java for data mining and analysis tasks [31].

We start our experiments by splitting the aforementioned dataset into two equal parts. One part is used for the feature selection process and to train the classifier while the other part is used for the final evaluation of the classifier.

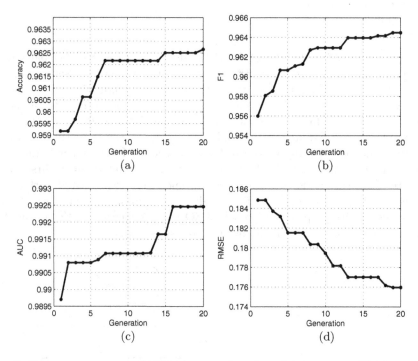

Fig. 2. Convergence curves of PSO wrappers based on (a) Accuracy, (b) F-measure, (c) AUC, and (d) RMSE fitness functions

In the feature selection process, the wrapper algorithm is applied with 5-folds cross validation in order to increase the reliability of its evaluation. GPSO is used as a heuristic search meathod in the wrapper to find the best subset of features. GPSO is applied with the tuned parameters listed in Table 2. Since PSO in general is computationally expensive search algorithm, the population size and the maximum number of generations are set arbitrarily to a small value of 20. Random Forest is selected as a base classifier for the wrapper to evaluate every candidate subset of features. Empirically we set number of trees to 100. Since the spam dataset is imbalanced, using only accuracy rate or RMSE for evaluating the classification models is not efficient [32]. Therefore, we experiment also the AUC and F-measure as cost functions since they are more appropriate evaluation

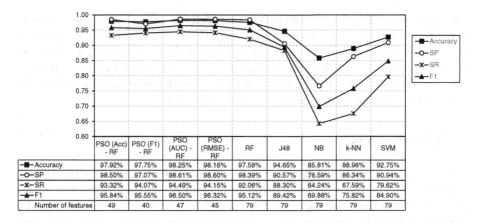

Fig. 3. Comparison of evaluation results

Table 2. GPSO parameters settings

Parameter	Value
Swarm size	20
Iterations	20
Mutation Type	bit-flip
Mutation Prob	0.01
Interia weight	0.33
Social weight	0.33
Individual weight	0.34

metrics for imbalanced datasets [32,33]. The convergence curves of the four cost functions are shown in Fig. 2.

In the second stage, RF is applied on the testing data using the best subset of features found by the PSO based wrapper. For final evaluation, the accuracy rate, SP, SR and F-measure are calculated. The evaluation results are compared with those obtained for RF without any feature selection and other four traditional classifiers: Decision Trees , NB, k-NN and SVM. The latter algorithms are used as follows: for Decision Trees the J48 implantation is applied. SVM is used with an RBF kernel, cost and gamma were tuned using 5-folds cross validation with cost=50 and gamma =0.1. Finally, k is set to 1 in k-NN.

The experimental results of all classifiers are shown in Fig. 3. The results of our developed approach are listed in the first four columns and indicated as PSO(cost function used)-RF. Firstly, we can notice that the RF in the fifth column which was applied without any feature selection got the best results compared to the other classifiers by means of all evaluation measures. On the other hand, we can notice also that the proposed PSO-RF approach enhanced the results with a small variation based on the cost function selected. In general,

PSO wrapper with all cost functions dramatically decreased the number of features to almost the half without any degradation in the evaluation results. It is also worth to mention that PSO with the AUC cost function obtained the best results among all other classifiers with an increment of 0.67 %, 0.22 %, 2.43 % and 1.38 % for Accuracy, SP, SR and F-measure, respectively over the standard RF approach with only 47 features out of 79.

5   Conclusions

In this paper, a hybrid machine learning approach was proposed for identifying e-mail spam. The proposed approach is applied in two stages. In the first stage, Particle Swarm Optimization is performed as a wrapper feature selection mechanism based on different cost functions. While in the second, a Random forest classifier is applied for classifying emails. The developed classification model is evaluated and compared with other traditional classification algorithms. Evaluation results show that Particle Swarm Optimization as a features selector and Random forest as a classifier proved to outperform the other techniques.

References

1. Su, M.C., Lo, H.H., Hsu, F.H.: A neural tree and its application to spam e-mail detection. Expert Syst. Appl. **37**, 7976–7985 (2010)
2. Carreras, X., Marquez, L.S., Salgado, J.G.: Boosting trees for anti-spam email filtering. In: Proceedings of 4th International Conference on Recent Advances in Natural Language Processing, RANLP 2001, Tzigov Chark, BG (2001)
3. Yang, J., Liu, Y., Liu, Z., Zhu, X., Zhang, X.: A new feature selection algorithm based on binomial hypothesis testing for spam filtering. Knowl.-Based Syst. **24**, 904–914 (2011)
4. Guzella, T.S., Caminhas, W.M.: A review of machine learning approaches to spam filtering. Expert Syst. Appl. **36**, 10206–10222 (2009)
5. Silva, R.M., Almeida, T.A., Yamakami, A.: Artificial neural networks for content-based web spam detection. In: Proceedings of the 14th International Conference on Artificial Intelligence (ICAI 2012), pp. 1–7 (2012)
6. Faris, H., Aljarah, I., Alqatawna, J.: Optimizing feedforward neural networks using Krill Herd algorithm for e-mail spam detection. In: IEEE Jordan Conference on Applied Electrical Engineering and Computing Technologies (AEECT), Jordan, Amman (2015)
7. Rodan, A., Faris, H., et al.: Optimizing feedforward neural networks using biogeography based optimization for e-mail spam identification. Int. J. Commun. Netw. Syst. Sci. **9**, 19 (2016)
8. Deshpande, V.P., Erbacher, R.F., Harris, C.: An evaluation of naive bayesian anti-spam filtering techniques. In: IEEE SMC Information Assurance and Security Workshopp, IAW 2007, pp. 333–340. IEEE (2007)
9. Sakkis, G., Androutsopoulos, I., Paliouras, G., Karkaletsis, V., Spyropoulos, C.D., Stamatopoulos, P.: A memory-based approach to anti-spam filtering for mailing lists. Inf. Retrieval **6**, 49–73 (2003)

10. Drucker, H., Wu, D., Vapnik, V.N.: Support vector machines for spam categorization. IEEE Trans. Neural Netw. **10**, 1048–1054 (1999)
11. Blanco, Á., Ricket, A.M., Martín-Merino, M.: Combining SVM classifiers for email anti-spam filtering. In: Sandoval, F., Prieto, A.G., Cabestany, J., Graña, M. (eds.) IWANN 2007. LNCS, vol. 4507, pp. 903–910. Springer, Heidelberg (2007)
12. Delany, S.J., Cunningham, P., Tsymbal, A.: A comparison of ensemble and case-base maintenance techniques for handling concept drift in spam filtering. In: FLAIRS Conference, pp. 340–345 (2006)
13. Al-Shboul, B.A., Hakh, H., Faris, H., Aljarah, I., Alsawalqah, H.: Voting-based classification for e-mail spam detection. J. ICT Res. Appl. **10**, 29–42 (2016)
14. Fernández-Delgado, M., Cernadas, E., Barro, S., Amorim, D.: Do we need hundreds of classifiers to solve real world classification problems? J. Mach. Learn. Res. **15**, 3133–3181 (2014)
15. DeBarr, D., Wechsler, H.: Spam detection using clustering, random forests, and active learning. In: Sixth Conference on Email and Anti-Spam, Mountain View, California (2009)
16. Rios, G., Zha, H.: Exploring support vector machines and random forests for spam detection. In: CEAS (2004)
17. Zitar, R.A., Hamdan, A.: Genetic optimized artificial immune system in spam detection: a review and a model. Artif. Intell. Rev. **40**, 305–377 (2013)
18. Fogel, D.B.: The advantages of evolutionary computation. In: BCEC, pp. 1–11. Citeseer (1997)
19. Gavrilis, D., Tsoulos, I.G., Dermatas, E.: Neural recognition and genetic features selection for robust detection of e-mail spam. In: Antoniou, G., Potamias, G., Spyropoulos, C., Plexousakis, D. (eds.) SETN 2006. LNCS (LNAI), vol. 3955, pp. 498–501. Springer, Heidelberg (2006)
20. Zhang, Y., Wang, S., Phillips, P., Ji, G.: Binary PSO with mutation operator for feature selection using decision tree applied to spam detection. Knowl.-Based Syst. **64**, 22–31 (2014)
21. Lai, C.C., Wu, C.H.: Particle swarm optimization-aided feature selection for spam email classification. In: ICICIC, p. 165. IEEE (2007)
22. Tan, Y.: Particle swarm optimization algorithms inspired by immunity-clonal mechanism and their applications to spam detection. In: Innovations and Developments of Swarm Intelligence Applications, p. 182 (2012)
23. Kennedy, J., Eberhart, R.C.: Particle swarm optimization. In: Proceedings of the IEEE International Conference on Neural Networks, Piscataway, NJ, USA, pp. 1942–1948 (1995)
24. Gandomi, A.H., Alavi, A.H.: Krill Herd: a new bio-inspired optimization algorithm. Commun. Nonlinear Sci. Numer. Simul. **17**, 4831–4845 (2012)
25. Aljarah, I., Ludwig, S.A.: A new clustering approach based on glowworm swarm optimization. In: 2013 IEEE Congress on Evolutionary Computation. Institute of Electrical & Electronics Engineers (IEEE) (2013)
26. Moraglio, A., Chio, C., Togelius, J., Poli, R.: Geometric particle swarm optimization. J. Artif. Evol. Appl. **2008**, 11 (2008)
27. Breiman, L.: Random forests. Mach. Learn. **45**, 5–32 (2001)
28. Guo, L., Chehata, N., Mallet, C., Boukir, S.: Relevance of airborne lidar and multispectral image data for urban scene classification using random forests. ISPRS J. Photogrammetry Remote Sens. **66**, 56–66 (2011)
29. Alqatawna, J., Faris, H., Jaradat, K., Al-Zewairi, M., Adwan, O.: Improving knowledge based spam detection methods: the effect of malicious related features in imbalance data distribution. Int. J. Commun. Netw. Syst. Sci. **8**, 118 (2015)

30. Kohavi, R., John, G.H.: Wrappers for feature subset selection. Artif. Intell. **97**, 273–324 (1997)
31. Hall, M., Frank, E., Holmes, G., Pfahringer, B., Reutemann, P., Witten, I.H.: The weka data mining software: an update. ACM SIGKDD Explor. Newsl. **11**, 10–18 (2009)
32. Burez, J., Poel, D.: Handling class imbalance in customer churn prediction. Expert Syst. Appl. **36**, 4626–4636 (2009)
33. Wang, S., Tang, K., Yao, X.: Diversity exploration and negative correlation learning on imbalanced data sets. In: International Joint Conference on Neural Networks, IJCNN 2009, pp. 3259–3266. IEEE (2009)

Fuzzy Logic and PD Control Strategies
of a Three-Phase Electric Arc Furnace

Loredana Ghiormez[1(✉)], Octavian Prostean[1], Manuela Panoiu[2],
and Caius Panoiu[2]

[1] Faculty of Automation and Computers, Politehnica University of Timisoara,
str. B-dul Vasile Parvan, No. 2, 300223 Timisoara, Romania
{loredana.ghiormez,octavian.prostean}@upt.ro
[2] Faculty of Engineering Hunedoara, Politehnica University of Timisoara,
str. Revolutiei, No. 5, 331128 Hunedoara, Romania
{manuela.panoiu,caius.panoiu}@upt.ro

Abstract. This paper presents a fuzzy control and a conventional proportional derivative control for the electrode positioning system of a three-phase electric arc furnace. Generally, it is necessary to maintain constant arc lengths for these kinds of furnaces. The two control strategies proposed in this paper regulates the current of the electric arc because arc length depends by the electric arc current. In order to do this, a new model of the electric arc developed by the authors of this paper was used. This paper illustrates a comparison of the performance analysis of a conventional PD controller and a fuzzy based intelligent controller. The fuzzy intelligent based controller has two inputs and one output. The whole systems are simulated by using of Matlab/Simulink software. These systems are tested when applying a disturbance in the process. The responses of the closed-loop systems illustrates that the proposed fuzzy controller has better dynamic performance, rapidity and good robustness as compared to the proposed PD controller.

Keywords: Modeling · Fuzzy logic controller · PD controller · Current control

1 Introduction

Electric arc furnace (EAF) is the equipment that converts the electrical energy in thermal energy, using graphite electrodes. Scrap is added in the furnace tank in order to be melted and obtain liquid steel [1]. EAF taken into consideration in this paper is a three-phase one with direct action, so, the electric arcs appear between each of the three electrodes and the metal that will be melted. In order to melt the metals and to reduce the energy consumption it is necessary to optimize the delivered power to the furnace. Arc power can be influenced in two ways: by modifying the position of the electrodes or by modifying the power delivered by the transformer to the EAF [2]. A new position of the electrodes is performed at an EAF by using hydraulic actuators [3] so, obtaining different arc lengths.

For a production cycle at an EAF can be found two stages: melting and refining stages. In the melting stage the metal is loaded in the furnace tank, it has solid shape

© Springer International Publishing Switzerland 2016
N.T. Nguyen et al. (Eds.): ICCCI 2016, Part I, LNAI 9875, pp. 509–519, 2016.
DOI: 10.1007/978-3-319-45243-2_47

and will be obtained the liquid state of this charging material. This stage is characterized by disturbances because the scrap loaded in the furnace tank will change the position during the melting process and arc length will be different. If there is not an automatic control of the position of the electrodes, these disturbances will be injected in the electrical power network or the electrode can break. In the refining stage, the metal is in the liquid state and the disturbances will be smaller, because arc length can be maintain constantly.

In the reference literature are illustrated several control strategies, these being executed through the electrode positioning system that consists of hydraulic actuators used to move the electrodes up or down [4–6].

In this paper is designed a new model of the electric arc which represents more accurately the voltage-current characteristic (VIC) of the electric arc. It is necessary to model the behavior of the electric arc, because it is the main cause of the nonlinearity of the EAF [7, 8]. This model is used than in the closed-loop system to regulate the electric arc current.

Also, in this paper are proposed two control strategies: fuzzy based intelligent control and PD control. This paper proposes to make a comparison between the two control strategies in order to notice which is better to use for the real installation.

This proposed fuzzy controller is like a PD fuzzy controller, so, its performance is compared with a conventional PD controller. In order to obtain the performance of the two proposed control strategies is injected in a moment of time a disturbance in the process.

2 Modeling of the EAF Electrical System

The electrical system of an electrode positioning system can be compound by two models. One model is for the modeling of the high current of the electric arc and one model for the modeling of the three-phase electrical supply system of the EAF.

2.1 Modeling of the Electric Arc

The VIC of the electric arc is the method used for the modeling of the statically and dynamically behavior of the electric arc for the three-phase EAF [9]. In Fig. 1 is presented the VIC of the EAF that will be obtained with the proposed model. In this model, the VIC was divided in five sections in order to obtain a characteristic similar with the one from the real installation. Because the EAF is supplied with alternative voltage, in each period the voltage and the current are crossing two times by 0. Each time that voltage and current change its polarities, the arc will be reignite.

In Fig. 1 v_{ig} represents the ignition voltage, v_{ex} is the extinction voltage, v_m is the average between the ignition and the extinction voltage of the arc, v_{arc} is the voltage of the electric arc, i_1 is the current of the electric arc corresponding for the v_{ig}, i_2 is the current of the electric arc corresponding for the v_m, i_3 is the current of the electric arc corresponding for the v_{ex} and i_{arc} is the current of the electric arc [9]. R_1, R_2 and R_3 represents lines slope for the linear section of the VIC.

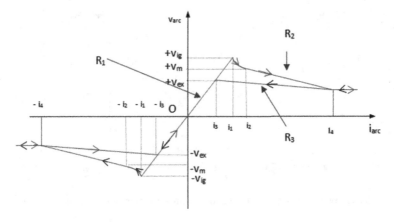

Fig. 1. Voltage-current characteristic divided by sections (Color figure online)

The first section of this characteristic is in the range $[-i_3, i_1]$ and electric current derivative is positive. The second section is in the range $[i_1, i_2]$ and electric current derivative is positive. The third section is in the range $[i_2, i_4]$ and electric current derivative is positive. The fourth section is greater than i_4 and electric current derivative is positive and the fifth section is greater than i_3 and electric current derivative is negative. These sections will be also for the negative semi-period of the electric parameters but the range will be different because the values of the electrical parameters are negative.

In Fig. 1 the arrows illustrate how will be designed the voltage-current characteristic of the electric arc. Red arrow illustrates the route when the electric arc current derivative is positive and black arrows illustrate the route when the electric arc current derivative is negative.

In (6) is presented the model of the electric arc that will be used in the closed-loop system. This model accurately represents the VIC of the electric arc. In order to validate this new model, it is performed a comparison, illustrated in Fig. 2, for the simulated data and the measured data from the real installation taken into consideration to study. In the model from (6) were chosen values for the model parameters in order to accurately represent the VIC of the electric arc, $R_1 = 0.05$, $R_2 = -0.0007$ and $R_3 = -0.0003$. Limit i_1 is computed as presented in (1), limit i_2 is computed as presented in (2), limit i_3 is computed as presented in (3), limit i_4 is computed as presented in (4) and voltage v_m is computed as presented in (5).

$$i_1 = v_{ig}/R_1 \tag{1}$$

$$i_2 = 2 \cdot i_1 \tag{2}$$

$$i_3 = v_{ex}/R_1 \tag{3}$$

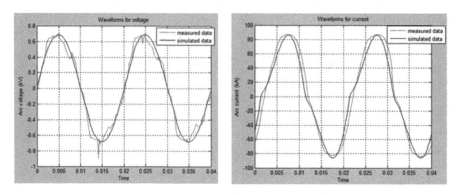

Fig. 2. Waveforms for voltage and current – measured and simulated data

$$i_4 = (v_m - v_{ex} + i_3 \cdot R_3 - i_2 \cdot R_2)/(R_3 - R_2) \tag{4}$$

$$v_m = (v_{ig} + v_{ex})/2 \tag{5}$$

$$v = \begin{cases}
-v_{ex} + (i_3 - i_4) \cdot R_3, & i < -i_4 \text{ and } di/dt \geq 0 \\
& \text{or } i < -i_4 \text{ and } di/dt < 0 \\
-v_{ex} + (i + i_3) \cdot R_3, & i \in [-i_4, -i_3) \text{ and } di/dt \geq 0 \\
R_1 \cdot i, & i \in [-i_3, i_1) \text{ and } di/dt \geq 0 \\
& \text{or } i \in [-i_1, i_3) \text{ and } di/dt < 0 \\
v_{ex} + (v_{ig} - v_{ex}) \cdot e^{(i_1 - i)/i_2} & i \in [i_1, i_2) \text{ and } di/dt \geq 0 \\
v_m + (i - i_2) \cdot R_2 & i \in [i_2, i_4] \text{ and } di/dt \geq 0 \\
v_{ex} + (i_4 - i_3) \cdot R_3 & i > i_4 \text{ and } di/dt \geq 0 \\
& \text{or } i > i_4 \text{ and } di/dt < 0 \\
v_{ex} + (i - i_3) \cdot R_3 & i \in [i_3, i_4] \text{ and } di/dt < 0 \\
-v_{ex} + (v_{ex} - v_{ig}) \cdot e^{(i_1 + i)/i_2} & i \in [-i_2, -i_1] \text{ and } di/dt < 0 \\
-v_m + (i + i_2) \cdot R_2 & i \in [-i_4, -i_2] \text{ and } di/dt < 0
\end{cases} \tag{6}$$

2.2 Modeling of the Three-Phase Electrical Supply System Circuit of the EAF

The power supply electric circuit of an EAF for a single phase can be considered as the one presented in Fig. 3a. This electric circuit is used in practical computations. The EAF has a resistive character [2], so, can be considered as a variable resistance. In Fig. 3a v_s represents the alternative voltage from the secondary side of the furnace transformer for a single phase, I is the electric current for a single phase in the secondary side of the furnace transformer, r is the total electrical resistance for a single phase of the secondary side of the furnace transformer, x is the total electrical reactance

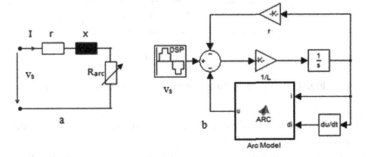

Fig. 3. (a) The power supply electric circuit of an EAF for a single phase and (b) Block diagram of the EAF system

for a single phase of the secondary side of the furnace transformer and R_{arc} stands for the resistance of the electric arc, being reported at the voltage from the primary side of the furnace transformer.

In Fig. 3b is presented the block diagram for the implementation of the electric arc furnace in Matlab/Simulink. This block diagram was implemented by applying second Kirchhoff's theory in the electric circuit from Fig. 3a. In this study were chosen the following values of the circuit parameters: $r = 0.47$ mΩ, $x = 5.5$ mΩ. Voltage from the secondary side of the furnace transformer was set as considered that the transformer has plot set on 894 V. These values are taken from a real industrial plant.

Total electrical resistance is obtained by adding the values of the source resistance and of the short network resistance. Total electrical reactance is obtained by adding the values of the source reactance and of the short network reactance.

Frequency considered of the circuit is 50 Hz. For the source sample time was set to 1/10000 s, meaning that are taken 10000 samples in 1 s. *ARC* block contains the mathematical model presented in (6).

3 Control Strategies of the EAF

In this paper are proposed, designed and implemented two control strategies. One control strategy is based on the fuzzy intelligent controller and the other is based on conventional PD controller. In both strategies, the control variable used in this paper is current of the electric arc and these control strategies are used to obtain constant arc length. In order to obtain the performance of the proposed control loops, the systems are tested when applying a disturbance in the process in order to observe if the controller will reject this disturbance and which of the controllers has better performance.

In Fig. 4 is presented the block diagram of the proposed control closed-loop. In this block diagram I_{ref} is the reference arc current, e_I is the error computed as the difference between the reference value and obtained value of the arc current, d_{eI} is the derivative of the error, c is the command for the hydraulic actuator in order to move the electrode up or down with a corresponding speed, x is a value that will be added in the position of the arc in this way obtaining the value of the reference current, because arc length influence the electric arc current. I is the current of the electric arc obtained by the

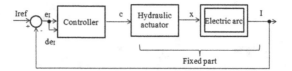

Fig. 4. Block diagram of the proposed control closed-loop

model of the electric arc. *Controller* block stands for the fuzzy logic controller (FLC) or the PD controller.

Electric arc block is the process in the control loop and the *hydraulic actuator* is the execution element. Hydraulic actuator and electric arc blocks compound the fixed part of the system. In both of the control strategies, the hydraulic actuator is considered as having a transfer function corresponding to a PT1. Chosen time constant for both of the control strategies is 0.75 s.

3.1 Fuzzy Logic Controller of the EAF

FLC is suitable for systems that present nonlinearity and strong random disturbances [10]. In this paper was design a PD-like fuzzy controller that has two inputs and one output. The parameter of the EAF system under control is the current of the electric arc. The inputs of the FLC are the error and the derivative of the error, and the output of the controller is the command for the hydraulic actuator that will modify the position of the electrode, in this way, obtaining a new arc length, so, a new arc current. The design of the fuzzy inference structure for FLC is illustrated in Fig. 5. The fuzzy inference method used is a Mamdani-type, so, the output membership functions used are distributed fuzzy sets. *Min-Max inference* engine is used for the inference and the *centroid* of a two-dimensional function is used for the defuzzification process. Membership functions for all the three variables of the fuzzy system are divided into seven fuzzy sets. These fuzzy sets are: *NL, NM, NS, Z, PS, PM* and *PL* which corresponds to *Negative Large, Negative Medium, Negative Small, Zero, Positive Small, Positive Medium, and Positive Large*. Trapezoidal membership functions are used for *NL* and *PL* and triangular for the rest. The variables can be positive of negative depending on the obtained current.

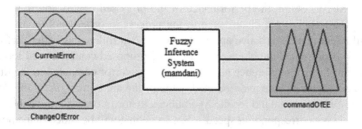

Fig. 5. Designing of the fuzzy inference structure for FLC

Figure 6 presents the membership functions for the error of the system, this being an input variable. The universe of discourse for this variable is in the range [−1700; 1700] amperes. Figure 7 presents the membership functions for the derivative of the error, this being an input variable. The universe of discourse for this variable is in the range [− 3e + 06; 3e + 06]. Figure 8 presents the membership functions for the command to the hydraulic actuator, this being an output variable. The universe of discourse for this variable is in the range [−40; 40].

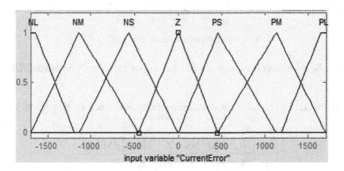

Fig. 6. Membership functions for the input variable *"CurrentError"*

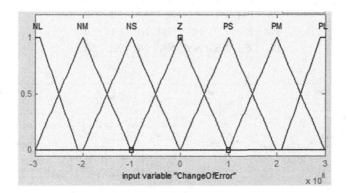

Fig. 7. Membership functions for the input variable *"ChangeOfError"*

The fuzzy system has two inputs and one output, each of the variables having seven fuzzy sets, so, are written 49 rules which are describing the command for the hydraulic actuator taking into account the values of the inputs of the fuzzy system. These rules are presented in Table 1.

In Fig. 9 is presented the Matlab/Simulink model for FLC. In the model of the electric arc it is set an initial length of the arc of *31* cm, which will correspond to the value of the reference parameter. Value that will be obtained from the hydraulic actuator will be added to the initial arc length. Figure 9 illustrates the case when the disturbance is injected in the process. Value of the reference current is set to *5650* A.

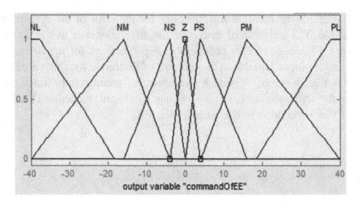

Fig. 8. Membership functions for the output variable "*CommandOfEE*"

Table 1. Fuzzy rules for developing fuzzy inference system for FLC.

e_I	d_{eI}						
c	NL	NM	NS	Z	PS	PM	PL
NL	PL	PL	PL	PL	PM	PS	Z
NM	PL	PL	PL	PM	PS	Z	NS
NS	PL	PL	PM	PS	Z	NS	NM
Z	PL	PM	PS	Z	NS	NM	NL
PS	PM	PS	Z	NS	NM	NL	NL
PM	PS	Z	NS	NM	NL	NL	NL
PL	Z	NS	NM	NL	NL	NL	NL

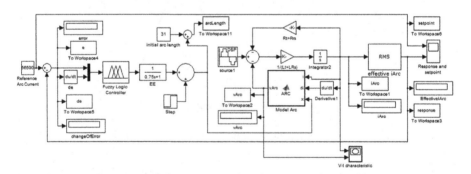

Fig. 9. Matlab/Simulink model for the FLC when is injected disturbance in the process

3.2   PD Controller of the EAF

In this paper is design a PD controller in order to make a comparison between the two proposed strategies because the FLC is PD-like fuzzy controller. In Fig. 10 is presented the Matlab/Simulink block diagram for PD controller for the case when is injected disturbance in the process. So, PD controller is used for the rejection of this

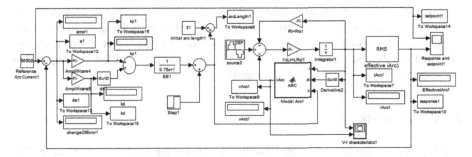

Fig. 10. Matlab/simulink model for the PD controller when is injected disturbance in the process

disturbance. This PD controller has two gains, one for the proportional component and one for the derivative component.

4   Comparisons of the Two Control Strategies

In order to make a comparison between the responses obtained for the FLC and for the PD controller, after the first second of the system functioning was applied a disturbance of 2 cm, which is corresponding for the real installation with the melting of some parts of the charging material and obtaining a new configuration of the scrap in furnace's tank. So, arc length is increasing and the arc current is decreasing. In Fig. 11 is illustrated the response of the systems, when is injected disturbance in the process.

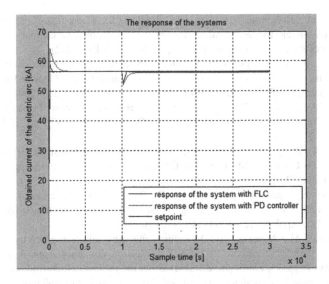

Fig. 11. Response of the systems when is injected disturbance in the process

For the supply source was set the sample time to 0.0001 s, so, in one second will be taken 10000 samples. One can notice that both responses reach the stable state but the response of the system with FLC is faster than with PD controller. Also, it can be observed that the response of the system with FLC has smaller overshot as compared to the response of the system with PD controller.

5   Conclusions

In this paper is proposed a new model of the electric arc for an electric arc furnace. The model is based on voltage-current characteristic of the electric arc and it is an exponential model. In order to validate this model comparisons were made for the simulated data and acquired data from the real installation. One can noticed that the model of the electric arc accurately follows the electric parameters acquired from the real installation. This model is used in the implementation of the two control strategies used in the current control of the electric arc. For both, fuzzy logic controller and PD controller, the systems were tested in the closed-loop. In order to make a comparison of both control strategies, a disturbance was injected in a moment of time in the process. It was demonstrated that system with fuzzy logic controller has better dynamic performance, rapidity and good robustness as compared to the PD controller. The response of the system is important because by using the electrode system regulation it is reduced the energy consumption and it is avoided the appearance of damage caused by the breakage of the electrodes, therefore, the productivity of steel will be increasing.

References

1. Taslimian, M., Shabaninia, F., Vaziri, M., Vadhva, S.: Fuzzy type-2 electrode position controls for an electric arc furnace. In: 2012 IEEE 13th International Conference on Information Reuse and Integration (IRI), pp. 498–501. IEEE (2012)
2. Panoiu, M., Panoiu, C., Deaconu, S.: Study about the possibility of electrodes motion control in the EAF based on adaptive impedance control. In: 2008 Power Electronics and Motion Control Conference, EPE-PEMC 2008, pp. 1409–1415. IEEE (2008)
3. Hong, H., Mao, Z.: Controller design for electrode regulating system of electric arc furnace. In: 2015 27th Chinese Control and Decision Conference (CCDC), pp. 864–867. IEEE (2015)
4. Li, X., Lu, X., Wang, D.: Arc furnace electrode control system design. In: 2010 International Conference on Computer, Mechatronics, Control and Electronic Engineering (CMCE), vol. 4, pp. 364–366. IEEE (2010)
5. Khoshkhoo, H., Sadeghi, S.H.H., Moini, R., Talebi, H.A.: An efficient power control scheme for electric arc furnaces using online estimation of flexible cable inductance. Comput. Math Appl. **62**(12), 4391–4401 (2011). Elsevier
6. Moghadasian, M., Alenasser, E.: Modelling and artificial intelligence-based control of electrode system for an electric arc furnace. J. Electromagn. Anal. Appl. **3**, 47–55 (2011)
7. Hui, Z., Fa-Zheng, C., Zhuo-qun, Z.: Study about the methods of electrodes motion control in the EAF based on intelligent control. In: 2010 International Conference on Computer, Mechatronics, Control and Electronic Engineering (CMCE), vol. 4, pp. 68–71. IEEE (2010)

8. Panoiu, M., Ghiormez, L., Panoiu, C.: Adaptive neuro-fuzzy system for current prediction in electric arc furnaces. In: Balas, V.E., Jain, L.C., Kovacevic, B. (eds.) Soft Computing Applications, Advances in Intelligent Systems and Computing, vol. 356, pp. 423–437. Springer International Publishing, Heidelberg (2016)
9. Panoiu, M., Panoiu, C., Ghiormez, L.: Modeling of the electric arc behavior of the electric arc furnace. In: Balas, V.E., Fodor, J., Várkonyi-Kóczy, A.R., Dombi, J., Jain, L.C. (eds.) Soft Computing Applications. AISC, vol. 195, pp. 261–271. Springer, Heidelberg (2012)
10. Yan, L.I., et al.: Model predictive control synthesis approach of electrode regulator system for electric arc furnace. J. Iron. Steel Res. Int. **18**(11), 20–25 (2011)

Hybrid Harmony Search Combined with Variable Neighborhood Search for the Traveling Tournament Problem

Meriem Khelifa$^{(\boxtimes)}$ and Dalila Boughaci

LRIA-FEI, Computer Science Department, USTHB, BP 32 El-Alia,
16111 Bab-Ezzouar, Algiers, Algeria
khalifa.merieme.lmd@gmail.com, dboughaci@usthb.dz

Abstract. In this paper, we are interested in the mirrored version of the traveling tournament problem (mTTP) with reversed venues. We propose a new enhanced harmony search combined with a variable neighborhood search (V-HS) for mTTP. We use a largest-order-value rule to transform harmonies from real vectors to abstract schedules. We use also a variable neighborhood search (VNS) as an improvement strategy to enhance the quality of solutions and improve the intensification mechanism of harmony search. The overall method is evaluated on benchmarks and compared with other techniques for mTTP. The numerical results are encouraging and demonstrate the benefits of our approach. The proposed V-HS method succeeds in finding high quality solutions for several considered instances of mTTP.

Keywords: Sport scheduling · Traveling tournament problems · Variable neighborhood search · Harmony search

1 Introduction

The Traveling Tournament Problem (TTP) is a core problem in sports scheduling. TTP is the problem of finding a feasible double round robin schedule that minimizes the overall distance traveled by all teams [8,11,17]. TTP is widely believed to be NP-hard [19] which makes difficult finding quality solutions even instances with just 8 teams due to strong feasibility issues together with a complex optimization. Various works in different contexts tackled the problem of traveling tournament scheduling. Among them, we give the following ones: the Branch-and-price-based approaches developed in [14]. The MaxminTTP variant of TTP that minimizes the longest distance traveled where both models are based on the detection of independent sets on conflict graphs [4]. Nitin and Choubey proposed an evolutionary approach based on genetic algorithm for TTP [6]. In [15], authors proposed a variable neighborhood search for TTP. A hybrid genetic algorithm (GA) with a simulated annealing (SA) for mTTP is proposed in [3]. In [12], a biogeography based optimization (BBO) combined with a simulated annealing (SA) is proposed for the mTTP. The authors in [9] proposed

© Springer International Publishing Switzerland 2016
N.T. Nguyen et al. (Eds.): ICCCI 2016, Part I, LNAI 9875, pp. 520–530, 2016.
DOI: 10.1007/978-3-319-45243-2_48

a tabu search based approach for TTP and Stephan and Westphal proposed an approximation algorithm for TTP [20]. Harmony search (HS) is an evolutionary meta-heuristic proposed by Geem *et al.* [10]. HS is inspired by the process of music players searching for a perfect state of harmony. In this paper, we propose an enhanced harmony search to solve mTTP. We use a largest-order-value rule to transform harmonies in the harmony memory from real vectors to abstract schedules and vice-versa. Further, we add a variable neighborhood improvement strategy into HS to enhance the quality of the generated solutions.

The rest of this paper is organized as follows: the second section gives a background on the traveling tournament problem. The third section presents in detail the proposed harmony search based approach for mTTP. The fourth section gives some numerical results. Finally, the fifth section concludes the work and gives some future works.

2   Problem Description

The Traveling Tournament Problem (TTP) is a well-known sport scheduling combinatorial optimization problem which involves n teams $T = (t_1, \ldots, t_n)$ (n is even). TTP is the problem of scheduling a double round-robin tournament (DRRT) that is a set of games in which every team plays every other team exactly once at home and once away and all teams must play only one match every round. Consequently DRRT has $n(n-1)$ games and $n/2$ games are played in every round. Thus exactly $2(n-1)$ slots (rounds) are required to schedule a Double round robin tournament. The distance between team cities are given by $(n \times n)$ symmetric matrix Dis, such that an element dis_{ij}, of Dis represents the distance between the homes of the teams t_i and t_j. The teams begin in their home city and must return there after the tournament. The aim of TTP is to find a double round-robin tournament that minimizes the distance traveled by all teams, and satisfy the following related constraints:

1. **Double Round Robin constraint** (DRRT): each team plays with every other team exactly twice, once in his home and once in the home of his opponent.
2. **AtMost Constraints**: each team must play no more than U and no less than L consecutive games at home or away. In general, L is set to 1 and U to 3.
3. **NoRepeat Constraints**: A game (t_i, t_j) can never be followed in the next round by the game (t_j, t_i) .

– **The TTP inputs** are: the number n of teams and the Dis distance matrix.
– **The output** will be a Double Round Robin Tournament on the n teams respected the three constraints: AtMost, NoRepeat and DRRT, and the total distance traveled by all the teams is minimized.

mTTP is the mirrored version of TTP. mTTP requires that the games played in round R are exactly the same as those played in round $R + n - 1$, for $R = 1, 2, \ldots n - 1$ with reversed venues. We called this additional constraint *mirrored constraint.* mTTP is then the problem of finding a schedule for a double round robin tournament with minimum cost satisfying the same constraints plus an additional constraint the mirrored constraint.

3   Proposed Approach

Harmony search (HS) is a new nature inspired meta-heuristic optimization developed by Geem *et al.* [10]. HS is based on the musical composition process in an orchestra where composers try to adjust their pitches to hunt for a fantastic harmony. In the optimization process, each decision variable takes values randomized within the possible array, forming a solution vector. In this section, we propose a harmony search based approach for mTTP. The details of the proposed approach are given in following.

3.1   Neighborhood Structures

We used three structures of neighborhood which are:

- **Swap Round**(s, R_i, R_j): this is a move that consists of swapping all games of a given pair of rounds.
- **Swap Home**(s, t_i, t_j): this move swaps the home/away roles of a game involving the teams t_i and t_j .
- **Swap Team**(s, t_i, t_j): this is a move that corresponds to swapping all opponents of a given pair(t_i, t_j) of teams over all rounds.

3.2   Creation of a SRRT

The well-known polygon method [7] is used to build a schedule for Single Round Robin Tournament (SRRT) with an abstract teams (we note an abstract schedule as A_SH), without assigning stadiums to the games.

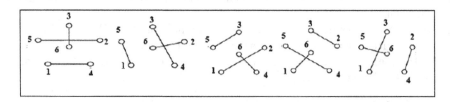

Fig. 1. Polygon method for n = 6 for the permutation [3 2 4 1 5 6]

In order to create a SRRT, we follow the followings steps:

- **(1)** We first generate an initial permutation of n random and unique positive integers from 1 to n.

- **(2)** We place integers from this permutation on the $n - 1$ nodes of a regular polygon consecutively as shown in Fig. 1 where team 1 is placed in node 1, team 2 in node 2, and so on. The nth integer is placed outside of the polygon.
- **(3)** We have $n * (n - 1)/2$ games in round $n - 1$. In each round k from 1 to $n - 1$, the abstract team placed at node 1 plays against the abstract team n, team $l = 2, 4 \ldots n/2$ plays with the team $n + l - 1$. Each team plays with a team located symmetrically opposite to it in the polygon. After a round is completed, we make a simple rotation of the polygon in the direction of clockwise in order to get the rest rounds.
- **(4)** Thus in $n - 1$ rounds, we get a Single Round Robin Tournament SRRT (noted S_SH) schedule. We complete this schedule to get a mirrored DRRT (noted as D_SH) as shown in Table 1.

Table 1. Mirror of Single Round Robin Tournament (DRRT) (n=6)

R_1	(t_3, t_6)	(t_5, t_2)	(t_1, t_4)	R_6	(t_6, t_3)	(t_2, t_5)	(t_4, t_1)
R_2	(t_4, t_3)	(t_2, t_6)	(t_5, t_1)	R_7	(t_3, t_4)	(t_6, t_2)	(t_1, t_5)
R_3	(t_3, t_5)	(t_2, t_1)	(t_6, t_4)	R_8	(t_5, t_3)	(t_1, t_2)	(t_4, t_6)
R_4	(t_1, t_6)	(t_5, t_4)	(t_2, t_3)	R_9	(t_6, t_1)	(t_4, t_5)	(t_3, t_2)
R_5	(t_6, t_5)	(t_3, t_1)	(t_2, t_4)	R_{10}	(t_5, t_6)	(t_1, t_3)	(t_4, t_2)

3.3   The Proposed Harmony Search for mTTP

Harmony Representation. A schedule is coded as a permutation (A_SH) of size n such that n is the total number of teams that are used to generate single round robin tournament (S_SH). However, due to the continuous nature of HS, the process of HS and their operator it is not suitable for our representations (integer permutation or abstract schedule A_SH).

In order to find an appropriate mapping between the permutation and harmony in HS, we use a largest-order-value (LOV) [1,16] and smallest position value (SPV) rules to convert the permutation to a harmony as shown in Figure 2.

The principle of the largest-order-value (LOV) rule is given as follows:
Let us consider the harmony $x_n = (x_{n,1}, x_{n,2}, x_{n,3} \ldots x_{n,n})$ n is the number of teams:

- **(a)** The LOV and SPV procedure consist in ordering all the elements of the vector x_n by descending order.
  We obtain: $Xord_i = (Xord_{n,1}, Xord_{n,2}, Xord_{n,3} \ldots Xord_{n,n})$. Table 2 gives an example for n=6 . ($x_6 = (0.31, 1.46, 0.11, 2.33, 4.6, 1.33)$ and $Xord_6 = (4.6, 2.33, 1.46, 1.33, 0.31, 0.11)$)

Table 2. Conversion of a harmony to an abstract schedule

Permut(n)	1	2	3	4	5	6
x_n	0.31	1.46	0.11	2,33	4.6	1.33
$Xord_n$	4.6	2.33	1.46	1.33	0.31	0.11
Ord_n	5	3	6	2	1	4
$Persh_n$	5	4	2	6	1	3

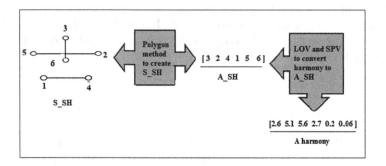

Fig. 2. Transformation of a harmony to SRRT

- **(b)** the vector $Ord_n = (Ord_{n,1}, Ord_{n,2}...Ord_{n,3})$ can be generated from $Xord(n)$ by the following formula:if $X_{n,k} = Xord_{n,j}$ then $Ord_{n,k} = j$.$(x_{6,5} = xord_{6,1} \rightarrow Ord_{6,5} = 1...Ord_6 = (5,3,6,2,1,4))$. **(c)** finally, we calculate the permutation of the abstract schedule by using the following formula : $Permsh_{n,Ord_{n,j}} = j$ $(Ord_{6,1} = 5, Permsh_{6,Ord_{6,1}} = 1)$.

Initialization of the Harmony Memory. The harmony memory (HM) is a set of randomly generated solution vectors (harmonies) where the corresponding abstract schedules are sorted according to the travel-cost values.

Parameter Setting. The HS parameters are as follows:

- The **HMCR** is the probability of choosing one value from HM. A large HMCR is not recommended because of the possibility that the best solution may be improved not stored in the HM and other harmonies are not explored, while a small HMCR value means every candidate value is chosen from the range of the variable space randomly and this may increase the diversity of the HM. Therefore, we use $HMCR = 0.70 \approx 0.95$.
- The pitch adjusting process **PAR** is the probability of mutating the chosen value from the HM memory. A small PAR value can slow the convergence of HS because the limitation in the exploration of only a small space of the total search space while a large PAR value can be risked losing the best solution. Thus, we usually use $RPA = 0.1 \approx 0.5$.

Generation of Novel Harmonies. After having initialized the HM memory and fixed the algorithm parameters, HS method launches the improvisation step where a novel harmony vector is generated by applying three possible rules: **(1)** memory consideration **(2)** pitch adjustment **(3)** random selection.

- **(1)** With probability HMCR, the new harmony is chosen from the current harmony memory (formula 1) according to its travel cost.
- **(2)** Pitch adjustment: if the decision variable of the harmony obtained from harmony memory should be mutated with probability PAR where bw is an uniform random number between 0 and 1 is the band width, control ling the local range of pitch adjustment.

$$hm_{new} = \begin{cases} hm_i, i \in (1,2...HMS) & \text{with a probability } (HMCR*(1-PAR)) \ (1) \\ hm_{new} \pm rand() * bw & \text{with a probability } (HMCR*PAR) \quad (2) \end{cases}$$

- **(3)** (1-HMCR) is the probability of generating the decision variable randomly by using the formula (3).

$$hm_{new} = LB + (UB - LB) * rand \quad \text{with a probability } (1 - HMCR) \ (3)$$

where LB and UB are the lower and upper bounds for the decision variables respectively and n is the number of teams.

In this paper, UB is set to n and $LB = 0$. The notation A_SH_{New} represents the abstract schedule round robin tournament getting by applying LOV rule. The latter permits to convert the harmony hm_{New} to the abstract schedule and S_SH_{New} is a Single round robin of the abstract A_SH_{New} which is obtained by polygon A_SH_{New}, here we used VNS [7,13] as a local search meta-heuristic. VNS is a systematic change of neighborhood combined within a local search (swap teams, Swap home, Swap round) to improve the cost travel of the schedule (D_SH_{New} Double Round Robin schedule) in each iteration. The VNS algorithm for TTP is sketched in Algorithm 1.

Algorithm 1. The VNS improvement strategy

Require: a current solution S, three Neighborhood structures are used $V_k:$, $k = 1$ for *Swap Round* structure, $k = 2$: for *Swap Home* and $k = 3$: for *Swap Team*, *Maxiter* is the maximum number of iterations.
1: $k \leftarrow 1$; $S_0 \leftarrow$ SwapRound (S, R_i, R_j); $S^* \leftarrow S_0$
2: Apply Local search with the neighborhood structure (N_1) on S_0) to obtain S.
3: **for** $I = 1$ to *maxiter* **do**
4:     $S' \leftarrow$ choose Random solution $N_k(S)$
5:     $S'' \leftarrow$ local search
6:     Apply SLS on S' with the neighborhood structure (N_k) to obtain S''
7:     **if** $f(S'') \ ; \ f(S)$ **then**
8:         $S \leftarrow S''$; $S^* \leftarrow S$
9:     **else**
10:         **if** $k < |K|$ **then**
11:            $k \leftarrow k + 1$
12:         **else**
13:            $k \leftarrow 1$
14:         **end if**
15:     **end if**
16: **end for**
    return the best schedule S found.

The HS algorithm for mTTP is sketched in Algorithm 2.

Algorithm 2. Harmony search algorithm

```
1: for I = 0 to Nbr_gener do
2:     if rand(0, 1) < HMCR then
3:         hm_New ← hm_i...i ∈ (1...P_bect)
4:         if rand(0, 1) < PAR then
5:             hm_New ← hm_New ± rand(0, 1) * bw
6:         end if
7:         A_SH_New ← Lov(hm_New)
8:         S_SH_New ← Polygon(A_SH_New)
9:         D_SH_New ← VNS(S_SH_New)
10:    else
11:        hm_New ← LB + (UB − LB) * rand(0, 1)
12:        A_SH_New ← Lov(hm_New)
13:        S_SH_New ← polygon(A_SH_New)
14:        D_SH_New ← VNS(S_SH_New)
15:    end if
16:    if Cost_travel(D_SH_New) < Cost_travel(D_SH_Worst) then
17:        Include(Lov(A_SH_New)) to the HM
18:        Exclude(Lov(A_SH_Worst)) from the HM
19:    end if
20: end for
       return the best schedule.
```
---

4 Experimental Results

The source codes are written in Java. The experiments were performed on a Intel(R) Core(TM)i5 4210UCPU @ (1.70 GHz) with 800 GB of RAM memory.

4.1 The Considered Benchmarks

We evaluated the proposed method (V-HS) on three different datasets of instances available at the website [5]. The considered datasets are: the NLx instances and the Constant distance instances, named CONx (the x stands for the dimension of the instance). We noted that t he NL_x and CON_x instances are probably the most well-researched TTP benchmark and virtually all researches studying the TTP publish their computational results with these instances.

4.2 Parameters Tuning

We use a hill climbing algorithm to set the different parameters of V-HS method. This choice is motivated by its simplicity and by its free parameters propriety. The adjusting of the parameters is automated by using Hill Climbing method given in Algorithm 3. The setting parameters are as follows: ($HMCR \approx 0,85$, $PAR \approx 0.3 \ HMsize = 100, Max - Iter = 1000, bw = 0.1$).

Algorithm 3. Hill climbing approach to set the parameters

1: Generate an initial random values $HM, HMCR, Max - Iter, PAR, bw$
2: Apply $V - HS(HM, HMCR, PAR, bw)$, to create schedule S
3: Generate a set of $NB - h$ neighbors $(HM_i, HMCR_i, PAR_i, bw_i)0 \leq i \leq NB - h$
4: **for** $I = 1$ to $NB - h$ **do**
5:     Apply $V - HS(HM_i, HMCR_i, PAR_i, bw_i)$ to create schedule(S')
6:     **if** $Cost - travel(S') < Cost - travel(S)$ **then**
7:         $HM \leftarrow HM_i, HMCR \leftarrow HMCR_i$
8:         $Max - Iter \leftarrow Max - Iter_i, PAR \leftarrow PARi \ bw \leftarrow bw_i$
9:     **end if**
10: **end for**
11: Go to 2 until no improvement in Cost-travel

4.3 Numerical Results

Table 3 gives the results found by V-HS method on the different considered instances of NL (respectively of CON). The first column gives the name of instance, the second gives the lower bound for the considered instance, the third, the fourth and the fifth give the results found by V-HS method (respectively the minimum maximum and the average) and the column sixth and seventh and eighth give the running time (Minimum, maximum, Average). The last column give Gap % which is the gap between the best solution of our method and the Lower-bound [5].

Table 3. Results found by (V-HS) on the NL family

Instance	LB	Distance			Time(secs)			Gap%
		Min	Aver	Max	Min	Aver	Max	
NL4	8276	8276	8276	8276	5.03	10.20	16.33	**0 %**
NL6	-	26588	26723	26905	30.14	62.05	80.04	**0 %**
NL8	41928	41928	44051	45044	928.62	1201.01	1853.19	**0 %**
NL10	58190	63832	66079	66877	18401.64	2688. 19	3305.34	8.83 %
NL12	110519	121454	128871	130496	2702.21	3816.50	4094.04	9.00 %
NL14	182996	208140	211240	295693	6987.00	8833.13	106045.14	12.08 %
NL16	253957	284629	290188	308743	9544.71	9903.11	208014.63	10.77 %
CON4	17	17	17	17	2.05	9.87	20.11	**0 %**
CON6	48	48	48	48	15.13	30.03	50.44	**0 %**
CON8	80	80	80	81	531 .81	614.82	891.34	**0 %**
CON10	130	130	130	130	714.23	821. 15	1044.00	**0 %**
CON12	192	192	195	199	1299.14	14533.32	19855.30	**0 %**
CON14	252	253	254	256	2554.26	3105.07	3855.63	0.39 %

The proposed method succeeds in finding the optimal solution for NL4, NL6 and NL8 $CON_4, CON_6, CON_8, CON_{10}, CON_{12}, CON_{14}$ and CON_{16}.

Table 4. Comparative study with other methods for NL

Instance	V-HS	GA-SA [3]		BBO-SA [12]		CSA [2]		HLSA [18]	
	best	best	Gap%	best	Gap%	best	Gap%	best	Gap%
NL4	**8276**	8276	0	_	_	8276	0	_	_
NL6	**26588**	26588	0	_	_	26588	0	_	_
NL8	**41928**	43112	2.82	42804	2.08	41928	0	41928	0
NL10	**63832**	66264	3.81	66331	3.91	65193	2.13	63832	0
NL12	**121454**	120981	−0.38	121070	−0.31	120905	−0.45	120655	−0.66
NL14	**208140**	208086	0.02	210132	0.95	208824	0.32	208086	−0.02
NL16	**284629**	290188	1.95	291394	2.37	287130	0.87	280174	−1.56

For the rest of instances the gap is small and between 0.23 and 0,95 %. However the gap between optimal solutions and our results in our strategies do not go down in general below 12,08 % We observe that our results is close to Lower bound. In order to quantify this, we calculate the performance ratio(PR) given by the formula $PR = \sum_{NBins}^{i=1} GAP_i/NBins$, where GAP_i is the gap between solutions for an instance i and NBins is the number of instances for each benchmark. The performance ratio for NL benchmark is equal to PR(NL)=5,81 %. This result indicates that the proposed approach finds near optimal solutions with deviation from optimality equal to 5,81 %, PR(CON)=0,15 %. The general deviation from optimality value is 2.98 %. The superiority of our method is due to the good combination between VNS and harmony search that improve the intensification and diversification mechanisms in HS.

4.4   Comparison with Other Techniques

We conducted a further comparison to other new well-known hybrid approaches for mTTP in order to demonstrate the performance of our method in solving mTTP. We considered the following methods for comparison: Hybrid clustering (CSA) [2] that uses a hybrid heuristic called clustering search to solve the mTTP. BBO-SA [12] that combines both Biogeography Based Optimization and Simulated Annealing for mTTP. GA-SA [3] that uses a genetic algorithm in association with a simulated annealing for solving mTTP and some interesting methods proposed in [18] that combines both GRASP and ILS to solve mTTP. The comparison shows a good performance in favor of our method. V-HS algorithm succeeds in finding better results for most checked instances compared to the other methods given in Tables 4. We notice that for the NL instances, V-HS improves the results obtained by $BBO - SA$ with an average 1.28 % (using the Gap operator defined in formula 4), and improves those of CSA with 0,41 %, V-HS also gives better average than $GA - SA$ 1,16 % and our approach finds near solutions with deviation from [18] equal to 0,32 %.

$$GAP_{(V-HS,BBO-SA)} = \frac{Cost - travel_{BBO-SA} - Cost - travel_{V-HS}}{Cost - travel_{V-HS}} * 100... \quad (4)$$

5 Conclusion

The well-known NP-Hard TTP problem is difficult to solve optimally due to the problem formulation containing both optimization goal and integer feasibility constraints. This work proposes a novel hybrid approach of harmony search and variable neighborhood search heuristics to generate schedules for mirrored version of TTP. The proposed method (V-HS) starts with an initial population generated by using polygon method. Then we use a largest-order-value rule to transform harmony in harmony memory from real vectors to the abstract schedule and vice-versa so that the harmony search can be applied for TTP. Further, we apply the variable neighborhood search as an improvement strategy to enhance the solution quality and improve the HS intensification mechanism. The overall method is evaluated on some datasets and compared with the lower bound (LB). The results shows that our approach is able to find optimal solutions for small instances and the general deviation from optimality equal to 2,98 %. This is due to the good combination between harmony search based population meta-heuristic with VNS which permits to ensure a good compromise between intensification and diversification in the search space.

References

1. Bean, J.C.: Genetic algorithms and random keys for sequencing and optimization. ORSA J. Comput. **6**, 154–160 (1994)
2. Biajoli, F.L., Lorena, L.A.N.: Clustering search approach for the traveling tournament problem. In: Gelbukh, A., Kuri Morales, A.F. (eds.) MICAI 2007. LNCS (LNAI), vol. 4827, pp. 83–93. Springer, Heidelberg (2007)
3. Biajoli, F.L., Lorena, L.A.N.: Mirrored traveling tournament problem: an evolutionary approach. In: Sichman, J.S., Coelho, H., Rezende, S.O. (eds.) IBERAMIA 2006 and SBIA 2006. LNCS (LNAI), vol. 4140, pp. 208–217. Springer, Heidelberg (2006)
4. Carvalho, M.A.M.D., Lorena, L.A.N.: New models for the mirrored traveling tournament problem. Comput. Ind. Eng. **63**, 1089–1095 (2012)
5. Challenge Traveling Tournament Problems. http://mat.gsia.cmu.edu/TOURN/
6. Choubey, N.S.: a novel encoding scheme for traveling tournament problem using genetic algorithm. IJCA Spec. Issue Evol. Comput. **2**, 79–82 (2010)
7. de Werra, D.: Some models of graphs for scheduling sports competitions. Discrete Appl. Math. **21**, 47–65 (1988)
8. Easton, K., Nemhauser, G.L., Trick, M.A.: The traveling tournament problem description and benchmarks. In: Walsh, T. (ed.) CP 2001. LNCS, vol. 2239, pp. 580–584. Springer, Heidelberg (2001)
9. Gaspero, L.D., Schaerf, A.: A composite-neighborhood tabu search approach to the traveling tournament problem. J. Heuristics **13**, 189–207 (2007)

10. Geem, Z.W., Kim, J.H.: A new heuristic optimization algorithm: harmony search. Simulation **76**, 60–68 (2001)
11. Guedes, A.C.B., Ribeiro, C.C.: A heuristic for minimizing weighted carry-over effects in round robin tournaments. J. Sched. **14**, 655–667 (2011)
12. Gupta, D., Goel, D., Aggarwal, V.: A hybrid biogeography based heuristic for the mirrored traveling tournament problem. In: 2013 Sixth International Conference on Contemporary Computing (IC3), pp. 325–330. IEEE, Noida (2013)
13. Hansen, P.: Mladenovi, N.: Variable neighborhood search: principles and applications. Eur. J. Oper. Res. **130**, 449–467 (2001)
14. Irnich, S.: A new branch-and-price algorithm for the traveling tournament problem. Eur. J. Oper. Res. **204**, 218–228 (2010)
15. Khelifa, M., Boughaci, D.: A variable neighborhood search method for solving the traveling tournaments problem. Electron. Notes Discrete Math. **47**, 157–164 (2015)
16. Qian, B., Wang, L., Hu, R., Huang, D.X., Wang, X.: A DE-based approach to no-wait flow-shop scheduling. Comput. Ind. Eng. **57**, 787–805 (2009)
17. Rasmussen, R.V., Trick, M.A.: The timetable constrained distance minimization problem. Ann. Oper. Res. **171**, 45–59 (2008)
18. Ribeiro, C.C., Urrutia, S.: Heuristics for the mirrored traveling tournament problem. Eur. J. Oper. Res. **179**, 775–787 (2007)
19. Thielen, C., Westphal, S.: Complexity of the traveling tournament problem. Theor. Comput. Sci. **412**, 345–351 (2011)
20. Westphal, S., Noparlik, K.: A 5.875-approximation for the traveling tournament problem. Ann. Oper. Res. **218**, 347–360 (2012)

Web Systems and Human-Computer Interaction

Utilizing Linked Open Data for Web Service Selection and Composition to Support e-Commerce Transactions

Nikolaos Vesyropoulos[1]([⊠]), Christos K. Georgiadis[1], and Elias Pimenidis[2]([⊠])

[1] University of Macedonia, Thessaloniki, Greece
{nvesyrop,geor}@uom.edu.gr
[2] University of the West of England, Bristol, UK
elias.pimenidis@uwe.ac.uk

Abstract. Web Services (WS) have emerged during the past decades as a means for loosely coupled distributed systems to interact and communicate. Nevertheless, the abundance of services that can be retrieved online, often providing similar functionalities, can raise questions regarding the selection of the optimal service to be included in a value added composition. We propose a framework for the selection and composition of WS utilizing Linked open Data (LoD). The proposed method is based on RDF triples describing the functional and non-functional characteristics of WS. We aim at the optimal composition of services as a result of specific SPARQL queries and personalized weights for QoS criteria. Finally we utilize an approach based on the particle swarm optimization (PSO) method for the ranking of returned services.

Keywords: Web services · Linked Open Data · RDF

1 Introduction

The Service Oriented Architecture (SOA) paradigm provides the means for the removal of interoperability barriers between heterogeneous systems. As a result, it has been widely adopted by enterprises for the fulfillment of their business needs. WS can both be used as stand-alone components or as parts of a value added composition, comprising of a number of (internal or external to the enterprise) services. Through the reuse of existing WS as parts of a composition, an enterprise can lower the cost of complex operations and adjust to specific business needs that rise during online transactions with partners or customers [8].

For the discovery and composition of WS, semantic rules and ontologies have widely been used over the past few years. Through the application of semantic technologies, metadata describing the functionality and properties of WS are utilized for the automated processing of orchestration rules, thus resulting in the rapid development of compositions.

A recent approach towards the identification of WS which fulfill specific requirements, as well as their selection and composition is the use of LoD [7]. This method,

© Springer International Publishing Switzerland 2016
N.T. Nguyen et al. (Eds.): ICCCI 2016, Part I, LNAI 9875, pp. 533–541, 2016.
DOI: 10.1007/978-3-319-45243-2_49

which has been widely adopted in various fields, refers to the utilization of structured data that can be machine understandable [13]. An important aspect of LoD is that data share connections, based on semantic rules. Through these rules linked data form RDF graphs, representing the ties between the interlinked data. Such data can be identified and retrieved by issuing specific queries through an appropriate querying language, such as SPARQL.

In this work, we describe a proposed framework for the selection and composition of WS. The main contribution of our approach lies within the architecture of the theoretical framework that utilizes RDF-based repositories, combined with a novel representation of WS related data as RDF graphs. In addition, we demonstrate a methodology for integrating a performance-based ranking algorithm, as a step towards the automation of WS compositions.

This work is organized as follows: In Sect. 2, we analyze related works and attempt to highlight the benefits of utilizing LoD for the optimal WS selection and composition. In Sect. 3, the proposed methodology is analyzed and a framework is demonstrated. In the following section, a case study is presented, based on the proposed framework and pertaining to the composition of services in the context of e-commerce transactions. Finally we conclude with a discussion on the lessons learned and point towards future research and areas of interest.

2 Related Work

The notion of utilizing LoD in the process of the discovery, selection and composition of WS has recently gained the attention of researchers. Nonetheless, in contrast to the corresponding semantic approaches, no specific standards have been established and the optimal utilization of this approach is still being debated.

The work presented in [16] pertains to the examination of the benefits of a LoD-based composition framework for RESTful services. The author discusses the similarities between LoD and RESTful services as both technologies rely on URI-described resources and highlights the potential of a proposed composition approach, though an implementation is lacking. The construction of value-added services based on RESTful resources, through the utilization of LoD is also examined in [12]. The authors discuss a methodology, where services are identified using graph patterns which also aid the construction of the appropriate RDF messages for the exchange of data between WS. This is a promising step towards automated compositions, but no elaboration is presented regarding the problem of selecting among WS with similar functionalities, which is a focal point in our approach.

In [14] the concept of Linked Open Services is presented, which is based on the utilization of LoD, RDF and SPARQ queries. Their approach is aiming at the discovery of WS that are described through interlinked data. In addition, in [15], authors elaborate on the applicability of SPARQL graph patterns for the description of such services. In more detail, they analyze the behavior of registries that are compatible with RESTful calls and LoD, and proceed to the implementation of such an approach using Hadoop. While authors also discuss composition methodologies, the focus of their approach is on the discovery of WS through RFD-based data.

In [10], authors describe the application of a LoD-as-a-service architecture. They highlight the challenges of applying domain specific datasets in a LoD format and developing appropriate architectures for the retrieval of these data. A dataset regarding WS descriptions can be added into such an implementation, thus providing WS discovery capabilities. We build on this notion, as we intent to provide enterprises with a methodology for the selection and composition of WS based on RDF descriptions. Our approach is differentiated, as we also include a methodology for the selection of optimal service compositions between a set of alternatives, utilizing those descriptions.

In [6], a service identification methodology is proposed, which enables the application of LoD and semantic links for the discovery of APIs. Authors describe the application of such an approach and demonstrate the affectability on two popular API repositories. While not tackling the problem of automated WS composition, it provides a cross-platform WS searching methodology.

Finally, in [1], Linked-OWL a modification to the OWL-S language is proposed which can describe both ontologies and LoD data. The proposed language takes advantage of REST-based services and is a promising approach towards the composition and dynamic reallocation of WS utilizing LoD. The approach, however, lacks a concrete selection mechanism. We believe that the integration of a ranking algorithm, as presented below, is an important step in the overall procedure, providing a performance-oriented solution.

3   Proposed Methodology

In this work, we propose a conceptual framework for the discovery of business oriented services, provided by a number of enterprises through an appropriate repository, based on descriptions stored in RDF format. Those descriptions provide details pertaining to both functional and non-functional characteristics and can be accessed through the corresponding SPARQL requests. Through the application of a proposed algorithm, the selection and composition process of these services can be aided. In such an approach, information regarding the available WS must be converted in an RDF-based format, such as the following tuple structure:

{WS_Name : Property1 : Value}
{WS_Name : Property2 : Value}
...
{WS_Name : PropertyN : Value}

By converting the aforementioned information into an RDF format, an RDF graph can be created, modeling all the available WS, their functional and non-functional characteristics and their interconnections. Such a graph enables the monitoring of the overall data and the identification of requested information, through appropriate queries.

The proposed framework includes a LoD-based repository containing descriptions of all available WS. Using a SPARQL endpoint the end-user is capable of issuing queries regarding to specific operations while a number of requested QoS characteristics can also be included in such a query. The framework's overall architecture is depicted in Fig. 1.

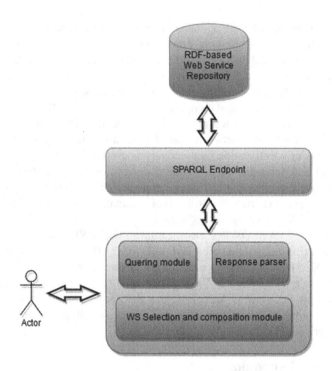

Fig. 1. The proposed framework architecture

An important module in the framework is the response parser. SPARQL responses can be in a number of predefined formats (such as XML, JSON, CSV and TSV) and the retrieved information will then be used as input for the composition methodology. In order to adhere to this, the parser is responsible for the transformation of the response to an appropriate format.

After the aforementioned transformation, a ranking of alternative solutions is required for the identification of the optimal composition. A number of researchers have utilized the PSO algorithm to adhere to similar problems in a variety of domain specific scenarios. The application of PSO for solving the WS selection and composition problem, based on QoS characteristics, is feasible as shown in [9, 18].

In this work we apply this algorithm as a part of the selection process, nonetheless other selection algorithms can also be applied, such as multicriteria decision analysis methods [17] or constraint algorithms [4]. Compared to these approaches, the application of the PSO algorithm is less effort demanding, while providing a more performance-oriented solution. Multicriteria and constraint methods, on the other hand, can be used in compositions where a higher level of personalization is required.

The PSO algorithm, was introduced in [11] and is described by authors as an algorithm simulating a social model, and in particular the behavior of bird flocking and fish schooling.

Two important values in the algorithm are the personal best and the neighborhood best, p_{best} and p_{gbest} accordingly. Each simulated particle (corresponding to a member of the swarm and to a potential candidate solution), is attracted to these two values, as it tries to identify a position better than its present. This attraction influences the particle's position and current velocity.

By a random alteration of the p_{best} and p_{gbest} values, all particles are in constant movement trying to converge to the point of the optimal values. Each repetition of this procedure represents in iteration i. For each iteration, the updated velocity v and current position x of a particle is being calculated according to the following formulas:

$$v(t+1)_i = wv(t)_i + c_1 r_1 \left(p(t)_{best} - x(t)_i\right) + c_2 r_2 \left(p(t)_{gbest} - x(t)_i\right) \qquad (1)$$

$$(t+1)_i = (t)_i + v(t-1)_i \qquad (2)$$

where w is the inertia weight, a variable changing values from a maximum of 0.9 to a minimum of 0.4, thus controlling the influence of a particle's current velocity to the one it will gain in the following iteration. In addition c_1 and c_2 pertain to the cognition and social weights, allocated to the p_{best} and p_{gbest} values accordingly while r_1 and r_2 refer to random values in the [0, 1] scale. After a number of iteration all particles are drawn near to the optimal values [20].

In order to minimize the time required for the selection and composition of requested services, we propose the following series of steps that include the application of PSO:

1. Execute SPARQL query regarding the requested functional characteristics
2. If the number of results is higher than one, issue request regarding to non-functional characteristics
3. Retrieve a list of appropriate compositions and the values of non-functional characteristics in the predefined format
4. Parse the list and construct the utility function
5. Fetch the above to the PSO method
6. Return the optimal composition along with the corresponding WSDL files.

4   Case Study

We demonstrate the effectiveness of the proposed work through the application of a case study. We have used the dataset described in [2, 3] which includes a number of real world WS and a value for their QoS attributes. We selected a subset of those services which can be used in the context of e-commerce transactions and converted their description and QoS values into appropriate RDF tuples. In real-life applications, the values of QoS parameters can be applied by the RDF-based repository. They can be attained by closely monitoring the behavior along with the invocation and response times of a WS (for parameters such as availability or response time), or can be assigned manually according to the corresponding attributes (for properties such as best practices and documentation). The structure of an e-commerce service in the converted RDF dataset is as follows:

- {WS1 : Functionality : OrderProductService}
- {WS1 : Reliability : 0.58} *(value is in a scale of 0–1)*
- {WS1 : Resp_time : 0.24275} *(value in seconds)*
- {WS1 : Best Practices : 0.84} *(value is in a scale of 0–1)*
- {WS1 : Documentation : 0.38} *(value is in a scale of 0–1)*

The sum of all involved services comprise the RDF graph. Using the appropriate SPARQL queries a number of services can be retrieved by the end-user, that fulfill his given functional requirements. Utilizing the series of steps provided in the previous section, a test case is presented below, as proof of concept of the proposed framework. Suppose that an end-user request a composition for an e-commerce application. The user requires services for product ordering, credit card validation and payment, as well as a sms service in order to be informed of any alterations in his order's state. After issuing the request, and based on the modeled RDF dataset, a list of 96 potential compositions is returned. The user is then prompted to select his requested values in the following QoS characteristics: Reliability, Response Time, Best Practices, and Documentation.

Suggesting that all services follow a serial execution path, the overall Response time, based on [19, 20], is:

$$RT_o = \sum_{i=1}^{n} RT_i \tag{3}$$

The corresponding best practices value is:

$$Bpr_o = \prod_{i=1}^{n} Bpr_i \tag{4}$$

The score pertaining to documentation is calculated based on:

$$Doc_o = \prod_{i=1}^{n} Doc_i \tag{5}$$

While the overall reliability, based on [5] is:

$$Rel_o = \prod_{i=1}^{n} Rel_i \tag{6}$$

Regarding the value of reliability, a multiplication aggregation type is applied, as in complex compositions a malfunction in one of the involved WS can greatly affect and compromise the whole value-added service. Thus, the overall reliability should be the product of the corresponding values of each involved service, instead of a simple mean. We opted to use a multiplication aggregation for the calculation of the overall documentation and best practices value as well, following the logic above, as those properties could play a pivotal role in the selection of services for some users.

By modeling the requested problem the optimal service composition can be returned. As a starting point, after the parsing of the response, the objective function must be constructed. As a serial execution flow is presumed this function receives the following form:

$$Of = Min_{((\frac{1}{Bpr_o}) + (\frac{1}{Doc_o}) + (\frac{1}{Rel_o}) + RT_o)} \tag{7}$$

In more detail, by inverting the values of the best practices, documentation and reliability attributes (where the higher the scores the better) and maintaining the state of the response time parameter, a minimum function is created. The optimal service composition, among the alternatives, is the one that produces the minimum value.

For the needs of this simplified case study, the values of the first three parameters were added and stored into a matrix (representing their summed value in all possible compositions), while a similar matrix was used for the response time parameter. The result of executing the, now less complex, objective function is shown Fig. 2.

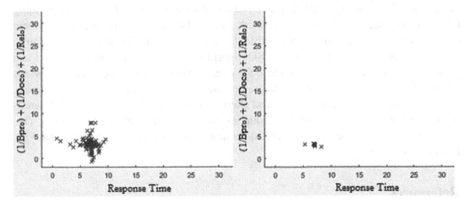

Fig. 2. Swarm converges to optimal values after 15 (left) and 30 (right) iterations

As demonstrated in the figure below, the velocity and position of each particle is altered between iterations, as particles gradually converge towards the solution. Using the returned values, a composition engine can easily determine which WS best fulfils the user's business needs. The number of required iterations and the overall execution time is based on the complexity of the objective function.

In this case study, most particles point towards the optimal composition after 30 iterations, proving both the feasibility of utilizing the PSO algorithm in the context of WS compositions, as well as its ability to provide results in reasonable time frames. In addition, the algorithm does not require as much human input as other similar approaches, such as multicriteria methods, that are more effort-demanding. While the case study demonstrates a scenario involving a few possible alternatives, in real life business scenarios a large number of WS can be involved and the size of required calculations rises exponentially. In order to provide scalability to the overall composition, the PSO algorithm can be used to enhance the performance of a composition engine, especially when there is a need for dynamic alteration of involved services (often a necessity in the rapidly changing business environments). In addition, the PSO algorithm provides users with the much desired flexibility to differentiate the composition, even in real-time, by using personalized user weights. Combining such an algorithm with service repositories that utilize the, machine understandable, RDF format is a step towards the automation of WS compositions and the minimization of required human intervention.

5 Conclusion and Future Work

We have demonstrated an approach for the selection and composition of WS, based on LoD and SPARQL endpoints. Our approach combines the benefits of utilizing semantic descriptions as well as those offered by the structured nature of RDF data. Through the application of a case study, we examined the feasibility of the proposed methodology and the integration of a performance-based ranking algorithm, with promising results. Future work involves the implementation of the framework along with a corresponding GUI, utilizing the Apache Jena framework and the Fuseki server for the development of SPARQL endpoints that are accessible through HTTP calls. In addition, we plan to investigate modified versions of the PSO algorithm in order to support both serial and parallel execution flows and to enhance the overall composition performance. Towards this direction, we plan to evaluate the framework in more complex business scenarios, involving numerous execution tasks and alternative services. Finally, we work towards the automated generation of SPARQL queries, by transforming plain text requests. This will allow the utilization of the proposed framework by non-programmers and ease the overall procedure.

References

1. Ahmad, H., Dowaji, S.: Linked-OWL: a new approach for dynamic linked data service workflow composition. Webology 10(1), 21–30 (2013)
2. Al-Masri, E., Mahmoud, Q.H.: Discovering the best web service. In: 16th International Conference on World Wide Web, pp. 1257–1258 (2007)
3. Al-Masri, E., Mahmoud, Q.H.: QoS-based discovery and ranking of web services. In: IEEE 16th International Conference on Computer Communications and Networks, pp. 529–534 (2007)
4. Alrifai, M., Risse, T., Nejdl, W.: A hybrid approach for efficient web service composition with end-to-end QoS constraints. ACM Trans. Web 6(2), 1–31 (2012)
5. Alrifai, M., Skoutas, D., Risse, T.: Selecting skyline services for QoS-based web service composition. In: 19th International Conference on World Wide Web, pp. 11–20 (2010)
6. Bianchini, D., De Antonellis, V., Melchiori, M.: A linked data perspective for effective exploration of web APIs repositories. In: Daniel, F., Dolog, P., Li, Q. (eds.) ICWE 2013. LNCS, vol. 7977, pp. 506–509. Springer, Heidelberg (2013)
7. Chen, W., Paik, I.: Improving efficiency of service discovery using linked data-based service publication. Inf. Syst. Front. 15(4), 613–625 (2013)
8. Fahad, M., Moalla, N., Ourzout, Y.: Dynamic execution of a business process via web service selection and orchestration. Procedia Comput. Sci. 51, 1655–1664 (2015)
9. Kang, G., Liu, J., Tang, M., Xu, Y.: An effective dynamic web service selection strategy with global optimal QoS based on particle swarm optimization algorithm. In: 26th IEEE International Parallel and Distributed Processing Symposium Workshops and Ph.D. Forum, pp. 2280–2285 (2012)

10. Kennedy, J., Eberhart, R.: Particle swarm optimization. IEEE Int. Conf. Neural Netw. **1995**, 1942–1948 (1995)
11. Kim, S., Berlocher, I., Lee, T.: RDF based linked open data management as a DaaS platform. In: First International Conference on Big Data, Small Data, Linked Data and Open Data, pp. 58–61 (2015)
12. Krummenacher, R., Norton, B., Marte, A.: Towards linked open services and processes. In: Berre, A.J., Gómez-Pérez, A., Tutschku, K., Fensel, D. (eds.) FIS 2010. LNCS, vol. 6369, pp. 68–77. Springer, Heidelberg (2010)
13. Neubert, J.: Linked data based library web services for economics. In: International Conference on Dublin Core and Metadata Applications, pp. 12–22. Dublin Core Metadata Initiative (2012)
14. Norton, B., Krummenacher, R., Marte, A., Fensel, D.: Dynamic linked data via linked open services. In: Workshop on Linked Data in the Future Internet at the Future Internet Assembly, pp. 1–10 (2010)
15. Norton, B., Stadtmüller, S.: Scalable discovery of linked services. In: 4th International Workshop on Resource Discovery, pp. 6–21 (2011)
16. Stadtmüller, S.: Composition of linked data-based RESTful services. In: Cudré-Mauroux, P. (ed.) ISWC 2012, Part II. LNCS, vol. 7650, pp. 461–464. Springer, Heidelberg (2012)
17. Vesyropoulos, N., Georgiadis, C.K.: QoS-based filters in web service compositions: utilizing multi-criteria decision analysis methods. J. Multi-criteria Decis. Anal. **22**(5–6), 279–292 (2015)
18. Wang, S., Sun, Q., Zou, H., Yang, F.: Particle swarm optimization with skyline operator for fast cloud-based web service composition. Mobile Networks Appl. **18**(1), 116–121 (2013)
19. Wang, W., Sun, Q., Zhao, X., Yang, F.: An improved particle swarm optimization algorithm for QoS-aware web service selection in service oriented communication. Int. J. Comput. Intell. Syst. **3**(1), 18–30 (2010)
20. Zhang, T.: QoS-aware web service selection based on particle swarm optimization. J. Networks **9**(3), 565–570 (2014)

Improved Method of Detecting Replay Logo in Sports Videos Based on Contrast Feature and Histogram Difference

Kazimierz Choroś[(✉)] and Adam Gogol

Faculty of Computer Science and Management, Wrocław University of Science and Technology, Wyb. Wyspiańskiego 27, 50-370 Wrocław, Poland
kazimierz.choros@pwr.edu.pl

Abstract. Sports videos are probably the most popular videos and the most frequently searched in the Web. Similarly to text documents for which abstracts can be automatically generated we need to automatically summarize the long videos presenting for example the whole soccer matches. The obvious solution to summarize a sports video is to detect and to extract replay segments as highlights which usually present the most interesting and exciting parts of a video containing important players and actions. Replays can be detected if replay shots are separated by special logo animations or replays are in slow motion. One of the automatic method of logo transition detection is a method based on contrast feature and histogram difference. The paper presents the improved method of replay logo detection and the results of the tests demonstrating the benefits of the proposed improvements.

Keywords: Content-based video indexing · Video summarization · Sports video highlights · Replay detection · Replay logo templates · Logo transition detection

1 Introduction

Content-based video indexing is a great challenge in computer science. The amount of digital videos stored and browsed in Web video collections, TV shows archives, documentary video archives, video-on-demand systems, personal video archives, etc. is very huge. Many indexing algorithms, techniques, approaches, solutions, and frameworks have been proposed for specific cases and for specific kinds of videos (see for example [1] or [2]). The automatic recognition of video structure is a basic process of video analyses. Content-based indexing of videos is based on the automatic detection of shots and scenes, i.e. of the main structural video units. Digital video is hierarchically structured. The great analogy [3] is observed between the structural elements of a text and the structural elements of an edited video. A video can be considered as a visual representation of a book. It should be noticed that a book is usually a sequential text. Traditional video has also a sequential nature, although semantically the video scenes of the same episode are not necessarily placed one after another in a given video. In a written text in a novel as well as in a movie two or several episodes can be

© Springer International Publishing Switzerland 2016
N.T. Nguyen et al. (Eds.): ICCCI 2016, Part I, LNAI 9875, pp. 542–552, 2016.
DOI: 10.1007/978-3-319-45243-2_50

presented in alternation. Similarly to text indexing and retrieval methods, we would like to be able to automatically index videos and to retrieve videos as efficiently as we can do it in the case of textual documents. However, the analyses of texts leading to the automatic generation of abstracts as well as to the extraction of key words used in retrieval process are relatively well managed nowadays. Whereas, the analogous process with digital videos, i.e. video indexing and summarization are not easy to perform. The processes of content-based analyses of videos are significantly more time-consuming, so in consequence their effectiveness as well as their efficiency are not yet satisfactory. Video summarization consists in detecting these parts of video which are the most representative for the whole videos. These are for example the highlights of sports events. The most interesting parts of sports videos are mainly those which are usually replayed in real time and which are broadcasted during the game. Replay video segments can be detected because they are broadcasted in slow motion or they begin and end with an overlapped digital animation called replay logo.

The characteristic of a given sports video strongly depends on the sports category. Some sports videos are characterized by a very dynamic background, others by a static background, close-up view of players, in-field medium view, far view, or out of field view of the audience, small or great objects of foreground, homogeneous type of playing field with one dominant colour or very diversified field. Also, the highlights for different sports categories are of different nature. The automatic summarization of sports videos is not simple. One of the practical solutions is to detect and to extract replay segments as highlights which usually present the most interesting and exciting parts of a video containing important players and actions.

There are four types of replays observed in sports videos [4]: replay of the same shot(s) recorded with the same camera (not in use in nowadays sports event broadcast), replay of the same shot(s) recorded with the same camera but in slow motion, replay of the same action(s) recorded with another camera, and replay of the same action(s) recorded with another camera and moreover in slow motion. The last two types of replays that is the replay of a given action recorded with another camera are not detectable. May be in the near future it will be possible but it will require very sophisticated techniques of action analyses with the recognition of players and their behaviour in a registered sports event. Nowadays only the replays of the same shots are detected unless the replay segments are separated by replay logo animations or replays are in slow motion. At that time there are two major approaches of replay detection: one is based on the detection of slow motion shots or scenes in a video, and the other is based on the detection of replay logo – logo transition detection.

To localize highlights in sports videos we need to automatically detect replay segments. Many methods of replay detection have been proposed. However, they are not efficient enough. In this paper an improved method of replay logo detection in sports videos is presented. The paper is structured as follows. The next section describes related work on replay detection. The method of a replay logo detection based on contrast feature and histogram difference is briefly described in the Sect. 3. The Sect. 4 discusses the influence of thresholds values used in this replay detection method. The improved method of replay logo detection and the results of the tests demonstrating the benefits of proposed improvements are presented in the Sect. 5. The final conclusions are discussed in the last section.

2   Related Work

A brief and categorized overview of video summarization techniques has been prepared in [5]. Also, a hierarchical classification of these techniques has been proposed. Video summarization techniques are based on features, clustering, event detection, shot selection, mosaic, and trajectory analysis. The first group of video summarization techniques based on feature was classified on the basis of motions, colours, dynamic contents, gestures, audio-visual features, speech transcript, objects. Video summarization based on clustering was classified into similar activities, K–means, partitioned clustering, and spectral clustering. The third group of techniques based on the detection of events is suitable for videos mostly captured from static camera. The next group of techniques which is based on shot selection consists in the determination of the most important shots in a video. The fifth kind of summarization techniques based on motion analyses is mainly applied for videos recorded in the surveillance systems where cameras are fixed and backgrounds do not change. In this case, the summarization consists in the analyses of behaviour of moving objects. And finally the last group of techniques is a group of summarization techniques generating a panoramic image from a large number of consecutive frames.

Many new proposed methods have been tested with soccer videos. In the case of a given sports category the highlights can be identified not only with replay shots but also with specific actions like goals in soccer [6]. Other authors emphasized that proposed solutions are general and adequate to most categories of sports game and that replays are a reliable clue to the highlight and the features they utilize are not limited to a specific sports category [7]. Some studies have been conducted on baseball videos [8], but the approach is applicable to other sports categories.

Replay extraction based on the detection of video segments broadcasted in slow motion has been proposed by many authors [4, 9–11]. However, not all replay shots are always slowed down. It happens that the rest of a replay scene is played in normal speed. Hence, the methods based only on slow motion detection could detect the boundary of replay scenes not accurately.

The detection of replays can be also performed if the logo transition is found. Logo transitions are detected similarly to other special digital video effects and they are inserted at the beginning and end of a replay scene [12]. Replay shots are usually placed between two replay logos. The automatic recognition of replay logo is difficult to locate [13] because the time of logo transition can be random, the position of logo in frame can be random, the shape of logo can be random and changing, and farther the texture and colour of logo can be random. In the first solutions replay logo was mark manually or semi-automatically [13]. First, logo template sequence is extracted, and then the replay segments are located by using the logo template sequence match method. Also automatic extraction of logo templates has been proposed [14–16].

In many approaches a combined solution has been proposed based on slow motion detection as well as logo transition detection to identify replay shots. In [17] logo template frames were found by extracting frames surrounding the detected slow motion shots. Then, replay shots were identified by grouping the detected logo frames and slow-motion segments. In [18] a general framework based on a Bayesian network has been proposed using different techniques, including shot structure, gradual transition pattern, slow motion, and sports scene detection. The other framework [19] used both inherent video characters and transition relations of replay and non-replay scenes based on annotation of the video. First, replay and non-replay shots were detected and simultaneously segmented basing on statistical inference and using a hidden Markov model and bi-grams. Then, the detected replay shots were further verified and their boundaries were adjusted to get more accurate replay segment considering probability distribution of lengths of replay and non-replay shots.

Whereas, in [20] replay detection method using scene transition structure analysis of broadcast soccer video has been tested and its computational efficiency has been shown. However, the method requires to predefine scene structures as well as to classify each shot into respective view type.

3   Detection of Replay Logo Transition

Several methods of detection of replay logo transitions have been tested in the AVI Indexer System [21]. One of the methods tested and implemented is the method presented in [22]. There are four main steps of this method of detecting replays: finding the set of candidates for logo images, calculating the contrast logo template, finding the beginning and the end of logo transition, and finally matching logo pairs.

Finding the candidates for replay logo is based on the observation that logo transitions contain logos of broadcasters or game organizers. In many sports games the logo colour is usually in strong contrast to the rest of a frame. Hence, for each frame, the contrast value is calculated, which is defined as the ratio of white pixels in the binary image of frame. To reduce noise, all the grass pixels belonging to the field (the soccer matches have been analyzed) are set to black.

A frame is added into the set of candidate logo images if its contrast value is greater than the defined threshold – contrast threshold. However, some wrong candidate logo images can be classified into the candidate set due to the player's bright colour and the appearance of white goal post, etc. Therefore, all frames classified as candidate logo should be verified by checking their histograms. A candidate is excluded if its histogram is not sufficiently different from the histograms of the images 15 frames before or from the histograms of the images 15 frames further, that is if the histogram difference is lower than a given threshold – histogram difference threshold. Logo transitions may contain a sequence of consecutive images that also have high contrast values. Hence, for each candidate, if there are also candidates in the next 15 frames, only the candidate with the largest contrast value is kept whereas all remaining candidates are discarded.

Then the contrast logo template Ic in the form of a binary image of the average image is calculated by averaging candidate logo images. For each candidate i in the template set S, the contrast difference of binary image Ii and logo template Ic of the width w and height h is defined as follows:

$$Diff_{contrast}(I_i, I_C) = \frac{1}{w*h} \sum_{u=0}^{w} \sum_{v=0}^{h} \frac{|I_i(u,v) - I_C(u,v)|}{255} \qquad (1)$$

The candidate will be discarded from the candidate set if the difference is greater than a given threshold – contrast logo template difference threshold (the authors of the method proposed 10 % of total pixels).

The beginning of logo transition is established by comparing the average contrast difference calculated for frames from $(i–k)$ to $(i–k + m)$ for each candidate logo image (the authors of the method proposed k = 20 and m = 5).

$$Diff_{Avg} = \frac{\sum_{l=i-k}^{i-k+m} |r_w(l+1) - r_w(l)|}{m} \qquad (2)$$

Then the difference of contrast values r of two successive frames was calculate. If it was greater than 0.5 of average contrast, the frame was marked as the beginning of logo transition. The same procedure was carried out for finding the end of logo transition, but the frame was examined from $(i + k)$ frame back to frame i.

The authors of the method observed that replay segment was usually approximately less than 800 frames. If the number of frames from current logo image to next logo image was larger than 800 frames, current logo image was considered as faulty detection, and it was discarded.

4 Setting the Thresholds

In the tests performed in the AVI Indexer 10 full matches have been tested. The results obtained for the tested 10 videos were not satisfactory. Moreover, some additional observations have been made. First of all, it happened seven times in these videos that the replay parts have been preceded by a replay logo but they have not ended by any replay logo as it could be expected. So, the replay segments could not be identified by pairing the beginning and the end of logo transition. Then, four replays have been accompanied by other replay logo comparing to the standard logo in a given video. Such inconsistencies make the method inefficient for these cases.

And finally, this method applied to basketball, ski jumping, or hockey match videos is not effective because of relatively bright or even white background. It leads to the conclusion that the detection of playing filed colour in the first step of this procedure is crucial [23].

Three parameters should be set for the method applied in the experiment: minimum contrast difference, minimum histogram difference, and maximum logo template difference. Unfortunately, the values of parameters used in the tests described in [22] were not adequate for videos tested in the AVI Indexer. Mainly the value of parameter to calculate the average contrast should be much smaller. Same as the threshold value of the histogram difference. It means that these parameters should be set and optimized for a given video. This is a strong inconvenience of the method.

And then it is very reasonable to introduce not only the maximum length of a replay segment but also a minimum length. It has been observed that replays are long at least a few seconds. In our experiments we assumed that the maximum length of replays is 40 s and their minimum length is 2 s.

To set the necessary thresholds used in this method a series of tests have been carried out in the AVI Indexer. The results of these tests helped to optimally set the three main parameters: contrast threshold – 20 %, the threshold of histogram differences – 25 %, and the contrast logo template difference threshold – 25 %. Despite that, the results were still not satisfactory (Table 1).

Table 1. Videos used in the experiments.

Video	Number of replays	Recall [%]
A Soccer game 2013	38	92.10
B Soccer game 2013	21	90.48
C Soccer game 2014	24	66.70
D Soccer game 2014	20	85.00
E Soccer game 2014	27	92.59
F Soccer game 2015	24	100.00
G Hockey match 2010	16	0
H Hockey match 2014	19	0
I Basketball match 2014	26	96.15
J Basketball match 2014	16	0
Total: 7 h 12 min	231	62.30

Let us to present the tests results helping to set the threshold values. To set the value of the contrast threshold parameter several values have been tested (Table 2). The optimal value of contrast threshold is about 20–25 %.

Table 2. Contrast threshold tests

Contrast threshold [%]	10	15	20	25	30
Replays detected	215	176	146	150	155
Correct detections	13	100	141	147	145
False detections	202	76	5	3	10
Recall [%]	5.63	43.29	61.04	63.64	63.20
Precision [%]	6.05	56.82	96.52	97.35	93.55
F-measure [%]	5.83	49.14	74.80	77.17	77.04

To establish the value of the histogram difference threshold parameter several values of histogram difference threshold have been then tested (Table 3).

The optimal value of histogram difference threshold is about 25–30 %. The best result of recall is obtained for the threshold equal to 25 %, whereas the best results of

Table 3. Histogram difference threshold tests.

Histogram difference [%]	15	20	25	30	35
Replays detected	147	146	165	158	143
Correct detections	122	141	161	155	139
False detections	25	5	4	3	4
Recall [%]	52.81	61.04	69.70	67.10	60.17
Precision [%]	82.99	96.52	97.58	98.10	97.20
F-measure [%]	64.55	74.80	81.31	79.69	74.33

precision is obtained for the threshold equal to 30 %. Because the recall seems to be the most important the optimal threshold value is 25 %.

To set the value of the contrast logo template threshold parameter its several values have been also tested (Table 4).

Table 4. Contrast logo template threshold tests.

Contrast logo template difference [%]	20	25	30
Replays detected	156	165	182
Correct detections	155	161	169
False detections	1	4	13
Recall [%]	67.10	69.70	73.16
Precision [%]	99.36	97.58	92.86
F-measure [%]	80.01	81.31	81.84

The optimal value of contrast logo template difference threshold is about 25 %. Although, the increase of this threshold to 30 % improves the recall, however, the precision has significantly decreased.

5   Detection Improvements

The analysis of tests with threshold values have signalled that the specificity of logo image has a great influence on the efficiency of replay detections. Bright logo, transition of more than two seconds, dark background, i.e. dark playing field, facilitate replay logo detection. Reply logo was not detected when logo image was relatively small not covering the whole frame, so in consequence the contrast was not significant, or the logo transition was too fast, very dynamic.

Wrong results of the method are also due to the fact that the logo template is generated using a very simple technique based on averaging logo candidates. It is the case mainly if between logo candidates there are many frames not belonging to logo transition or if a background is far more dominant in the frame than appearing logo. Two proposed remedies are as follows.

First of all the logo template can be generated iteratively. In each iteration step a new template is generated but at the same time the accepted difference between the template and logo candidate, i.e. the contrast difference threshold is decreased.

The proposed and tested values were 35, 30, and 25 %. If any logo candidate is eliminated a new template is generated and the procedure is restarted. In the tests performed on the video H nine iterations were completed resulting in the reduction of the number of logo candidates from 92 to 24. But the most important was that the contrast logo template was much more similar to the binary image of real replay logo comparing to the simple average of logo candidates (Fig. 1).

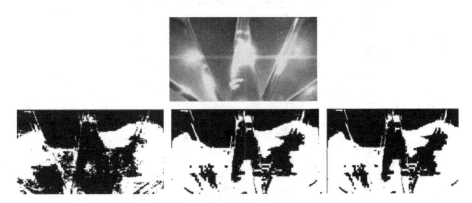

Fig. 1. Logo frame and binary logo templates generated by the original algorithm, improved algorithm using iterations, and improved algorithm taking into account logo candidate similarities.

The second proposed alternative solution is based on the assumption that correctly detected transition logos are similar to one another, whereas false detections are different from each other. Each candidate is compared with each other from the set of logo candidates and for each candidate we calculate how many times it is sufficiently similar (75 % of pixels – formula 1 is applied) to any other candidate. The logo template is generated using only those candidates which are similar to more than the average received for all logo candidates.

The iterative algorithm does not work well for low values of thresholds. It can be applied when the contrast threshold as well as the histogram threshold parameters are set close to optimum values.

Whereas, the logo template generation taken into account similarities between logo candidates significantly improves the recall as well as precision of replay detection for different values of thresholds (Table 5).

It should be also noticed that these proposed improvements has unfortunately an influence on the processing time. The processing time may increase even by one third.

Table 5. Efficiency of replay detections using the original and improved method of logo template generations.

Contrast threshold	15			20			20		
Histogram threshold	20			20			25		
Logo template generation algorithm	Original	Iterations	Similarities	Original	Iterations	Similarities	Original	Iterations	Similarities
Replays detected	176	196	191	146	170	176	165	172	175
Correct detections	100	94	170	141	161	171	161	169	171
False detections	76	102	21	5	9	5	4	3	4
Recall [%]	43.29	40.69	73.59	61.04	69.70	74.03	69.70	73.16	74.03
Precision [%]	56.82	47.96	89.01	96.58	94.71	97.16	97.58	98.26	97.71
F-measure [%]	49.14	44.03	80.57	74.80	80.30	84.03	81.31	83.87	84.24
Relative average time of logo template generation [%]	100	138	746	100	140	646	100	138	345
Relative average time of replay detection [%]	100	107	137	100	107	131	100	107	114

6 Conclusions

The tests performed in the AVI Indexer have clearly shown that the efficiency of the method of detecting replay logo in sports videos based on contrast feature and histogram difference is strongly dependent on the thresholds used in the method. Because of the diversified specificity of logo image these thresholds may be different for different sports categories. Bright logo, transition of more than two seconds, dark background, for example green playing field in soccer games, facilitate replay logo detection. Whereas, reply logo is difficult to detect when logo image is relatively small not covering the whole frame, so in consequence the contrast is not significant, or the logo transition is very dynamic.

The proposed improvements of the methods consist in generating logo template using an iterative approach or taking into account logo candidate similarities. The results of the tests described in the paper have confirmed the usefulness of these improved techniques of logo template generation.

References

1. Hu, W., Xie, N., Li, L., Zeng, X., Maybank, S.: A survey on visual content-based video indexing and retrieval. IEEE Trans. Syst. Man Cybern. Part C Appl. Rev. 41(6), 797–819 (2011)
2. Schoeffmann, K., Hudelist, M.A., Huber, J.: Video interaction tools: a survey of recent work. ACM Comput. Surv. (CSUR) 48(1), 1–34 (2015). Article no. 14
3. Choroś, K.: Video structure analysis for content-based indexing and categorisation of TV sports news. Int. J. Intell. Inf. Database Syst. 6(5), 451–465 (2012)
4. Gu, L., Bone, D., Reynolds, G.: Replay detection in sports video sequences. In: Correia, N., Chambel, T., Davenport, G. (eds.) Multimedia 1999, Proceedings of the Eurographics Workshop in Milano, pp. 3–12. Springer, Vienna (2000)
5. Ajmal, M., Ashraf, M.H., Shakir, M., Abbas, Y., Shah, F.A.: Video summarization: techniques and classification. In: Bolc, L., Tadeusiewicz, R., Chmielewski, L.J., Wojciechowski, K. (eds.) ICCVG 2012. LNCS, vol. 7594, pp. 1–13. Springer, Heidelberg (2012)
6. Ekin, A., Tekalp, A.M., Mehrotra, R.: Automatic soccer video analysis and summarization. IEEE Trans. Image Process. 12(7), 796–807 (2003). IEEE
7. Zhao, Z., Jiang, S., Huang, Q., Zhu, G.: Highlight summarization in sports video based on replay detection. In: Proceedings of the IEEE International Conference on Multimedia and Expo, pp. 1613–1616. IEEE (2006)
8. Su, P.C., Lan, C.H., Wu, C.S., Zeng, Z.X., Chen, W.Y.: Transition effect detection for extracting highlights in baseball videos. EURASIP J. Image Video Process. 27(1), 1–16 (2013)
9. Pan, H., Van Beek, P., Sezan, M.I.: Detection of slow-motion replay segments in sports video for highlights generation. In: Proceedings of the IEEE International Conference on Acoustics, Speech, and Signal Processing (ICASSP), vol. 3, pp. 1649–1652. IEEE (2001)
10. Wang, L., Liu, X., Lin, S., Xu, G., Shum, H.Y.: Generic slow-motion replay detection in sports video. In: Proceedings of the International Conference on Image Processing (ICIP 2004), vol. 3, pp. 1585–1588. IEEE (2004)

11. Chen, C.M., Chen, L.H.: A novel method for slow motion replay detection in broadcast basketball video. Multimedia Tools Appl. **74**(21), 9573–9593 (2015)
12. Duan, L. Y., Xu, M., Tian, Q., Xu, C.S.: Mean shift based video segment representation and applications to replay detection. In: Proceedings of the IEEE International Conference on Acoustics, Speech, and Signal Processing (ICASSP), vol. 5, pp. V-709–V-712. IEEE (2004)
13. Dang, Z., Du, J., Huang, Q., Jiang, S.: Replay detection based on semi-automatic logo template sequence extraction in sports video. In: Proceedings of the Fourth International Conference on Image and Graphics (ICIG 2007), pp. 839–844. IEEE (2007)
14. Tong, X., Lu, H., Liu, Q., Jin, H.: Replay detection in broadcasting sports video. In: Proceedings of the Third International Conference on Image and Graphics (ICIG 2004), pp. 337–340. IEEE (2004)
15. Xu, W., Yi, Y.: A robust replay detection algorithm for soccer video. IEEE Signal Process. Lett. **18**(9), 509–512 (2011)
16. Zhao, F., Dong, Y., Wei, Z., Wang, H.: Matching logos for slow motion replay detection in broadcast sports video. In: Proceedings IEEE International Conference on of the Acoustics, Speech and Signal Processing (ICASSP), pp. 1409–1412. IEEE (2012)
17. Pan, H., Li, B., Sezan, M.I.: Automatic detection of replay segments in broadcast sports programs by detection of logos in scene transitions. In: Proceedings of the IEEE International Conference on Acoustics, Speech, and Signal Processing (ICASSP), vol. 4, pp. IV-3385–IV-3388. IEEE (2002)
18. Han, B., Yan, Y., Chen, Z., Liu, C., Wu, W.: A general framework for automatic on-line replay detection in sports video. In: Proceedings of the 17th ACM International Conference on Multimedia, pp. 501–504. ACM (2009)
19. Yang, Y., Lin, S., Zhang, Y., Tang, S.: A statistical framework for replay detection in soccer video. In: Proceedings of the IEEE International Symposium on Circuits and Systems (ISCAS), pp. 3538–3541. IEEE (2008)
20. Wang, J., Chng, E., Xu, C.: Soccer replay detection using scene transition structure analysis. In: Proceedings of the IEEE International Conference on Acoustics, Speech, and Signal Processing (ICASSP), vol. 2, pp. II-433–II-436. IEEE (2005)
21. Choroś, K.: Video structure analysis and content-based indexing in the automatic video indexer AVI. In: Nguyen, N.T., Zgrzywa, A., Czyżewski, A. (eds.) Advances in Multimedia and Network Information System Technologies. AISC, vol. 80, pp. 79–90. Springer, Heidelberg (2010)
22. Nguyen, N., Yoshitaka, A.: Shot type and replay detection for soccer video parsing. In: IEEE International Symposium on Multimedia (ISM), pp. 344–347. IEEE (2012)
23. Choroś, K.: Automatic playing field detection and dominant colour extraction in sports video shots of different view types. In: Zgrzywa A. et al. (eds.) Multimedia and Network Information Systems. AISC, vol. 506, pp. 39–48, Springer, Heidelberg (2016)

Dynamic MCDA Approach to Multilevel Decision Support in Online Environment

Jarosław Jankowski[1,2(✉)], Jarosław Wątróbski[1], and Paweł Ziemba[3]

[1] West Pomeranian University of Technology, Szczecin,
Żołnierska 49, 71-210 Szczecin, Poland
jaroslaw.jankowski@pwr.edu.pl,
jwatrobski@wi.zut.edu.pl
[2] Department of Computational Intelligence, Wrocław University
of Technology, Wybrzeże Wyspiańskiego 27, 50-370 Wrocław, Poland
[3] The Jacob of Paradyż University of Applied Sciences in Gorzów Wielkopolski,
Teatralna 25, 66-400 Gorzów Wielkopolski, Poland
pziemba@pwsz.pl

Abstract. Effective online marketing requires technologies supporting campaign planning and execution at the operational level. Changing performance over time and varying characteristics of audience require appropriate processing for multilevel decisions. The paper presents the concept of adaptation of the Multi-Criteria Decision Analysis methods (MCDA) for the needs of multilevel decision support in online environment, when planning and monitoring of advertising activity. The evaluation showed how to integrate data related to economic efficiency criteria and negative impact on the recipient towards balanced solutions with limited intrusiveness within multi-period data.

Keywords: Online marketing · Intrusiveness · Decision support · MCDA methods

1 Introduction

It does not raise doubt that marketing science and commercial aspects are the base of marketing activity in both online and offline environments. However, especially for online marketing, the technological background and supporting technologies play the key role. They are developed in several directions including campaign planning [12] or real time optimization towards better conversions [22]. New methods are implemented in the area of algorithms for computational advertising including adaptive approaches [20] or linear mathematical models [9] with their extensions [23]. Attempts to increase the effectiveness of online commercial activity often leads to negative side effects such as growing intrusiveness of online marketing content [33] and, as a result, physical or cognitive avoidance [14]. Searching for compromise between content intrusiveness and its influence on user experience within a web system is one of directions of research in this area [18, 34]. The approach proposed in this paper integrates data related to effectiveness of online marketing content together with the evaluation of its intensity and negative impact on user experience. Changes of online environments are taken into

© Springer International Publishing Switzerland 2016
N.T. Nguyen et al. (Eds.): ICCCI 2016, Part I, LNAI 9875, pp. 553–564, 2016.
DOI: 10.1007/978-3-319-45243-2_51

account and multistage decision support with the use of MCDA methods is introduced. Direct application of the MCDA approach in this class of problems is hampered as the MCDA methodology is based on assumptions of stability of parameters forming part of the decision-making support process, e.g. datasets, criteria, decision variants and evaluations. In online planning parameters may change dynamically and are conditioned on the changes in audience characteristics, variable efficiency of advertising message or competitors' activity [5]. Employing the classic MCDA approach without considering time evolution is the way to oversimplify the problem [17]. In this context the paper presents an adaptation of the MCDA approach for the needs of dynamic decision-making support in the online environment in the process of multi-stage planning of marketing activity. The solutions were verified on the basis of data from real advertising campaigns. The paper is organized as follows: section two includes the review of literature, section three presents the conceptual framework and assumptions for the proposed approach. In the next section empirical results are presented followed by a summary in section five.

2 Literature Review

The development of electronic media and the growing role of online advertising create the area for searching for new solutions both in the practical and scientific dimension. The main purpose is usually to increase efficiency of marketing activity in multiple dimensions. On the level of advertising message, tasks which include the use of persuasion, colors, animation and call to action images [32] as well as identification of factors affecting efficiency [30] are realized. On the operational level, optimization in real time [10] and factorial methods [6] are used. Other areas include the use of available broadcasting resources [7], personification of message [20] or choice of message content on the basis of context [29]. The basis for implementing marketing activity is planning and scheduling ad expositions with the participation of available broadcasting resources. Plans are implemented using advertising servers which carry out the selection of marketing content as a response to a request coming from an internet browser [2]. The problem of optimization of the selection of advertisements was formulated as a task of linear programming with maximizing the number of interactions under given constraints which include the number of times an ad was displayed in a given period [22]. The basic model of linear optimization was developed towards a compromise between searching for and exploration of decision-making solutions [23] and a balanced distribution of broadcasts [9]. Other methods on the operational level are based on the monitoring user activity and maximizing likelihood of interaction [16]. In other solutions the selection of advertising content in based on user profiles created during internet sites browsing [13]. Another model takes into account pricing strategies in the process of managing advertising space and maximization of revenue from the sale of advertising space [12]. The literature review shows that majority of the available optimization systems and models is oriented towards increasing the number of interactions within a webpage and automatic selection of advertisements so that it is maximized. Even though maximizing broadcasts of invasive forms of advertising may increase the financial result, in a short while it may also result in the decrease of user experience and negatively affect the brand perception. In a typical process of designing websites user experience should be taken into account for better functionality and creating solutions oriented on internet users' needs. A question arises here about the level to which it is worth increasing the intensity of the marketing message in order to draw users' attention and maintain profit at an acceptable level without disturbing user experience excessively. Excessive use of video, audio and animation within online content results in the problem of excessive burdening with commercial content and brings side effects negatively affecting the user [26]. The solution proposed in this paper integrates parameters related to effectiveness of message and its negative impact on the recipient resulting from the intensity of employed persuasive mechanisms. Taking into account several stages of decision-making support gives the opportunity to reflect the changeability of preferences and measurement data. The basis of earlier-proposed linear models are Pareto-optimal solutions where bringing tasks down to one function makes it more difficult to take into account decision-makers' preferences and criteria. In MCDA approaches the basis are nondominated solutions and they feature the possibility of taking into account qualitative factors subject to subjective assessment as well. The next section presents assumptions for the proposed solutions.

3 Conceptual Framework

The conceptual framework presented in this stage is the continuation of earlier research based on searching for compromise between marketing effectiveness and user experience [18]. In the proposed approach data from multi-stage campaign is integrated with measurements of intensity of message and subsequently global and local objectives are employed in the evaluation of results from various perspectives. The proposed model allows obtaining compromise solutions on the basis of measurements from a real environment and a decision-maker's preferences. Methodologically, framework is based in on the Dynamic Multi-Criteria Decision Analysis (DMDCA) which is an extension of static approach [5] along with an indication including the taking into account the dynamics of domains in the MCDA subject matter [1, 21]. At present majority of work done in this field is oriented towards the expansion of the classic MCDA model allowing its areas oriented application for various aspects of dynamic decision making. They are focused on changes in the MCDA domain such as variable sets of decision variants or criteria for their assessment and expand the classical MCDA paradigm (see [15]) with additional components of the decision making process such as changeable spaces or domain analyses [31]. The synthesis of approaches can be found, i.e., in the following works: [5, 19]. Due to the specific nature of online marketing-related issues and great dynamics of environment the research assumes a constant form of the set of decision variants as well as the family of criteria for their evaluation. It is at the same time in line with the specific nature of the discussed decision-related problems. The aspects of dynamics in multicriteria decision making process which were highlighted includes changeability of partial evaluation of decision variants in time (performance tables and global performance of variants – see [15]). The analysis of impact of this changeability on the final result of the decision making process is taken into an account. The presented procedure is based on the classic MCDA framework (see [15, 27]). Additionally, assumptions allowing of the dynamic decision situation modelling were introduced. The research procedure itself was based on a five-stage course of the process of multiple-criteria decision-making proposed by Guitouni [15]. It is composed of: (I) the structuring of the decision situation, (II) the preferences articulation and modelling, (III) the aggregation of these preferences, (IV) the exploitation of this aggregation, (V) the recommendation. The multiple-criteria procedure itself covers stages (III) and (IV), whereas the decision problem can be characterized by stages (I), (II) and (V), where stages (I) and (II) address the input data of the decision process and stage (V) defines the output data. The introduction here of the dynamics of modelled decision situations requires expanding the classical MCDA model with consequences of implementation of the time factor. Let t_k denote k-th time period for which the multiple-criteria decision model is built, and, let T denote a set of time periods $T = \{t_1,...,t_k...,t_p\}$. The intention of the decision maker in k-th multiple-criteria decision problem is to select the alternative that best meets their preference for a specific set of criteria. Further consideration was adopted as a solution to the problem of decision-making to maximize the outcome of the transformation F designating the degree of fulfillment of the criteria selected by the successive decision variants as shown in the equation:

$$G\left(a_k^b\right) = \max F(C(A)) \tag{1}$$

where a_k^b is the most preferred alternative selected from a set of decision-making variants A in period t_k and $G\left(a_k^b\right)$ is a performance variant a_k^b denoted also as an assessment of the fulfillment of criteria C. The course of a DMCDA process formulated in this way is presented in Fig. 1.

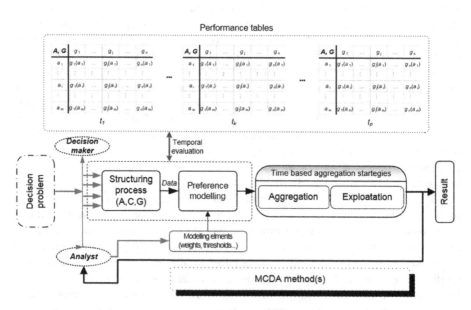

Fig. 1. Integration of MCDA methods with multi-stage measurement of performance in online environment

The presented procedure expands classical paradigms presented in works [15, 27] with the aspect of modelling the dynamics of modeled domain. In the presented model, particularly, it has the form of a cyclical process of generation of structure of individual performance tables and their aggregation. The research presented in the paper assumes a balanced impact of each performance table on the final evaluation support result. It's worth to notice, that he choice of an aggregation strategy itself can be realized in accordance with the specifics of a given decision problem and the following can be examples of such [8]: Time Appreciated Aggregation, Time Depreciated Aggregation, Time Period Mostly Appreciated Aggregation. As shown above, the framework was formulated using classical assumptions (set of alternatives, set of criteria, outcome of each choice, preference structures of decision makers and stakeholders are fixed and steady). However, it may be expanded also where the assumptions are not fulfilled. For instance, the changeable spaces listed above can be effectively modelled using the theory of Dynamic Multiple-Criteria Decision Analysis, habitual domains, and competence set analysis presented by Po-Lung Yu and Yen-Chu Chen [31].

4 The Empirical Study

Structuring of the decision problem was carried out in stage I. For this purpose the set of decision variants (A), the set of criteria (B) and performance tables of individual variants were defined. The set of decision variants (A) consisted of advertisements located on a web site. Five advertisers were taken into account here and for each of them ten ad units differing in the level of intensity of impact on the recipient. Therefore, the total of 50 ad units formed the discussed set of variants. Each variant was examined in terms of 3 criteria: C1 - conversion rate, C2 - intensity of advertising message, which may negatively influence user experience and C3 - profit of an portal operator which can be treated also as advertising costs covered by the advertiser. Conversion rate (CR) is a basic measure of effectiveness of advertising expressed by the ratio of the number of desired user interactions to the number of contacts with the advertising content in which they can be potentially realized [18]. In the case of online advertising a desired interaction may be e.g. user's "clicking" on an ad unit, e.g. a banner, and the number of contacts is equal to the number of times a given ad is displayed. The CR coefficient was designated *a posteriori* in the performed analyses on the basis of real data. In turn, intensity of an ad was specified in a subjective study. The broadcaster's profit was calculated on the basis of the number of interactions and costs covered by the advertiser. Due to the fact the research focused on the changeability of preferences in time, it looked separately at the efficiency of variants in three equal time periods for which different conversion rates were obtained due to audience characteristics and interest in promoted services. Moreover, aggregation of three efficiency rankings into one was carried out using a group procedure and the efficiency of variants averaged from three time periods was examined. The obtained efficiencies for projected variants represented by conversion rate (CR) for each from three periods for first advertiser are presented in Table 1.

Stages II and III, i.e. modelling and preference aggregation, require in particular selecting a calculating procedure (MCDA methods) [27]. The research applied the

Table 1. Examples of criteria efficiencies of variants for selected advertisers.

Ad unit	CR Period1	CR Period2	CR Period3	Mean	Intensity	Profits
A1.1	0.0015	0.0008	0.0027	0.0017	0.0026	0.0015
A1.2	0.0017	0.0011	0.0024	0.0018	0.2584	0.0025
A1.3	0.0036	0.0019	0.0056	0.0037	0.1843	0.0072
A1.4	0.0018	0.0005	0.0028	0.0017	0.8474	0.0045
A1.5	0.0027	0.0014	0.0034	0.0025	0.7028	0.0081
A1.6	0.0028	0.0020	0.0032	0.0027	0.8386	0.0098
A1.7	0.0026	0.0014	0.0044	0.0028	0.4392	0.0104
A1.8	0.0023	0.0014	0.0022	0.0020	0.5785	0.0103
A1.9	0.0035	0.0012	0.0023	0.0024	0.5000	0.0175
A1.10	0.0036	0.0012	0.0060	0.0036	0.6481	0.0198

Promethee method based on the outranking relation [35]. The method allows the application of six preference functions: usual criterion, U-shape criterion, V-shape criterion, level criterion, V-shape with indifference criterion, gaussian criterion [3]. Promethee allows obtaining a total preorder of decision variants (Promethee II) and carrying out the aggregation of individual rankings into a group ranking (Promethee GDSS - Group Decision Support System) [4]. Therefore, it is suitable for the above discussed structure of a decision problem. In stage II, for the Promethee method, the following needed to be done: defining the weight of the criteria and directions of preferences, selection of criteria preference functions and defining values of thresholds for the criteria. The selection of preference functions and values of thresholds greatly affect the order of the variants in a ranking [24, 25]. Moreover, the type of preference functions applied depends of the type of criteria. For quantitative criteria it is recommended that functions using the following are applied: V-shape criterion, V-shape with indifference criterion or gaussian criterion [11]. The developed decision model applied the V-shape criterion. This function uses preference threshold p, whose value should fall within reliable min and max values taken by a given criterion [28]. For threshold p, the developed model adopted the value of two times standard deviation. When it comes to the weights of criteria, for the purpose of our analysis it was assumed that all criteria are equally significant. Full preference model with assigned weights and direction for each from three periods is presented in Table 2.

Table 2. Preference model in the discussed decision problem.

Criterion	Direction	Weight [%]	Preference function	Preference threshold			
				Period 1	Period 2	Period 3	Mean
Conversion rate	Max	33.3	V-shape	0.0016	0.0010	0.0046	0.0024
Invasiveness	Min	33.3	V-shape	0.5330	0.5330	0.5330	0.5330
Profit	Max	33.3	V-shape	0.0079	0.0079	0.0079	0.0079

Five rankings of preference variants were obtained in stage II: 3 individual rankings from subsequent time periods, a ranking based on averaged values and a ranking obtained using the Promethee GDSS group procedure based on three individual rankings. These ranking are presented in Table 3.

Table 3. Obtained variant rankings for top twenty ad units.

Rank	Period 1 Variant	ϕ_{net}	Period 2 Variant	ϕ_{net}	Period 3 Variant	ϕ_{net}	Mean Ad unit	ϕ_{net}	Group Ad unit	ϕ_{net}
1	A1.9	0.6121	A1.3	0.4881	A5.10	0.4431	A1.3	0.4991	A1.3	0.4689
2	A1.3	0.5646	A2.3	0.3537	A3.3	0.3656	A5.10	0.4642	A1.10	0.3822
3	A1.10	0.5394	A1.9	0.3295	A1.10	0.3573	A1.10	0.4620	A1.9	0.3730
4	A1.7	0.4296	A1.7	0.3171	A1.3	0.3542	A3.3	0.4179	A1.7	0.3300
5	A5.10	0.4026	A3.1	0.3118	A5.3	0.2948	A1.7	0.3233	A3.3	0.3260
6	A3.3	0.3550	A3.3	0.2599	A4.3	0.2714	A1.9	0.3183	A5.10	0.3178
7	A2.9	0.3305	A1.10	0.2500	A3.8	0.2572	A5.3	0.2864	A5.3	0.1851
8	A1.8	0.2863	A1.6	0.2386	A1.7	0.2432	A3.8	0.2691	A3.8	0.1792
9	A5.3	0.2362	A1.8	0.2291	A5.4	0.2271	A5.4	0.2267	A1.8	0.1734
10	A1.6	0.2158	A3.8	0.2147	A3.7	0.1991	A3.7	0.1859	A2.9	0.1662
11	A1.5	0.2026	A2.10	0.2062	A3.10	0.1846	A4.3	0.1556	A2.3	0.1601
12	A2.7	0.1997	A3.9	0.1919	A1.9	0.1773	A3.1	0.1503	A3.7	0.1539
13	A2.3	0.1811	A3.7	0.1678	A3.1	0.1563	A3.10	0.1437	A3.1	0.1430
14	A2.2	0.1621	A5.7	0.1226	A3.2	0.1249	A3.2	0.1164	A2.10	0.1377
15	A2.10	0.1532	A2.7	0.1177	A5.2	0.1222	A3.9	0.1160	A1.6	0.1223
16	A2.1	0.1265	A5.10	0.1077	A5.1	0.1155	A5.2	0.1153	A5.1	0.0811
17	A5.2	0.1039	A3.2	0.1018	A3.9	0.0969	A5.1	0.0918	A2.7	0.0724
18	A5.1	0.1037	A2.1	0.0929	A2.9	0.0867	A2.9	0.0877	A3.9	0.0709
19	A3.7	0.0949	A2.9	0.0813	A3.5	0.0850	A1.8	0.0725	A1.5	0.0681
20	A1.1	0.0801	A1.5	0.0731	A5.7	0.0672	A3.5	0.0712	A5.7	0.0665

Stage IV of the research procedure is based on the exploitation of the obtained solution. For exploitation of individual rankings the analysis of their changeability in time was carried out. This analysis shows that in a dynamic environment such as the Internet, and in particular an internet ad, user preferences may be subject to constant, significant changes. It may be proven by the fact that out of 10 best variants of the first individual ranking, only 6 variants featured in the top of the third ranking (A5.10, A3.3, A1.10, A1.3, A5.3, A1.7). The position of variant A1.6 will serve as another example, which fell from position 10 in the first ranking and position 8 in the second ranking to position 32 in ranking 3. Great changeability of preferences obtained for individual variants in subsequent time periods are shown by rankings' scatter graphs presented in Fig. 2. This is why the analysis of individual rankings shows the need for permanent preference research in DSS systems operating on dynamic data.

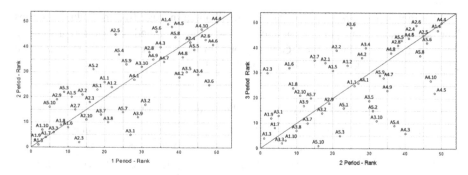

Fig. 2. Scatter graphs of rankings of variants in individual periods with pairwise period comparison and showed localization of each design variant for each advertiser.

Exploitation of rankings: based on averaged values and aggregated to group evaluation, was performed through the application of the analysis of rankings' sensitivity to changes of criteria weights. The purpose behind performing it was to find guidelines for optimal decision variants depending of decision makers' preferences. Findings of the sensitivity analysis for an averaged ranking and for a group ranking were presented respectively in Figs. 3(a) and (b). The comparison of results of the sensitivity analysis for both rankings shows that aggregation of preferences from three individual rankings into a group ranking gives more transparent results that a ranking based on averaged criteria values. In a group ranking, across the entire field of criteria values, there is a smaller number of dominant variants, which allows obtaining more transparent recommendations. It can be observed above all in the case of the Conversion Rate criterion for which along with the increase in its weight, for the averaged ranking, variants A1.9, A1.3, A5.10 and A5.4. dominate subsequently. In turn, in the group ranking, only two variants are dominant: A1.9 and A1.3.

Stage V, i.e. drawing up the recommendation, is based in the results of stages III and IV. The sensitivity analysis carried out for the group ranking obtained using the Promethee GDSS methods indicates high stability of the obtained solution for dominant variants. Variant A1.3 remains the best in terms of weights: between 15 % and 100 % for the Conversion Rate criterion, between 27 % and 72 % for the invasiveness criterion and between 0 % and 40 % for the profit criterion. If the weight of the profit criterion is greater than 40 %, then the best variant may be A1.10. In turn, when the weight of the invasiveness criterion is lesser that 27 % then variant A1.10 may be assumed as optimal. The following variant dominations resulting for the sensitivity analysis may be assumed as doubtful: A1.9 for the weight of the Conversion Rate lesser that 15 % and A1.1 and A3.1 for the weights of the invasiveness criterion greater than approx. 75 %. This doubt results from the fact that in individual rankings the position of these variants is characterized by great dynamics and in the most current ranking (Period 3) they take remote positions.

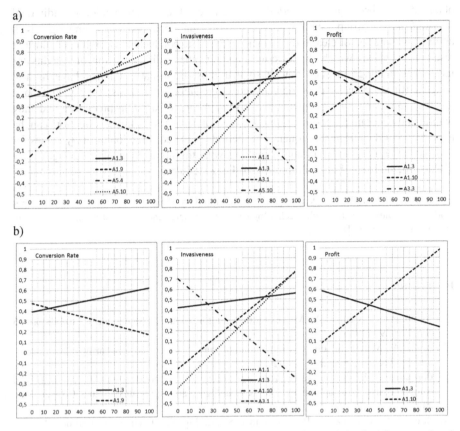

Fig. 3. Results of the analysis of sensitivity to changes in criteria weights in (a) averaged and (b) group ranking

5 Summary

The presented results confirm the ability of application of the proposed procedure for obtaining decision solutions given changeability of measurement data and the presence of multiple criteria. The obtained results show discrepancies in the application of the averaged and group approaches. Making decisions based on averaged values may lead to simplifications which lower the quality of decision. A significant element introduced in the model was the taking into account of both characteristics relating to the effectiveness of message represented by the Conversion Rate as well as parameters relating to the impact intensity. Research results indicate the dynamics of obtained solutions in the field of internet advertising and its effectiveness. This is why there is a need of constant evaluation of the effectiveness of advertising in relation to its other aspects, such as invasiveness and profits for the service owner or costs borne by the advertiser. A certain way to capture this changeability allowing the drawing up of the recommendations valid for a slightly longer period of time is averaging individual solutions obtained for subsequent time periods. In the research case, aggregation of individual rankings into a group ranking proved a more functional way of averaging which allowed preparation of clearer recommendations. The conducted research opens further research directions which should cover, among others, differentiation of weights for individual rankings in such a way so that most up-to-date rankings have highest weights (time appreciated or depreciated aggregation [8]). Thus, they would have greatest impact on the aggregated ranking by means of a group procedure and they would allow obtaining more up-to-date user preferences. Another direction can be application of data from subsequent time periods in the construction of fuzzy values for criteria preferences and building a ranking with the application of a selected fuzzy MCDA method, e.g. Fuzzy Promethee or Fuzzy TOPSIS.

Acknowledgments. The work was partially supported by European Union's Seventh Framework Programme for research, technological development and demonstration under grant agreement no 316097 [ENGINE].

References

1. Agrell, P.J., Wikner, J.: An MCDM framework for dynamic systems. Int. J. Prod. Econ. **45** (1), 279–292 (1996)
2. Amiri, A., Menon, S.: Efficient scheduling of internet banner advertisements. ACM Trans. Internet Technol. **3**(4), 334–346 (2003)
3. Behzadian, M., Kazemzadeh, R.B., Albadvi, A., Aghdasi, M.: PROMETHEE: a comprehensive literature review on methodologies and applications. Eur. J. Oper. Res. **200**, 198–215 (2010)
4. Brans, J.P., Mareschal, B.: Promethee methods. In: Figueira, J., Greco, S., Ehrgott, M. (eds.) Multiple Criteria Decision Analysis, pp. 163–195. Springer, Boston (2005)
5. Campanella, G., Ribeiro, R.A.: A framework for dynamic multiple-criteria decision making. Dec. Supp. Syst. **52**(1), 52–60 (2011)
6. Chakrabarti, D., Kumar, R., Radlinski, F., Upfal, E.: Mortal multi-armed bandits. In: Koller, D., Schuurmans, D., Bengio, Y. (eds.) Advances in Neural Information Processing Systems 21, pp. 273–280 (2008)
7. Chakrabarti, D., Agarwal, D., Josifovski, V.: Contextual advertising by combining relevance with click feedback. In: Proceeding of the 17th International Conference on World Wide Web, pp. 417–426 (2008)
8. Chen, Y., Li, K.W., He, S.: Dynamic multiple criteria decision analysis with application in emergency management assessment. In: IEEE International Conference on Systems Man and Cybernetics (SMC), pp. 3513–3517 (2010)
9. Chickering, D.M., Heckerman, D.: Targeted advertising with inventory management. In: Proceedings of the 2nd ACM Conference on Electronic Commerce, pp. 145–149 (2000)
10. Cookhwan, K., Sungsik, P., Kwiseok, K., Woojin, Ch.: How to select search keywords for online advertising depending on consumer involvement: An empirical investigation. Expert Syst. Appl. **39**(1), 594–610 (2012)
11. Deshmukh, S.C.: Preference ranking organization method of enrichment evaluation (Promethee). Int. J. Eng. Sci. Invent. **2**(11), 28–34 (2013)

12. Du, H., Xu, Y.: Research on multi-objective optimization decision model of web advertising —takes recruitment advertisement as an example. Int. J. Adv. Comput. Technol. **4**(10), 329–336 (2012)
13. Giuffrida, G., Reforgiato, D., Tribulato, G., Zarba, C.: A banner recommendation system based on web navigation history. In: Proceedings of the IEEE Symposium on Computational Intelligence and Data Mining (CIDM), pp. 291–296 (2011)
14. Goldstein, D.G., McAfee, R.P., Suri, S.: The cost of annoying ads. In: Proceedings of the 22nd International Conference on World Wide Web, pp. 459–470 (2013)
15. Guitouni, A., Martel, J.M., Vincke, P.: A Framework to Choose a Discrete Multicriterion Aggregation Procedure. Defence Research Establishment Valcatier (1998)
16. Gupta, N., Khurana, U., Lee, T., Nawathe, S.: Optimizing display advertisements based on historic user trails. In: Proceedings of the ACM SIGIR, Workshop: Internet Advertising (2011)
17. Hashemkhani Zolfani, S., Maknoon, R., Zavadskas, E.K.: An introduction to Prospective Multiple Attribute Decision Making (PMADM). Technol. Econ. Develop. Econ. **22**(2), 309–326 (2016)
18. Jankowski, J., Ziemba, P., Wątróbski, J., Kazienko, P.: Towards the tradeoff between online marketing resources exploitation and the user experience with the use of eye tracking. In: Nguyen, N.T., Trawiński, B., Fujita, H., Hong, T.-P. (eds.) ACIIDS 2016, Part I. LNCS (LNAI), vol. 9621, pp. 330–343. Springer, Heidelberg (2016)
19. Jassbi, J.J., Ribeiro, R.A., Varela, L.R.: Dynamic MCDM with future knowledge for supplier selection. J. Dec. Syst. **23**(3), 232–248 (2014)
20. Kazienko, P., Adamski, M.: AdROSA - adaptive personalization of web advertising. Inf. Sci. **177**(11), 2269–2295 (2007)
21. Kornbluth, J.S.H.: Dynamic multi-criteria decision making. J. Multi-Criteria Dec. Anal. **1**(2), 81–92 (1992)
22. Langheinrich, M., Nakamura, A., Abe, N., Kamba, T., Koseki, Y.: Unintrusive customization techniques for web advertising. Comput. Netw. **31**(11–16), 1259–1272 (1999)
23. Nakamura, A., Abe, N.: Improvements to the linear programming based scheduling of web advertisements. Electron. Comm. Res. **5**, 75–98 (2005)
24. Podvezko, V., Podviezko, A.: Dependence of multi-criteria evaluation result on choice of preference functions and their parameters. Technol. Econ. Develop. Econ. **16**(1), 143–158 (2010)
25. Podvezko, V., Podviezko, A.: Use and choice of preference functions for evaluation of characteristics of socio-economical processes. In: 6th International Scientific Conference on Business and Management, pp. 1066–1071 (2010)
26. Rosenkrans, G.: The creativeness and effectiveness of online interactive rich media advertising. J. Interact. Advertising **9**(2), 18–31 (2009)
27. Roy, B.: Multicriteria Methodology for Decision Aiding. Springer, Dordrecht (1996)
28. Roy, B.: The outranking approach and the foundations of ELECTRE methods. In: Bana e Costa, C.A. (ed.) Readings in Multiple Criteria Decision Aid, pp. 155–183. Springer, Heidelberg (1990)
29. Teng-Kai, F., Chia-Hui, Ch.: Blogger-centric contextual advertising. Expert Syst. Appl. **38**(3), 1777–1788 (2011)
30. Tsai, W.H., Chou, W.Ch., Leu, J.D.: An effectiveness evaluation model for the web-based marketing of the airline industry. Expert Syst. Appl. **38**(12), 15499–15516 (2011)
31. Yu, P.-L., Chen, Y.-C.: Dynamic MCDM, habitual domains and competence set analysis for effective decision making in changeable spaces. In: Greco, S., Ehrgott, M., Figueira, J.R. (eds.) Trends in Multiple Criteria Decision Analysis. International Series in Operations Research & Management Science, vol. 142, pp. 1–35. Springer, New York (2010)

32. Yun, Y.Ch., Kim, K.: Processing of animation in online banner advertising: the roles of cognitive and emotional responses. J. Interact. Market. **19**(4), 18–34 (2005)
33. Zha, W., Wu, H.D.: The impact of online disruptive ads on users' comprehension, evaluation of site credibility, and sentiment of intrusiveness. Am. Commun. J. **16**(2), 15–28 (2014)
34. Jankowski, J., Watróbski, J., Ziemba, P.: Modeling the impact of visual components on verbal communication in online advertising. In: Núñez, M., Nguyen, N.T., Camacho, D., Trawiński, B. (eds.) ICCCI 2015. LNCS, vol. 9330, pp. 44–53. Springer, Heidelberg (2015). doi:10.1007/978-3-319-24306-1_5
35. Wątróbski, J., Jankowski, J.: Guideline for MCDA method selection in production management area. In: Różewski, P., Novikov, D., Bakhtadze, N., Zaikin, O. (eds.) New Frontiers in Information and Production Systems Modelling and Analysis. ISRL, vol. 98, pp. 119–138. Springer, Switzerland (2016)

Usability Testing of a Mobile Friendly Web Conference Service

Ida Błażejczyk, Bogdan Trawiński[✉], Agnieszka Indyka-Piasecka,
Marek Kopel, Elżbieta Kukla, and Jarosław Bernacki

Faculty of Computer Science and Management,
Department of Information Systems, Wrocław University of Science
and Technology, Wybrzeże Wyspiańskiego 27, 50-370 Wrocław, Poland
i.blazejczyk@hotmail.com, {bogdan.trawinski,
agnieszka.indyka-piasecka,marek.kopel,elzbieta.kukla,
jaroslaw.bernacki}@pwr.edu.pl

Abstract. Results of usability study of the conference website developed using responsive web design approach are presented in the paper. Two variants of the application architecture were examined using laptops and smartphones. The testing sessions with users took place in laboratory conditions, whereas three expert inspection methods including cognitive walkthrough, heuristic evaluation, and control lists were accomplished remotely. The list of 111 recommendations for improving the website was formulated. In consequence, a new version of the website was developed and the second round of usability testing was planned.

Keywords: Responsive Web Design · Usability Testing · Web applications · Mobile applications

1 Introduction

Responsive Web Design (RWD) is an approach to designing and implementing web pages which are optimized for various devices with diverse screen sizes, resolutions and proportions. The approach is based on fluid grids, flexible images, and media queries [1]. It presents a series of opportunities and challenges for web designers [2–5]. Despite the basic principles of mobile user experience (UX) are similar to desktop UX, the best practices of mobile UX are still appearing due to vigorous growth of mobile technologies [6]. Usability is key for success of any software product including both desktop systems and mobile applications. A usable system allows its users to achieve specific goals effectively and efficiently with a high level of satisfaction. Usability is described in literature by means of models specifying a number of usability attributes. The most popular are ISO 924-11 model comprising three attributes: effectiveness, efficiency and satisfaction [7] and Nielsen's model which consists of five attributes: efficiency, satisfaction, learnability, memorability, and errors [8]. In turn, the PAC-MAD model devised for mobile applications encompasses seven attributes: effectiveness, efficiency, satisfaction, learnability, memorability, errors and cognitive load [9]. The consolidated QUIM model decomposes usability into factors, criteria and metrics

© Springer International Publishing Switzerland 2016
N.T. Nguyen et al. (Eds.): ICCCI 2016, Part I, LNAI 9875, pp. 565–579, 2016.
DOI: 10.1007/978-3-319-45243-2_52

taken from other models and standards [10]. A number of works reporting usability evaluation of mobile and responsive applications employing users as well as experts have been published recently [11–14].

Usability testing of a conference website based on the ISO Standard 9241-11 attributes, namely effectiveness, efficiency, and satisfaction, is presented in the paper. The conference website was developed to assist with the organization of the ACIIDS 2016 conference [15]. The target users were researchers, ranging from PhD Students to professors, potentially interested in contributing their works to the conference as well as the conference attendees whose papers were accepted for the presentation at the conference. The main purpose of the service was to supply the users with all organizational information about the event including instructions how to submit papers and take part in the event. Therefore, the textual form of information was prevailing on the website. One subpage contained forms where the users could enter data necessary to register at the conference. The service did not support paper submission and reviewing which were carried out using the external system EasyChair.

The website was worked out using RWD approach to address the needs of the users accessing it with diverse devices including desktop computers, tablets, and smartphones. The website was designed according to the standard requirements and possessed a typical structure of conference services to be familiar and easy to use to all scientists interested in the conference. We limited the number of options in the main menu to 10 following items: Home, News, Dates, Committees, For Authors, Program, Participants, Venue, Previous, Contact.

This paper extends our earlier work [16] by providing in-depth analysis of the results of usability testing referring to effectiveness and efficiency. The usability metrics such as correctness of task completion, number of requests for assistance and percent of proper options selected as well as time of task completion, number of actions to achieve task and wandering time during task fulfilment were considered. Users' perceived satisfaction was also measured using the Post-study System Usability Questionnaire (PSSUQ). Complementary inspection of the website was carried out by 19 experts who employed three methods, namely cognitive walkthrough [17], heuristic evaluation [18], and control lists [19]. The results of satisfaction survey and experts' inspection are presented in [16]. Moreover, the Google Analytics provided statistics of the website usage from the beginning of organizational activities till the end of the conference.

2   Google Analytics Statistics

The Google Analytics [20, 21] was applied to collect statistics of the ACIIDS 2016 website usage. We tried to find answers for the following questions: (a) how many people visited our website, (b) what countries and cities the visitors came from, (c) whether the decision to develop a mobile friendly website was correct, (d) what our activities and events drove the most traffic to the website, (e) which pages on our website were the most popular. Selected metrics based on data gathered during ten months from July 9th, 2015 till May 8th, 2016 are presented in the rest of this section. The statistics encompass the period from the moment we started promoting the conference via email till the point of time when the current content was moved to the archive and replaced by information of the next edition of the conference.

Basic metrics provided by Google Analytics for three types of devices, namely desktop, smartphones, and tablets are shown in Table 1 and then illustrated in Figs. 1, 2 and 3. The majority of visitors utilized desktops, they accounted for 88.4 % of sessions and 91.6 % of pageviews. On the other hand, the number of visits and pageviews with mobile devices, such as smartphones and tablets, was 3,111 and 7,933, respectively. Despite these figures made up 11.6 % of all sessions and 8.4 % of all pageviews, they justify our decision to develop the website using the responsive web design approach.

Table 1. Basic metrics provided by Google Analytics

Metrics	Desktop	Phone	Tablet	All
No. of visits (sessions)	22,999	2,452	569	**26,020**
No. of page views	86,853	6,167	1,766	**94,786**
Avg. visit duration [min]	3:24	2:03	3:39	**3:39**
Pages/Session	3.78	2.52	3.10	**3.64**
New visitors	9,337	1,188	238	**10,763**
New sessions	40.60 %	48.45 %	41.83 %	**41.36 %**
Bounce rate	38.18 %	53.51 %	51.49 %	**39.92 %**

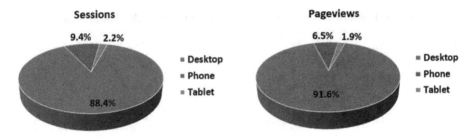

Fig. 1. Distribution of sessions and pageviews by device types

At first glance it is clear that smartphones were more popular than tablets. The average visit duration was the shortest as well as the number of pages viewed per session was the smallest for users utilizing the smartphones. The percentage of new sessions with smartphones was greater by 7.9 % and 6.6 % than visits with desktops and tablets respectively. The bounce rate for smartphones and tablets, although relatively low, exceeded the one for desktops by 15.3 % and 13.3 % respectively.

The Map Overlay report in Fig. 4 illustrates that 26 thousand visits came from 110 countries and 6 continents and the most frequent places accounted for Vietnam and Poland. 20 top countries where the most visits came from are listed in Table 2. Representatives of all these countries attended the conference. In turn, the top 14 most

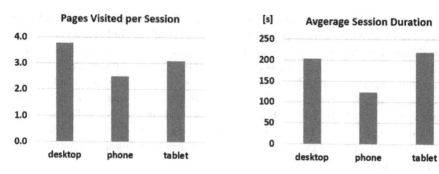

Fig. 2. Number of pages per session and average session duration for individual devices

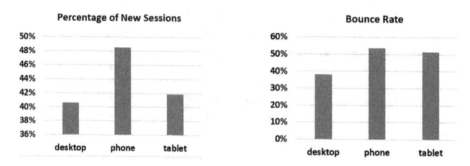

Fig. 3. Percentage of new sessions and bounce rate for individual devices

Fig. 4. Map Overlay report on countries where site's visitors came from

Table 2. 20 top countries where the most visits came from

	Country	Visits		Country	Visits
1	Vietnam	7425	11	Czech Republic	610
2	Poland	5294	12	France	543
3	United States	919	13	Singapore	468
4	Japan	875	14	Algeria	438
5	Taiwan	781	15	Australia	430
6	South Korea	764	16	Brazil	395
7	Russia	664	17	Germany	391
8	India	628	18	China	371
9	Malaysia	616	19	Brunei	278
10	Thailand	614	20	Indonesia	247

Table 3. Top website pages viewed

	Country	Views		Country	Views
1	Main Page	26219	8	Venue	4211
2	For Authors	11912	9	Admin Panel	3671
3	Dates	8944	10	Program	3407
4	News	6308	11	Participants	2813
5	Committees	5819	12	Registration Form	2549
6	Registration	5530	13	Previous Conferences	2010
7	Log In	4921	14	Edit Your Data	907

viewed pages are presented in Table 3, and we can observe that all more important pages of the conference website were visited more than two thousand times.

The daily distribution of ACIIDS 2016 site visits is presented in Fig. 5, where peaks of the curve reflect the increased number of visits caused by our planned activities, deadlines, and events. The number of sessions soared when either mailing actions were carried out and deadlines for paper submission, notification of acceptance, registration, camera-ready papers, and payment expired. There was a rise in visits also during the conference and when the photos taken in the course of the conference were published.

3 Setup of Usability Testing

The purpose of our research was to evaluate the Internet conference service in terms of fundamental attributes of usability, stated by the ISO 9241-11 standard [7], namely effectiveness, efficiency and satisfaction. The conference website was worked out based on responsive design concept. For the purpose of experiment two versions of the conference website were developed. The first one possessed the classic architecture based on a hierarchical menu containing options and sub-options with links to content

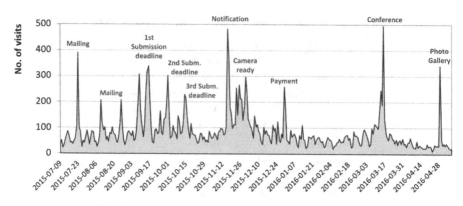

Fig. 5. Daily distribution of ACIIDS 2016 website visits

pages. In turn, the second version was built on the basis of a flat main menu and loaded the stack of subpages when a given option was selected. The content was then displayed in the form of scrolling vertical columns. Thus, the first version was designed for the users of laptops and desktop computers whereas the second one for those who utilize mobile devices with small vertically oriented touch screens.

The first version was named *Classic* and the second one *Vstack*. Different groups of research participants used these versions with laptops (*L*) or smartphones (*S*). In consequence, we obtained the results of usability tests for four version/device combinations denoted *Classic_L*, *Classic_S*, *Vstack_L*, and *Vstack_S*, respectively.

Both versions were implemented using HTML5, CSS3 and Bootstrap [22] according to the responsive web design paradigm. Dynamic components of the service, like forms, were implemented with PHP.

The participants were asked to accomplish 15 tasks presented in Table 4. The pages and subpages where the user could find the requested data or fulfil requested activity are shown in column *Page*. The tasks reflected the most common usage of the website and covered the whole functionality of the service. 13 tasks were to find specific information on the website, whereas tasks T11 and T13 assumed that participants enter data to the service through online forms.

32 participants of usability testing were enlisted from among the researchers and PhD students of Wrocław University of Science and Technology. All of them were interested in the field of intelligent information and database systems and more than half, 53 %, of them took part in at least one of the previous ACIIDS conferences. A very large majority of users, 88 %, were males, more than half were below 35 years and the fraction of persons aged 35 to 55 was equal to a third. A significant proportion of participants, 69 %, held the degree of PhD. Thus we might state that they represented well the target audience of the website. The participants were split into four disjoint groups. Each group was asked to fulfil the tasks with the *Classic* or *Vstack* applications run on either laptops (L) or smartphones (S). Thus, the testing was performed using four variants of application/device combinations denoted *Classic_L*, *Classic_S*, *Vstack_L*, and *Vstack_S*, respectively. Baseline characteristics of the participants are presented in Table 5.

Table 4. 15 tasks to accomplish by the participants

Task	To complete	Page
T1.	Find when the authors will be notified of paper acceptance.	Dates
T2.	Find what is the registration fee for authors of accepted papers.	Participants > Fees
T3.	Find what is the page limit for a paper submitted to the conference.	For Authors > How to Submit
T4.	Check whether John Smith from USA is the member of Program Committee.	Committees > Program Committee
T5.	Find the topic of the plenary lecture delivered by Professor John Smith from USA.	Program > Keynotes
T6.	Find where the conference proceedings will be published.	For Authors > How to Submit
T7.	Display the photo gallery from the previous ACIIDS 2014 conference held in Bangkok.	Previous
T8.	Find to what special session you could submit a paper on traffic optimization in a smart city.	For Authors > Special Sessions
T9.	Find the date and time of the conference banquet.	Program > Outline
T10.	Proceed to the EasyChair paper submission system.	For Authors > How to Submit
T11.	Register for the conference as an author of two papers.	Participants > Registration > Register
T12.	Download the detailed program of the conference in the pdf file.	Program > Detailed
T13.	Log in as the user John Smith and download the invoice for the registration fee.	Participants > Registration > Log in
T14.	Find how to book hotel room at the Pullman Danang Beach Resort 5*.	Venue > Accommodation
T15.	Download the conference flyer in the pdf file.	For authors > Downloads

Usability evaluation was done by individual users independently with the participation of a facilitator. Users' activities during sessions with laptops and smartphones were captured by the Morae [23] and Lookback [24] software tools, respectively. Having finished testing sessions the users were asked to fill the Post-study System Usability Questionnaire (PSSUQ). The questionnaire was applied to measure users' perceived satisfaction with the conference website. PSSUQ belongs to the most popular standardized questionnaires, it contains 16 items grouped in four sub-scales: overall, system, information, and interface quality. The final score is obtained by averaging all sub-scales [25, 26]. PSSQU is free of charge and is characterized by high reliability with Cronbach's alpha coefficient equal to 0.94. The results of our study of users' satisfaction are shown in [16].

Table 5. Characteristics of users taking part in usability testing

Appl/Device		Classic_L	Vstack_L	Classic_S	Vstack_S	Total
No. of users		11	8	6	7	32
Age	<35	1	4	5	7	17
	35-55	6	4	1	–	11
	>55	4	–	–	–	4
Gender	M	9	6	6	7	28
	F	2	2	–	–	4
Academic degree	Prof.	1	–	–	–	1
	PhD	9	6	4	3	22
	MSc	1	2	2	4	9
ACIIDS participant	Yes	5	4	5	3	17
	No	6	4	1	4	15

4 Analysis of Experimental Results

Several measures referring to efficiency, effectiveness, and satisfaction, which are the main attributes recommended by ISO Standard 9241-11, were gathered and analysed.

Effectiveness: Binary tasks completion rate. This basic metrics of effectiveness is defined as the ratio of successfully completed tasks to all tasks undertaken. In [16] we showed that average binary tasks completion rate for individual combinations of application/devices was 83.6 %, 95.6 %, 95.8 %, 98.1 % for **Classic_L**, **Classic_S**, **Vstack_L**, and **Vstack_S**, respectively. It is a good result with respect to Sauro's research [27] who, based on 1189 tasks, determined an average completion rate equal to 78 %.

Correctness of task completion. This metrics of effectiveness refers to the errors committed by the users while accomplishing individual tasks. The moderators rated each task completion using a 4-point scale with scores from 0 to 3 as well as standard scenarios and predefined criteria. The score of 0 was given when a user completed a task correctly, 1 – when the user committed only a minor error, 2 – when the user accomplished a task making a number of errors, and 3 – when the user quitted a task or completed it wrongly. An error was counted when the user tried to accomplish a task using an option different from the one recommended in a standard scenario or performed incorrect action which did not bring him closer to the goal. The distribution of errors for individual tasks is illustrated in Figs. 6, 7, 8 and 9 for **Classic_L**, **Classic_S**, **Vstack_L**, and **Vstack_S** variants, respectively. These figures give a deeper insight into the users' effectiveness. If you take into account the scores of 0, i.e. *none* errors, then you attain the percentage of users who completed a task with no problems. In turn, after discarding the scores of 3, the success rate is obtained. It can be seen in Figs. 7, 8, 9 and 10 that the users had difficulty completing tasks *T2, T4, T8, T11, T13, T14, T15*. This can be explained in the following way:

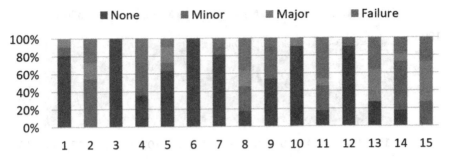

Fig. 6. Errors committed during individual task completion for *Classic_L*

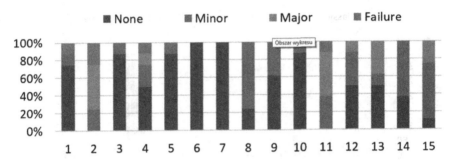

Fig. 7. Errors committed during individual task completion for *Vstack_L*

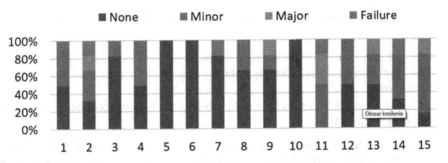

Fig. 8. Errors committed during individual task completion for *Classic_S*

T2 - the users confused the options *For Authors* with **Participants.** Moreover, they could not select the right type of fees, because they were described unclearly,

T4 - the users confused the options *Program committee* with *Organizing committees*,

T8 - users confused the options *For Authors* with *Program* to find the *Special Sessions* sub-option,

T11 - the users mixed up the option *Registration* with the external system *Easy-Chair* supporting paper submission and reviewing,

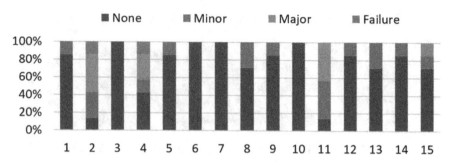

Fig. 9. Errors committed during individual task completion for *Vstack_S*

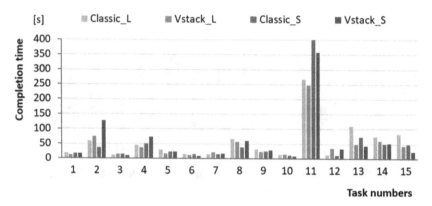

Fig. 10. Average completion time of individual tasks (geometric mean)

T13 - the users could not find the right option where they could log in to the system,
T14 - the users could not find the right sub-option with information how to book hotel rooms,
T15 - the users could not find the right sub-option where the conference flyer was available for download.

Other effectiveness measures were collected, namely average number of requests for assistance while accomplishing individual tasks and percent of proper options selected as the first when the users started performing a task. The results are shown in Table 6. The reasons for asking for assistance can be explained as follows:

T2 - the users often started searching the fees from the option *For Authors* and when they could not find them there, they tried to get a hint from the moderator.

T11 - the users could not distinguish the option *Registration* from the external system *EasyChair* supporting paper submission and reviewing. 63 % of users did not realize the option *Registration* and asked the moderator if they should register in *EasyChair*. 22 % of users did not request for assistance and were unable to complete the task. Only 16 % of users managed to register in the service without any assistance.

Table 6. Number of requests for assistance and percent of proper options selected

Task	Requests for assistance				Percentage of proper options			
	Cl_L	Cl_S	Vs_L	Vs_S	Cl_L	Cl_S	Vs_L	Vs_S
T1	0	0	0	0	82	80	50	80
T2	0	0.5	0.2	0.2	0	0	40	0
T3	0	0	0	0	100	100	83	83
T4	0	0	0	0	88	100	83	75
T5	0	0.1	0	0.1	50	100	100	60
T6	0	0	0	0	100	100	100	100
T7	0	0	0	0	100	100	100	83
T8	0.1	0.1	0	0	14	80	83	17
T9	0	0	0.2	0	90	67	83	100
T10	0	0	0	0	91	83	100	100
T11	1	0.9	1.2	0.7	50	100	40	67
T12	0	0	0	0	100	80	100	100
T13	0	0	0.2	0	71	100	83	100
T14	0	0.1	0	0	67	83	100	83
T15	0	0.1	0	0	0	20	20	17

In turn, the percentages of correctly selected options as the first indicate the users had the biggest problems with tasks T2, T4, T8, T11, T13, T14, T15. This coincides with the results of correctness of task completion presented above.

Efficiency. Time of task completion, number of actions to achieve task and wandering time during task fulfilment were used to measure the *Efficiency*.

The geometric mean of completion time for individual tasks is shown in Fig. 10. The geometric mean was employed because according to Sauro and Lewis [28] task times tend to be positively skewed. Therefore, the geometric mean provides more accurate estimation of the population center for small size samples than the sample median and arithmetic mean.

The number of navigation actions including clicks, taps, scrolls, and swipes to accomplish individual tasks, expressed in terms of the median, is depicted in Fig. 11.

In turn, the wandering time during task fulfilment was measured when the user was not using an option recommended in a standard scenario or was performing actions which did not progress toward the target. The results are illustrated in Fig. 12.

The main observations and issues could be described as follows:

T2 - information about fees was placed at the bottom of a scrolling vertical column in the ***Vstack*** versions so that it took longer to scroll the content especially on smartphones. Similarly, the users had to perform greater number of navigation actions to reach the target. Some users could not distinguish between the options *For Authors* and Participants. They began searching information about fees with the option *For Authors* and then moved to other options.

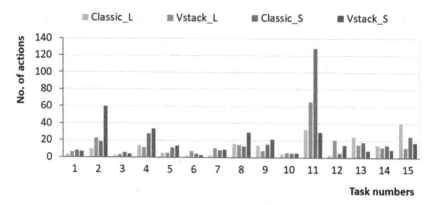

Fig. 11. Number of actions to complete individual tasks

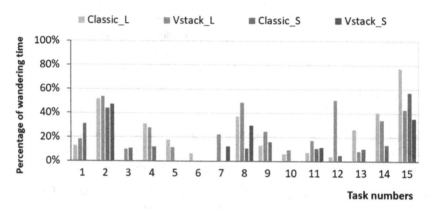

Fig. 12. Percentage of wandering time for individual tasks

T4 – the users utilizing smartphones should perform more navigation actions and it took longer time to find a given name in the vertically displayed list.

T5 – the users utilizing smartphones should perform more navigation actions and it took longer time to find the information about the fourth keynote speaker placed at the bottom of the vertical column.

T8 – the users complained that **Special Sessions** could be often found in the main menu of other conference websites and therefore they had difficulty in discovering where this option is placed.

T11 - registration in the service was the longest task, because it consisted in entering some personal and invoice data into online forms. It took much more time on smartphones due to their smaller screen sizes compared to laptops. The number of navigation actions with the **Classic_L** version was excessive because the users had to scroll the whole page after having filled a given field.

T12 - the detailed program of the conference was accessible through a barely visible link placed at the bottom of a scrolling vertical column in the **Vstack** versions. Therefore, it took longer and required more navigation actions to download the program with these versions on both laptops and smartphones. Moreover, some users missed it while scrolling quickly the content of the column.

T14 – information how to book a hotel room was not clearly emphasized on the page.

T15– All users have problems with finding the conference flyer. The users utilizing the *Vstack* versions often applied a search engine whereas there were no possibility to search the whole content of the service in the *Classic* versions. Some users began searching the conference flyer with the option *Home* and then were wandering the service having no idea where it could be placed.

Summary of results. In total 111 usability problems were detected as a result of usability testing (*UT*) with users and three methods of expert inspection, namely cognitive walkthrough (*CW*), heuristic evaluation (*HE*), and control lists (*CL*). The problems were grouped into nine categories as shown in Table 7. Then, the problems were evaluated by the moderators according to their severity. The scores of 0, 1, 2, and 3 were given to mark none, minor, major, and critical usability problems, respectively. The average severity of problems for individual categories are placed in the column *Score* in Table 7. The number of problems detected by individual groups of participants and methods is presented in Fig. 13.

Table 7. Usability problems detected by individual methods

Category	UT	CW	HE	CL	Total	Score
Data entry forms	16	0	4	12	22	2.9
Information architecture	9	9	10	2	16	2.6
Navigation	6	3	5	4	7	2.6
System behaviour	4	4	7	8	15	2.4
Content	3	3	10	2	15	2.2
GUI elements	2	0	1	4	6	2.1
Page layout	10	2	3	8	16	1.9
Text formatting	0	5	2	2	7	1.7
Visual design	2	4	1	1	7	1.7

No. of problems detected

- 17
- 59
- 35

■ only experts
■ only users
▦ both groups

No. of problems detected

60
52
40
43   43
30
20
0
UT   CW   HE   CL

Fig. 13. Number of problems detected by groups of participants and methods

5   Conclusions and Future Work

The outcome of usability study of the conference website developed using responsive web design approach is presented in the paper. Usability testing with the participation of potential users and expert inspection were conducted to identify usability problems of a responsive application. Classic and vertical stack variants of the application were examined using laptops and smartphones. The testing sessions with users took place in laboratory conditions and were controlled by facilitators. In contrast, three inspection methods including cognitive walkthrough, heuristic evaluation, and control lists were accomplished remotely by the experts alone. The experiments conducted with 32 participants and 19 experts resulted in the list of 111 recommendations for improving the website.

Expert inspections detected greater number of issues than the tests with users did. The most important problems identified by the users were also disclosed during the cognitive walkthrough. Quantitative results revealed differences among individual combinations of version/device. However, there were no sufficient evidence to choose the best one. Nevertheless, the analysis of quantitative data led to discover some defects which could not be found during observation and interview with the user. To sum up, our study revealed that the methods we employed are complementary and all should be applied to detect as many usability issues as possible.

As a result of our research an improved version of the website was developed. The improvements can be partitioned into three main groups, solving major problems found by users. They are as follows: (1) redesign of the navigation menu, (2) make registration process more user friendly, (3) make login and user session more transparent. Redesigning the menu also meant restructuring the content. In the new version the menu is hierarchical, having 11 main options and 16 more items hidden at the second level. With desktop computers and laptops one can navigate from any content section to any other section with a single click. In turn, on mobile devices, another click/touch is needed to open a submenu. The registration form was equipped with more validation using HTML5 and JavaScript. So, the format constraints on fields like e-mail or zip code are enforced now on a client side, before clicking the submit button. And this minimizes the risk of returning an invalid form after submission. Login was placed as the 12th item in the top level menu and moved to the right top corner. When user is logged in the item converts into the username.

The second round of usability testing is planned with the improved version of the conference website.

Acknowledgments. This paper was partially supported by the statutory funds of the Wrocław University of Science and Technology, Poland.

References

1. Marcotte, E.: Responsive Web Design, 2nd edn. A Book Apart, New York (2014)
2. Wroblewski, L.: Mobile First. A Book Apart, New York (2011)

3. McGrane, K.: Content Strategy for Mobile. A Book Apart, New York (2012)
4. Jehl, S.: Responsible Responsive Design. A Book Apart, New York (2014)
5. Clark, J.: Designing for Touch. A Book Apart, New York (2015)
6. Krug, S.: Don't Make Me Think, Revisited: A Common Sense Approach to Web Usability, 3rd Edition. New Riders (2014)
7. ISO 9241-11:1998 Ergonomic requirements for office work with visual display terminals (VDTs) - Part 11: Guidance on usability
8. Nielsen, J., Budiu, R.: Mobile Usability. New Riders Press, Berkeley (2012)
9. Harrison, R., Flood, D., Duce, D.: Usability of mobile applications: literature review and rationale for a new usability model. J. Interact. Sci. **1**, 1 (2013). doi:10.1186/2194-0827-1-1
10. Seffah, A., Donyaee, M., Kline, R.B., Padda, H.K.: Usability measurement and metrics: A consolidated model. Softw. Qual. J. **14**, 159–178 (2006)
11. Hussain, A., Hashim, N.L., Nordin, N., Tahir, H.M.: A metric-based evaluation model for applications on mobile phones. J. ICT **12**, 55–71 (2013)
12. Lestari, D.M., Hardianto, D., Hidayanto, A.N.: Analysis of user experience quality on responsive web design from its informative perspective. Int. J. Softw. Eng. Appl. **8**(5), 53–62 (2014)
13. Saleh, A., Isamil, R.B., Fabil, N.B.: Extension of PACMAD model for usability evaluation metrics using Goal Question Metrics (GQM) Approach. J. Theor. Appl. Inf. Technol. **79**(1), 90–100 (2015)
14. Hussain, A., Mkpojiogu, E.: The effect of responsive web design on the user experience with laptop and smartphone devices. J. Teknologi **77**(4), 41–47 (2015)
15. ACIIDS 2016 conference website. https://aciids.pwr.edu.pl/2016/. Accessed 1 June 2016
16. Bernacki, J., Blazejczyk, I., Indyka-Piasecka, A., Kopel, M., Kukla, E., Trawinski, B.: Responsive web design: testing usability of mobile web applications. In: Nguyen, N.T., et al. (eds.) ACIIDS 2016. LNCS, vol. 9621, pp. 257–269. Springer, Heidelberg (2016). doi:10. 1007/978-3-662-49381-6_25
17. Mahatody, T., Mouldi Sagar, M., Kolski, C.: Sate of the art on the cognitive walkthrough method, its variants and evolutions. Int. J. Hum. Comput. Interact. **26**(8), 741–785 (2010)
18. Nielsen, J., Molich, R.: Heuristic evaluation of user interfaces. In: Proceedings of the SIGCHI Conference on Human Factors in Computing Systems: Empowering People (CHI 1990), pp. 249–256. ACM, New York (1990)
19. 247 web usability guidelines. http://www.userfocus.co.uk/resources/guidelines.html. Accessed 1 June 2016
20. Clifton, B.: Advanced Web Metrics with Google Analytics, 3rd edn. Wiley, Hoboken (2012)
21. Google Analytics website. http://www.google.com/analytics/. Accessed 1 June 2016
22. Bootstrap website. http://getbootstrap.com/. Accessed 1 June 2016
23. Morae website. https://www.techsmith.com/morae.html. Accessed 1 June 2016
24. Lookback website. https://lookback.io/. Accessed 1 June 2016
25. Lewis, J.R.: Psychometric evaluation of the PSSUQ using data from five years of usability studies. Int. J. Hum. Comput. Interact. **14**(3&4), 463–488 (2002)
26. Sauro, J., Lewis, J.R.: Quantifying the User Experience. Practical Statistics for User Research. Morgan Kaufmann, San Francisco (2012)
27. Sauro, J.: What Is A Good Task-Completion Rate? (2011). http://www.measuringu.com/ blog/task-completion.php. Accessed 1 June 2016
28. Sauro, J., Lewis, J.R.: Average task times in usability tests. In: Proceedings of the SIGCHI Conference on Human Factors in Computing Systems (CHI 2010), pp. 2347–2350. ACM, New York (2010)

Latent Email Communication Patterns

Miloš Vacek[✉]

Faculty of Informatics and Management, University of Hradec Králové,
Rokitanského 62, 500 03 Hradec Králové, Czech Republic
milos.vacek@uhk.cz

Abstract. The paper introduces a framework to analyze company internal email communication. In some cases, users tend to use email for various purposes where other tools would be more appropriate. Such behavior patterns cause high workload to other users who have to keep doing the same. A sample from Enron email corpus is used to discover such latent patterns using Latent Class Analysis method and strategies how to improve communication are proposed.

Keywords: Email overload · Communication patterns · LCA · Enron

1 Introduction

Email is still the most frequently used tool in both, internal as well as external communication in many companies. It seems that email survives despite a huge development of new technologies with more or less subtle ambitions to substitute it and there are only a few big companies that fight excessive email overuse with an attitude sometimes called zero-email policy. Such companies simply use other means of communication, for example social networks, instant messengers and other collaboration tools. This brings them freedom in not having to deal with hundreds of incoming emails which need to be read, answered, classified or otherwise processed. It can be assumed that there are emails with some common characteristics which could be used to identify communication patterns within a company. Then, perhaps a more challenging task is to assess whether another communication channel would be more appropriate in a given situation. In this work, a real process is to be introduced that helps to classify company emails based on specific email metadata. Thus, the main objective of this paper is to describe a framework that can be adopted by any company interested in reducing internal email traffic and to provide some general guidelines on how such identified patterns can be substituted with another existing collaboration software.

This paper is structured in the following way. After this short introduction, there is a section Related work which puts results of the author in the context of previously published papers either related directly or validating methods and processes used hereby. The section Methodology describes 3 steps of the framework: data selection, transformation and analysis, using clustering method LCA. The next section Overall results provides descriptive statistics obtained

© Springer International Publishing Switzerland 2016
N.T. Nguyen et al. (Eds.): ICCCI 2016, Part I, LNAI 9875, pp. 580–589, 2016.
DOI: 10.1007/978-3-319-45243-2_53

from Enron email corpus as well as results from LCA classification. Potential improvement strategies for internal communication are given in the section Discussion and the paper is concluded with the final section that summarizes what has been achieved.

2 Related Works

Research on the field of email has been in progress for almost as long as development of email itself. While the SMTP protocol, still used for sending emails today, was defined for the first time in 1982, in RFC 821 and 822, the first broadly acknowledged study [16] defining the problem known as email overload was published in 1996. Since then, there has been a large number of papers that observed this problem from different perspectives. For example, [11] brings own definition of email overload which *describes the situation where possible business disruption due to email use may significantly harm the well-being of users and impair their productivity.* Another fundamentally different approach was taken in [5] who correctly claims that *anyone who knows your email address can send email to you* and marks all email users as owners of their data who ought to be able to manage which emails they will be getting, when and from whom by providing a special license. In addition, each user might build their *whitelist containing emails addresses that are allowed to send messages to the user without attaching a license.* One of a few attempts to innovate the actual email protocol, IMAP in this case, was described in [13] where authors propose collaborative sharing of tags for better inbox organization compared to the classic tree structure of folders. None the less, they conclude that *a single strategy alone will not be sufficient to solve the problems with email.*

In the meantime, several other papers and studies appeared that estimated how the problem would get even worse, predicting the number of emails sent and received widely further ahead. From these, let us pay attention at least to a large information study [8] that in 2003 predicted that 31 billion of emails sent daily worldwide would double by 2006. [4] further developed a method to predict email stress concluding that *email volume significantly predicted email stress amongst the participants* of their study.

The truth is that email overload is not caused primarily by spam and other unsolicited emails but by real people with who knowledge workers often cooperate on daily basis, such as colleagues in the same organization. Spam filters with self-learning algorithms rarely let commercial offers and emails with abusive and potentially harmful content into email clients, stopping them already on the server side. The main amount of regular email workload comes from verified domains and known email addresses. It can be assumed that this workload has several potential causes, introduced and analyzed in author's previous work [15]. Also, other researchers are continuously working on new interfaces to help users better manage their emails. An interesting plug-in for email clients was developed and described by [10]. He introduces semantics of actions an email sender can require from a recipient. Thus, the recipient is automatically offered an appropriate reaction. Email was also attempted to integrate with social networks such as

Facebook or LinkedIn in [12]. Another interesting tool that fills data from generated emails such as account statements, invoices or flight arrangements into prepared templates is described in [7]. Some of these also find their way to real usage, for example the latest is now implemented in Inbox by Google.

The problem, nevertheless, is much wider to be solved just by a couple of new features in email clients. The author believes that the problem is focused too much on email itself and less to electronic communication in general. Instead of overcoming the problem at its very beginning, engineers tend to solve it at the end, for example by installing features for better organizing and pre-selecting important messages first. This paper intends to carry out a root cause analysis, discovering what the purposes for sending emails are and classifying them in order to understand how people communicate.

Classification of emails in general is not an entirely new approach. Grouping with different expectations and outputs was attempted, for example, in [2] who focused on creating clusters among email users based on symmetric relationships between senders and increasing number of recipients. One of the conclusions was that *above a certain threshold, emails sent to a great number of people (approximately 13) make them less likely to know each other*. This may sound already like internal spam. Another study [1] had used a similar set of email metadata as was used for research described hereby, for clustering in order to provide better navigation in one's inbox. The approach, nevertheless, was not meant to be applied globally on a real dataset.

Thus, to the best of author's knowledge, this is the first attempt that uses a statistical clustering method LCA to find groups of emails based on email metadata to identify communication patterns. To justify LCA method for this purpose, see variable types and use cases in [14] which quite well fit the dataset described in the following section. It aims to describe the theoretical approach for which Enron email corpus was used. This archive had provided the basis for numerous researches in communication as well as linguistic fields. Some that can be considered as related to this research are, for example, [17] who used it to analyze keywords that could be used to identify email threads. Extracted data from Enron dataset was also used in [9] to compare with another sample of real population in Social Network Analysis. A linguistic approach was taken in [6] where email matrix was analyzed for frequency of certain keywords to point out user's personality, role in a company or likelihood of actions to come. Among these, clustering Enron emails to learn about behavioral patterns is indeed an innovation.

3   Methodology

Framework described in this section consists of 3 different phases with the common objective to find out as much information as possible about email usage patterns in a company. Each phase is explained below in consequent subsections.

3.1 Data Selection

As it is usual in quantitative research, the right sample of data had to be selected first. In email communication, this meant to consider several factors, such as position of an employee in the company (some use email more than others), different communication preferences or seasonal peaks. To achieve the ability to generalize the results to the whole population and avoid differences caused by the first two factors, all users from Enron dataset were selected in the sample. To avoid the problem with seasonal peaks a generous timeframe, i.e. a whole year, had to be chosen that would suggest some real findings. Nevertheless, data from 150 users for 12 months would have generated and incredible amount of emails and made it very difficult even for computer processing. Therefore, some reduction had to be done. To keep the overall structure of the sample, it was decided that a systematic selection of 1 from every 10 emails from each user would be taken in the work dataset. Thus, it can be generalized that all absolute results are in reality 10 times bigger while keeping nearly the same explaining values. This approach provides a complex picture of all email users and eliminates peaks which was the right decision as seen on Fig. 2.

3.2 Data Transformation

The technical process of collecting emails for this research from Enron email corpus is not different from the principles described in author's previous work [15]. There was the same choice of variables, same software tools and the same structure of database used for analyzing the data. Obviously, observed timeframe, number of users and the method used to select the previous sample were different by the means described in previous paragraph. After all, the end of this phase was a MySQL database instance full of data in a desired structure for the concerned period of year 2001. Therefore, it was no surprise that with a significantly bigger data sample the descriptive statistics were different as well.

3.3 Data Analysis

For data analysis, which would discover hidden features that Enron employees use their email clients for, a classification method called Latent Class Analysis (LCA) was used. This method is implemented in a number of commercial and free tools. As an example of commercial tools, LatentGold was tested whereas AutoClass provided an open-source option. Both tools generated slightly different yet comparable results. In order to make this framework widely reusable, AutoClass was given the preference in all further described steps. It is easy to use and results can be quickly reproduced with the same or slightly changed input parameters. AutoClass creates suspected latent classes in two steps *search* and *report*, each executed by a command with its own parameters passed in referenced input files. More about AutoClass classification can be found in [3]. In the first step, the algorithm tries to reach the maximum of likelihood function for a given statistical model. The second step prints results into output txt files. These were imported into MS Excel and using standard features, such as sorting, ordering and pivot tables, the results were formatted into profiles and charts presented in the next section.

4   Results

This section brings a summary of achieved results and points some interesting statistics which are discussed in more detail in Sect. 5.

4.1   Descriptive Statistics

Using a database query, a dataset with the total number of 12 999 unique emails was generated from Enron dataset. This actually represents approximately 130 000 emails received by users from Enron email corpus in 2001, however, as mentioned before only every 10th email was selected to the sample in this research. Despite this selection, it was possible to create the following charts representing daily routines in sending emails (Fig. 1) and seasonal peaks (Fig. 2). Especially the second chart proves the previous statement that a monthly dataset would not represent the real image of email communication in the company. The daily routines suggest that Enron employees were rather slow starters who turned into some heavy late workers. It is quite surprising that more emails were sent between 8pm and midnight compared to morning period between 8am and noon. Supported by the fact that 74,6 % of all emails were internal ones (9691), it can be stated that Enron email users tended to postpone their email communication after their regular office hours. There are two eye-catching peaks at monthly received emails report which are quite difficult to explain. At energy companies it is common that summer months are easier and the season culminates just before the heating season begins which would be October in this case. The other aspects of the chart are unknown and probably do not bring any new behavioral patterns in communication.

The email dataset contained the total number of 2072 unique attachments. However, only file names could be fetched from emails as the physical files were never present in the corpus. Therefore, the variable email size (in bytes) is a

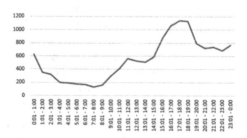

Fig. 1. Total numbers of emails received during a day in 2001

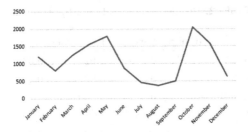

Fig. 2. Total numbers of emails received during the year 2001

Table 1. Most frequent file extensions received by Enron employees in 2001

doc	xls	pdf	jpg	ppt	tif	vcf	rtf	gif	bmp	dtf	wpd	htm	txt	zip
1278	360	248	104	53	40	33	28	27	25	24	24	23	17	12

bit misleading in this research as it does not reflect the real size of emails and actual volumes of data flows. Some emails with one or more attachments could be significantly bigger in size than those without attachments but with plenty of text. Still, a closer look at attachment types in Table 1 shows that users mainly shared working documents or pictures which is what users nowadays do too.

4.2 Classification Case Study

This section describes found classes using clustering method LCA and explains latent communication patterns that these classes can actually represent. Using the data provided in input files, AutoClass finished searching procedure in 200 cycles with following top results on Fig. 3, taken from the log file.

LCA belongs in a group of fuzzy clustering methods, which assign to each instance (email) probability from interval <0,1>, that it belongs to a certain cluster. The sum of probabilities of each instance always equals to 1. Nevertheless, fuzzy clustering methods always produce intersections where two or more classes overlap each other. Elements from such intersections have membership probability distributed among all such classes and sometimes it is better to exclude them from reporting. Thus, emails considered as real members of classes discussed below had membership probability equal or greater than 0,8. There

```
------------------  SUMMARY OF 10 BEST RESULTS  ------------------
PROBABILITY  exp(-163139.499) N_CLASSES   9 FOUND ON TRY  128 *SAVED*
PROBABILITY  exp(-163151.885) N_CLASSES   8 FOUND ON TRY  185 *SAVED*
PROBABILITY  exp(-163164.093) N_CLASSES   8 FOUND ON TRY   84
PROBABILITY  exp(-163171.530) N_CLASSES   7 FOUND ON TRY  171
PROBABILITY  exp(-163271.820) N_CLASSES   8 FOUND ON TRY  117
```

Fig. 3. Saved results of clustering Enron data in AutoClass

Table 2. Profiles of classes found by AutoClass

Class	Class size	Internal	Replies	Forward	Recipients	Attachments	Email size*
0	32,3 %	96,2 %	11,7 %	0,1 %	2,02	0,03	956
1	14,7 %	88,3 %	99,9 %	0 %	5,92	0,46	3489
2	15,5 %	98,9 %	0 %	0 %	12,69	0	1912
3	11,3 %	0 %	3,3 %	0,6 %	11,82	0,19	8384
4	8,6 %	90,6 %	0 %	100 %	4,93	0,44	2688
5	6,5 %	97,5 %	0,5 %	3,9 %	34,75	0,12	3985
6	5,7 %	99,0 %	1,6 %	18,1 %	64,82	0,03	9379
7	3,6 %	0 %	0 %	0,3 %	6,98	2,19	1548
8	1,8 %	97,5 %	1,9 %	5,6 %	151,59	0,32	35689

were 8672 of such emails (66,7 %). Other intersection elements were excluded. Profiles of classes found during clustering are summarized in the Table 2.

Based on characteristics of variables in each of these classes, some hidden patterns may be observed and described. The first attempt ever of this effort is listed below.

Class 0 - very small emails without attachments sent to a small number of recipients in many aspects remind instant messaging. Users usually write several short messages in such conversations which may be supported by the greatest size of this class.

Class 1 - this class may represent replies with or without attachments sent to members of a workgroup. It is a typical practice even nowadays which does not mean that some collaboration tool oriented on sharing information in teams would probably serve this purpose better.

Class 2 - exclusively original emails (no replies or forwarded ones) coming almost entirely from internal addresses and shared with rather average number of recipients might be looking like blog posts or intranet announcements. Senders do not share files, probably write only a couple of paragraphs of text.

Class 3 - all external emails sent to a reasonable number of recipients. Due to private company networks, email is still a good tool for B2B communication. Even if a new collaboration platform was put in place, administrators would not want to provide access to external users. In this communication pattern, email might survive.

Class 4 - very interesting group of emails, all of which are forwarded ones. Forwarded emails always involve extra persons into an existing communication. These newcomers often need to read the whole stack of previous emails to learn about the topic. However, there are modern tools which enable creating topic threads and users can follow them upon invitation and stop following when a thread is no longer relevant. This seems to be a better alternative then forwarding emails which endlessly pile up in inboxes.

Class 5 - this class seems to be the least interesting from all as it lacks some characteristic properties. The number of recipient is high above the average but there are groups with even much greater numbers. Let us assume that these are emails representing standard mostly internal communication.

Class 6 - internal mail spread and forwarded to a large number of recipients. Once again, this pattern might be substituted with threads where users would follow only such they would be involved or interested in. It is quite common that emailing groups are artificial (like departments or workgroups) and need to be often updated with people coming and leaving. On the other hand, threads can be created as private and accessible upon invitation only or public which anyone can join and leave as they wish.

Class 7 - again all external emails only, but this time with the highest number of attached files. Potentially, this pattern might be improved by creating shared folders accessible even for authorized external users or, considering the number of Word, Excel and Powerpoint files, by using services such as Google Drive or OneDrive.

Class 8 - this class differs from the others by the highest number of recipients and the biggest average email size (which may not be relevant due to missing attachments). The high number of recipients can always be complicated, though. It can be assumed that amongst hundreds (sometimes almost 800 recipients) there are always some who suffer this bulk messages as spam. If some information is meant for almost the whole company it can be distributed through certainly more efficient channels.

5 Discussion

Enron was a company which came to the end in January 2002. Therefore, there is no recommendation to use modern communication and collaboration tools which might have improved rather old-fashioned email patterns discovered in this research. Nevertheless, there is no reason to believe that the same or similar patterns do not exist in companies nowadays. The author is ready to support any private or public company that would be interested in carrying out this research and learning about their latent email communication patterns making employees' work difficult. Of course, it is obvious that a different dataset would probably yield different classes and conclusions about ineffective patterns would have to be adapted accordingly. The last step of evaluating patterns and proposing solutions will always have to be individual. Let us briefly introduce a few strategies how to improve internal communication in companies.

Strict email policies - one of the ways how to reduce email workload is restricting certain email attachments. To make users sending less attachments, it is better to install a central document repository with versioning features, standard in most document management systems nowadays. Another often applied policy in companies is limiting the maximum number of recipients in group emails. Nevertheless, this precaution brings two disadvantages at the same time: a/ such an email is still sent to users for who the content is not relevant, b/ a recipient list does not include users who might be interested in the news spread by such emails.

New communication channels - with respect to [Joung] all users should be given the opportunity to decide what information topics they want to follow themselves. In modern communication tools there are channels covering different topics and user can either start following them or unsubscribe from them at anytime. If there is new channel created, users can be invited to follow the information shared there which equals to forwarding an email and again, each user can decide whether they want to read it or they do not. There could also be an official instant messaging tool, promoted by the company, integrated within the main communication platform, if possible.

Intranet message board - last but not least important is the possibility to share internal news with all employees at once or within certain teams or departments only. This option is often overlooked, especially in small and middle-sized companies that do not see its advantages, postpone it on later or save the expenses on licenses or maintenance. Sure, neither of these strategies come for free but satisfaction of employees and increased utilization of their working time should outweigh in long term the initial investments.

The global solution of email overload is not yet known and it is unlikely to appear in short time horizon either. Another good model pattern for a number of followers might be approach named zero-email policy applied even in large corporations such as Atos. This approach shows that there are always other ways then sending tens of emails every day, most of which land in inboxes of colleagues sitting at the next table. Nevertheless, it is a subject of author's future work, to be broadly described in his dissertation, to design features of a cutting-edge communication and collaboration platform that would integrate the most common agendas in a company. Above usual one-to-one and one-to-many communications, these would also cover collaboration on documents, planning events and managing calendars in general, sharing contacts, integration with social networks and other important aspects of daily work. Everyday experience from a digital agency where the author is employed brings new ideas and challenges even in 21st century.

6   Conclusion

This paper introduced a new 3-step semi-automated procedure which analyses metadata from email users in a company to learn about their communication patterns. For proof of concept, Enron email corpus was used from which emails received in 2001 were selected and clustered by LCA method in AutoClass. In total, 9 classes were found and based on their profiles several characteristics were described, resulting in a potentially increased workload caused by incoming emails. The framework does not reflect individual attitudes but it tends to outline the overall image of internal communication in the company. Finally, a few strategies are proposed which might decrease email workload using modern and more complex technologies.

Acknowledgments. The support of the FIM UHK Project "Solution of production, transportation and allocation problems in agent-based models" and Czech Science Foundation GAČR 15-11724S DEPIES is gratefully acknowledged.

References

1. Frau, S., et al.: Dynamic coordinated email visualization. In: Visualization and Computer Vision, pp. 182–196 (2005)
2. Engel, O.: Clusters, recipients and reciprocity: extracting more value from email communication networks. Procedia Soc. Behav. Sci. **10**, 172–182 (2011)
3. Cheeseman, P., Stutz, J.: Bayesian Classification (AutoClass): Theory and Results, NASA (1996)
4. Jerejian, A., et al.: The contribution of email volume, email management strategies and propensity to worry in predicting email stress among academics. Comput. Hum. Behav. **29**, 991–996 (2013)
5. Joung, Y., Yang, C.: Email licensing. J. Netw. Comput. Appl. **32**, 538–549 (2009)
6. Keila, P., Skillicorn, D.: Structure in the enron email dataset. Comput. Math. Organ. Theor. **11**(3), 183–199 (2005)
7. Laclavík, M., Šeleng, M., Dlugolinsky, Š., Gatial, E., Hluchý, L.: Tools for email based recommendation in enterprise. In: Quintela Varajão, J.E., Cruz-Cunha, M.M., Putnik, G.D., Trigo, A. (eds.) CENTERIS 2010, Part I. CCIS, vol. 109, pp. 209–218. Springer, Heidelberg (2010)
8. Lyman P., Varian H.: How Much Information? (2003). http://groups.ischool. berkeley.edu/archive/how-much-info-2003/
9. Michalski, R., Palus, S., Kazienko, P.: Matching organizational structure and social network extracted from email communication. In: Abramowicz, W. (ed.) BIS 2011. LNBIP, vol. 87, pp. 197–206. Springer, Heidelberg (2011)
10. Scerri, S.: Semantics for enhanced email collaboration. In: Fred, A., Dietz, J.L.G., Liu, K., Filipe, J. (eds.) IC3K 2010. CCIS, vol. 272, pp. 413–427. Springer, Heidelberg (2013)
11. Sumecki, D., et al.: Email overload: exploring the moderating role of the perception of email as a business critical tool. Int. J. Inf. Manag. **31**(5), 407–414 (2011)
12. Tran, T., Rowe, J., Wu, S.F.: Social email: a framework and application for more socially-aware communications. In: Bolc, L., Makowski, M., Wierzbicki, A. (eds.) SocInfo 2010. LNCS, vol. 6430, pp. 203–215. Springer, Heidelberg (2010)
13. Tungare, M., Prez-Quiones, M.: You scratch my back and I'll scratch yours: combating email overload collaboratively. In: Extended Abstracts on Human Factors in Computing Systems, pp. 4711–4716 (2009)
14. Uebersax, J.: LCA Frequently Asked Questions (2009). http://www.john-uebersax. com/stat/faq.htm#data
15. Vacek, M.: Email overload: causes, consequences and the future. Int. J. Comput. Theor. Eng. **6**, 170–176 (2014)
16. Whittaker, S., Sidner, C.: Email overload: exploring personal information management of email. In: Proceedings of the SIGCHI Conference on Human Factors in Computing Systems, pp. 276–283 (1996)
17. Zajic, D., et al.: Single-document and multi-document summarization techniques for email threads using sentence compression. Inf. Process. Manag. **44**(4), 1600–1610 (2008)

Spectral Saliency-Based Video Deinterlacing

Umang Aggarwal[1], Maria Trocan[1(✉)], and Francois-Xavier Coudoux[2]

[1] Institut Superieur d'Electronique de Paris,
28 rue Notre Dame des Champs, Paris, France
{umang.aggarwal,maria.trocan}@isep.fr
[2] Valenciennes University, EMN (UMR CNRS 8520) Department OAE,
59313 Valenciennes Cedex 9, France
francois-xavier.coudoux@univ-valenciennes.fr

Abstract. A spectral saliency-based motion compensated deinterlacing method is proposed in the sequel. We present a block-based deinterlacing method wherein the interpolation strategy is taken upon both field texture and viewer's region of interest, for ensuring high quality frame interpolation. The proposed deinterlacer overpasses the classical interpolation approaches for both objective and subjective quality results and has a low complexity in comparison with the state of the art deinterlacers.

1 Introduction

In deinterlacing, a stream of interlaced fields in video footage in transformed into progressive frames [1] for guaranteeing the playback on nowadays devices. This process can be a very costly one, as it needs to store and use the information in several fields in order to interpolate the missing field in the full frame. Many different deinterlacing methods have been proposed so far, but due to the absence of half vertical resolution within the frames there is no error-prone recovery method till the date.

Based on their complexity and need in both temporal and spatial information in order to recover the progressive frames, the deinterlacers can be classified in three main groups: the ones [2] using the current field information to interpolate the missing field lines, thus only the spatial information, the ones using both the neighboring fields and the current field in the reconstruction process [1] and finally, the group of deinterlacers trying to perform the interpolation on the motion direction [4]. The complexity associated to these classes varies from one algorithm to another, but the best performance in both terms of reconstructed frame quality and complexity is given by the adaptive deinterlacers, e.g., the methods which adapt the interpolation strategy based on some spatial and/or temporal criterion. In order to remove the flickering artifacts, motion-compensation is a classical technique for performing the temporal interpolation, however, it requires substantial computation resources. In order to alleviate this complexity, block-based implementation solutions are preferred for the motion estimation, however at the expense of important visual artifacts in the recovered video sequence [4].

© Springer International Publishing Switzerland 2016
N.T. Nguyen et al. (Eds.): ICCCI 2016, Part I, LNAI 9875, pp. 590–598, 2016.
DOI: 10.1007/978-3-319-45243-2_54

In order to alleviate the short-comings of the existent methods, we propose in the followings an interpolation scheme which takes into consideration both the information which might be sensitive to the human visual system as well as humans' preference for sharp image quality. Knowing that the deinterlacing complexity might be an issue for real-time processing as required by nowadays display devices, we propose a block-based recovery approach, which differentiates between salient (e.g. important to the human visual system) and non salient regions, as well as between textured and smooth areas, in order to obtain the best trade-off in term of reconstructed quality and overall complexity.

Given that visual saliency [3] is used for deciding on the interpolation type: spatial, i.e. low-complexity or motion-compensated one, having average complexity, has several advantages: the non-important areas, e.g. non-salient blocks are interpolated only in a spatial manner, requiring no further buffering or computational power, whether the salient information is treated more carefully, in a motion-compensated manner, all insuring that the sharpness of the recovered image is preserved. Different from our previous deinterlacing method in [5], in which a high-complexity saliency detector is used (e.g. Graph-based Visual Saliency) along with an optical-flow-based motion estimation, in this paper a low-complexity saliency map generator is deployed [3] coupled with simple block-based estimation, such that the proposed system can be further implemented on real-time format converters.

Our paper is organized as follows: Sect. 2 presents the benefits of the visual information for interpolation systems and introduces the spectral saliency model which has been chosen for the proposed deinterlacer. Further, Sect. 3 presents our spectral saliency spatio-temporal video deinterlacing approach which is validated by the simulation framework results on different video sequences in Sect. 4. A summary of the presented method and conclusions are given in Sect. 5.

2   Saliency Information

Visual saliency can be defined as a property which makes some object attract the attention of the viewers and appear important from its neighbors [6]. Visual information also helps us to find a particular object or visual stimuli in a scene by analyzing the scene for some semantic features like motion, orientation or color.

Visual saliency is a very important characteristic of the human visual system (HVS). It can be applied in many digital image and video processing applications including image and video compression, segmentation or automatic cropping of images [7,8]. Due to the increasing number of applications of visual saliency, several visual models that try to simulate the human visual system have been proposed recently in the literature [9–11]. Generally, a visual saliency map is generated by these models that highlights the visually salient regions that appear to be most important and of interest in a scene. Out of the different frequently used saliency models, the most popular model is the one proposed by Itti et al. [6,12].

In this paper we use spectral saliency [3] method to find the spectral saliency map that is used to classify the block. Spectral residual saliency detection is a approach that imitates the behavior of a visual search where attention is not focused on a particular visual stimuli (see Fig. 1). Spectral residual is computed by analyzing the log spectrum of each image. The saliency map is generated by transforming the spectral residual back to spatial domain. This method explores the properties of the backgrounds, rather than the target objects. The main advantage of this method is its very low complexity.

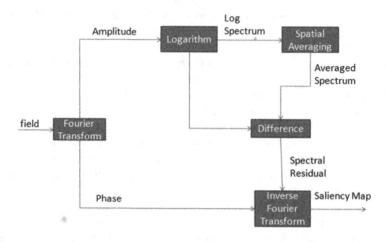

Fig. 1. Block diagram of spectral residual saliency algorithm.

Given the luminance component of field Y, we find the amplitude spectrum $A(f)$ and the phase spectrum $P(f)$ as:

$$A(f) = \Re(Fourier(Y)), \tag{1}$$

$$P(f) = \Im(Fourier(Y)). \tag{2}$$

The logarithmic spectrum $L(f)$ is obtained by:

$$L(f) = \log(A(f)). \tag{3}$$

Further, the average spectrum $As(f)$ can be approximated by convoluting the logarithmic spectrum with a 3×3 convolutional kernel h_n,

$$As(f) = h_n * L(f) \tag{4}$$

where h_n is $a = 1/n^2$ unit matrix (i.e. $n = 3$).

The spectral residual $R(f)$ is the statistical singularity that is specific to the input image and it is obtained as difference between the logarithmic spectrum and the average spectrum, as:

$$R(f) = L(f) - As(f). \tag{5}$$

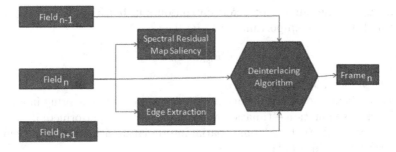

Fig. 2. Spectral saliency-based deinterlacer processing chain.

The spectral residual is then converted to the saliency map $S(x)$ using the inverse Fourier transform. The saliency map contains mainly the regions of interest(ROI) of the scene. To indicate the map in terms of estimation error the saliency map is squared. The saliency map is convoluted with a Gaussian filter $g(x)$ ($\sigma = 6$) to obtain better visual representation.

$$S(x) = g(x)IFT[\exp(P(f) + R(f))]^2, \tag{6}$$

where IFT denotes the inverse Fourier transform.

3 Spectral Saliency-Based Deinterlacing

Figure 2 presents the block diagram of the proposed spectral saliency-based deinterlacer. The proposed algorithm works by classification of blocks as per their saliency and texture information, and then using a interpolation method adapted to this classification. Hence the first step is the generation of saliency map for the current field of the interlaced video sequence, using the spectral residual saliency model discussed in Sect. 2. The saliency map obtained is denoted by S and consists of positive gray values represented on 8 bits. Also, to evaluate the texture information, edges mask C is obtained by applying a Canny edge detector on the current field.

To implement block based interpolation, the current field is divided into blocks of fixed size B^2, and each block is classified as mentioned above. To categorize based of saliency, the mean S_{b_n} of the entire collocated block s_n is computed as:

$$S_{b_n} = \Sigma_{i=1}^{B}\Sigma_{j=1}^{B}s_n(i,j)/B^2, \tag{7}$$

The block s_n belongs to the salient region if its mean S_{b_n} is greater than a empirically found threshold T_s, otherwise the block belongs to non-salient region.

Also, to classify the blocks with respect to texture information as textured or smooth, the mask field C is obtained for the current field f_n with the Canny filter. Then the number of edges for each block b_n is found by counting the number of pixels on local maxima in the collocated block c_n by using the Eq. (8).

$$CE_{b_n} = \Sigma_{i=1}^{B}\Sigma_{j=1}^{B}c_n(i,j) \tag{8}$$

where, CE_{b_n} is the number of pixels on contours in block b_n. b_n is categorized as textured if CE_{b_n} is important w.r.t. B^2, i.e.:

$$CE_{b_n} > T_b \tag{9}$$

where the value of T_b depends on B^2.

If b_n is salient and smooth, thus motion estimation will not bring in any artifacts, block based motion estimation and compensation is performed; if the block is salient but highly textured, only spatial interpolation using a 5-coefficients filter is used for the interpolation.

In the case that b_n is found not to be salient, only the spatial 5-coefficients filter as presented in (Fig. 3) is applied in order to obtain the deinterlaced block $\hat{b}_n(i,j)$ as:

$$\hat{b}_n(i,j) = \frac{b_n(i-1, j+x_0) + b_n(i+1, j-x_0)}{2}, \tag{10}$$

x_0 being triggered to minimize the intensity value on the 5 directions:

$$|b_n(i-1, j+x_0) - b_n(i+1, j-x_0)| = \min_{x_0 \in \{-2,-1,0,1,2\}} |b_n(i-1, j+x_0) - b_n(i+1, j-x_0)|. \tag{11}$$

For the case wherein b_n belongs to a salient area and is not textured, the motion estimation is performed in a bidirectional way. We consider that the displacement of the objects within the field is linear, so firstly a direct forward motion estimation is performed between the just adjacent fields, f_{n-1} and f_{n+1}. This direct vector constitutes the initialization for the bidirectional search, by splitting it into 2 vectors (forward and backward). For a given block in the current field there is the possibility to have several or even no vector MV passing through, therefore the selected MV for the current block b_n is given by the closest MV^n, in terms of Euclidean norm w.r.t. the block center.

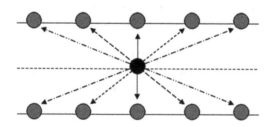

Fig. 3. Spatial interpolation scheme.

As mentioned above, for each block, the motion estimation (ME) refinement in a bidirectional way starts with the initialization:

$$MV_B^n = \frac{-MV^n}{2}, \quad MV_F^n = \frac{MV^n}{2}. \tag{12}$$

and refines the backward MV_B^n and forward MV_F^n MVs in an additional ME step.

Further, the left

$$\mathcal{F}_{MV_B}^n$$

and right

$$\mathcal{F}_{MV_F}^n$$

temporal compensations for the current block b_n are given by:

$$\mathcal{F}_{MV_B}^n(i,j) = b_{n-1}(i + MV_{By}^n, j + MV_{Bx}^n), \qquad (13)$$

$$\mathcal{F}_{MV_F}^n(i,j) = b_{n+1}(i + MV_{Fy}^n, j + MV_{Fx}^n), \qquad (14)$$

and the final prediction of b_n, i.e. $\hat{b}^n$ is simply the mean of these bidirectional compensations:

$$\hat{b}^n(i,j) = \frac{\mathcal{F}_{MV_B}^n(i,j) + \mathcal{F}_{MV_B}^n(i,j)}{2}. \qquad (15)$$

The missing pixels in b_n are finally obtained by:

$$\hat{b}_n(i,j) = \frac{b_n(i-1, j+x_0) + b_n(i+1, j-x_0) + k\hat{b}^n(i,j)}{k+2}, \qquad (16)$$

where k is the empirically found constant miximizing the reconstruction quality (in our framework its value its 20), and the value of x_0 is given by (11).

4   Simulation Framework

Our spectral-saliency based deinterlacer has been tested on seven video sequences, namely "Foreman", "Hall", "Mobile", "Stefan" and "News" - in CIF resolution (352×288) and "Salesman" and "Carphone" in QCIF resolution (176×144). These sequences, highly used by the video community, have been chosen for their heterogeneous dynamics and texture. Moreover, full progressive videos have been used in order to perform a fair qualitative and subjective comparison between the originals and the deinterlaced sequences.

The above mentioned videos have been transformed into interlaced format by alternating the removal of odd and even line in consecutive frames (see Fig. 4).

The setup of our simulation framework used $4 \times 4(B = 4)$ blocksize and a motion research window of 8×8 pixels (see Sect. 3) The saliency trigger, e.g. T_s threshold was empirically set up to 2, this value given the best trade-off between the salient and non-salient areas (i.e. the number of pixels in the salient areas $\approx$ the number of pixels in non-salient regions). The texture trigger have been set $T_b = 8$, such that a block is declared to be textured whether at least half of the pixels within are situated on edges.

The qualitative efficiency of our deinterlacing method is measured in terms of peak signal to-noise ratio (PSNR), computed on the first 50frames for every sequence and taking into consideration only the luminance (Y) component from

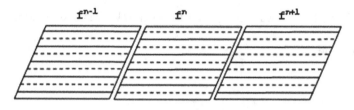

Fig. 4. Conversion from full frames to interlaced fields - dashed lines have been removed.

Table 1. PSNR results

	Foreman	Hall	Mobile	Stefan	News	Carphone	Salesman
ELA	33.14	30.74	23.47	26.04	32.19	32.33	30.51
VA	32.15	28.26	25.38	27.30	34.64	32.17	31.52
TFA	34.08	37.47	27.96	26.83	41.06	37.39	45.22
MCD	35.42	34.23	25.26	27.32	35.49	33.55	33.16
AME	33.19	27.27	20.95	23.84	27.36	29.63	28.24
$SRSGD$	**37.32**	**36.98**	**29.48**	**30.78**	**41.66**	**37.80**	**41.48**

Table 2. Estimated time

	Foreman	Hall	Mobile	Stefan	News	Carphone	Salesman
$SaliencyTime$	0.45	0.41	0.41	0.39	0.40	0.43	0.40
$TotalTime$	**6.57**	**19.67**	**20.09**	**19.90**	**19.15**	**4.49**	**4.75**

the YUV 4:2:0 sampling format. Table 1 presents the PNSR comparison of our proposed method (SRSGD) with the state of the art deinterlacers like Edge Line Average (ELA), Vertical Average (VA), Temporal Field Average (TFA), Motion-Compensated Deinterlacing (MCD) and Adaptive Motion Estimation (AME).

The computational complexity is estimated by the average time it takes to deinterlace a frame added with the average saliency map generation time for a field. Table 2 shows the average time of saliency map generation and average total time to make a frame for all the sequences.

Our spectral-saliency deinterlacer outperforms the state-of the art deinterlacers with $\approx 4.5dBs$. Moreover, a complexity analysis, reported as average deinterlacing time for a frame and thus including the computation of the saliency map is proposed in Table 2. Considering that our deinterlacer has been implemented in Matlab (8.0.0.783 (R2012b)) and run on a quad-core Intel PC@1.8 GHz, the time results in Table 2 can still be optimized, both from the implementation point of view, programming language as well as processing platform. Given that the proposed method is block-based, further optimizations can be done, as parallel processing, in order to obtain a real-time deinterlacing framework.

Fig. 5. Subjective evaluation of the proposed method on (a) CIF frame from Foreman and (b) QCIF frame from Salesman video sequences.

Further, the block dimension considered for the reported results is well suited for the considered resolutions (CIF and QCIF), but it can be viewed also as a trigger for lowering the complexity: increasing the block-size along with a constant search window will result in complexity reduced by the ratio between the new blocksize and the original one.

Moreover, in Fig. 5 we propose a subjective evaluation of our deinterlacing framework: the visual quality of several reconstructed frames is presented.

5   Conclusion

A spectral saliency-based block-based motion-compensated strategy is proposed in this paper, wherein the best trade-off between computational complexity and human visual system is achieved by adapting the interpolation based on texture and area of interest for the viewer. Our simulation framework for the proposed scheme overpasses the classical deinterlacing methods with 4.5dBs in average, at fair complexity. Given the block-based implementation of our algorithm, the proposed deinterlacer can be further optimized and triggered either to lower the complexity or to increase the recovered frame quality by a simple tweak of saliency/edges thresholds and block-size parameters.

References

1. Haan, G.D., Bellers, E.B.: Deinterlacing - an overview. Proc. IEEE **86**(9), 1839–1857 (1998)
2. Atkins, C.B.: Optical image scaling using pixel classification. In: International Conference on Image Processing (2001)
3. Hou, X., Zhang, L.: Saliency detection: a spectral residual approach. In: Proceedings of the IEEE CVPR, pp. 1–8, June 2007
4. Trocan, M., Mikovicova, B., Zhanguzin, D.: An adaptive motion compensated approach for video deinterlacing. Multimedia Tools Appl. **61**(3), 819–837 (2011)
5. Trocan, M., Coudoux, F.X.: Saliency-guided video deinterlacing. In: Núñez, M., et al. (eds.) ICCCI 2015, Part II. LNCS, vol. 9330, pp. 24–43. Springer, Switzerland (2015)
6. Itti, L., Koch, C., Niebur, E.: A model of saliency-based visual attention for rapid scene analysis. IEEE Trans. PAMI **20**(11), 1254–1259 (1998)
7. Itti, L.: Automatic foveation for video compression using a neurobiological model of visual attention. IEEE Trans. Image Process. **13**(10), 1304–1318 (2004)
8. Rahtu, E., Kannala, J., Salo, M., Heikkilä, J.: Segmenting salient objects from images and videos. In: Daniilidis, K., Maragos, P., Paragios, N. (eds.) ECCV 2010, Part V. LNCS, vol. 6315, pp. 366–379. Springer, Heidelberg (2010)
9. Zhang, L., Tong, M., et al.: SUN: a bayesian framework for saliency using natural statistics. J. Vis. **9**(7), 1–20 (2008)
10. Lu, S., Lim, J.-H.: Saliency modeling from image histograms. In: Fitzgibbon, A., Lazebnik, S., Perona, P., Sato, Y., Schmid, C. (eds.) ECCV 2012, Part VII. LNCS, vol. 7578, pp. 321–332. Springer, Heidelberg (2012)
11. Seo, H.-J., Milanfar, P.: Static and space-time visual saliency detection by self-resemblance. J. Vis. **9**(12), 1–12 (2009)
12. Itti, L., Koch, C.: Computational modeling of visual attention. Nat. Rev. Neurosci. **2**(3), 194–203 (2001)

Author Index

Printed in the United States
By Bookmasters